IET POWER A.

Series Editors: Professor A.T. Johns
G. Ratcliff
Professor A. Wright

Power Circuit Breaker Theory and Design

Other volumes in this series:

Power Circuit Breaker Theory and Design

Edited by C.H. Flurscheim

The Institution of Engineering and Technology

Published by The Institution of Engineering and Technology, London, United Kingdom

First edition © 1975 Peter Peregrinus Ltd
Revised edition © 1982 Peter Peregrinus Ltd
Reprint with new cover © 2008 The Institution of Engineering and Technology

First published 1975
Second impression 1977
Revised edition 1982
Reprinted 1985 with minor corrections
Reprinted 2008

The Institution of Engineering and Technology
Michael Faraday House
Six Hills Way, Stevenage
Herts, SG1 2AY, United Kingdom

www.theiet.org

While the authors and the publishers believe that the information and guidance given in this work are correct, all parties must rely upon their own skill and judgement when making use of them. Neither the authors nor the publishers assume any liability to anyone for any loss or damage caused by any error or omission in the work, whether such error or omission is the result of negligence or any other cause. Any and all such liability is disclaimed.

The moral right of the authors to be identified as authors of this work have been asserted by them in accordance with the Copyright, Designs and Patents Act 1988.

British Library Cataloguing in Publication Data
Power circuit breakers. – Rev. edn – (Power and energy series; 1)
  1. Electric circuit-breakers
  I. Flurscheim, C.H.   II. Series
  621.31'7   TK2482

ISBN (10 digit) 0 90604 870 2
ISBN (13 digit) 978-0-90604-870-2

Printed in the UK by Lightning Source UK Ltd, Milton Keynes

Authors

AMER, D. F., F.I.E.E.

formerly
Manager of Engineering,
A. Reyrolle & Co. Ltd.

1981 Revision
FAWDREY, C. A., B.Sc., C.Eng., M.I.E.E.
Chief Development Engineer
High Voltage Switchgear Unit
GEC Switchgear Ltd.

FLURSCHEIM, C. H., B.A., F.Eng., Fel.I.E.E.E., F.I.Mech.E., F.I.E.E.
Consultant to Merz and McLellan
formerly
Director of Engineering,
Associated Electrical Industries Ltd.

Chief Engineer, Switchgear,
Metropolitan-Vickers Ltd.

Chairman, Power Division,
Institution of Electrical Engineers

GONEK, S. M., C.Eng., F.I.E.E.
1975 Technical Manager,
A. Reyrolle & Co. Ltd.
1981 Revision
Technical Executive Engineering Division
Mitsubishi Electric Company
Croydon

JOHN, M. N., B.S.C., F.Eng., Sen.Mem.I.E.E.E., C.Eng., F.I.E.E.
Partner,
Kennedy & Donkin,
Consulting Engineers

KENNEDY, M. W., B.Sc., Ph.D., Mem.I.E.E.E., C.Eng., M.I.E.E.
Associate of Merz and McLellan, and
Assistant to the Chief Electrical Engineer (Transmission)

LEGG, D., B.Sc., C.Eng., M.I.E.E.
Switchgear and Production Manager
A. Reyrolle & Co. Ltd.

Visiting Professor,
University of Strathclyde
formerly
> Head of Research,
> A. Reyrolle & Co. Ltd.

MAGGI, ERNESTO, Dr.Ing.
Technical Director,
Magrini SPA until December 1962

Chairman, Large Switchgear Commission,
Italian Electrotechnical Committee

Chairman, Technical Committee TC-17, and
> Sub-Committee SC-17A of
> IEC Switchgear, 1962–71

MORTON, J. S., B.Sc., C.Eng., F.I.E.E., Sen.Mem.I.E.E.E.
Technical Director
Whipp & Bourne (1975) Ltd.

REECE, M. P., B.Sc., Ph.D., C.Eng., F.I.E.E.
Technical Director,
Vacuum Interrupters Ltd.

SULLIVAN, J. A., C.Eng., F.I.E.E.
Head of Electrical Power Equipment Department,
Merz and McLellan

Contents

13   Cost effective design   563
C. H. Flurscheim, B.A., Fel.I.E.E.E., C.Eng., F.I.Mech.E., F.I.E.E.

Appendix A   Virtual current chopping   591
M. P. Reece, B.Sc., Ph.D., F.Eng., F.I.E.E.

Foreword

The objectives of this book are to discuss in some depth the wide range of technologies that are involved in power circuit breaker design, and consequential space limitations have made it necessary to omit related but interesting fringe subjects, such as aircraft circuit breakers and 'metalclad'.

I am, then, grateful to the authors who have contributed the specialist chapters; and I recognise that this approach leads to differences in philosophy and presentation, for which I make no apology, because I believe this can be of value with a product in which the physics of the arc and engineering are combined in the 'art of design'. I hope that inadequacies in editorial continuity will be overlooked in relation to the authority a somewhat individualistic approach permits.

I am also grateful to T. W. Wilcox, Partner, and to Merz and McLellan, Consulting Engineers, who have lent me their support throughout this project and have made available the extensive services it has required.

Charles H. Flurscheim

Foreword – Second Edition 1982

In the seven years since this work was compiled there have been significant advances in theory and design of circuit breakers, and the opportunity afforded by the production of a second edition has been used to update relevant chapters.

I am grateful to the authors, and again to the partners of Merz & McLellan, Consulting Engineers, for their continued support.

Charles H. Flurscheim

Chapter 1

Development of circuit breakers

C. H. Flurscheim, B.A., F.Eng., Fel.I.E.E.E., F.I.Mech.E., F.I.E.E.

1.1 The function of circuit breakers

Circuit breakers are required to control electrical power networks by switching circuits on, by carrying load and by switching circuits off under manual or automatic supervision. The character of their duty is unusual and they will normally be in the closed position carrying load, or in the open position providing electrical isolation. They are called on to change from one condition to the other only occasionally, and to perform their special function of closing on to a faulty circuit or interrupting short-circuit current only on very rare occasions. They must, therefore, be reliable in the static situations, but be effective instantaneously when they are called on to perform any switching operation, often after long periods without movement.

The most arduous of these duties is the interruption of short circuits, and the severity of this duty imposed on circuit breakers has increased immensely during the past 50 years as the result of growth in network size. Network voltages have risen from 132 to 750 kV in this period and experimental systems are now being built for 1000 kV. Short-circuit ratings have risen from the order of 1×10^6 kVA on networks with low circuit severity factors and associated with ill-defined proof testing techniques, to 50×10^6 kVA on networks that involve very high circuit severity factors, and associated with elaborate proof testing.

During this period the total break time required to interrupt short circuits has also been reduced radically in the interests of system stability. The early plain break oil circuit-breaker designs required a rather variable time of the order of 10—20 cycles because of long arc durations. The introduction of arc controlled systems quickly reduced this to 6—8 cycles, and with improved techniques a number of designs can now operate within 2 cycles.

Looking to the future the continuing growth in demand for electricity in the major industrial countries may be expected in the next 25 years to require compact closely interconnected synchronous networks of the order of 200×10^6 or even 300×10^6 kW, the limit depending on the network voltage adopted, with several such networks being interconnected by supervoltage ties which might be required to operate at between 750 and 1500 kV. This, together with the siting problems, could lead to very large concentrations of power generation, perhaps of the order of 30×10^6 kW, geographically sited so as in effect to be coupled to a single busbar, and this in turn may result in short-circuit rating levels of the order of $100 \times 10^6 - 200 \times 10^6$ kVA over the range 400–1500 kV. Although short-circuit limitation techniques now in an embryo state of development could modify the rate of growth in short-circuit levels there will remain very considerable difficulties in the design of future circuit breakers needed to meet ratings of this general order, in association with 1 cycle interruption which may be necessary for system stability reasons.

The major design problem in a circuit breaker is to provide the ability to carry out its function of controlling any current which may flow in the circuit of which it constitutes a part of the conductor system, under both normal and abnormal conditions. The function of a circuit breaker is then to possess two stable conditions 'close' in which it has ideally zero, and in practice a very small impedance; and 'open' in which it has ideally an infinite and practically an extremely high impedance. The circuit breaker must be able to change from either condition to the other when so instructed, at a rate of change of impedance which is compatible with the parameters of the circuit being controlled.

The switching element in such a device, the impedance of which is caused to change, may take the form of a resistance or inductance or capacitative impedance. In all actual circuit breakers the switching element is resistive, although in some other switching systems, such as 'resonance links', it is reactive.

The rate at which impedance changes is important. If the element is resistive a change in resistance occurring over a relatively long time will involve the release of considerable thermal energy, which the element must absorb and then dissipate, making it large and expensive. However, for other reasons related to the response of the circuit, it must not change too rapidly. For as all realistic circuits contain inductance, a voltage of $-L(di/dt)$ will be generated during switching as the current falls, and to keep this voltage to a level at which it will not damage insulation of the circuit, $-(di/dt)$ must be limited to some acceptable value.

In direct-current circuits this puts an absolute limit on the rate at which impedance of the circuit breaker may increase, since, if it is too large, $-(di/dt)$ will be excessive and result in the generation of destructive overvoltages across the system inductance. The d.c. circuit breaker must therefore be designed so that its impedance changes at some maximum predetermined rate compatible with the insulation levels employed.

In alternating-current circuits the parameters are different. The current passes through two zero periods in each cycle automatically, when for an instant there is no current. The impedance may in theory now change instantaneously from zero to infinity without generating dangerous overvoltages, providing it does so precisely at one of these moments of current zero. Should rapid change of impedance occur more than a few microseconds away from current zero, disastrous overvoltages may be generated.

Of course, the option of using the slowly changing impedance is available also to designers of a.c. switchgear, and to some extent this system is employed in air-break circuit breakers. In general however in circuit breakers for large a.c. power systems, a resistor which changes gradually over a period of time even as short as 10 ms, would dissipate an unacceptably large amount of thermal energy, and consequently for these large circuit breakers a rapidly changing impedance becomes essential, the changes being synchronised with the current zero.

How then can such a resistive system changing rapidly at current zero be built? Fortunately the answer is that the electric arc can be constrained to provide just the properties required, that is not merely that it can change its resistance very rapidly, but that the change can be automatically and synchronously invoked by the changing value of the alternating current itself. A similar condition can exist with the current flowing in the crystalline lattice of a semiconductor slice.

There are then two basic requirements for alternating current interruption. The first is to increase arc plasma resistance exponentially in the region of the natural current zero so that any subsequent flow of current is suppressed and interrupted. The second is to ensure that the dielectric strength of the arc path beyond this point increases faster than the electric stress imposed on it by the network, so avoiding reignition. All circuit breakers are therefore designed with the objective of optimising their response to these requirements, using a wide variety of arc control devices based on parameters such as liquid, gas, pressure, vacuum, turbulence, arc and contact movement. In addition, some circuit breakers embody features designed to reduce the severity of the network restriking voltage imposed on them.

The arc is therefore an essential component of the interrupting process. To the physicist it appears in many forms and is always a subject of great scientific interest and difficulty. To the switchgear designer the arc all too often appears to have only one objective, self preservation, for power arcs evade deionisation systems by swerving, by striking out in new and unexpected directions, and by reigniting after they appear to have been successfully interrupted, if the design has any weakness which permits this.

Nevertheless, the arc possesses inherent features which are helpful. The arc plasma has no upper limit in current carrying capacity, is resilient and can be stretched, and possesses resistance which can be increased both by length and by confinement. Then the mobility of the arc, which can in some circumstances aid it to escape from the deionising means provided, can be turned to advantage by using its own magnetic field to react on the arc plasma itself and so promote controlled high-speed movement of the arc, which can be harnessed to aid the interrupting process.

The scientific development of circuit breakers dates from the construction of special short-circuit test plants about 1925, followed by the development of the cathode ray oscillograph, which made possible controlled testing in association with accurate measurement of performance, not previously attainable with the *ad hoc* and sometimes spectacular arrangements customary before that time, using production turbogenerators and test beds on the open shop floor. Resulting from this more sophisticated approach a wide range of systems for arc extinction have been developed using air, oil, compressed air, SF_6 gas and vacuum as interrupting medium and a brief history of some of this background is given in the following pages.

1.2 Development of circuit breakers: Oil

Historically the most successful of the arc interrupting systems in the earlier years was undoubtedly the oil circuit breaker in its many forms some of which still retain their place among the leaders of present day practice. The oil circuit breaker employs the properties of the arc by using its energy to crack the oil molecules and generate gas, principally hydrogen, which with properly designed control systems can be used to sweep, cool and compress the arc plasma and so deionise it in a self extinguishing process. The early forms of high-voltage oil circuit breakers in use until about 1925 were of plain-break construction usually with two or more breaks, with very little if any formal arc extinction system other than that provided by extending the arc in an enclosed oil filled tank capable of withstanding the considerable pressures resulting from the large quantities of gas generated by long arcs.

Because of the fortuitous nature of the interruption process the plasma leakage current conditions following the final current zero tended to be unstable in character, and instead of increasing in resistance towards infinity a reversal in trend sometimes occurred with resultant restrike of the full power arc through the hot plasma. In extreme cases the energy released could damage the circuit breaker. But this leakage current was also one of the principal factors that contributed to the considerable degree of success these plain-break designs enjoyed for many years, for the leakage resistance damped the restriking voltage oscillations, thereby reducing circuit severity and making interruption easier. At the same time it provided a crude form of resistance grading which assisted in equalising the distribution of the restriking voltage across the multi-break system, overcoming the inherent voltage distortion caused by stray capacitance of the moving contact system to the metal tank, which would otherwise have rendered most of the interrupting gaps virtually useless because they carried so small a share of the overall recovery voltage duty. The combined effect of these actions on a double-break plain-break circuit breaker could reduce the peak restriking voltage of the most highly stressed of the two breaks to about one third of that which would have occurred without leakage.

It becomes evident there was a need for circuit breakers which possessed a more positive system of interruption than the fortuitous deionisation associated with uncontrolled oil and gas flow. An early and notable step forward was the General

Electric (USA) H Type circuit breaker introduced in the 1920s, which employed two metal explosion pots per phase, oil filled and with insulating nozzles through which the moving contacts were withdrawn vertically upwards, the explosion pots being mounted on ceramic insulators within an air-insulated cubicle structure. Later, Slepian (Westinghouse) proposed a 'deion grid' in which the arc was forced electromagnetically into a narrow insulation slot within an assembly of fibre plates submerged in the OCB tank, thus increasing the 'effectiveness of the means for preventing the escape of gases generated in the vicinity of the arc without passing through the arc stream' (Baker and Wilcox, 1930). Another approach was to use the arc to generate high pressures within a small insulating chamber immersed in the oil, such as that developed by General Electric (GE) (Prince and Skeats, 1931) in the USA (Fig. 1.1a) which restricted oil and gas escape to an axial flow surrounding the arc plasma in the throat of the interrupter; and later in the cross flow interrupter developed by the British Electrical Research Association (Whitney and Wedmore, 1930), which forms the basis of many modern designs (Fig. 1.1b). The controlled

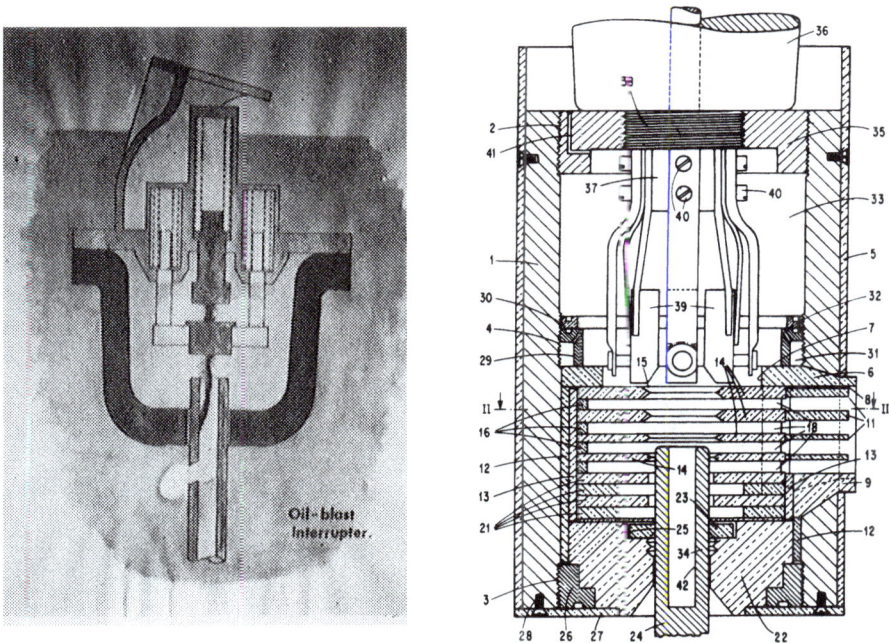

Fig. 1.1 Oil Interrupters

    a Axial flow oil-blast interrupter [GE-USA, 1930]
    b Crossblast interrupter, 1930 [ERA, 1930]

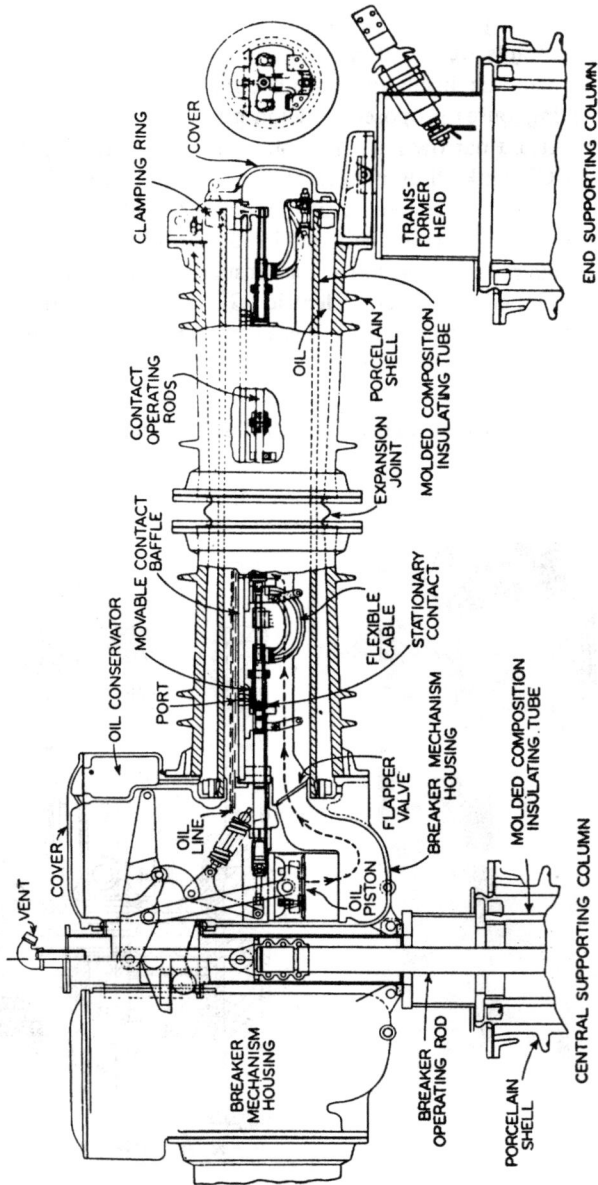

Fig. 1.2 Multibreak 'impulse' oil circuit breaker 1934 [GE-USA, 1934]

Labels on figure:

CLAMPING RING
COVER
CONTACT OPERATING RODS
OIL CONSERVATOR
MOVABLE CONTACT BAFFLE
PORT
COVER
VENT
OIL LINE
OIL PISTON
FLAPPER VALVE
FLEXIBLE CABLE
STATIONARY CONTACT
BREAKER MECHANISM HOUSING
EXPANSION JOINT
PORCELAIN SHELL
OIL
MOLDED COMPOSITION INSULATING TUBE
TRANS-FORMER HEAD
END SUPPORTING COLUMN
MOLDED COMPOSITION INSULATING TUBE
CENTRAL SUPPORTING COLUMN
BREAKER MECHANISM HOUSING
BREAKER OPERATING ROD
PORCELAIN SHELL

turbulence and high pressure and resultant rapid deionisation in these systems eliminated the erratic operation of the plain-break by virtually eliminating the leakage current, but with this it also eliminated the useful voltage damping and voltage control function this current had performed in earlier designs, voltage division then reverting to the capacitance controlled distribution.

Development at this period took account of this better understanding of the voltage division problem and single-break circuit breakers were introduced for duties previously associated with multibreak construction such as the single-break metalclad system designed by Metropolitan–Vickers, an arrangement inspired by layout considerations but justified by the fact that when two breaks were used, and depending on the neutral earthing methods, one was subjected to of the order of 76–85% of the voltage in any case (Davies and Flurscheim, 1936). On the other hand the multibreak impulse circuit breaker conceived by Prince (1935) at GE (Fig. 1.2) was equally justified because the distorting stray capacitance was reduced by live tank construction, and the voltage duty was distributed with a reasonable degree of parity across each of the breaks employed using external capacitance shielding. Typical of a third approach was the multibreak e.h.v. dead tank circuit breaker, developed by Cox and Wilcox (1947) at BTH, (Fig. 4.21) fitted with positive voltage grading and damping devices in the form of shunt resistors. These additions were expensive, but made the e.h.v. oil circuit breaker reliable by performing consistently the function the leakage current had previously carried out in a primitive way.

A desirable compromise would be to retain the advantages that leakage current can afford but eliminating the erratic nature of this control. No means of achieving this have as yet been suggested and this may remain insoluble, because of the difficulty of the control problem it postulates. For this phenomenon takes place in an environment in which dielectric stress imposed by the network is changing at several thousand volts per microsecond and in which arc plasma conductivity changes approximately a billion times as fast as temperature in the critical range of 1000–3000 K associated with thermal ionisation, as shown in Fig. 1.3.

The idea of a single break carrying out the whole duty was however extended too high in voltage in some designs in terms of contemporary techniques at this period, and some difficulties occurred in situations such as switching long open ended transmission lines. These limitations were associated with the electrical and mechanical strength of the insulation materials then available, which neither permitted the circuit breakers to be designed with the acceleration necessary to ensure restrike free switching, nor to have their jet assemblies restricted sufficiently to prevent the arc, in unfavourable circumstances, from flashing through the jets and along the outside of the interrupter, thus by-passing the interrupting mechanism provided.

The advances in performance of present day e.h.v. dead tank oil and low oil circuit breaker construction have been brought about by using multibreak designs, but with the added complication of positive voltage control; by reducing the inertia

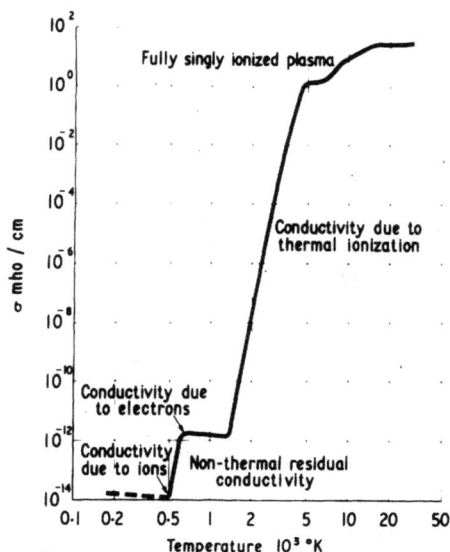

Fig. 1.3 Electrical conductivity against temperature of air at atmospheric pressure [Kreichbaum, AEG]

of the moving parts through the use of new high tensile materials or by eliminating mechanical linkages by the use of high pressure oil drives; by improved containment of the arc within the interrupter as the result of the higher pressures that can be sustained through the use of materials such as thread wound fibreglass; and by improved techniques for arc control, including limited forced oil flow and pressurisation of the interrupter. The overall complication of low oil circuit breakers has also been reduced by the elimination of the former external automatic isolating switches, made possible by the improved internal dielectric conditions following shorter arc durations.

The multibreak (Prince, 1935) impulse circuit-breaker already referred to was a special case as it relied entirely on oil flow produced by a piston driven by external energy. The best known example of this type is the 8-break 287 kV 2500 MVA General Electric Boulder Dam installation commissioned in 1935, which afforded a 3-cycle interruption under all conditions of switching. These circuit breakers were also the first to be proved by means of realistic high power synthetic testing using current and voltage supplied from different circuits and synchronised within a few microseconds at current zero, using a system devised by Skeats (1936). These tests were carried out without failure to an equivalent short-circuit level in excess of 4000 MVA, and it is of historic interest to the world of synthetic testing, on which modern high power breakers rely largely for proof of rating, to note that these circuit breakers were still operating successfully, after 35 years of service, in a network with a fault capacity of the order of 7000 MVA.

The high costs of the powerful mechanisms needed to drive the oil in both American and British derivatives of this system discouraged further development, at a time when the modular construction that the airblast circuit breaker made possible began to be apparent. This, together with a change in fashion away from oil – and it should be recognised that even engineering is not free from the influence of fashion – encouraged a swing to airblast construction. Nevertheless, the difficulties inherent in deciding on such long term development policies in switchgear are exemplified by the decade or more which passed before the high-voltage airblast circuit breaker matched the best oil circuit-breaker practice in either performance or reliability.

1.3 Development of circuit breakers: Air

Air is perhaps the obvious environment to consider as an arc extinguishing medium for it has good insulation properties at normal temperatures and is freely available. The problems of using air at atmostpheric pressure for power circuit breakers are however difficult, as once a power arc has been drawn in air a considerable volume of air surrounding the arc is raised to high temperatures, which makes cooling of the actual ionised conducting path difficult, and introduces a risk of flashover across the contacts and to earth through leakage of hot plasma.

One of the earliest successful approaches to high power airbreak circuit breakers was that developed by Slepian (1929) at Westinghouse, who recognised that below a certain voltage equal to the cathode voltage drop an arc in air between two metal electrodes will extinguish automatically however small the gap and however large the current, provided melting and generation of metal vapour from the electrodes was suppressed. From this background he developed the "deion" airbreak circuit breaker which consisted of a large number of airgaps between flat metal plates, (Fig. 1.4a and b) insulated from each other, the arc being blown into the gap electromagnetically. The metal electrodes were kept cool during arcing by electromagnetic rotation of the arc, and voltage across the series gaps was controlled by capacitive shielding.

An alternative approach was to use arc chutes made from insulation materials, of which the GE design using phosphorous asbestos mouldings was an early example (Boehne and Linde, 1940). The contours of the slots in these arc chutes was such that as the arc was blown into the chute it was very considerably lengthened and being in close contact with the chute material it was also cooled. Arc resistance was therefore raised and very considerable energy was released, to the extent that the

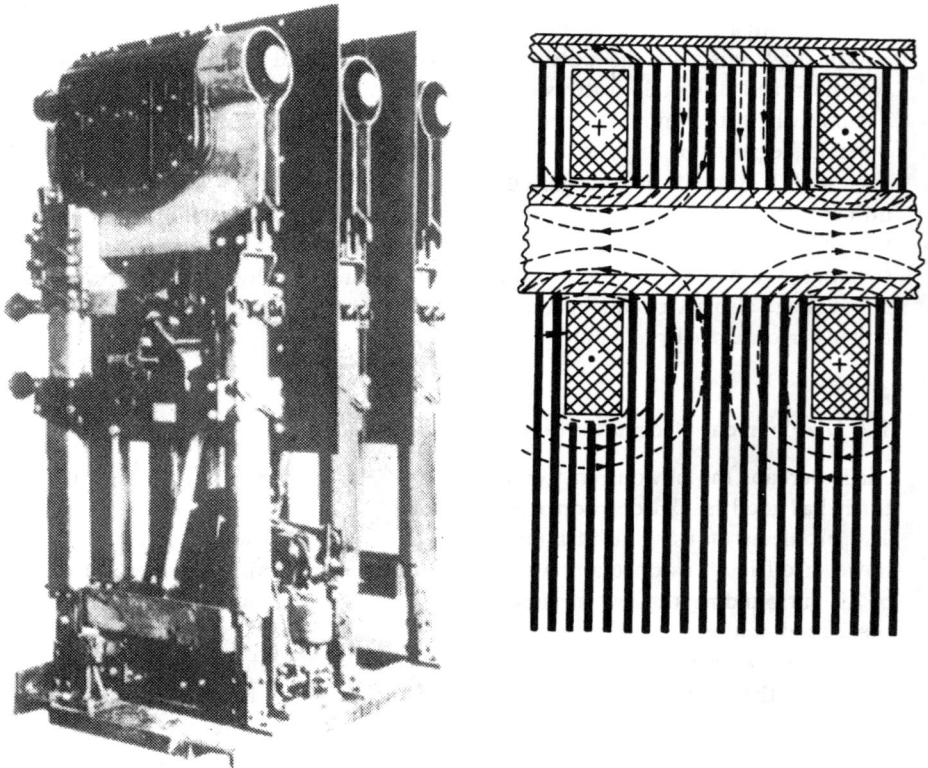

Fig. 1.4 De-ion air circuit breaker [Westinghouse, 1929]

    a 15 kV circuit breaker
    b Schematic arrangement of arc chute

network phase angle during arcing was significantly reduced and the circuit severity at current zero was therefore eased.

1.4 Development of circuit breakers: Airblast

The airblast circuit breaker was another invention of Whitney and Wedmore (1926) (Fig. 1.5). This was developed experimentally by the British Electrical Reasearch Association and was put into production in Germany (Biermanns, 1929) and Switzerland (Walty, 1935).

In the higher-voltage field, a wide divergence of practice developed in the early period 1935—45. This was between the German AEG Freistrahl designs (Biermanns, 1938), which operated at up to 110 kV per break and employed nozzles made of

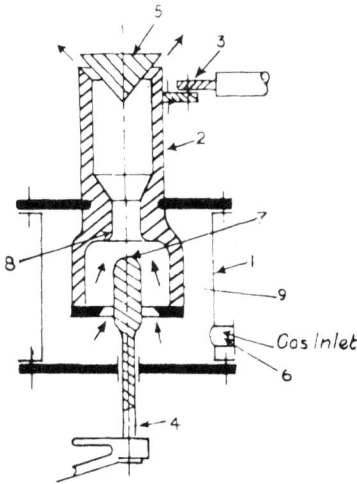

Fig. 1.5 Air-blast circuit breaker nozzle system 1926 [ERA, 1926]

insulation developed by Ruppel, a system still in effective use (Fig. 6.17), and the metal nozzle technique, which was used by most other manufacturers working at about 35 kV per break, in which a number of interrupters were mounted in series in one vertical column exemplified by a Brown Boveri concept (Thomen, 1941) (Fig. 1.6). Advances in airblast engineering since that period have resulted in many changes in design configuration. One may compare the earlier designs with interrupters arranged in vertical array with the later trend to mount these horizontally, exemplified by a Metropolitan–Vickers range of designs with two, four or six breaks in parallel per blast pipe, the primary objective being to ensure symmetrical aerodynamic conditions and identical performance from each interrupter (Flurscheim and L'Estrange, 1949) (Fig. 1.7).

Another major change has been in the method of providing the insulation strength in the open position, which has seen the automatic external airbreak isolating switch used with earlier types replaced in most e.h.v. designs by more sophisticated systems using internal pressurised isolation. The first production design of the internal isolation system was produced by Brown Boveri (Thomen, 1950) with an arrangement in which the contact system was held open under pressure, the contacts being closed by releasing the entrapped compressed air to atmosphere (Fig. 1.8). This concept not only eliminated the series isolator, but it also eliminated the need for any mechanical coupling between the control mechanism and the live contact system.

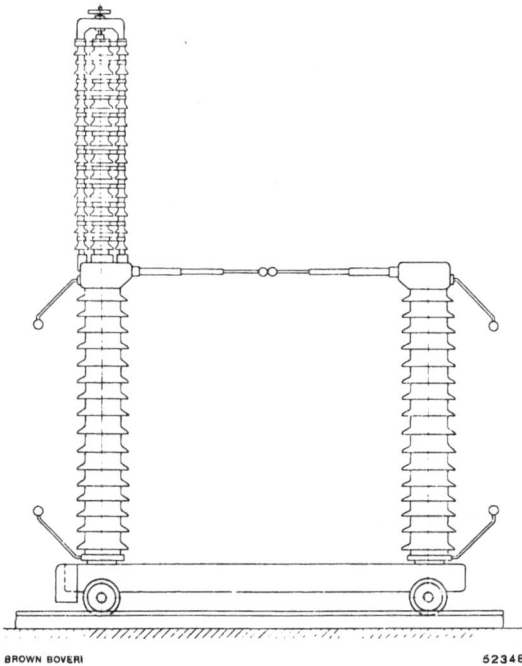

BROWN BOVERI   52348

Fig. 1.6 Multibreak air-blast circuit breaker with 12 interrupters in series aerodynamically 1940
[Brown Boveri, 1940]

More recent development has been directed towards maintaining the insulation pressure vessel containing the interrupting system under full pressure both when closed and when open, this being required to obtain higher breaking capacity and shorter interrupting times. These later systems have the advantage of avoiding compressed air supply tubes between live metal and earth, which operate part of the time at atmospheric pressure and which therefore depend on air conditioning to prevent internal condensation under unfavourable conditions.

The criteria for the number of breaks to be used are no longer based entirely on interrupting characteristics because the omission of the external series isolation has made it necessary to provide the required impulse level internally within the interrupter gap assemblies, and the number of breaks used may therefore be determined either by interruption or by impulse level requirements. Gasblast interruption is dependent on cooling, and is therefore influenced by aerodynamic configuration including both nozzles and gas flow passages, as well as by mass flow.

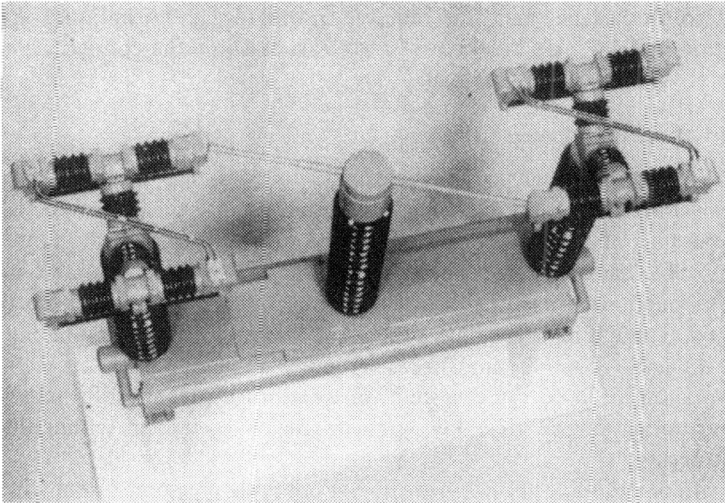

Fig. 1.7 Multibreak air-blast circuit breaker with eight interrupters in parallel aerodynamically [Metropolitan-Vickers, 1950]

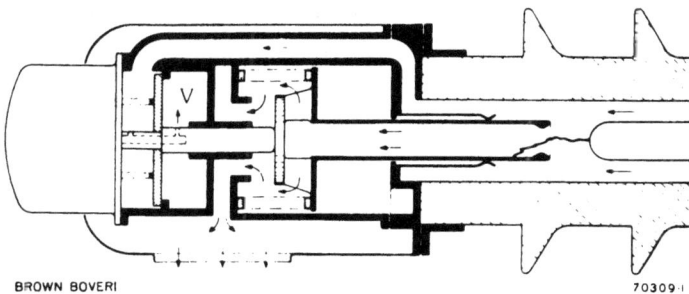

BROWN BOVERI   70309·1

Simplified longitudinal section through an arc extinction chamber

Fig. 1.8 Air-blast interrupter with internal insulation [Brown Boveri, 1950]

Restriking voltage stress gradients in the arc plasma and impulse breakdown stress in the open position are controlled by the field form and are also dependent on contact and interrupter shape, and this requires careful design to optimise the overall interruption and impulse breakdown performance since the configuration requirements for these may not be identical.

1.5 Development of circuit breakers: SF$_6$

Air is of course the most easily available gas for interrupting purposes, but alternatives have attractions, of which sulphur hexafluoride has so far proved the most effective. Because of its electronegative attraction for electrons and its high ionisation energy, SF$_6$ has an interrupting characteristic at 15 atm comparable to compressed air at around 50 atm so that it offers the possibility of working at much lower pressures or at much higher ratings per break. The necessity for containing the exhaust gas on both economic and health grounds requires that SF$_6$ circuit breakers shall work on a closed cycle system, in which high pressure SF$_6$ passes through the interrupting nozzle to a low pressure region contained within the circuit-breaker structure.

The more usual construction has been to provide a double pressure system with the gas held at two pre-determined pressures by a small electrically driven pump, an arrangement first successfully developed in production by Westinghouse (Leeds, *et al* 1957 and Friedrich and Yeckley 1959) (Fig. 7.4).

The alternative developed in different form by Westinghouse, (Easly and Telford, 1964) by Delle Alsthom in France and by Magrini in Italy (Fig. 7.9) is to provide the high pressure on a transient basis by compressing the SF$_6$ gas adiabatically during the opening operation, the energy to do this being provided by an external mechanism. This is the Prince impulse breaker principle applied to a new media and the lower mechanical energies required with SF$_6$ as compared with oil have made this system an attractive possibility. In recent years this type has in fact made quite remarkable progress, and the system voltage per break that can now be met is of the order of 250 kV, made possible not only by the interrupting capability of SF$_6$ gas, but because of its very high dielectric strength.

1.6 Development of circuit breakers: Vacuum

Contrary to all the trends the newest and most sophisticated of designs, the vacuum circuit breaker is the simplest yet proposed, if this is judged by mechanical construction, for it consists, in principle, of only a fixed and movable contact located in a vacuum vessel. When the contacts are separated, the arc is supported by ionised metal vapour derived from the contacts instead of by ionised gas as in other forms of interrupter, and at current zero collapse of ionisation and vapour condensation is very fast, ensuring interruption virtually independent of the rate of rise of restriking voltage.

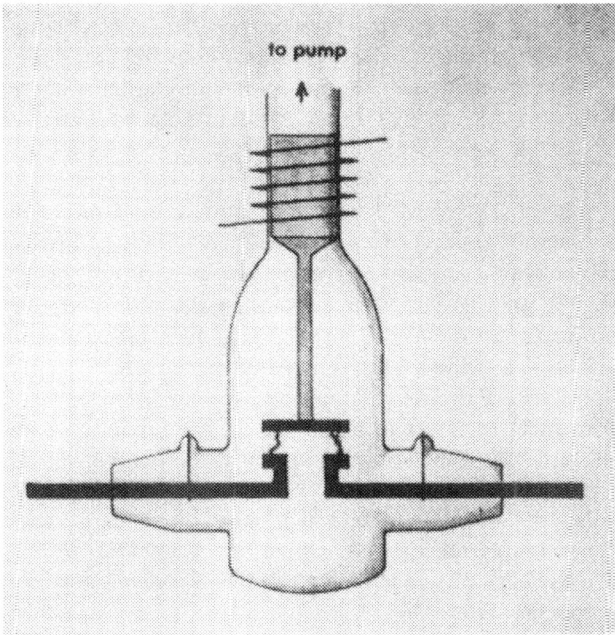

Fig. 1.9 Vacuum circuit breaker [Sorenson and Mendenhall, 1926]

The development of vacuum interruption which in concept seems so simple has, however, proved by an order of magnitude to be the most difficult exercise in the switchgear world. Early work was originated in the USA as far back as 1926 (Sorenson and Mendenhall, 1926) (Fig. 1.9), and they succeeded at that time in interrupting small currents at voltages up to 40 kV, but the techniques did not then exist to make it possible to build high power interrupters. It was not until 20 years later that the extensive programmes of research in the field of pure metals made further work in vacuum power interruption worthwhile. Development was recommenced in the early 1950s in a number of laboratories, but there were many difficult problems to resolve. These included the necessity to remove completely the gas molecules absorbed in the metal of the components or attached as a thin layer on exposed surfaces, and the prevention of recontamination during assembly operations, requiring development of a number of elaborate processing techniques; the necessity to provide efficient cooling of the contact surfaces to prevent residual pools of molten and over-heated metal persisting at current zero with consequential restrikes in the resulting metal vapour atmosphere, attacked by electromagnetic arc rotation; problems associated with cold welding of the contacts as the result of extreme cleanliness, and the production of metal whiskers from the contacts which reduced impulse breakdown, and current chopping with resultant overvoltages, all difficult questions of metallurgy; problems of voltage distribution and deterioration of surfaces caused by metal vapour deposition on the internal insulation, a question of shielding; and problems of metal to glass seals and flexible metal bellows, which again required new processing techniques to attain the high degree of reliability required.

The first interrupters capable of really heavy short-circuit duty were built around 1960 by General Electric, USA (Lee *et al.,* 1962). A decade of intensive research and development was then necessary to convert the early prototypes into production designs, which required sustained confidence in the objectives. Both vacuum contactors and power circuit breakers then went into quantity production, largely as the result of work by Lee in the USA and by Reece in the UK.

Vacuum interruption now offers the nearest approach so far achieved to the ideal circuit-breaker performance, in terms such as freedom from maintenance of the contact system for life, and consistency of interruption at the first current zero after contact separation. However the maximum service voltage per interrupter is still restricted to the order of 36 kV, and although multi-break e.h.v. circuit breakers have been demonstrated as technically feasible, the economic problems associated with such assemblies have not as yet been adequately solved in relation to the competition, more especially from SF_6.

1.7 Development of circuit breakers: special systems

The increases in short-circuit rating and reduction in total break time suggested here as being required in the future will present difficult problems for all classes of design. With conventional tripping we cannot avoid the possibility of an arc duration of up to 1 cycle in the case of the asymmetrical short circuit, with the consequential problems of contact deterioration and excessive residual metal temperatures at current zero, which could interfere with interruption and make 1-cycle performance unreliable.

This has resulted in the development of selective tripping systems to prevent the contacts opening until the ideal moment for contact separation is approached. This envisages an electronically controlled locking device on the trip gear which delays initiation of contact movement until the latest permissible point in the current wave is reached, which still leaves adequate time for the contact system to open sufficiently to guarantee interruption at the following current zero. It also requires a circuit breaker which provides very high acceleration and needs only a small travel for interruption and which has extreme mechanical and electrical consistency. High acceleration is essential so that the time within which the character of a short circuit can change once the trip sequence has been set in motion is kept to a minimum. Development is moving towards using point-on-wave closing for extra-high-voltage transmission lines with the objective of reducing the overvoltages than can be generated when long lines are energised at an unfavourable point on the voltage wave, and with improved reliability such electronic techniques may become acceptable as an aid to interruption.

The rate of increase in system short-circuit levels could be slowed down by a number of techniques, some of which are relevant here as being circuit interrupting systems. One is the use of high-voltage direct-current interconnections, which avoid the fault contribution provided by a.c. ties, but which present a new set of switching problems. Although d.c. circuit breakers have not so far proved essential they are desirable, and proposals have been made to base such a design on high arc resistance systems, or alternatively to produce an artificial current zero in the d.c. network by discharging the energy of a condenser into the system and then opening the circuit by means of a high-speed isolator synchronised with the artificial current zero thus produced.

The second interesting and relevant system is the a.c. resonance link first proposed by Kalkner (1966) at AEG (Fig. 9.3) which is basically an automatic switching device, and which introduces a new concept in network design. The resonance link is a 3-phase tuned LC circuit coupling which has negligible effect on normal system operation but which intervenes in the case of system short circuit or over current to limit the first half cycle and subsequent values of through current. Detuning of the link results in the circuit assuming a high impedance, and this is arranged to occur as a function of the level of through current by using a capacitor bypass circuit having a characteristic similar to that of a saturating reactor.

Since the introduction of this concept a number of new developments have been made including the use of superconductivity, which have expanded their possibilities, but the overall economics of such installations is often marginal, which still restricts their application.

Moving into an entirely new form of circuit breaking, static switching by means of thyristors may also have a place in this field of system control (Fig. 9.17). The large array of series parallel units needed together with their complicated control circuitry does, however, make them uneconomic for normal service in their present stage of development but they may nevertheless have a number of specialised but useful applications.

1.8 Summary

In this chapter the historical background of circuit-breaker design has been outlined, and some of the problems that may require to be solved in the next 25 years have been identified.

Circuit-breaker design has not reached finality, and there is scope both for new systems and for advancing existing systems directed towards reducing total break time and switching overvoltages, increasing the maximum available voltage and short-circuit ratings, and improving reliability and cost effective performance. The design spectrum involved is very wide and the principal theoretical and practical design approaches which are peculiar to circuit breakers are considered in the following chapters. In all engineering development however there is in addition a component of 'engineering art', as opposed to the more definable scientifically based technologies, and in providing systems to control so complicated a phenomenon as the power arc the contribution made by 'art' is perhaps unusually significant.

1.9 References

BAKER, B. P., and WILCOX, H. W. (1930): 'The use of oil in arc rupture', *Trans. Am. Inst. Electr. Eng.,* 49, pp. 431-446

BIERMANNS, J. (1929): 'Hochleistungschalter ohne OL', *Elektrotech. Z.,* 1929-2 pp. 1073-1079 and 1114-1119

BIERMANNS, J. (1938*a*): 'Fortschritte im Bau von Druckgasschaltern', *ibid.,* 1938-1, pp. 165-168

BIERMANNS, J. (1938*b*): 'The AEG airblast breaker'. AEG Progress Report 1938-2

BOEHNE, E. W., and LINDE, L. J. (1940): 'Magne blast air circuit-breaker for 5000 volt service', *Trans. Am. Inst. Electr. Eng.,* 59, pp. 202-209

BOEHNE, E. W. (1941): 'The geometry of arc interruption', *ibid.,* 60, pp. 524-532 and 1944, 63, Pt II, pp. 375-386

COX, H. E., and WILCOX, T. W. (1947): 'The performance of high-voltage oil circuit-breakers incorporating resistance switching', *J. IEE,* 94, Pt. II, pp. **351-361**

DAVIES, D. R., and FLURSCHEIM, C. H. (1936): 'Development of the single break metalclad circuit breaker', *J. IEE,* 79, pp. 129-178

EASLY, C. J., and TELFORD, J. M. (1964): 'A new design 34.5 kV to 69 kV intermediate capacity SF$_6$ circuit-breaker', *Trans. Am. Inst. Electr. Eng.,* 83, pp. 1172-1177

FLURSCHEIM, C. H. and L'ESTRANGE, E. L. (1949): 'Factors influencing the design of high voltage air-blast circuit breakers', *J. IEE,* 96, **Pt. II, pp.** 351-361

FLURSCHEIM, C. H. (1956): 'Switchgear − a review of progress', *Proc. IEE,* 103A, pp. 239-262

FRIEDRICH, R. E., and YECKLEY, R. N. (1959): 'A new concept in power circuit-breaker design utilising SF$_6$', *Trans. Am. Inst. Electr. Eng.,* 78, Pt. IIIA, pp. 695-706

GARRARD, C. J. O. (1960): 'High-voltage switchgear − a review of progress', *Proc. IEE,* 107B, pp. 1523-1539

GARRARD, C. J. O. (1969): 'High voltage distribution switchgear − a review of progress', *ibid.,* 116, pp. 1031-1045

GERSZONOWICZ, S. (1953): 'High voltage a.c. circuit-breakers'

KALKNER, B, (1966): 'Short-circuit current limiter for coupled high power systems'. CIGRÉ Paper 301

LEE, GREENWOOD, CROUCH and TITUS (1962): 'Development of power vacuum interrupters', *Trans. Am. Inst. Electr. Eng.*, 81, Pt. III, pp. 629-639

LEEDS, W. M., BROWNE, T. E., and STRAM, A. P. (1957): 'The use of SF$_6$ for high power arc quenching', *ibid.*, 76, pp. 906-909

MORTLOCK, J. R. (1956): 'AC switchgear − a survey of requirements'

PRINCE, D. C., and SKEATS, W. F. (1931): 'The oil blast circuit-breaker', *ibid.*, 50, pp. 506-512

PRINCE, D. C. (1935): 'Circuit-breakers for Boulder Dam line", *ibid.*, 54, pp. 366-372

SKEATS, W. F. (1936): 'Special tests on impulse circuit-breakers', *ibid.*, 55, pp. 710-716

SLEPIAN, J. (1929): 'Theory of the De-ion circuit-breaker', *ibid.*, 48, pp. 523-553

SORENSON, R. W. and MENDENHALL, H. E. (1926): *ibid.*, 45, pp 1102-1107

THOMEN, H. (1941): 'An a.c. high speed circuit-breaker for very high service duty and breaking capacity', *BBC Eng.*, 28, pp. 292-294

THOMEN, H. (1950): 'Simplified outdoor type high speed air-blast circuit-breaker with voltage rating up to 380 kV', *ibid.*, 37, pp. 123-136

WALTY, W. (1935): 'The Brown Boveri air-blast circuit-breaker', *ibid.*, 22, pp. 199-207

WHITNEY, W. B. (1930): British Electrical & Allied Industries Research Association Report G/XT15 and *J. IEE*, 1929, 67, pp. 557-578

WHITNEY, W. B. and WEDMORE, E. B. (1926): British Patent 278764

WHITNEY, W. B. and WEDMORE, E. B. (1930): British Patent 366998

ZAJIC, V. (1957): 'High voltage circuit-breakers'

Chapter 2

Physics of circuit-breaker arcs

M. P. Reece, B.Sc., Ph.D., C.Eng., F.I.E.E.

2.1 Introduction: arc as switching element in circuit breakers

In chapter 1 it has been shown that the heart of a circuit breaker is a switching element having a variable impedance. In most circuit breakers, this variable impedance is a high pressure arc, although in some it may nowadays be a vacuum arc; in most hv dc convertors it is a low-pressure mercury arc, and in certain other switching applications, which include some modern hv dc convertors, it may be a 'discharge' in the crystal lattice of a solid-state device. The arc acts, in all circuit breakers, as a variable resistor, and the art and science of circuit breaker design is to control the arc so that its resistance varies with current in such a way that satisfactory current interruption will be obtained.

When interrupting direct current (Rudenberg 1922, and Von Engel, 1929) the resistance of the arc must be increased rapidly enough to force the current down to zero in a reasonably short time, but not so rapidly that high overvoltages are generated in the inductance of the circuit.

For a.c. circuit breakers (Rudenberg, 1922), there are two current zeros every cycle, and in the majority of a.c. circuit breakers the designers aim is to maintain a high-conductance arc during the period of current flow, to minimise the energy dissipated in the circuit breaker and to cause the conductance of the arc to decrease extremely rapidly a few microseconds before the current zero. The designers problem is to control the arc so that the energy-removal processes at the current zero are sufficiently intense to raise the resistance of the arc rapidly enough at the current zero to prevent reignition when interrupting large fault currents, and yet are not so intense that small currents are 'chopped' before the current zero, giving rise to high overvoltages.

Some relatively low-voltage (15 kV and below) a.c. circuit breakers, arc-chute (Babakov, 1947, and Dickinson *et al.*, 1946) and air magnetic breakers are designed to maintain a high arc resistance, so that the fault current is reduced to about one third of the prospective short-circuit current, and the arc is finally extinguished at a current zero. The rate of energy dissipation in the arc chute of the circuit breaker is almost the maximum power the circuit can develop, and this technique is therefore limited to relatively low-power circuits.

The physics of discharges in circuit breakers is only understood qualitatively; many of the processes occurring are too complex for analytical treatment at present. When it is considered that a rigorous mathematical analysis of turbulence in an anisotropic flow is not yet possible, without the effect of adding to the flowing-gas problem an electric arc, which involves temperature gradients of thousands of degrees per millimetre (with our knowledge of gas properties not very accurate), it can be seen that large uncertainties exist. Nevertheless, progress in circuit breaker-arc physics is probably more rapid now than at any time in the past, owing mainly to advances in aerodynamics and computational techniques.

It follows that the circuit breaker is quite unlike other electrical machines; in no sense can it be designed from first principles. Our knowledge of the arcing process merely guides us when making changes to a design, and that is all. We rely very largely on empirical knowledge, experience and on testing, in arriving at a final design. For these reasons, it will be clear that it is not possible to give accurate numerical accounts of the physics of circuit-breaker arcs; however, in this chapter, an attempt is made to show the bases of the physical processes controlling arc interruption.

2.2 Types of arc used in circuit breakers

Arcs used* in circuit breakers can be divided into high-pressure arcs, with pressures between one and some hundreds of atmospheres, and vacuum arcs, with ambient pressures below 10^{-4} torr. Since neither mercury-arc convertors nor semiconductor junctions have been used in normal circuit breakers, as opposed to certain specialised switching applications, these types of discharge are not discussed.

2.3 High-pressure arcs
2.3.1 Characteristics of high-pressure arcs
The arcing and extinguishing medium in all blast-type circuit breakers is a flowing gas, commonly either air or SF_6 in blast-type breakers, and hydrogen (Wedmore *et al.*, 1929) in oil circuit breakers. Table 2.1 shows the actual composition of the gases involved in different circuit breakers, and also the nominal composition used for theoretical studies and the pressure ranges involved. From the Table, it is clear that the gases of most interest to theoreticians are air and nitrogen, hydrogen and sulphur hexafluoride.

---

*The word *used* is employed advisedly; the arc is one of the most essential features of the circuit breaker, and not just a troublesome and unwanted phenomenon

Table 2.1   Pressure and composition of gases in circuit breakers

Gas		Approximate pressure range	Applications
Composition	Nominal composition assumed by theoreticians		
Air, 100%	Air, 100%, often	1 atm	Arc-chute circuit breakers
	N_2, 100%	5 atm — a few hundred atmospheres	Air-blast circuit breakers and e.h.v. load-breaking switches
H_2, 70% acetylene, 25% CH_4, 5% and many trace hydrocarbons	H_2, 100%	1 — 3 atm	Oil switches for load breaking
		3 — 100 atm	Oil circuit breakers, bulk oil and small oil volume
SF_6, 100% but may contain small quantities of air	SF_6, 100%	3 — 15 atm	'Single' and 'double' pressure SF_6 circuit breakers and e.h.v. load-breaking switches
'Hard' gas, H_2, CO_2, H_2 and many others	No detailed[†][‡] theoretical studies	1 — 10 atm	Load-break switches and expulsion fuses

*Rudenberg (1922)
†Trencham (1953)
‡Petermichl (1940)

Figs. 2.1 and 2.2 show the two arc geometries most commonly employed. Fig. 2.1 is the air-blast and SF_6 breaker basic geometry, with a flow of gas predominantly along the axis of the arc, but with a radial inward flow upstream and a radial outward flow downstream. This geometry also applies in the 'double' or 'divided' (Trencham, 1953) blast arrangements. Fig. 2.2 shows the oil-circuit-breaker geometry, in which the hydrogen and hydrocarbon gases provided by the thermally decomposed oil and fibre appear, at first sight, to flow transversely to the arc; however, when gas production from the decomposing surface of the baffle plates is allowed for, and the looping of the arc into the slots between the baffles, it will be seen that the flow is substantially axial.

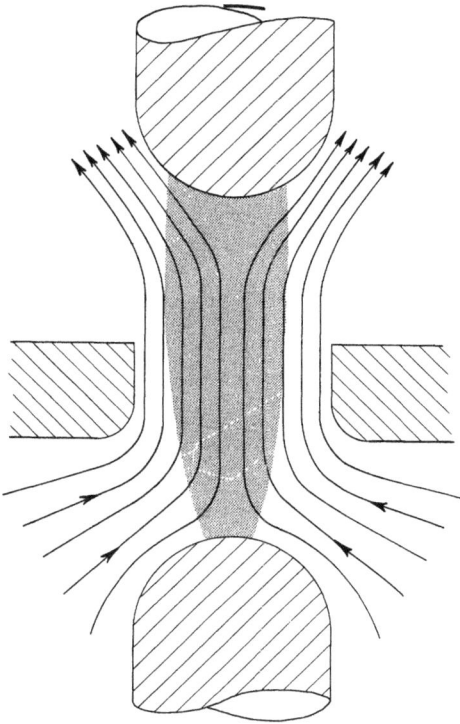

Fig. 2.1 Flow of gas through the arc in a blast-type circuit breaker

It is, therefore, not necessary to consider arc geometries other than axial flow, and the high-pressure arc thus may be taken to consist of a cathode region where electrons and metal vapour are emitted from the metal electrode into the plasma, either by 'field' emission (Fowler, 1928, and Lee 1959) or by thermionic (Jones, 1936) emission, an arc column, subjected to axial flow, where the current is carried by moving electrons and ions, and an anode region where electrons from the discharge enter the electrode, and where metal vapour may also enter the discharge.

2.3.2 Electrode regions
The electrode regions are usually considered to be relatively unimportant in relation to arc extinction in most high-pressure circuit breakers. Arcing electrodes in blast-type circuit breakers are sometimes highly refractory (pure tungsten) and sometimes semirefractory (sintered materials). Non- or semirefractory electrodes show a very rapid (in microseconds) dielectric recovery at current zero, but only to

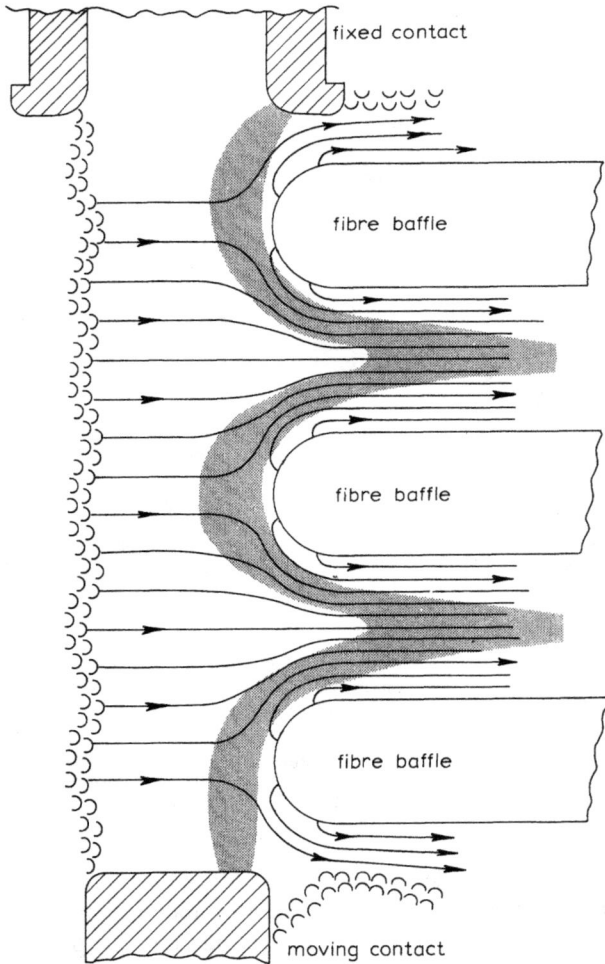

Fig. 2.2 Generation and flow of gas through the arc in an oil circuit breaker

a level of a few hundred volts. This recovery of electric strength is due to the relatively cold surface of the electrode cooling a very thin layer of plasma adjacent to it from temperatures of many thousands of degrees (Trencham, 1953), where the plasma is a good conductor, to a few thousand degrees only, where, as shown in Fig. 2.7 it forms an insulating layer. The voltage level to which these thin gas layers recover is determined by a number of parameters, and is related to the minimum sparking potential of Paschen (Loeb, 1939) curves, in which 150 V is typically the minimum possible breakdown voltage, whatever the condition of the gap. Theoretical studies of interruption processes in blast-type circuit breakers usually neglect this effect, since the electric strength of the layer is so low compared with practical recovery voltages in blast-type circuit breakers.

There is, however, one type of arc-chute circuit breaker that uses metal arc-splitter plates and operates wholly on this principle. It employs about 15 plates per kilovolt of recovery voltage. An early type of metal-splitter-plate arc-chute circuit breaker was developed by Slepian (1929) and described by Jamieson (1929) but this had complex magnetic circuits and was expensive to manufacture, and so there is only a simple type of metal-splitter-plate arc-chute in wide use (Reece, 1956, and Babakov, 1947) at present, and only for small ratings.

Another process that occurs at electrodes in high-pressure arcs is the formation of magnetically constricted jets (Maecker, 1955). The gas in these jets may have velocities of 10^5 cm/s, and these velocities are due to axial gradients in the current density, causing axial gradients in magnetic-field strength and axial pressure gradients. These jets can affect the performance of arc chutes and arcing horns in arc-chute circuit breakers, and can influence the injection of metal vapour into the arc column in blast-type breakers. The most important effect of the electrode region in high-pressure blast-type circuit breakers is probably this injection of quantities of metal vapour into the arc, with significant effects on ionisation and thermal conditions in the arc column.

2.3.3 Arc column

The arc column in the high-pressure arc is a cylindrical region in which ionised gas gives almost exact equality of positive- and negative-charge density, so that very high current densities may be supported with relatively low axial electric fields. Heat produced by the $I^2 R$ losses in the arc column maintains the ionisation, and flows axially towards the electrodes and radially outwards from the arc. In many low-current-arc studies, the axial flow of heat is neglected, but it is important in blast-type circuit breakers.

The ionisation in the arc column is maintained by the energy dissipated in the arc, IE watts per centimetre, and, as heat is continually being lost from the periphery and ends of the arc, the arc is all the time adjusting its conditions so that the heat input per unit time equals the heat stored per unit time plus the heat lost per unit time. Steenbeck (1940) showed that the diameters and temperatures in the steady state take values such that the power loss from an arc in any particular environment is a minimum.

2.3.4 Ionisation in arcs

Ionisation in arc columns (Von Engel, 1955, Von Engel *et al.*, 1932, and Cobine, 1941) takes place when any process occurs in which sufficient energy can be supplied to neutral particles to detach one or more electrons. This energy may be supplied by the direct impact of electrons, neutral atoms, positive ions or photons. It need not be all supplied in one encounter; a particle may be excited by one collision and ionised by the next.

In circuit-breaker arcs, two major ionisation processes are of interest. During the conduction period, ionisation is almost wholly thermal ionisation, owing to the thermal agitation of the very hot gas particles, but, very near to the current zero, ionisation by direct electron impact in the electric field of the discharge can occur.

2.3.4.1 Thermal ionisation

Thermal ionisation occurs when a mass of gas is heated sufficiently for the random thermal velocities of the particles to cause ionisation. The degree of ionisation depends on the pressure, the temperature and the ionisation potential of the gas or vapour. The ionisation potential is the work done in removing an electron from an atom or molecule, and is measured in electron volts; more than one electron may be removed per particle in highly ionised plasmas.

The detailed collision processes that cause thermal ionisation vary according to the gas temperature (Von Engel, 1955). At the lower temperatures, where the degree of ionisation is small, photons and molecules impacting with molecules or atoms are responsible for the largest proportion of the ionising collisions, but, at the higher temperatures, electron collisions are the dominant ionising process. For these, the energy of the particles colliding has been obtained thermally because of the high temperature of the gas, and not because of any high electric field accelerating the particles.

The expression for the proportion of particles in a gas thermally ionised was first derived by Saha (1920), and is given by

$$\frac{x^2}{1-x^2}p = 3{\cdot}16 \times 10^{-7} T^{5/2} \exp\left(-eV_i/kT\right) \tag{2.1}$$

Typical ionisation potentials (Von Engel, 1955, 1932 and 1934, Cobine, 1941, and Meek and Craggs, 1953) are as follows:

C_s = 3·9 eV	F_2 = 17·8
K = 4·3 eV	N_2 = 15·5 eV
Cu = 7·7 eV	He = 24·5 eV
Xe = 12·1 eV	H_2 = 15·4 eV
O_2 = 12·2 eV	S = 10·3 eV

SF_6 dissociates before ionising. The relationship between the fraction of particles ionised and the temperature is a sigmoid curve, and typical curves are shown in Fig. 2.3.

2.3.4.2 Ionisation by collision

The term Ionisation by collision is usually taken to mean ionisation occurring in a gas by electrons that are directly accelerated by high electric gradients in the arc,

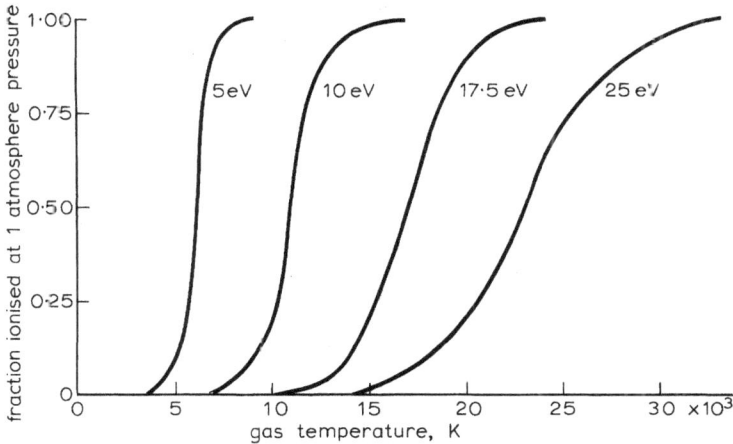

Fig. 2.3 Fraction ionised as a function of temperature for a range of ionisation potentials

and whose velocities are not only much higher than they would be if they were in thermal equilibrium with the surrounding gas at arc temperatures, but are no longer random in direction and are directed by the electric field. The probability of ionisation by collision (Von Engel, 1955, and Meek 1953) is, of course, zero when the impacting electron has an energy less than the ionisation potential, and rises to a maximum for most gases at about 100 V, falling slowly as the energy of the impacting electron is further increased. A typical curve of ionisation probability is shown in Fig. 2.4.

2.3.4.3 Formation of negative ions

Not only may positive ions be formed by the removal of electrons from a previously neutral atom or molecule, but, for some gases, negative ions can be formed by the attachment of an additional electron, so that the particle now has a negative charge (Von Engel, 1955, Meek, 1953, and Cobine 1941). This affinity of an already neutral atom or molecule for an additional electron depends on quantum-mechanical considerations of the stability of the electron-shell arrangements in the atom. It is expressed in electron volts, and varies from zero for the noble gases (which have filled electron shells), to a maximum of 3:9 eV for fluorine; some values are as follows:

F = 3·9 eV	O = 2 eV
Cl = 3·8 eV	O_2 = 1 eV
Br = 3·6 eV	H = 0·7 eV
I = 3·2 eV	

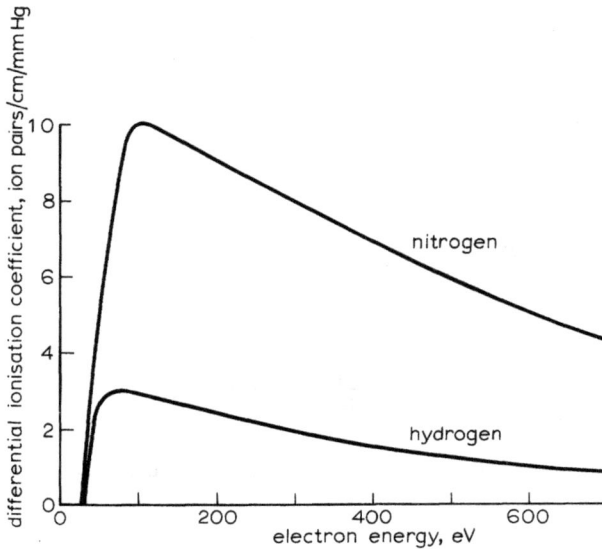

Fig. 2.4 Effect on ionisation efficiency of electron energy in ionisation collision (Meek and Craggs)

It will be found that the elements with the highest affinities have electron-shell structures in which the addition of a single extra electron changes the shell structure into that of the noble gas next to the element in the periodic table, and that noble gases have electron-shell structures of great stability. Elements with this property of attaching an electron are termed electronegative.

When a negative ion is formed, the kinetic energy of the incident electron and the electron affinity must somehow be dissipated. This almost always occurs by the emission of a photon, a process termed called radiative capture, but may also involve the transfer of kinetic energy to a third particle. The lower the velocities of the incident electrons, the greater is the probability of the formation of a negative ion. The formation depends on the neutral particle spontaneously emitting a quantum of energy during the period when the electron is within the atom. Now, since an electron of 1 eV passes through an atom in about 10^{-15} s, and the coefficient of spontaneous emission is of the order 10^8/s, the probability that a negative ion will be formed in any one encounter is about 10^{-7}. Measurements vary from 10^{-4}, for highly electronegative gases, downwards.

In considering the existence and movement of an electron in an electronegative gas, it is possible to define an 'attachment time' as the average time for which an electron will exist free in a gas at s.t.p. before being absorbed by a neutral particle to form a negative ion. Very approximate attachment times are as follows:

Gas	Attachment time, seconds
Noble gases	∞
CO	1×10^{-3}
NH_3	3×10^{-4}
Air	6×10^{-7}
O_2, H_2O	2×10^{-7}
Cl_2	5×10^{-9}
F	4×10^{-9}
SF_6	4×10^{-9}

It will be appreciated that negative ions are more easily formed at low temperatures, where the electrons are slower and therefore linger longer in the vicinity of the atom or molecule, than at high temperature.

2.3.5 Recombination

Where positively and negatively charged particles exist in a gas, there is a rate of recombination, pairs of oppositely charged particles recombining to form neutral particles (Von Engel 1955 and 1932). In doing this, they must emit energy, either as radiation or as kinetic energy; the amount is the original ionisation energy of the particles. Thus the degree of ionisation attained in a gas subjected to some ionising influence, such as a high temperature or an X-ray flux, is an equilibrium situation, where the rate of formation of ion pairs exactly balances the rate of recombination. In any process of recombination, the particles must remain close enough to one another for a sufficient time for recombination to occur, since the emission of excess energy is not a process that can occur instantly, but may take a few nanoseconds. This means that recombination occurs more rapidly at low temperatures, where particles have lower velocities, than at high temperatures.

In studying recombination, it is necessary to consider three main processes, although recombination can occur in a number of other ways. These three processes are as follows:

(*a*) recombination on solid surfaces
(*b*) negative-ion − positive-ion recombination
(*c*) electron − positive-ion recombination

Recombination on surfaces is highly effective in that, when a positive ion impinges on a surface, it is neutralised very rapidly and leaves the surface as a neutral particle. In high-pressure circuit breakers, the process is not important, because the mean free path, of the order of 10^{-5} cm and below, is so short that ions in the critical region of the discharge channel almost never meet a solid surface. In the vacuum interrupter, however, where the mean free path is many centimetres, this is the dominant deionisation process, the ions being mainly metallic, and, being deionised, as they condense on shields and electrodes.

Negative-ion — positive-ion recombination has a higher probability than electron — positive-ion recombination, mainly because the particles are moving more slowly at any given temperature, and thus remain adjacent for a longer time. The recombination coefficient, defined by the expression

$$C_i = - \frac{1}{N_i^2} \frac{dN_i}{dt} \tag{2.2}$$

has values typically of the order of 10^{-6} cm^3/s.

Electron-ion recombination is a process that is of most importance in discharges in noble gases, e.g. high-pressure zenon lamps and similar, where negative ions cannot occur. The coefficient, similarly defined by

$$C_e = - \frac{1}{N_e^2} \frac{dN_e}{dt} \tag{2.3}$$

is typically some two orders smaller than C_i, i.e. about 10^{-8} cm^3/s. C_i and C_e vary greatly from reference to reference, and little reliability can be placed on many of the published figures, because of the difficulties involved in making measurements.

Recombination rates are very sensitive to temperature and pressure, and, of course, very small quantities of electronegative impurities in the gas will form negative ions and will greatly increase the recombination rate.

2.3.6 Electron and positive-ion velocities in arc column

In the high-temperature arc column, the particles are in thermal equilibrium. They are all in rapid random motion, and each particle describes a random walk, but, when no overall electric field is present, no net movement occurs. The time-averaged energy of all the particles is the same, since they are in thermal equilibrium, and so $mu^2/2$ is the same for every particle and is equal to $3kT/2$. At room temperature, u for nitrogen and oxygen molecules is about 4×10^4 cm/s and, for hydrogen, about 16×10^4 cm/s. These velocities increase, of course, as the square root of the absolute temperature.

When an electric field is applied to the gas, the charged particles gain a drift velocity superposed on their random walk. The drift velocity is proportional to the electric gradient, the coefficient of proportionality being termed the particle mobility. The heavier the particle, the lower the mobility, since the particle loses its directed velocity at each collision of the random walk, and has to be reaccelerated in the direction of the field again. Similarly, the higher the pressure, the lower the mobility, since collisions occur more frequently.

Many different values can be found in the literature for mobilities of ions and electrons, since the mobility depends on the species of ion in a given gas, and, for electrons, on the nature of the gas. However, at atmospheric pressure, it is possible to take two figures, one for ions, positive and negative, and the other for electrons, and these at least give an order of magnitude. Ion mobilities are usually in the range $1-10$ cm^2 s^{-1}V^{-1}, and electron mobilities are typically in the range $10-10^4$ cm^2 s^{-1}V^{-1} however, wide variations can occur.

2.3.7 Current flow, conductivity and power dissipation in arc column

The current flowing per unit area of column is proportional to the number of charges crossing the plane perpendicular to the axis of the column per second. Although positive ions move in the opposite direction to electrons, the positive-ion current is added to the electron current, because of the opposite sign of the charges. The particle velocities are equal to their respective mobilities multiplied by the electric-field strength, and, since the mobilities of ions are typically one-thousandth of electron mobilities (because of their large masses), the ion current is only about $0 \cdot 1\%$ of the total current, and is neglected. Therefore

$$J = EeN_e(\mu_e + \mu_i) \cong EeN_e\mu_e \tag{2.4}$$

and

$$\sigma = eN_e(\mu_e + \mu_i) \cong eN_e\mu_e \tag{2.5}$$

where

$\quad\quad J$ = current density
$\quad\quad E$ = electric gradient
$\quad\quad e$ = electronic charge
$\quad\quad N_e$ = electron density
$\quad\quad \sigma$ = electrical conductivity
$\quad\quad \mu_e$ = electron mobility
$\quad\quad \mu_i$ = ion mobility

When the arc burns in an electronegative gas, or where electronegative impurities exist in sufficient quantity, the electrons become attached to form negative ions as the temperature falls before the current zero. The positive- and negative-ion currents are now about equal, and each is approximately one-thousandth of the electron current that would flow if attachment had not taken place. Therefore

$$\sigma = 2 \, eN_i\mu_i$$

and, since, $\mu_i = 10^{-3}\mu_e$, the conductivity drops to $1/500$ of its previous value. In practice, of course, falling temperatures reduce the degree of ionisation, as well as permitting attachment, so conductivities fall very rapidly with temperature. Attachment is particularly important in sulphur-hexafluoride circuit breakers during and after the current-zero period.

It is possible, therefore, if the nature of the gas is known, to calculate many of the electrical properties of the plasma as a function of temperature, since the ionisation density can be calculated from the temperature by the Saha equation. Fig. 2.7 shows electrical conductivity against absolute temperature for a number of different gases (Yos, 1967, and Frost *et al.*, 1971). The conductivities of these gases are rather similar above 5000 K by SF_6 is lower at temperatures in the 2000 K region, particularly when temperatures are falling rapidly.

2.3.8 Magnetic phenomena in arcs: circumferential fields

Since an arc is merely the flow of an electric current in a gas, it sets up magnetic fields and experiences forces due to magnetic fields, as does a current flow in any conductor. The practical difference between a gaseous plasma and a solid conductor, so far as magnetic forces are concerned, is that, in the solid conductor, the 'positive ions' are fixed and the forces, which act on the moving electrons, are transferred to the 'ions', and thus to the structure of the solid, electrostatically. For the plasma, of course, the ions can move. The axial current flow in an arc sets up a circumferential field. This circumferential field, interacting with the axial current, sets up a pressure acting radially inwards. The pressure on the axis is given, approximately, by $P = JI \times 10^{-8}$ atmospheres. If the cross-section of an arc changes along the length of the arc, J changes and therefore the pressure at the axis changes, and an axial pressure gradient exists.

This situation occurs at the electrodes of arcs, where the current densities are commonly higher than in the arc column. A high pressure exists therefore just in front of the electrode and this high pressure can set up a jet of plasma (Maecker, 1955, and Barrault *et al.*, 1972) moving away from the electrode at velocities of up to 10^5 cm/s.

2.3.9 Magnetic phenomena in arcs: transverse fields

When an arc is burning in a transverse magnetic field, the sum of the lateral forces on the charged particles in the arc is the same as the force on a solid conductor carrying the same current. The lateral momentum of the charged particles will be transferred to the neutral particles, and so powerful lateral flows will occur. The transverse magnetic field will, in part, be due to the current flowing in the arc itself and in its associated metallic circuit, and thus an initially straight arc will almost always bow under the influence of its self magnetic field, if the current is high enough. Because the self-generated magnetic forces on any circuit tend to increase the inductance of the circuit, the arc will bow out into a large loop and a long arc may even take up a helical shape under the influence of these forces. These long high-current high-pressure arcs, such as the arcs resulting from flashovers on h.v. systems, tend to be very mobile and show whorls and loops when high-speed films of them are taken.

If an arc subject to a transverse magnetic field (its own or a separate field) is prevented from moving transversly by being pressed against a stack of refractory plates, as occurs in certain arc-chute circuit breakers (Dickinson *et al.*, 1946) then the transverse pressure gradient pumps gas through the arc (magnetohydrodynamic pumping), and, when some arc-chute circuit breakers interrupt fault currents, the flows set up by the transverse magnetic field may reach sonic velocity.

In general, it may be said that magnetic phenomena in high-pressure arcs are not very significant in blast-type circuit breakers, where the pressures due to flowing gases are in general greater than magnetic pressures. However, in arc-chute-type circuit breakers, magnetic forces are very important, and, in vacuum interrupters, essential to the design, since it is only the magnetic force on the arc that causes the arc to enter the arc chute from the contacts, and, in the vacuum interrupter, only the correct magnetic design of the contact structure can prevent the formation of electrode spots with long thermal time constants.

2.3.10 Thermal phenomena in arcs
Two main problems are of interest in the transfer of heat in the arc. These are

(*a*) the transfer of heat within the arc
(*b*) the loss of heat from the 'surface' of the arc.

Although it is not possible to separate these two phenomena so completely as it is, for example, for a heated metal bar in air, considerable separation is possible for the arc in still gas. For the arc in flowing gas, the position is rather more complex and will be considered later.

The thermal conductivities of the very hot gases are so very much greater than those of cold gases, and their viscosities so high, that an arc does, to some extent, behave like a heated metal bar. Typical figures for thermal conductivities are shown in Fig. 2.5. It will be seen that the curves show peaks in the thermal conductivity at various temperatures; these peaks occur at those temperatures

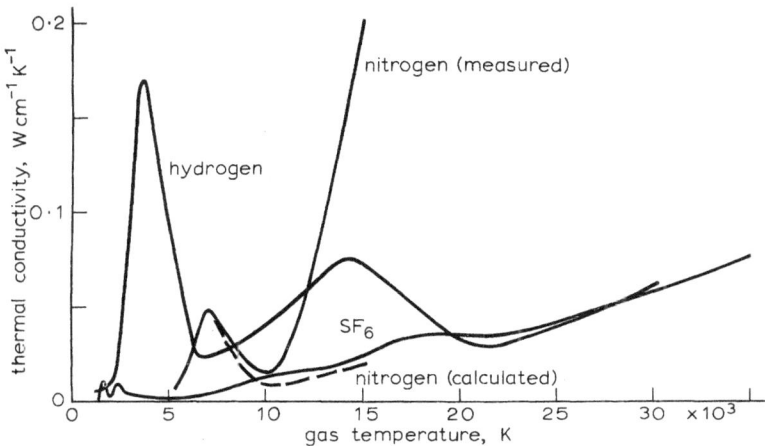

Fig. 2.5 Thermal conductivity

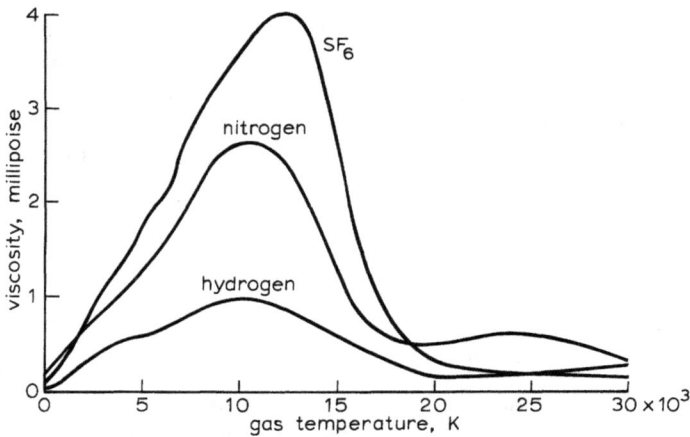

Fig. 2.6 Effect of temperature on viscosity

where the rate of change of dissociation or ionisation with temperature is a maximum. The cause of the dissociation peak, for example, at the temperature when the rate of dissociation against temperature is a maximum, is dissociated particles diffusing to the cold regions carrying their energy of dissociation with them, while undissociated particles diffuse from cold to hot regions carrying no dissociation energy; similarly for ionisation peaks. Peaks occur for second and higher ionisations, as well as for the first ionisation. The quality of the published data on thermal conductivity of hot gases is not good, and it is not uncommon to find an order of magnitude difference between the figures from various sources.

The viscosity of gases (Yos, 1967, and Frost *et al.*, 1971) at arc temperature is very much higher than at room temperature (Fig. 2.6), and this increase in viscosity means that, within the arc, the flow may frequently be laminar, when the flow surrounding the arc is turbulent. It is now considered by some workers that, even under the extreme convection occurring around the arc in blast-type circuit breakers, the core of the arc, at least during the high-current period, is experiencing laminar flow.

The electrical conductivity of gases as a function of the temperature has been shown to depend on the degree of ionisation and on the particle mobilities, and the conductivities of relevant gases at arc temperatures are shown in Fig. 2.7.

2.3.10.1 Temperature distribution in arcs

Now, with the data of Figs. 2.5 – 2.7, it is possible to compute the temperature distribution across a steady-state arc, providing that an external boundary condition

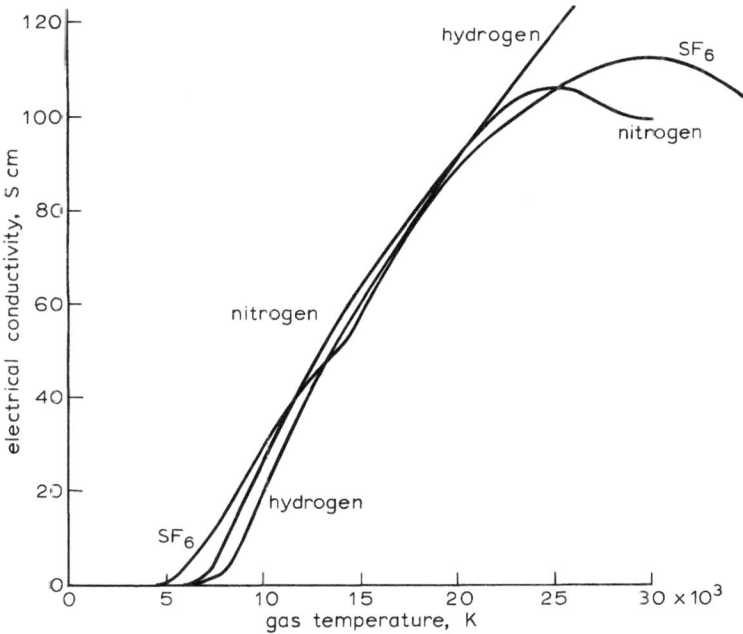

Fig. 2.7 Effect of temperature on electrical conductivity

(a known temperature at a known radius) and a voltage gradient can be given. The Elenbaas–Heller (Elenbaas, 1951) equation for a cylindrically symmetrical arc is

$$\sigma E^2 = \frac{1}{r}\frac{d}{dr} r\left(-K\frac{dT}{dr}\right) \tag{2.6}$$

This equation is derived by merely considering that the heat flowing radially out of any elementary annulus in the arc is the sum of the heat flowing into that annulus from the area of arc contained within it plus the heat generated within the annulus itself. It will be appreciated that a numerical solution of this equation is necessary, since σ and K are both varying rapidly with temperature (Figs. 2.7 and 2.5). A solution of the equation gives rise to a temperature profile across the arc, and a typical profile is shown in Fig. 2.8. In those temperature ranges where the thermal conductivity of the gas is very high, the temperature gradient is small, since heat easily flows radially outwards; where the thermal conductivity is low, as near the edge of the arc, where gas temperatures are low, the temperature gradients are high. It should be remembered that the thermal conductivity includes a transfer of

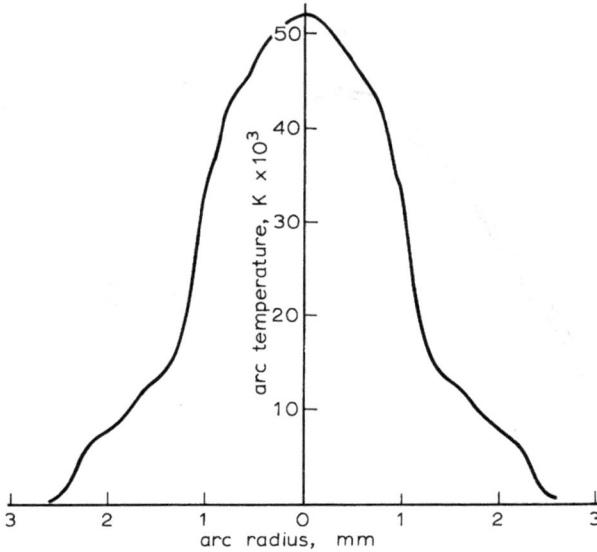

Fig. 2.8 Theoretical T/r distribution for 1 kA arc in axial N_2 flow [Shrapnel and Siddons]

energy from atom to atom by emission and absorbtion of radiation, as well as by physical collision between the atoms. Further, the longer-wavelength radiation can escape entirely from the arc, and this form of heat loss may account for up to 30% of the total heat loss of a very high temperature arc.

2.3.10.2 Arc boundary
A much more difficult problem emerges when we attempt to deal analytically with the boundary of the arc. At the boundary, the temperature falls steeply, and the thermal conductivity, electrical conductivity and the viscosity of the plasma all fall very rapidly. At this point, conduction is replaced by convection as the dominant heat-transfer process, and so far this region has defied satisfactory analytical treatment; at the boundary, the student of arcs generally falls back on measurements. A considerable amount of work (Topham, 1971) is being done on conditions at the boundary of the arc, and it is conceivable that an analytical treatment will be possible within the next ten years.

2.3.11 Static arc characteristics
The relationship between arc current and arc voltage is known as the arc characteristic: the static characteristics apply when changes of current impressed on

the arc take place slowly, and the dynamic characteristics when the changes take place rapidly.

Typical static characteristics (King, 1964) are shown in Fig. 2.9. The shape of the static characteristic depends on the mode of heat loss from the arc, as can easily be seen if we consider very simple models. Considering, first, a simple model arc with a constant current density and a constant heat loss per unit area of the cylindrical surface of the arc. The circumference of the arc column, and therefore, by definition, the heat loss, vi watts, is proportional to i, since, for a constant current density, $D^2 \propto i$. Hence

$$v \propto \frac{1}{\sqrt{i}}$$

It can be seen from Fig. 2.9 that this does give a shape not unlike the curve in the lower-current range.

We can now consider a slightly different arc model: again let the current density remain constant, but now assume that the loss vi is proportional to the cross-section area of the arc, such as might occur with gas flowing through the arc, removing heat from the whole volume of the arc. In this case, $vi \propto i$, and thus v remains constant, and does not change as i changes. It can be seen from the Figure that this gives a tolerably good reproduction of the higher-current regions of the curve. In practice, of course, as previously indicated, the heat-loss processes are much more complex, and so complex vi arc characteristics result.

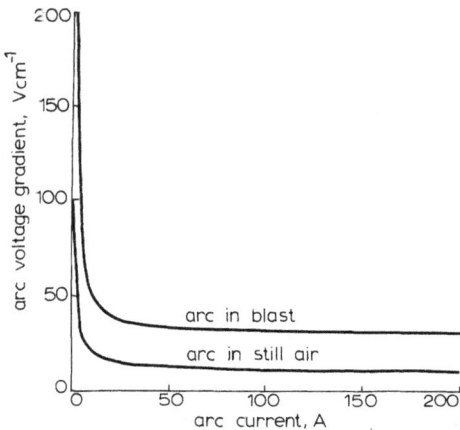

Fig. 2.9 Static *V/I* characteristics of arcs in forced and free convection [King]

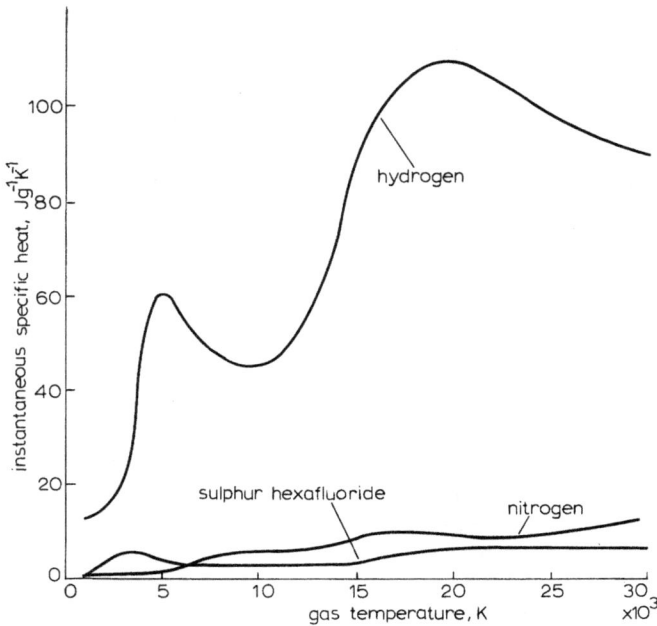

Fig. 2.10 Effect of temperature on specific heat

2.3.12 Arc in transient state

Section 2.3.10 shows, how, in principle, the temperature distribution across the arc in the steady state can be calculated. If we know the specific heat of the gas, we can also plot the energy stored in the various parts of the arc. Like the other properties of high temperature gases, specific heats vary steeply with temperature, and some specific heats (Yos, 1967, and Frost *et al.*, 1971) are shown in Fig. 2.10. Using these data, and data for density of the gases, the stored energy of every part of the arc can be computed. If now the current changes in some given way (usually falling linearly, as in the latter part of a sinewave), it is, in principle, possible to calculate a whole series of different temperature profiles, one for every element of time, by taking account not only of the heat generated in any annulus of the discharge by the current flow, but also of the heat released in the same annulus by its falling temperature, and so the problem of the transient state can also in principle be solved numerically. In practice, the problem is once again the boundary of the arc. To solve the heat-flow problem in the arc core, a boundary condition is needed; this boundary condition is not yet amenable to calculation, and so no complete analytical arc solution is yet possible.

Fang, Fung and Edels (1973) and Bowman and Edels (1968) have shown how, provided that the problem of the arc boundary is circumvented by using a well-stabilised arc, i.e. one where heat loss is wholly by conduction, even the transient behaviour of low-current arcs can be accurately calculated, and they show results where experimental results agree closely with theory.

Because of the heat stored in the arc, the static arc *vi* characteristics (Fig. 2.9) only apply when the changes of current occur slowly. If the current changes virtually instantly, in step-function form, the arc initially acts as a linear resistor, the voltage change being proportional to the change in current. Because the conductance of the arc depends on the stored energy, it cannot change instantly. This state may only last for a fraction of a microsecond, however, and then the voltage begins to change (though the current remains constant at its new value) and exponentially approaches the level determined by the static characteristic for the new value of current. If the arc is in the throat of an air-blast breaker, with a very large ratio of power loss to energy stored, the voltage will return to the static characteristic in a few microseconds, since the arc time constants are short. If, however, the arc is unconfined and not subjected to strong forced convection, such as the arc resulting from a flashover of an overhead-line insulator, the voltage may take a millisecond or more to return to the value given by the static characteristic.

2.3.13 Dynamic arc equations and arc extinction

It became apparent in the early 1930s that the current interrupted and the magnitude of the recovery voltage were not, in themselves, sufficient to specify the duty on a circuit breaker, and that the rate of rise of the transient-recovery voltage could also affect the performance of circuit breakers. This led to the development of the concept of a 'race' between the growing electric strength of a circuit-breaker gap after arc extinction (Slepian, 1930), and the transient recovery voltage of the circuit, the results of the race determining whether the gap interrupted or reignited.

It was at first propounded (Prince and Poitras, 1931) that the arc was 'severed' at the instant of current zero, and that the electric strength was merely the breakdown voltage of the gap in the arc column. This hypothesis, which may describe conditions in the circuit-breaker nozzle reasonably well at long times, say 100 μs after zero, when reignitions are due to 'dielectric' breakdown, was originally conceived as applying at all times after the current zero, even a few microseconds, when we now know that 'thermal' reignitions involving energy-balance processes usually occur.

Later in the 1930s, however, it became obvious from aerodynamic studies that it was not possible to 'sever' an arc in a gas blast, at least, not in the few tens of microseconds in which interruption effectively occurs, and that, even if it had been possible, the velocity of the gas, limited to the speed of sound, of the order of 1 mm/μs, could not possibly account for the initial rate of growth of electric strength between the broken ends of the 'severed' arc. Consideration of the flow lines soon showed that, in these short times, the arc could be stretched lengthwise and thinned down, but could not be cut.

These arguments were developed during the 1930s, and led Cassie (1939) to put forward the first energy-balance theory of arc extinction, and Slepian and Browne (1941) to suggest that turbulence in the gas flow was the means by which the very rapid dissolution of the arc was obtained. A considerable amount of work around the world, both experimental and theoretical, on the manner in which the resistance of the arc changed during the current zero period, commenced at this time.

Some of the first rough estimates of arc resistance during the zero period were made by measuring the degree of damping of the transient-recovery voltage introduced by the postzero resistance of the circuit breaker; this was followed by a method where a constant amplitude of radio-frequency current was superposed on the power-frequency current, and the radio-frequency voltage across the arc measured during the zero period; this gave a measure of the arc resistance. Finally, in the late 1940s and early 1950s, developments in oscillography, forced on during the Second World War, enabled the small currents flowing in the arc during the zero period to be measured directly (Trencham, 1953), in spite of the high fault current immediately preceding the current zero.

The finite conductivity after the current zero indicated by the damping measurements was at last beginning to show the effect of stored energy in the arc gap, leading to a finite time for the arc resistance to rise, and researchers now turned their attention to formulating mathematical models of the arc that would explain the measurements.

In 1938, Cassie (1939) postulated an arc that had a constant current density, so that its area varied directly with the current; it also had a constant resistivity and stored energy per unit volume. This model was intended to represent an arc in an air-blast circuit breaker, and Cassie assumed that the air flow penetrated the whole cross-section of the arc, carrying away heat, and thus made the dissipation per unit volume constant as well.

These assumptions led to the following differential equation:

$$R\frac{d(1/R)}{dt} = \frac{1}{\theta}\left\{\left(\frac{v}{v_0}\right)^2 - 1\right\} \tag{2.7}$$

where

R = arc resistance
v = arc voltage at any instant
v_0 = arc voltage in steady state
θ = time constant = $\dfrac{\text{energy stored per unit volume}}{\text{energy loss rate per unit volume}}$

If a constant current were suddenly cut off,

$$R = R_0 \exp(t/\theta) \tag{2.8}$$

the resistance increasing exponentially as the stored energy is removed at a finite rate by the air flow. Although this model indicated, in a general way, the properties of time constants in arcs, it had a serious weakness, in that such an arc could never sustain, for any appreciable period, a voltage greater than its steady-state burning voltage, and thus such an arc could not interrupt. This was because the loss of power from the arc was a volume effect, and the voltage that would inject just sufficient power to supply the losses of the arc was independent of arc diameter. This type of arc model in fact described the performance of the arc in the air-blast breaker very well under conditions when the current was large, but it did not really seem applicable to an arc near current zero.

In 1943, Mayr (1943) proposed a somewhat improved model, in which the loss occurred from the periphery of the arc only, and the conductance of the arc varied exponentially with the energy stored in it. This model gave a differential equation

$$R \frac{d(1/R)}{dt} = \frac{1}{\theta} \left\{ \left(\frac{vi}{W_0} \right) - 1 \right\} \tag{2.9}$$

the time constant is, of course, different to Cassie's in its derivation, although it is still a ratio of energy stored to rate of energy loss. This equation, it is clear, has certain advantages. In the steady-state condition, when currents and voltages are changing slowly

$$\frac{d(1/R)}{dt} = 0$$

and therefore $vi = W_0$. Thus the steady-state characteristics are hyperbolic, and this is a moderately good representation of what happens in a circuit breaker in the low-current region (Fig. 2.9), and therefore it was hoped that it would represent what would happen near the current zero. This equation does allow the arc to interrupt, since, after current zero, R can be very high and v can be very large, since, with R high, i, will still be very small and vi/W_0 can still be less than 1, so that $d(1/R)/dt$ is negative and the resistance of the arc continues to increase towards the final interruption.

Many other arc models have been proposed since, and, although they have become more complex with time, none of them has truly been able to represent the extreme complexity of the processes occurring in circuit breakers at current zero. In 1948, Browne (1948) developed a mathematical model of the arc that combined many of the features of previous models, but, once again, absolutely no progress was made in numerically relating the terms of the equation with the physical properties of the interrupting medium.

In 1958 Browne extended his mathematical model combining both the Cassie and Mayr models and suggested the use of the Cassie model before current zero and the Mayr model after zero.

The 1960s started a new look at the physics of the arc-interruption processes and some doubt was expressed by Rieder and Urbanek (1966) and Lee and Greenwood (1968) regarding the assumption of thermal equilibrium used in most models up to that time. The paper of Rieder and Urbanek extends the previous models to include non-equilibrium effects in the post-arc region of the interruption process.

The aerodynamic effects on the arc and the effects of the high-current arc on the gas flow were studied by Zuckler (1969) and Topham (1970). The paper of Zuckler shows how the main gas flow could be reversed at high currents and that normal flow may not be recovered at the instant of current zero.

During the 1970s, a number of investigators have published papers on a new approach to arc modelling.

It was realised that the Cassie-Mayr models inadequately described the physical processes occurring in gas-blast circuit breakers, and that in some cases severe departures from these two models occurred.

It appears that for a satisfactory understanding one is compelled to resort to the numerical solution of the arc conservation equations in full differential form. These highly nonlinear simultaneous partial differential equations are notoriously difficult and costly, in terms of computing time, to solve, and the discharge conditions are extremely complicated and not easily defined mathematically.

The design engineer today asks questions which cannot be answered by the Cassie-Mayr models. 'How do we optimise the mechanical design of the interrupter?' 'What nozzle diameter and electrode gap is required for a given rating?' 'What radius should the entry to the nozzle be?' Thus a method of arc analysis which blends empiricism with formal analysis will have to be found.

The arc models of Swanson and Roidt in a series of seven papers from 1970 to 1977 start from the basic conservation equations and deal with the different aspects of arc modelling. The main theme of Swanson and Roidt's papers is the dominant role played by turbulence in the energy removal.

In 1972 Butler and Whittaker proposed an arc model which showed very good agreement between theoretical predictions and measurements, although it still did not start from first principles. Butler and Whittaker have taken an arc model that, in the centre of the conducting region of the arc, takes cognisance of energy stored, as well as electrical and thermal conductivity, and in which the only significant loss process in the centre region is radial thermal conductivity. So far this is the common basic assumption made by most workers, but the boundary conditions imposed are quite new. An arbitrary loss process, giving rise to a loss of q per unit area, is assumed to occur at the radius at which the electrical conductivity falls effectively to zero, r_c. It is assumed by the authors that $q = Qr_c{}^n$. Values for Q and n can be chosen so that the steady-state arc analysis gives a static vi characteristic similar to an experimentally determined characteristic. Using this concept, the arc equations are normalised, by using simple relationships between gas properties and temperatures, and four basic equations are derived that describe the arc model. The arc that results from this model has more sophisticated properties than previous models, and properties that appear realistic to those who have worked on current-zero processes.

First the normalised shape of the temperature distribution changes during the zero period, and so does the arc radius. There is no longer a single arc time constant, but the time constant varies with time throughout the zero period. Secondly, when $Rd(1/R)/dt$ is plotted against vi, the instantaneous arc voltage and current during the current-zero period, the plot may have peaks and troughs, or even loops. This is exactly the behaviour of similar plots made from measured values of v and i in circuit breakers during the current-zero period, and previous arc equations have never been able to produce plots similar to the measured ones. Finally, it is of interest to note that Butler and Whittaker are at present attempting, in collaboration with workers who have specialised in the study of the arc-boundary conditions, to replace the empirical portions of their theory, i.e. Q and the exponent n, by data taken from basic arc theory. If this is achieved, a very real advance will have been made, in which the true behaviour of a gas-blasted arc at current zero can be calculated from first principles.

Further extensive investigation on arcs in gas flow have been carried out at Brown Boveri and in 1977 a model was proposed by Hermann and Ragaller (1977). In this work turbulance plays the dominant role; the authors also argue that the arc behaviour is dominated by the section upstream of the throat during the high-current phase and by the section downstream of the throat during the current-zero period. Thus it is considered that the magnitude of turbulence at some distance downstream of the throat is the reason for arc extinction.

Lowke and Ludwig (1975) developed a simple arc model for convection-stabilised arcs that rigorously applies only to a certain current regime. The current must first be large enough so that thermal conduction and turbulence losses do not influence the central arc temperature. The current must not cause blocking in the nozzle being studied. Tuma and Lowke (1975) applied the Lowke-Ludwig model to analyse the data of Hermann *et al.*, and besides the adiabatic approximation they also consider clogging and isothermal approximations for pressure and Mach number. El-Akari and Tuma (1977) extended the Tuma-Lowke model and applied it to analyse transient arc behaviour.

The desirable features of these models are that they account for nozzle geometry and are relatively simple. Their undesirable feature is that the equations for temperature and arc diameter are not so satisfactory near current zero.

The desirable features of the general model of Hermann *et al.* are that it is satisfactory for all currents and information on the effect of nozzle dimensions on the arc behaviour can be obtained. The undesirable feature is that this general model requires the difficult solution of six simultaneous partial differential equations that are used to define the two zone model.

In 1978 Fang and Brannen produced a simplified arc model for the current-zero period on the general formulation of boundary-layer integral method of arc analysis by Cowley (1974). The model takes enthalpy and kinetic energy transport into consideration, and is based on the following assumptions,

(i)   Temperature profile (magnitude and shape but not the radial extent) is characterised by the dynamic power loss, which is reduced to the power input per unit arc length for steady state. Thus, in order to generate the temperature profile, it is assumed that the temperature in the conducting core is the same as the wall-stabilised arc with the same dynamic power loss, and a parabolic profile is assumed for the thermal region, matching the two profiles at the core boundary. The method is the same as that used by Topham (1971) and by Chan *et al.* (1975), and is successful in predicting steady-state *E/I* characteristics. The validity of this method is based on a comparison of core temperature profiles under different arcing conditions (Topham, 1971).

(ii)   The axial velocity profile ω is generated by assuming constancy of kinetic energy density across the arc, $\rho\omega^2 = \rho_\infty \omega_\infty^2$, where subscript ∞ refers to the quantities in the external flow (Cowley, 1974). This assumption is equivalent to constant Mach number and has been verified as a good approximation during quasi-steady state (Malghan *et al.*, 1977). It should be noted that after the breakdown of quasi-steady state the velocity usually lags behind the temperature variation (Fang *et al.*, 1973). Thus, $\rho\omega^2 =$ constant tends to underestimate the velocity before current zero and to overestimate after current zero. It is difficult to assess quantitatively the effect of this assumption because of the scarcity of experimental information on arc velocities.

(iii) Arc radiation is neglected. This is justified by choosing a current ramp (i.e. I_0 and di/dt) before current zero such that a quasi-steady state is assured while radiation can be neglected. For most gas-blast circuit breakers this choice can be realised.

With the foregoing assumptions, the governing equations are reduced to:

$$\frac{\partial}{\partial T}\left(\frac{\Theta_\delta}{a}\right) + \Omega_h\, V_\infty \frac{\partial}{\partial z}\left(\frac{\Theta_\delta}{a}\right) + V_\infty\left(\frac{\Theta_\delta}{a}\right)\frac{\partial \Omega_h}{\partial z} = \frac{V_\infty\, i^{\cdot 2}\ (T)}{a_t^2\ \Omega c\left(\frac{\Theta_\delta}{a}\right)\alpha}$$

Where Θ_δ = thermal area, a = nozzle area, V_∞ = axial velocity, a_t = throat area, i = current, α = nozzle shape, z = axial co-ordinate, and T = time.

All quantities are non-dimensional (Fang and Brannen, 1978). This equation can easily be solved numerically by the method of characteristics. The conductance shape factor Ω_c and the enthalpy flux shape factor Ω_h are uniquely related to the local dynamic power loss by assumption (i) and (ii). For the first time, a quantitative definition of the current-zero period has been given. The model gives detailed information on the axial field strength, thermal area and the arc temperature. The effects of different nozzle shapes on the arc behaviour can easily be investigated, and for the first time the model is refined enough to investigate the effects of current-waveform distortion around current zero due to arc-circuit interaction. The model predicts satisfactorily without resort to turbulence and to the adjustment of parameters the critical r.r.r.v. $\sim di/dt$ and r.r.r.v. $\sim p_0$ relations for an orifice arc in air flow, where detailed current waveforms near current zero are available (Chapman, 1977). There are probably two reasons for the success of this model:

(*a*)   Only overall energy balance is considered. Thus the effects of turbulence (if it exists) at the core boundary does not affect overall energy balance;

(*b*)   Turbulence does not seriously affect the temperature and velocity profile, and the two shape factors are not very sensitive to the detailed structure of the profiles.

The assumptions used in various recent arc models have been discussed in detail. The models proposed by Swanson and Roidt, and by Hermann and Ragaller involve assumptions which cannot easily be justified on either experimental or theoretical grounds. Thus, it is by no means clear whether turbulence plays the dominant role in energy transfer during current-zero period. The most self-consistent and rigorous arc model seems to be that proposed by Fang and Brannen. Their model does not involve adjustable parameters, and has achieved quantitative agreement without introducing turbulence. The model is still simple enough and economical in computing time.

It is still (in 1981) not possible to design a circuit-breaker nozzle from first principles. However, modern computing power has enabled more complex arc equations to be solved, and these solutions are in some cases capable at least of guiding the circuit-breaker designer in a qualitative way.

Arc theory is, nevertheless, no substitute for experience, and this remains as true in 1981 as it was in 1950, and many empirical relationships still have to be used in design.

2.3.14 Post-zero currents and arc time constants

The fact that energy is stored in the arc column means that a curve of arc conductance must fall to zero sometime after the current zero, even in a circuit with a very small recovery voltage. This means that, at high currents, some 'post-zero' current will flow, which may or may not be sufficient to modify noticeably the shape of the inherent transient-recovery voltage. Figs. 2.11 and 2.12 show post-zero currents recorded (Irwin, 1971) in air-blast circuit breakers interrupting various currents. The magnitude and duration of the postzero current depend on the parameters of the circuit and the design of the circuit breaker, usually being greater in those circuit breakers with relatively long nozzles. It is necessary to view many of the published oscillograms of post zero currents with considerable suspicion, since sophisticated measurement techniques are involved

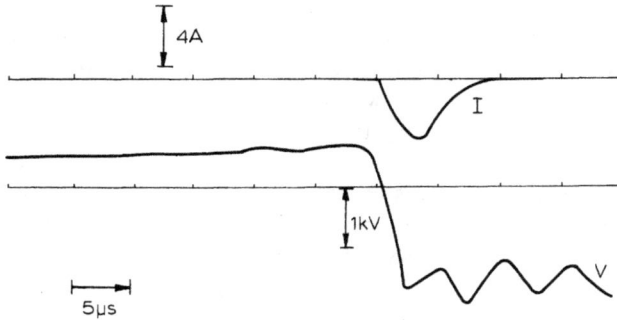

Fig. 2.11 Post-zero current following interruption of 8 kA r.m.s. at 1 kV r.m.s. (air blast)

and spurious results are easily obtained from inductance in shunts, zero shift and overshoot in amplifiers. Improper earthing of the recording circuits and isolating of the power supplies to the recording equipment is also likely to affect results adversely.

Many figures will be found in the literature quoting arc 'time constants'. It has already been seen that the properties of the arc are highly nonlinear, and, in practice, the decay of conductance is never strictly exponential. Some writers have

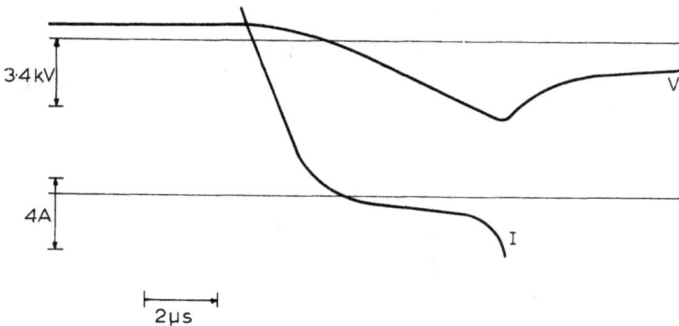

Fig. 2.12 Post-zero current leading to re-ignition by thermal failure

referred to 'time constants' in the mathematical sense, i.e. the time required for the arc conductance to fall by $1/e$; others tend to talk of it as, for example, the time between the passage of the current through zero and the conductance of the circuit breaker becoming substantially zero. Many authors talk about arc time constants without properly defining them. In any case, the time constant, according to most definitions, will vary with time throughout the zero period, with the magnitude of the current being interrupted and with the wave shape of the early part of the transient-recovery voltage. The term should thus not be uncritically accepted.

2.3.15 Current chopping

'Current chopping' is used to describe the sudden reduction of the current in a circuit breaker (Young, 1952) to zero at a time other than the 'natural' instant of current zero. Careful measurements have elucidated a process by which blast-type breakers may chop. The circuit breaker is in series with an inductive circuit, usually having a large inductance. There are, however, local, often distributed, LC circuits, having frequency responses in the 10-100 kHz range, between the circuit breaker and its load, and these have a resonant frequency and a dynamic resistance at that frequency. As the power-frequency current in the arc falls towards zero, the arc diameter shrinks and the time constant shortens, so that the arc can exhibit the negative resistance of the static characteristics at higher and higher frequencies. The negative resistance of the arc (at the resonant frequency of the local LC circuit), which is effectively zero at large currents, becomes larger as the power-frequency current falls. At some stage, the negative resistance of the arc becomes larger than the positive dynamic resistance of the local LC circuit, and thus the resistance of the whole circuit (arc and LC circuit) becomes negative. This is exactly the process used in the Poulsen arc transmitter in the early days of continuous-wave radio. When the total circuit resistance becomes negative in sign, high-frequency oscillations of the current in the arc build up, and, when the peak value of this h.f. oscillating current is equal to the instantaneous value of the power-frequency current, an 'artificial' current zero has been obtained, and the circuit breaker interrupts, diverting the residual power-frequency current into the shunt capacitance.

2.4 Vacuum arc

2.4.1 Definition of vacuum arc

The arc that occurs in vacuum interrupters (Selzer, 1971, and Reece, 1959a and 1959b) is very different from both high-pressure arcs and the low-pressure arc used, for example, in mercury-arc convertors. The differences are so great that it must be considered separately from the previous sections. At first sight it may be thought that 'vacuum arc' is a contradiction in terms, since an arc, by definition contains positive ions; however, a vacuum arc can justifiably be defined as an arc where the vapour to produce positive ions is generated from the electrodes by the arcing process itself (Reece, 1956). Thus, after the arc is extinguished, the vapour density between and around the electrodes falls substantially to zero, leaving a good vacuum.

2.4.2 The two vacuum-arc regimes

The vacuum arc can exist in two forms: the 'diffuse' arc and the 'constricted' arc (Reece, 1963a and 1963b). Ordinarily, a low-current, less than some few thousands of amperes, vacuum arc burns in the diffuse mode, and a high-current vacuum arc in the constricted mode, but the current at which the transition occurs varies with the shape, size and materials of the electrodes and with the rate of change of current. The diffuse vacuum arc has an extraordinarily high interrupting ability, measured in terms of the product

$$-\frac{di}{dt} \times \frac{dv}{dt}$$

that is, the product of the rate of fall of current and the rate of rise of the transient-recovery voltage, greater than that of other types of circuit interrupter (Goodwin *et al.*, 1960, Rich and Farrall, 1964, and Cobine and Farrall). The constricted arc has a negligible interrupting ability.

2.4.3 Diffuse vacuum arc

A vacuum arc carrying a few hundred to a few thousand amperes between two simple disc-shaped electrodes exists as a number of entirely separate arcs in parallel (Reece, 1956). Each arc consists of a cathode spot on the negative electrode (Tanberg, 1930) and, issuing from the spot towards the anode, a diverging arc plasma column. Each separate arc carries a current of a few tens to a few hundred amperes (Reece, 1959, and Rondeel, 1971), depending on the cathode material; the separate arcs exist in parallel because they each have a small, but finite, positive incremental resistance (Reece, 1959). The cathode spots move on the electrode surface at speeds of up to 10 m/s, and repel one another, so that a diffuse vacuum arc tends to spread across the cathode surface. The magnetic forces between the individual plasma columns are Ampèrian, so that they attract one another, but, at low currents, the retrograde-motion effect (Kesaev, 1968) at the cathode spots predominates.

2.4.3.1 Cathode spot

The properties of cathode spots depend on the material of the cathode and its physical form. They have been extensively investigated, but there is still some divergence of opinion about the details of the processes involved. The cathode spot is the region where electrons and metal vapour leave the cathode and enter the arc. The current density depends on the cathode material, being very high on metals with high products of boiling point and thermal conductivity, and lower on metals with low products. Values given in the literature range from 10^6 to 10^8 A/cm^2. In the absence of magnetic fields, spots move randomly on the cathode, but tend to favour ridges or other places where their heat loss by conduction into the body of the cathode is reduced. In magnetic fields, they exhibit 'retrograde' motion, and move in the direction opposite to that dictated by Ampères law.* The track of a spot on a clean metal surface always exhibits melting, and the spot temperature gives fairly rapid evaporation rates. Just in front of the cathode spot is a region where, because of the high current density, the vapour pressure must be high (see page 32), and this is a region of intense light output and ionisation. From this region there is emitted a blast of metal vapour at high velocities, 10^6 cm/s for copper (Reece, 1959, Tanberg 1930, and Davis *et al.*, 1969). The net rate of metal–vapour emission from the cathode spot varies from metal to metal (Table 2.2). For copper, therefore, when the value is 80 μg/c, this corresponds to about one copper atom per ten electrons.

Table 2.2   Metal-vapour emission for different metals

Cathode metal	Evaporation rate	Evaporation rate
	μg/C	atoms/electron
Copper	70	0·1
Silver	35	0·03
Magnesium	70	0·28
Lead	1,100	0·5
Zinc	230	0·33
Cadmium	400	0·35
Mercury	250	0·12

2.4.3.2 Plasma region
The plasma region, although ill defined, is roughly conical in shape, the vertex of the cone being at the cathode spot and in the base on the anode (Reece, 1956). Current is carried in the plasma by the electrons emitted from the cathode spot, and spread into the cone. The electrons in the plasma cone of a diffuse copper arc have a velocity towards the anode of about 10^8 cm/s. The metal vapour emitted from the spot also expands into the cone from the high-pressure region in front of the spot, the velocity for copper being about 10^6 cm/s. Positive metal ions formed in the high-pressure region in front of the spot also enter the plasma cone, accelerated against the electric field by momentum transferred from the particles in the high-pressure region. We thus have an arc in which the positive ions travel from cathode to anode because of their high initial energy. A consequence is that the voltage gradient in the plasma region of a low-current diffuse vacuum arc is effectively zero.

*It is thought that retrograde spot motion occurs because the emission process moves continuously to that region in the spot where electron emission is most efficient, and this occurs where the total magnetic field is highest. This is on the side of the spot where the self field of the spot current is in the same direction as the perturbing field, and this, of course, is on the side of the spot away from the effective geometrical centre of the remainder of the spots

Table 2.3   **Properties of diffuse vacuum arc on copper cathode**

Current per cathode spot, A	100
Total arc voltage, V	21
Cathode drop, V	8
Apparent anode-voltage drop, V	13
Metal-vapour emission rate, μg/C	80
	(0·1 atom/electron)
Mean velocity of vapour stream, cm/s	10^6
Mean kinetic energy of vapour particles, eV	33
Mean velocity of positive ions, cm/s	10^6
Mean kinetic energy of positive ions, eV	33
Positive-ion emission rate, ion/electron	0·01

Current range 100–1000 A

Table 2.2 (Reece, 1959) shows some parameters of a low-current diffuse copper vacuum arc. The gradient in the plasma region is zero because there are negligible losses, and the ions moving in the same direction neutralise the charge of the electrons up to and into the anode surface. For the same reason, there is no voltage drop at the anode, but the 'apparent' voltage drop heats the anode, owing to the kinetic energy originally transferred to the particles near to the cathode. If arc voltage is measured for low-current vacuum arcs on a number of different metals, a relationship is found between the arc voltage and the product of boiling point and thermal conductivity, as shown in Fig. 2.13 (Reece, 1963). The detailed physics of this relationship are obviously very complex, but the explanation, in general terms, is that a certain quantity of metal vapour must be evaporated to produce sufficient positive ions for the discharge. To provide this vapour, the surface of the metal must be hot, and the higher the boiling point, the hotter it must be. The heat loss from the spot into the surrounding cold metal is obviously a direct function of the temperature at the spot and the thermal conductivity of the metal, and this heat has to be supplied by the voltage drop at the cathode; the higher the heat loss, the higher the arc.voltage needed to dissipate the necessary power to maintain the spot temperature.

The stability of a vacuum arc at very low currents is also related to the thermal properties of the cathode metal. The cathode spot with a high rate of heat loss is less stable than one with a low rate of heat loss. Arc stability is thus an inverse function of arc voltage, and, although it strictly requires a statistical definition, it can be said that a 1 A d.c. arc on a bismuth cathode will be stable, and 100 A d.c. arc will be required to give similar stability on a tungsten cathode. The phenomenon of arc stability is closely related to 'current chopping' in vacuum interrupters (Reece, 1959a and 1959b), and 'chopping' may be controlled by the proper choice of contact materials.

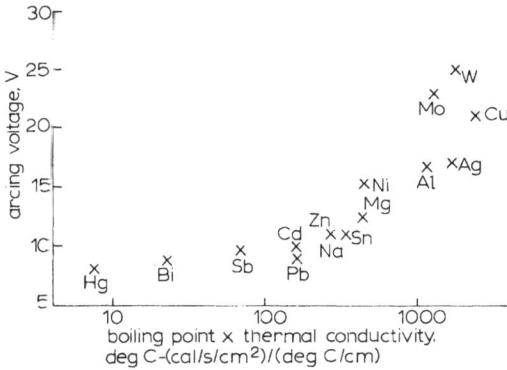

Fig. 2.13 Effect on arcing voltage of thermal properties of cathode material

2.4.4 Transition from diffuse to constricted arc

If the current in a diffuse arc is steadily increased, the number of cathode spots and associated plasma columns increases, as spots divide into two, and, at about 1000 A, the exact value depends on the electrode size and material, the arc voltage, which until this current has been constant, begins to increase (Rich, 1957, and Mitchell, 1967 and 1970). This increase is due to the magnetic field of the current deflecting the initially straight paths of electrons and ions, so that collisions, causing loss of axial momentum, previously rare, begin to become more frequent. The voltage continues to rise smoothly to about 40 V at about 7000 A, above which current the voltage suddenly rises to several hundred volts, but with continued very large transient disturbances. The explanation of this process is that, when momentum dissipating collisions have increased to the extent that the arc voltage has risen to 40 V, the arc voltage is greater than the initial kinetic energy of the positive ions, and they are brought to a standstill by the electric field before they reach the anode. Thus a region just in front of the anode becomes starved of positive ions, and a high voltage suddenly develops to overcome the electron space charge. The process occurs suddenly, because a positive-feedback effect takes place; as the voltage rises, positive ions from near the cathode are repelled from the ion-starved region. As the voltage across the newly formed electron sheath at the anode rises, neutral metal vapour in the sheath region undergoes ionisation by collision, and the current concentrates in those regions where the metal-vapour density allows sufficient ionisation. This concentration or constriction is also a self-sustaining process, since the high power now being applied to the anode in this small region heats the anode until it emits copious metal vapour, so increasing the ionisation density and lowering the voltage. At this stage, the arc is constricted, and has a voltage of some hundred or more volts, with a very irregular waveshape. The phase lag necessary to set up the oscillatory system giving rise to the irregular voltage is provided by the heating and cooling time of the anode spot.

2.4.5 Fully constricted vacuum arc

Once the arc has become constricted, its future development depends on the thermal conditions at the electrodes. If intense cooling is applied to the electrodes by, for example, moving the arc rapidly over the surface of the electrode by a magnetic field, so that the arc is all the time moving onto cold metal, the high and irregular arc voltage will persist. If however, the arc is allowed to remain in one place and the temperature of the electrodes to rise over a relatively large area, the arc voltage may fall again, as the local vapour density becomes so high that the arc, remaining constricted, becomes virtually a high-pressure arc with thermal ionisation occurring in the arc column.

2.4.6 Interrupting ability of vacuum arc

The interrupting ability of the vacuum arc depends on the vapour density between and around the contacts varying virtually synchronously with current (Reece, 1959). Since metal vapour is emitted from the hot spots on the electrodes, the rate of fall of vapour density at the current zero depends almost entirely on the cooling time constant of the heated spot. A cathode spot with an area of, say, 10^{-6} cm^2 has a time constant well below 1 μs, whereas a superheated spot on an anode, say, 1 cm^2 in area and a millimetre or so in depth may have a time constant of many milliseconds. Although the diffuse arc therefore has a very short time constant and a very high interrupting ability, the constricted arc has a much longer time constant, which may vary from a few hundred microseconds or so to many milliseconds, depending on the size of the hot spots, and, consequently, the interrupting ability of the constricted arc is virtually negligible, since the contacts continue to emit metal vapour copiously after the current zero. It is therefore necessary to design power vacuum interrupter contacts so that the arc at, and for some time before the current zero, is in the diffuse regime, and it is thus only necessary to study the theory of arc interruption in the diffuse arc.

2.4.7 Current-zero processes in diffuse arc

The model we take is for a diffuse arc on a copper electrode. As the current falls, the number of spots falls, as spots extinguish, until, when the current is 100 A, only a few spots remain. By assuming that one spot persists down to the current zero, assuming metal-vapour emission with an emission rate of 1 atom per 10 electrons and a mean velocity of 10^6 cm/s for the metal vapour, and assuming that the metal-vapour atoms have a Maxwellian velocity distribution, it is possible to calculate the vapour density as a function of position in the arc gap and of time after the current zero. The results of such a calculation are shown in Fig. 2.14, where it can be seen that, even at the instant of current zero, the metal-vapour

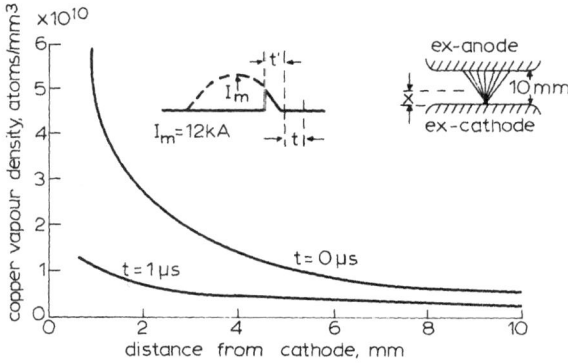

Fig. 2.14 Vapour density at and after current zero

density is so low throughout the gap that the gap is substantially an insulator (Reece, 1963). A somewhat different theory has been proposed by Farrell (1968), in which the reduction of vapour density occurs by radial diffusion of the metal vapour at thermal rates, and not by condensation of the metal vapour onto the anode; however, this theory gives the same sort of results as the previous one.

In fact, both theories neglect the fraction-of-a-microsecond delay in the cooling of the cathode spot, and they also neglect that, even when interrupting large fault currents, the last cathode spot is usually extinguished just before the true current zero, at an instaneous current of a few amperes. These two effects tend to neutralise each other, and the evidence is that figures calculated from the first theory are in good agreement with the measured values of interrupting ability. That the interrupting ability of a diffuse vacuum arc is so high is indicated by the oscillogram in Fig. 2.15, where, in an experiment (Goodwin *et al.*, 1960) for the interruption of direct current by the artificial commutation of a vaccum arc, the current, falling at a rate of 500 A/μs from a peak of 23 kA, is extinguished, although the recovery voltage rises at 18 kV/μs, both values being many times greater than occur in power systems. Other experiments have been reported in which a vacuum arc has been extinguished by artificial commutation in which the arc current has been reduced to zero in 10 ns, and the cathode spot has still been extinguished (Kesaev, 1968).

2.4.8 Post-zero currents
Prevention of the reignition of the vacuum arc after the current zero is achieved by ensuring that the vapour density between and around the contacts is sufficiently low that breakdown cannot occur in the tenuous vapour (chapter 8). Nonetheless,

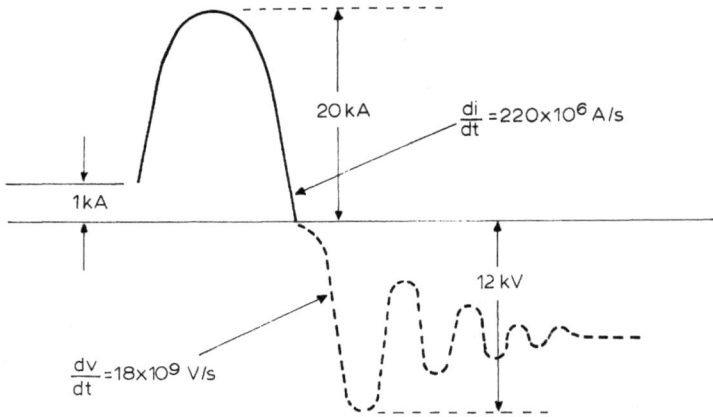

Fig. 2.15 Interruption in vacuum

although breakdown, with its associated cathode spots, may not occur, some post-zero current can flow in a vacuum interrupter. This current flows as a result of the movement of the few ions and electrons surrounding the contacts at the current zero (density $\simeq 10^{12}$/cc) i.e. a mean free path of 100 cm, much greater than the dimensions of the interrupter. Provided that the density of charged particles is known, and that the geometry of the interrupter can be represented by a simple 1-dimensional model (Reece, 1959b), the post-zero current flowing can be calculated for any given recovery-voltage waveshape by an equation derived from Child's law:

$$i_t = \{1 + f(V)\}\, eN_+ \frac{d}{dt} \left\{ \frac{\int_0^t V^{3/2} dt}{3\pi \sqrt{\left(\frac{em}{2}\right) N_+}} \right\}^{1/3} \tag{2.10}$$

where

  i_t = instantaneous post-zero current at time t after the current zero
  V = instantaneous transient recovery voltage
  $f(V)$ = secondary-emission coefficient, electrons per positive ion, as function of V
  e = electronic charge
  N_+ = positive-ion density within interrupter-shield volume
  m = mass of positive ions

Equation 2.10 merely takes account of the formation of a positive-ion sheath that sweeps through the tenuous plasma and removes all the charges. The current flow

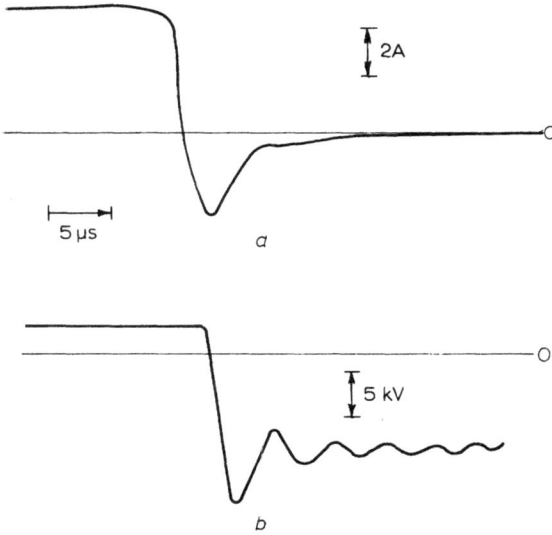

Fig. 2.16 Post-zero current in vacuum interrupter after simulated 15 kA clearance

    a Post-zero current record
    b Restriking-voltage record

ceases when all the charges have been swept to the electrodes. Fig. 2.16 shows a post-zero current wave in a vacuum interrupter. The post-zero current flows to almost all the electrically exposed regions of the contact, and not merely to those regions where the power arc was previously burning. Since the areas of the electrodes are usually quite large, the post-zero power density is relatively small, and the flow of post-zero current does not usually lead to the re-establishment of an arc, for which cathode spots are necessary.

2.5 Methods of measurement of arc properties

The study of the properties of arcs, so far as switchgear is concerned, can be divided into two fields:

(*a*) Investigations of arc properties typically made by switchgear engineers on circuit breakers or development rigs that fairly closely resemble circuit breakers. These investigations are almost always made at switchgear-testing stations.

(*b*) Studies of a more fundamental nature made on discharges by arc physicists, which may be applicable to any technological-device using discharges, and not merely to circuit breakers.

Although the two categories tend now, more than in the past, to merge into one another, it is possible, in practice, to draw a fairly sharp dividing line; this depends, perhaps rather oddly, on the place where the studies are made. Commercial pressures on high-power testing plants are so large that only the rather conventional measurements fitting into category (*a*) can be made there, and measurements falling into class (*b*) are made using small generators or capacitor sources.

2.5.1 Category (*a*): measurements

2.5.1.1 Voltage
The voltage across the circuit-breaker arc may be measured by electromagnetic oscillographs, having frequency responses up to a kiloherz or so, by drum-type cathode-ray oscillographs with sweep speeds up to about 10 μs/mm (or some ten times higher if a time base on the tube is used as well as the drum rotation to give a triangular zero line on the film with the signal along the rotational direction of the drum). These drum-type cathode-ray oscillographs have the advantage that fairly high frequencies, typically up to between 100 and 1 MHz, may be recorded for relatively long periods, i.e. not less than 20 ms. For very-high-speed recording, single-shot multibeam cathode-ray oscillographs are used, with writing speeds up to at least 1 μs/cm. Records may be taken of the arc voltage, as opposed to the transient recovery voltage, in which case diode clipping or clamping circuits often backed up by gas-discharge tubes, must be used to ensure that the oscillograph is not overloaded by the recovery voltage, which may sometimes be a thousand times higher than the arcing voltage.

For all these types of recording, except electromagnetic oscillography, great care must be taken with the installation and equalisation or balancing of the whole potential divider and signal cabling from the test area to the recording equipment, and also with the earthing and power supply to the recording apparatus. The voltage records obtained by these methods may be used to deduce a number of features of the behaviour of the circuit breaker; The arcing-voltage record can show the instant of contact separation, and can be used to determine where the arc is in the nozzle or other interrupter. If the circuit breaker failts to interrupt, the instant and the shape of the collapsing recovering voltage will show whether the failure is thermal, i.e. due to an energy-balance process, as described in Section 2.3.4, or whether it is dielectric, due simply to breakdown of the gas, which may be caused by different processes than those leading to thermal reignition.

2.5.1.2 Current
The situation for the recording of the current flowing in the circuit breaker is similar to that for voltage. At the low-frequency end of the recording range, current

transformers and electromagnetic oscillographs are used. Suitably designed current transformers may also be used for the drum-type oscillograph, and, for the most accurate work, shunts must be used. At the very highest frequencies, when single-shot cathode-ray oscillographs are used for recording, highly noninductive shunts must be used, and, since this technique is usually used to measure small postzero currents immediately after the flow of the full short-circuit current, wideband direct-coupled amplifiers with very effective clamping at the input, and often also at later stages of the amplifier, are needed. Measurements of postzero current were pioneered in Britain by Mason of the Electrical Research Association Ltd. (Mason, 1952), who had to develop his own amplifiers and oscillographs, but suitable commercial equipment, except for shunts and diode clamps, is now available (Irwin, 1971). Sweep speeds in the same range as for voltage records, i.e. up to 1 μs/cm are commonly used. The low-frequency electromagnetic-oscillograph current records are used merely to check the magnitude and general shape of the short-circuit current, and, for arc-chute circuit breakers, to study the forcing down of the short-circuit current (Boehne, 1941) by the high arc voltage developed in the arc chute. The very-high-frequency postzero current records are used, in conjunction with the same type of voltage records, to determine the postzero resistance of the circuit breaker and its energy-balance processes, and to determine the arc time constants. This type of measurement can assist the designer to make suitable modifications to the contact and nozzle structure should the breaker be unable, for example, to interrupt short-line faults.

2.5.1.3 Pressure
The pressure in the arc core is not usually measured in commercial switchgear-testing stations, since this can only be done by advanced optical techniques. However, the pressure of the gas surrounding the arc is measured by use of capacitive or inductive pressure transducers (Trenchan, 1953), and the pressure is recorded on slow- or medium-speed oscillograms. The recording of pressure in vacuum interrupters is not normally practised under commercial testing-station conditions.

2.5.1.4 Arc geometry
The position of the arc in the interrupter is commonly studied by high-speed cinematography (Bagshaw, 1959). Although cameras are available with framing rates of up to 10^8 frames per second, using rotating mirrors or image-convertor techniques, the maximum framing rates normally used in commercial testing stations are 10 000 frames/s on 16 mm film or 20 000 frames/s on 8 mm film. The cinematography used in the high-power testing stations almost always uses the light emitted by the arc, and the more complex optical systems involving laser and similar light sources are usually reserved for the arc-physics laboratories. High-speed colour cinematography at framing rates of 10 000 frames/s can give the switchgear engineer a large amount of information about the behaviour of the arc in

the circuit breaker during the period of current flow, and this is usually sufficient information for the designer.

These are the measurements commonly made on the arcs in production or developmental circuit breakers, and are normally the only measurements available to, or used by, the majority of switchgear engineers. Such measurements have carried the design techniques used in circuit breaker development forward from the plain-break oil circuit breaker of the 1920s to the very-high-power and sophisticated units in use today.

2.5.2 Category (*b*): fundamental studies

The arc physicist uses more complex and delicate measurements in his studies of discharges (Lochte—Holtgreven, 1968), and these are not usually suited to commercial testing stations, because of the time they take to set up, or the delicacy of the optical systems involved. There are, however, a number of arc-physics laboratories (usually operated by major switchgear manufacturers, but sometimes by electricity-supply authorities or universities), where the more fundamental measurements may be made on fairly-high-power arcs in circuit breaker type rigs, but where the power sources are sufficiently cheap, smaller generators arranged for synthetic testing, or capacitor banks, to allow these more involved techniques to be used without interruption by commercial testing. It is not possible to do more than mention briefly the type of techniques used for these more fundamental measurements, but details of the techniques may be obtained from the literature, and most of them represent a very considerable study in themselves.

2.5.2.1 Arc temperature

Gas temperatures are usually measured spectroscopically (Burhorn, 1951) by measuring the relative intensities of various spectral lines. The probabilities of transitions of the atoms from one energy state to another depend on the gas temperature, and therefore the relative intensities of particular spectral lines from the arc can be used as an absolute measure of the gas temperature in the arc. Other spectroscopic methods include the use of high-dispersion spectrographs to make accurate measurements of the shape of given spectral lines, since the lines are broadened by, amoung other causes, the Doppler effect of the random thermal motion of gas atoms, and the shape of the broadened line can be used to determine the temperature. At the lower end of the temperature range, measurements may be made by absorption rather than emission, when light from a higher and known-temperature source is passed through the arc, and the absorption of particular spectral lines can be measured, again, of course, spectroscopically. All these spectroscopic methods normally require large and expensive spectrographs. A particularly useful technique for measuring the lower temperatures associated with the outer regions of the arc has been developed by a number of workers using shock waves (Jones *et al.*, 1971). Shock waves are passed through the arc, and their velocity recorded by suitable high-speed cinematographic techniques, such as schlieren optics or interferometry. The velocity of the shock front depends on the gas temperature, and this method works in the outer regions of the arc, where convection is the main energy-removal source and where gas is too cool for normal optical methods to be useful.

2.5.2.2 Gas Density

Gas density can be measured in a number of ways. Absorption of soft X rays and of low-energy electron beams has been used in the past, but by far the most common methods used today employ optical techniques (de Silva, 1970). Schlieren, interferometric and, with the advent of the laser, holographic optical systems are used to measure densities, or strictly, density gradient, and, by integration of the gradient across the arc, the absolute density may be obtained at any point (Friedrich *et al.*, 1970). It must be remembered that, in all these methods, a beam of radiation or particles is passed through a cylindrical arc for measurement purposes, and account has to be taken of the fact that the beam is passing through a nonuniform arc, owing to the circular symmetry, and allowance must be made to obtain the temperature or density at a particular radius.

2.5.2.3 Electron densities

Electron densities may be measured by probing the discharge with millimetric radiation, and also by the use of physical probes. The probe methods developed for measuring ionisation conditions in high-pressure arcs, which are usually very hot, have involved sweeping refractory probes very rapidly through the arc so that they do not melt. Other probe measurements can be made by using sections of the wall of the arc chamber, when the chamber is composed of stacks of metal washers with insulation between the washers. In general, it can be said that probe methods are probably not as satisfactory in high-pressure arcs as in lower pressure and cooler arcs, since they have a greater tendency to affect the arc being measured than do those methods that involve radiation or particle beams.

2.6 Acknowledgement

The author acknowledges the assistance of many friends and colleagues, in particular Mr. L.H.A.King, and wishes to thank Dr. M.T.C.Fang for permission to quote from his paper, and Vacuum Interrupters Ltd., for permission to publish this chapter.

See also Appendix A, Virtual current chopping, p. 591.

2.7 References

BABAKOV, N. A. (1947): 'Passage of the electric arc into the arc quenching grid', *Vestn. Elektroprom.*, 18, p. 15

BAGSHAW, J. R., ELLIS, N. S., and FITTON, D. (1969): 'Photographing the air flow in an airblast circuit-breaker interrupter during arcing', *Reyrolle Rev.* 192

BARRAULT, M. R., BLACKBURN, T. R., EDELS, H., and SATYANARAYANA, P. (1972): 'Investigations of a high current free burning electric arc'. *IEE Conf. Publ.* 90, pp. 221–223

BOEHNE, E. W. (1964): 'The Geometry of arc interruption', *GE Rev.*, **44**, p. 207

BOWMAN, B., and EDELS, H. (1968) 'Radial temperature measurements of alternating current arcs', *J. Phys. D.*, 2, pp. 53–63

BROWNE, T. E. Jun. (1948): 'A study of ac arc behaviour near current zero by means of mathematical models', *Trans. Am. Inst. Elec. Eng.*, 67, Pt. 1, pp. 141–153

BROWNE, T.E. Jnr. (1958): 'An approach to mathematical analysis of AC arc extinction in circuit breakers', AIEE Fall Meet, Pittsburgh, Paper 58-1155

BUTLER, T. F., and WHITTAKER, D. (1972): 'A normalised moving-boundary circuit-breaker model in severe circuits'. *IEE Conf. Publ.* 90, pp. 302–303

BUTLER, T. F., and WHITTAKER, D. (1972): 'Distributed moving-boundary circuit-breaker model', *Proc. IEE*, 119, pp. 1295–1300

BURHORN, F., MAECKER, H., and PETERS, T. (1951): 'Temperature measurements on water stabilised high current arcs', *Z. Phys.*, 131, pp. 28–40

CASSIE, A. M. (1939): 'Arc rupture & circuit severity'. CIGRÉ paper 102, pp. 1–14

CHAN, S.K., FANG, M.T.C., and COWLEY, M.D. (1975): 'Dynamic behaviour of an A.C. arc column in a steady laminar accelerating flow' *Proc. IEE*, 122, pp.585-590

CHAPMAN, A. (1977): 'The electrical conductance of a gas blast interrupter arc near current zero.' Ph.D. thesis. Univ. of Liverpool, June

COBINE, J. D. (1941): 'Gaseous conductors' (McGraw-Hill)

COBINE, J. D., and FARRALL, G. A. (1962): 'Recovery characteristics of vacuum arcs'. AIEE Conference Paper 62–139

COWLEY, M.D. (1974): 'Integral method of arc analysis – Part I,' *J.Phys. D:Appl. Phys.*, 7, pp.2218-31

DAVIS, D. W., and MILLER, H. C. (1969): 'Analysis of the electrode products emitted by DC arcs in a vacuum ambient', *J.A.P.*, 40, pp. 2212–2221

DICKINSON, R. C., and FRINK, R. (1946): 'Size reduction and rating extension of magnetic air circuit-breakers up to 500,000kVA, 15kV', *Trans. Am. Inst. Electr. Eng.*, 65, pp. 220–223

EL-AKKARI, F.R., and TUMA, D.T. (1977): 'Simulations of transient and zero current behaviour of arcs stabilized by forced convection', IEEE *Trans.*, PAS-96, pp.1784-1788

ELENBAAS, W. (1951): 'The high pressure mercury vapour discharge', *Electronics*, 25, pp. 304–306

von ENGEL, A. (1929): 'Length and duration of arcs in air by switching off direct current'. *Wiss. Veroff Siemens*, 7, Pt. 2 pp. 50–66

von ENGEL, A. (1955): 'Ionised gases' (Clarendon Press, Oxford)

von ENGEL, A., and STEENBECK, M. (1932, 1934): 'Elektrische Gasentladungen', 1 and 2 (Springer, Berlin)

FANG, M. T. C., FUNG, H., and EDELS H. (1973): 'AC-arc-column and current-zero properties', *Proc. IEE*, 120, (6), pp. 709–714

FANG, M.T.C., and BRANNEN, D. (1978): 'Arc modelling and performance prediction of gas blast circuit breakers'. Univ. of Liverpool Arc Descharge Research Project, Report ULAP-T53, January

FARRALL, G. A. (1968): 'Decay of residual plasma in a vacuum gap after forced extinction of a 250 ampere arc', *Proc. Inst. Electr. Electron. Eng.*, pp. 2137–2145

FOWLER, R. H., and NORDHEIM, L. (1928): *Proc. Roy. Soc. A.* 119, pp. 173–181

FRIEDRICH, O. M., WEIGL, F., and DOUGAL, A. A. (1970): 'Plasma diagnostics utilising optical interferometry and holographic techniques' in 'Gas discharges'. *IEE Conf. Publ.* 70

FROST, L. S., and LIEBERMANN, R. W. (1971): 'Composition and transport properties of SF_6 and their use in a simplified enthalpy flow arc model', *Proc. Inst. Electr. Electron. Eng.,* **59**, pp. 474–485

GOODWIN, A., REECE, M. P., and SMITHERS, B. W. (1960): 'A preliminary study of direct current interruption by vacuum switches, using the artifical current zero injection method'. Report UR:CON.HAR.EMR/1085

HERMANN, W., and RAGALLER, K. (1977): 'Theoretical description of the current interruption in H.V. gas blast circuit breakers' *IEEE Trans.,* **PAS-96**, pp.1546-1555

IRWIN, C. (1971): 'The static and dynamic arc'. MSc. Thesis, University of Manchester Institute of Science & Technology

JAMIESON, B. G. (1929): 'Deion grid circuit-breaker', *Trans. Am. Inst. Electron. Eng.,* **48**, pp. 101–104

JONES, G. R., FREEMAN, G. H., and EDELS, H. (1971): *J. Phys. D.,* **4**, p. 236

JONES, T. J. (1936): 'Thermionic emission' (Methuen, London)

KESAEV, I. G. (1968): 'Cathode processes in electric arcs' (Hayka, Moscow)

KING, L. H. A. (1964): 'The voltage gradient of an arc column under forced convection in air or nitrogen'. Report 5072

LEE, T. H. (1959): 'T-F theory of electron emission in high current arcs', *J. Appl. Phys.,* **30**, p. 166

LEE, T.H., and GREENWOOD, A. (1968): 'The mechanism of thermal breakdown of a high temperature gas and its implication to circuit breakers'. CIGRE Paper 13-06

LOCHTE-HOLTGREVEN, H. (Ed.) (1968): 'Plasma diagnostics' (North Holland Publishing Company)

LOEB, L. B. (1939): 'Fundamental processes of electrical discharges in gases' (Wiley)

LOWKE, J.J., and LUDWIG, H.C. (1975): 'A simple model for high current arcs stabilized by forced convection' *J. Appl. Phys.,* **46**, pp.3352-3360

MAECKER, H. (1955): 'Plasmastromungen in Lichtbogen in Folge eigenmagnetischer Kompression', *Z. Phys.,* **141**, pp. 198–216

MALGHAN, V.R., FANG, M.T.C., and JONES, G.R. (1977): 'Investigation of quasi-steady state high current arcs in an orifice air-flow' *J. Appl. Phys.,* **48**, pp.2331-2337

MASON, F. O. (1952): 'The recording of current in ac power arcs around the period of current zero', *Engineering,* **193**, pp. 686–689

MASON, F. O. (1960): 'Air blast circuit-breakers', *Electr. Rev.,* **166**, pp. 571–572

MAYR, O. (1943): 'Beitrage zur Theorie des Statischen und des dynamischen Lichtbogens', *Arch. Elect.,* **37**, pp. 588–608

MAYR, O. (1943): 'On the theory of the electric arc and its extinction', *ETZ,* **64**, pp. 645–652

MEEK, J. M., and CRAGGS, J. D. (1953): 'Electrical breakdown of gases' (Clarendon Press, Oxford)

MITCHELL, G. R. (1967) 'The high current vacuum arc and its relevance to circuit interruption' Ph.D. Thesis, CNAA London

MITCHELL, G. R. (1970): 'High-current vacuum arcs', *Proc. IEE,* **117**, (12), pp. 2315–2332

PETERMICHL, F. (1940): 'Druckgasschalter bau in Schweiz', *ETZ,* **61**, pp. 1192–1193

PRINCE, D. C., and POITRAS, E. J. (1931): 'Oil-blast breaker theory proved experimentally', *Electr. World,* **97**, pp. 400

REECE, M. P. (1956): 'Arc chutes for a.c. contactors: some notes and data for experimental designs'. ERA Report G/T 309

REECE, M. P. (1959): 'Vacuum Switching I – Review of literature and scope of ERA research'. ERA Report G/XT 166

REECE, M. P. (1959): 'Vacuum switching II – Extinction of an a.c. vacuum arc at current zero'. ERA Report G/XT 167

REECE, M. P. (1956): 'Circuit interruption in high vacuum'. ERA Report G/XT 156

REECE, M. P. (1963a) 'The vacuum switch – pt. 1 – Properties of the vacuum arc', *Proc. IEE,* 110, pp. 793–802

REECE, M. P. (1963b) 'The vacuum switch – Pt. 2 – Extinction of an a.c. vacuum arc', *ibid.,* 110, pp. 803–811

RICH, J. A., and FARRALL, G. A. (1964): 'Vacuum arc recovery phenomena', *Proc. IEEE,* 52, pp. 1293–1301

RIEDER, W., and URBANEK, J. (1966): 'New aspects of current-zero research on circuit-breaker reignition. A theory of non-equilibrium arc conditions'. CIGRE Paper 107

RICH, J. A., PRESCOTT, L. E., and COBINE, J. D. (1971): 'Anode phenomena in metal-vapour arcs at high currents', *J. Appl. Phys.,* 42, pp. 587–601

RUDENBERG, R. (1922): 'The switching of direct and alternating current in inductive heavy current circuits', *Bull. Schweiz. Elek. Verein.,* 13, pp. 248–263

RUMA, D.T., and LOWKE, J.J. (1975): 'Prediction of properties of arcs stabilized by forced convection,' *ibid.,* 46, pp.3361-3367

RONDEEL, W. G. J. (1971): 'Electrode errosion and energy balance in a metal vapour arc'. Thesis, Department of Electrical Engineering, Norwegian Institute of Technology, Trondheim, Norway

SAHA, M. N. (1920); 'Ionisation in the solar chromosphere', *Phil. Mag.,* 40, pp. 72–78

SELZER, A. (1971): 'Switching in vacuum: a review', *IEEE Spectrum,* 8, pp. 26–37

DE SILVA, A. W. (1970): 'Plasma diagnostics by light scattering' (Plasma Physics VolA PartA Academic Press)

SLEPIAN, J. (1929): 'Theory of the deion circuit-breaker', *Trans. Am. Inst. Electr. Eng.,* 48, pp. 523–527

SLEPIAN, J. (1930): 'Extinction of a long a.c. arc', *ibid.,* 49, pp. 421–430

SLEPIAN, J., and BROWNE, T. E. Jun. (1941): 'Photographic study of a.c. arcs in flowing liquids'. *ibid.,* 60, p. 823

STEENBECK, M. (1940): 'Eine prufung der minimumprinzides', *Wiss. Veroff. Siemens* 19, pp. 59–67

SWANSON, B.W., and ROIDT, R.M. (1971): 'Boundary layer analysis of an SF6 circuit breaker arc,' *IEE Trans.,* PAS-90, pp.1086-1093

SWANSON, B.W., and ROIDT, R.M. (1971): 'Arc cooling and short line fault interruption,' *ibid.* PAS-90, pp.1094-1102

SWANSON, B.W., and ROIDT, R.M. (1971): 'Some numerical solutions of the boundary layer equations for an SF6 arc,' *Proc. IEEE,* 59, pp.493-501

SWANSON, B.W., and ROIDT, R.M. (1972): 'Thermal analysis of an SF6 circuit breaker arc,' *ibid.,* PAS-91, pp.381-389

SWANSON, B.W., ROIDT, R.M., and BROWNE, T.E.(1972): 'A thermal arc model for short line fault interruption,' *ETZ-A,* 93, pp.375-380

SWANSON, B.W. (1974): 'A thermal analysis of short line fault interruption,' IEEE Winter Power Meeting, Jan-Feb Paper C-74, 186-3

SWANSON, B.W. (1977): 'Nozzle arc interruption in supersonic flow,' *IEEE Trans.,* PAS-96, pp.1697-1766

SUCKLER, K. (1969): 'Contribution to the back-pressure problem in high-voltage circuit breakers,' **ETZ-A, 90,** pp.711-714

TANBERG, R. (1930): 'On the cathode of an arc drawn in vacuum', *Phys. Rev.,* **35,** pp. 1080–1089

TOPHAM, D.R. (1970): 'The aerodynamics of electric arcs in axial flow.' Ph.D. thesis, Loughborough University of Technology

TOPHAM, D. R. (1971): 'The electric arc in constant pressure axial gas flow', *J. Phys. D.,* **4,** p. 1114

TRENCHAM, H. (Ed.) (1953): 'Circuit-breaking' (Butterworths Scientific Publications)

WEDMORE, E. B., WHITNEY, W. B., and BRUCE, C. E. R. (1925): 'Circuit-breaking' *JIEE,* **67,** pp. 557–578

YOS, J. M. (1967): 'Revised transport properties for high temperature air and its components'. AVCO Space Systems Division

YOUNG, A. F. B. (1953): 'Some researches on current chopping in high-voltage circuit breakers', *Proc. IEE,* **100,** Pt. II pp. 337–353

Chapter 3

Network switching conditions

M. W. Kennedy, B.Sc., Ph.D., Mem.I.E.E.E., C.Eng., F.I.E.E.

3.1 Introduction

The duties required from a circuit breaker can be summarised as follows:

(*a*) It should be capable of making and breaking any possible load or fault condition envisaged for the circuit that it has to control.

(*b*) It should be capable of carrying fault currents for short times and load currents continuously without damage.

This chapter is concerned with the duties listed alongside (*a*) above and in particular, with the network voltage conditions during the switching period. The determination of current magnitudes and waveforms in the period prior to current interruption is comprehensively covered in the literature which includes those references given in Section 3.7 of this chapter.

The switching conditions described in this chapter can be conveniently considered as either short-circuit switching or load switching (including capacitor and reactor banks, unloaded transformers etc) which, in turn, demand a recognition of both power frequency and natural frequency phenomena.

3.2 Short-circuit switching: power-frequency phenomena

3.2.1 Three-phase short-circuit switching:

In considering the switching of 3-phase short circuits, it should be noted that, because the currents in each of the three phases pass through zero at different points in time, the interrupters clear the fault in a definite sequence. The analysis of this problem is facilitated by the theory of symmetrical components, and some knowledge of the use of sequence impedances is assumed. For a treatment of the theory of symmetrical components, reference should be made to Wagner *et al.* (1933) and Lackey (1951*b*).

Fig. 3.1 illustrates the equivalent circuit of a source feeding a 3-phase fault F through a circuit breaker. It should be noted that the use of this equivalent circuit presupposes that the negative-sequence source impedance is equal to the positive-sequence source impedance. This assumption is valid for a power system which consists largely of static plant, and is approximately true for rotating plant during the short period of time that a circuit breaker takes to clear a fault.

The voltage across the first pole to clear, say A, is given by

$$E_a \left(\frac{3Z_{s0}}{2Z_{s0} + Z_{s1}} \right) \tag{3.1}$$

For a 3-phase earth-fault in a solidly earthed system, i.e. where $Z_{s0} = Z_{s1}$, the recovery voltage is equal to E_a. For an insulated neutral system, i.e. where $Z_{s0} \gg Z_{s1}$, or where the 3-phase fault is clear of earth, the recovery voltage is $1 \cdot 5E_a$.

The voltage across the second pole to clear, say B, is given by

$$\sqrt{(3)}E_a \frac{\sqrt{\{(Z_{s1}^2 + Z_{s0}Z_{s1} + Z_{s0}^2)\}}}{2Z_{s1} + Z_{s0}} \tag{3.2}$$

For a 3-phase earth fault in a solidly earthed system, the recovery voltage is Ea. For an insulated neutral system, or where the 3-phase fault is clear of earth, the

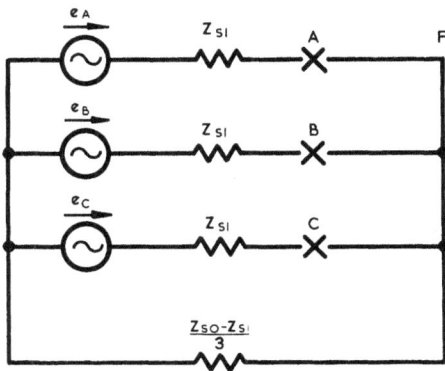

Fig. 3.1 Equivalent circuit for determination of power frequency recovery voltage

recovery voltage is $1 \cdot 732 E_a$. In this case, however, interruptor C must clear simultaneously with interruptor B since the same current flows through each. The recovery voltage, per pole, for this case, therefore, is $0 \cdot 866 E_a$.

For a fault involving earth, the voltage across pole C, if it does not clear simultaneously with pole B, is E_a, since this is the only driving voltage in the circuit at that instant.

3.2.2 Asymmetric short-circuit switching

As mentioned earlier, the theory of symmetrical components is of great assistance in the calculation of the power frequency recovery voltage following asymmetric faults, provided that the phase impedances and voltages of the network, as viewed from the point of fault, are symmetrical. Since this condition is almost true, the technique can be used to determine circuit breaker power frequency recovery voltages. It should be noted that for all asymmetric faults involving earth in a solidly earthed system, the power frequency recovery voltage on the faulted phase is equal to the system maximum phase-neutral voltage E. For a phase-phase fault, the power frequency recovery voltage on each of the two faulted phases is $0 \cdot 866E$, since for both earthed and unearthed phase-phase faults, the interruptors on the faulted phases are likely to clear simultaneously. An extensive coverage of the determination of transient recovery voltage for asymmetric faults is given by Geng *et al.* (1966).

In general, the most probable type of fault is the single-phase-earth type. This is particularly true in h.v. and e.h.v. systems, and it should be noted that on some long transmission lines, the magnitude of the voltage on the healthy phases following a single-phase-earth fault can be as high as 2 p.u.

Some aspects of this phenomena have been covered by Glavitsch (1964) and Kimbark *et al.* (1968).

Earth faults can also cause the current in the interruptors on the faulted phases to be greater than the rated 3-phase short-circuit current if the zero-sequence source impedance behind the circuit breaker is less than the positive-sequence source impedance. An important example of this occurs with the h.v. circuit breaker on a generator-transformer unit, where the transformer has a delta winding on the generator side and is star connected on the h.v. side. In this circuit, the total positive-sequence impedance of the source consists of the generator impedance plus the transformer impedance, whereas, the zero-sequence source impedance is that of the star-connected transformer winding only. This results in the current interrupted in the h.v. circuit breaker for close-up earth faults, typically 20—30% higher than for a similarly located 3-phase fault.

3.2.3 Effect of neutral earthing

The choice of the method of power-system earthing depends on the type and voltage of the system, and on the way in which the system designer wishes to control the short-circuit current of the system.

The following are examples of the types of neutral earthing commonly used:

(a) solid earthing
(b) reactive earthing
(c) resistive earthing
(d) Petersen coil earthing.

As mentioned above, the insertion of impedance in transformer neutrals, for example, determines to a significant extent the flow of fault current for asymmetric faults involving earth. The general equations governing the recovery voltage across a circuit breaker following system faults are given in Section 3.2.1, and from these it can be seen that the recovery voltage is to a certain extent dependent upon the zero-sequence source impedance. To this extent, therefore, the choice of neutral earthing not only determines the fault current but also may affect the power factor of the fault current. A full analysis for the recovery voltage across a circuit breaker, in a system where Petersen coil earthing is used, is given by Carter (1950).

3.3 Short-circuit switching: natural frequency phenomena

3.3.1 Transient recovery voltage
The treatment of short-circuit switching given in Sections 3.2.1 and 3.2.2 has been restricted to power-frequency considerations only. The natural frequency analysis which introduces the concepts of 'rate-of-rise' and amplitude of the recovery voltage across a circuit breaker, are now introduced. These factors are amongst the most important affecting the severity of the duty of a circuit breaker. The natural frequency analysis of circuit breaker performance has been extensively covered in published work, and mention may be made of the contributions by Park *et al.* (1931), Harle *et al.* (1944), Hammarlund (1946), Lackey (1951*a*), Rittenhouse *et al.* (1963*a*), and Mazza *et al.* (1972).

The three basic important aspects of transient recovery voltage are:

(a) frequency of the recovery voltage transient
(b) amplitude of the power frequency recovery voltage
(c) damping constant of the circuit.

Single-frequency recovery voltage transients
Consider the circuit shown in Fig. 3.2*a*. The source voltage is expressed by

$$e = E \cos w_s t \tag{3.3}$$

Therefore, when the circuit breaker is closed, if $w_s L \gg R$,

$$i = \frac{E}{w_s L} \sin w_s t \tag{3.4}$$

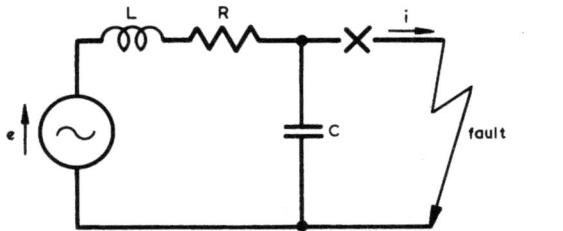

(a)

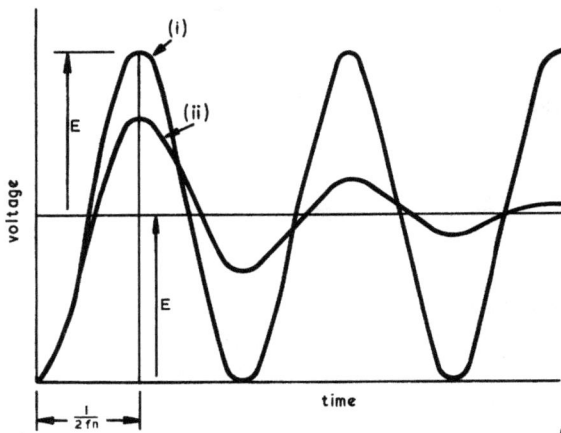

(b)

Fig. 3.2 Single-frequency recovery-voltage transients

Eq. 3.4 establishes the initial conditions in the circuit. Immediately after the circuit breaker opens

$$L\frac{di}{dt} + Ri + \frac{1}{C}\int idt = E \cos w_s t \tag{3.5}$$

The solution of eqn. 3.5 yields an expression for the recovery voltage across the circuit breaker as follows:

$$\text{recovery voltage} = E\left\{\cos w_s t - \exp\left(\frac{-Rt}{2L}\right)\cos w_n t\right\} \tag{3.6}$$

where

$$w_n = \frac{1}{\sqrt{(LC)}}$$

With no damping in the circuit, i.e. $R = 0$, and, therefore, with a short-circuit current having zero power factor, Fig. 3.2b(i) shows a simple, single-frequency recovery voltage transient of the type yielded by eq. 3.6. The peak value of the recovery voltage is $2E$. indicating the so-called 'voltage doubling' effect for small values of R. With resistance in the circuit, or other losses present, the transient is damped so that it diminishes exponentially. The peak value is then less than $2E$, as shown by Fig. 3.2b(ii).

By only considering this simple case, several difficulties are overlooked, e.g. the determination of the correct value of the power-frequency recovery voltage in a 3-phase circuit and the values of L and C. The problem is also complicated by the fact that, in most practical cases, the recovery voltage transient is the sum of two recovery voltage transients, one associated with each side of the circuit breaker. In general, each of these transients has a different frequency determined by the circuit parameters.

Double-frequency recovery voltage transients

Fig. 3.3a shows two separate, single-frequency sources of transient voltage, one on each side of a circuit breaker. The power-frequency voltage on the generator side increases at the instant of clearance from a value equal to iZ, where i is the short-circuit current and Z the transformer impedance, to e, the source voltage, whilst that on the transformer side falls from iZ to zero. If L_s and L_t are, respectively, the inductances of the source and the transformer, the voltage at the transformer side of the circuit breaker at current zero and zero power-factor, prior to fault clearance is

$$e_t' = \left(\frac{L_t}{L_s + L_t} \right) e \tag{3.7}$$

The change in voltage on the generator side after interruption is

$$e_s' = e - \left(\frac{L_t}{L_s + L_t} \right) e$$

$$= e \left(\frac{L_s}{L_s + L_t} \right) \tag{3.8}$$

and the frequency of the oscillation is

$$f_n = \frac{1}{2\pi} \frac{1}{\sqrt{L_s C_s}} \text{ hertz} \tag{3.9}$$

On the transformer side of the circuit breaker, the change in voltage is

$$e_t' = - \left(\frac{L_s}{L_s + L_t} \right) e \tag{3.10}$$

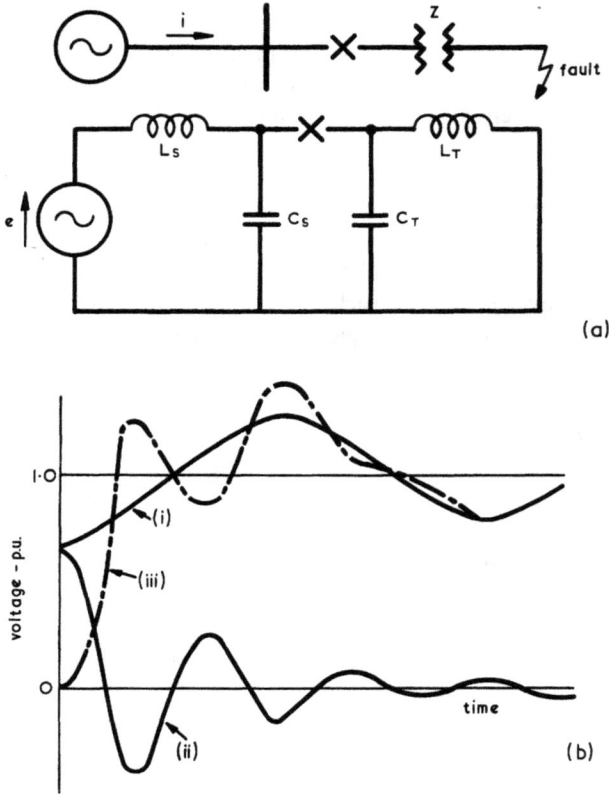

(a)

(b)

Fig. 3.3 Double-frequency recovery-voltage transient

and the frequency of the oscillation is

$$f_n = \frac{1}{2\pi} \frac{1}{\sqrt{L_t C_t}} \quad \text{hertz} \tag{3.11}$$

Curves (i) and (ii) of Fig. 3.3b show two typical transients illustrating these conditions. The transient recovery voltage is, of course, the vector difference between the voltages on the generator and transformer sides of the circuit breaker, and is shown by curve (iii) in Fig. 3.3b.

The case of a 3-phase fault, clear of earth, was considered in Section 3.2.1, insofar as power frequency conditions were concerned, this type of fault may,

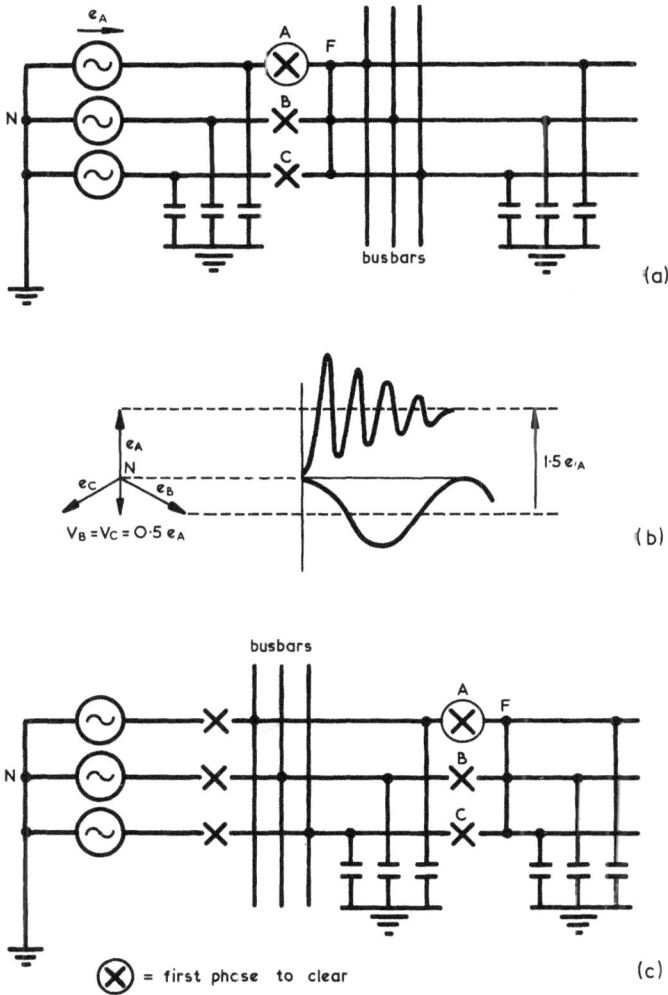

Fig. 3.4 Double-frequency transient arising from a 3-phase fault

however, give rise to another double-frequency condition. Fig. 3.4a shows a generator feeding a 3-phase busbar fault at F. Assuming that phase A of the circuit breaker is the first phase to clear, the voltage on the generator side of this phase will recover from zero to the phase-to-neutral voltage e_A, and on the busbar side from zero to $-0·5e_A$, as shown by the vector diagram in Fig. 3.4b. The steady-state value of the power-frequency recovery voltage of phase A is therefore $1·5e_A$ (see also the analysis in Section 3.2.1). The capacitance of phase A is very small on the generator side, however, being equal to that of the generator and its main connections only. As a result, the recovery from zero to e_A is by way of a high-frequency transient. On the busbar side, however, the capacitance is that associated with all the circuits connected to that busbar, and the recovery from zero to $-0 \cdot 5e_A$ is by way of a relatively low-frequency transient. These conditions are shown at the righthand side of Fig. 3.4b. It follows, therefore, that a high-frequency recovery voltage transient based on a power-frequency recovery voltage of $1 \cdot 5e_A$ may be unnecessarily severe for some conditions. However, a recovery voltage transient based on the high-frequency component and on a power-frequency recovery voltage of e_A only, may not be sufficiently severe. The short-circuit MVA .with which the circuit breaker has to deal under these conditions, is, however, only that corresponding to the short-circuit current of the generator and this is usually small compared with its rating.

If the fault is on an outgoing circuit, as in Fig. 3.4c for example, the short-circuit MVA may be equal to the circuit-breaker rating. In this case, the time function of the transient recovery voltage is quite different from the previously considered example. The capacitances associated with each side of the circuit breaker are now more nearly equal, so that the natural frequencies are similar and the recovery voltage transient must be based on a power-frequency recovery voltage having a value approaching $1 \cdot 5e_A$.

Another example giving rise to double frequency recovery-voltage transients concerns series reactors. If, for example, a reactor limits the faults MVA to 50% of that for a busbar fault, a fault on the circuit side of the reactor would give rise to a relatively high-frequency transient based on a change of voltage of $0 \cdot 5e$ to zero at the reactor side of the circuit breaker, and a low-frequency oscillation of $0 \cdot 5e$ to e on the busbar side. The conditions are, therefore, similar to those for faults on the low-voltage sides of transformers, with the exception that the fault MVA is usually much higher and, consequently, the conditions more severe.

A short length of cable, say 10–20 m, inserted between the circuit breaker, and the reactor would reduce the frequency of the higher-frequency component and would make switching conditions somewhat easier.

Damping

As has been previously mentioned, power losses in the circuit or the presence of shunt-connected loads damp recovery voltage transients. The main effect, so far as circuit breakers are concerned, is the reduction of the amplitude of the first peak and the consequent reduction of the rate of rise of the recovery voltage transient. For a circuit with resistance included (Fig. 3.5), one of three equations may be appropriate for the transient recovery voltage appearing across the circuit breaker. The choice depends on whether the resistance is greater than, equal to, or less than that which would give critical damping.

The equations describing these three conditions have been given by Lackey

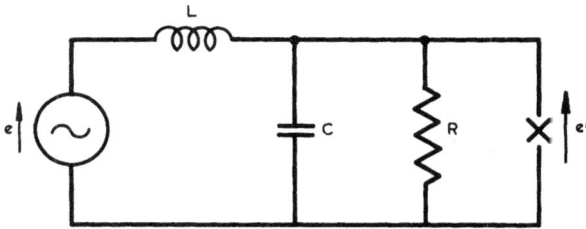

Fig. 3.5 Basic equivalent circuit with damping

(1951*a*), they are:

(*a*)   where R is greater than the critical damping resistance (underdamped)

$$e' = E \sin w_s t \left\{ 1 - \exp \frac{-t}{2RC} \left(\cos mt + \frac{1}{2Rm} \sin mt \right) \right\} \qquad (3.12)$$

where

$$m = \sqrt{\frac{1}{LC} - \frac{1}{4R^2 C^2}}$$

(*b*) where R is equal to the critical damping resistance

$$e' = E \sin w_s t \left\{ 1 - \exp \left(\frac{-t}{2RC} \right) \left(1 + \frac{t}{2RC} \right) \right\} \qquad (3.13)$$

(*c*)   where R is less than the critical damping resistance (overdamped)

$$e' = E \sin w_s t \left[1 - \frac{1}{2} \left(1 + \frac{1}{n} \right) \exp \left(\frac{1-n}{2RC} \right) t - \frac{1}{2} \left(1 - \frac{1}{n} \right) \exp \left\{ \frac{-(1+n)t}{2RC} \right\} \right]$$

$$\qquad (3.14)$$

where

$$n = \sqrt{\left(1 - \frac{4R^2 C}{L} \right)}$$

The critical damping resistance is given by

$$R_d = \sqrt{\left(\frac{L}{4C} \right)} \qquad (3.15)$$

Results yielded by eq. 3.12–3.14 have been plotted in Fig. 3.6 which clearly shows the importance of taking due account of the damping of the circuit, when assessing

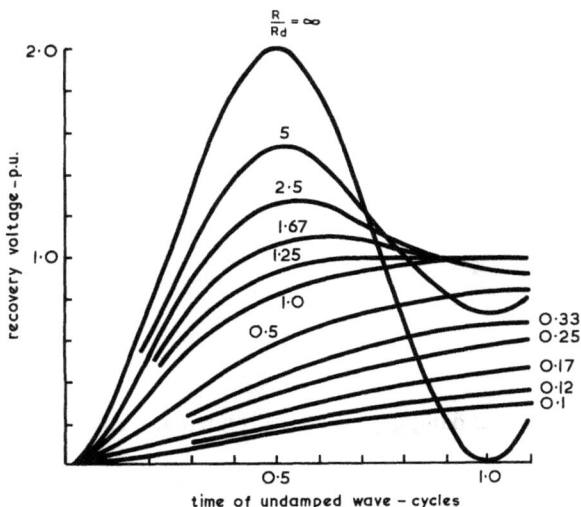

Fig. 3.6 Effect of shunt resistance in damping a recovery-voltage transient

the rate-of-rise of the recovery voltage. The curves can most easily be applied in practice by measuring the reduction of the rate-of-rise of the recovery voltage from that appropriate to an undamped transient, and plotting this against the ratio R/R_d. This has been done in Fig. 3.7 which gives the so-called damping factor for any known value of R/R_d.

Damping resistance can, of course, be provided by purpose-designed resistors connected across the interruptors to give so called 'resistance switching'. This technique has been described by Cox *et al.* (1944), Allan *et al.* (1947) and Cox *et al.* (1947). The value of the resistor can be determined by substituting the general expression for system current in eq. 3.15, i.e.

since

$$I = \frac{E}{w_s L} \text{ or } L = \frac{E}{w_s I} \tag{3.16}$$

therefore

$$R_d = \sqrt{\left(\frac{E}{4w_s IC}\right)} \tag{3.17}$$

and on this basis the critical damping resistance is inversely proportional to I, that is, low values of resistance are required for large currents and high values for small

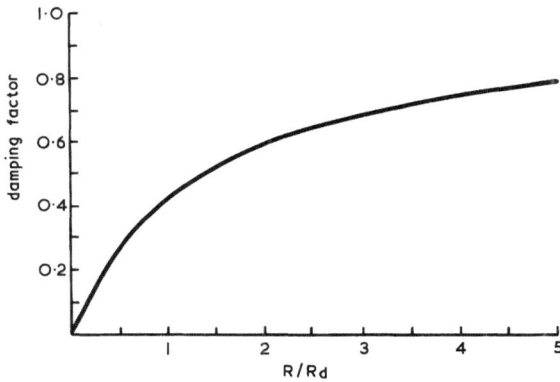

Fig. 3.7 Effect of shunt resistance in reducing rate of rise of recovery voltage

currents. It follows, therefore, that the ohmic value of the resistor must be chosen to give critical damping at a particular value of short-circuit current. Because of the inherently different characteristic of, say, oil and air blast circuit breakers, significantly different values of resistor would be chosen to provide critical damping at different values of short-circuit currents (Section 3.4.1).

Loads also provide damping resistance, but the effect is generally small during the period immediately after fault clearance, because of the delay time in the transmission system and this contribution to damping is usually ignored in calculation.

The most important natural damping resistance is provided by overhead lines and cables. At fault clearance, travelling waves of voltage and current are set up along the line, and in the initial period of time before the reflected wave returns, the line can be considered to be equivalent to a resistor of magnitude equal to its natural impedance. It can be shown that the reflections from the ends of a line have the effect of progressively increasing this equivalent resistance, from which it follows that the damping effect of short lines, say less than 10 km of overhead or less than 5 km of cable, is relatively small. It is then best to treat the line or cable as a lumped-impedance network for the purpose of calculation. Section 3.2.5 covers the treatment of the so called short line or kilometric fault in more detail.

Defining the severity of transient recovery voltage
The two important aspects of transient recovery voltage which help define its severity are:

(a) The transient peak voltage which is generally expressed as a per unit multiple of the peak power-frequency recovery voltage.

(b) The time function of the voltage as it rises from zero to the peak and its effect on the energy dissipation rate from the circuit breaker interruptor.

The peak voltage is easily defined, but precise definition of the rate of change of transient recovery voltage is more difficult.

The large variety of transient recovery-voltage waveforms which characterise particular circuits have raised doubts concerning the adequacy of the previous simplification in specifying the voltage/time characteristic of transient recovery voltage, merely as an average rise from zero to peak voltage.

Single frequency transients are uncommon in practice and Fig. 3.8 shows the initial part of a typical transient recovery voltage waveform. In practice, most transients tend to separate into a fast-rising initial component, $M-P$, which flattens out after a period, typically, some $300-1200$ μs on a 300 kV system, to reach a peak value S. The transient then gradually decays into the normal power-frequency rate of change. In general, gas-blast circuit breakers are sensitive to the shape of the initial portion of the transient, whereas oil-break circuit breakers are more likely to be affected by the amplitude of the peak S.

The rate of rise of recovery voltage is generally used as a measure of severity, but there are a number of ways of defining it. For example, Fig. 3.8(i) shows a rate of rise defined as the average rate of rise from zero to the maximum, Fig. 3.8(ii) as the average rate of rise from zero to the first peak, Fig. 3.8(iii) as maximum average

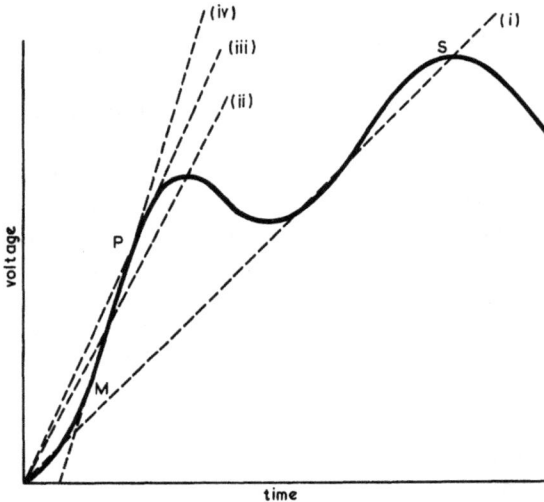

Fig. 3.8 Methods of defining rate of rise of recovery voltage

rate cf rise expressed as a tangent to the recovery voltage wave shape, and Fig. 3.8(iv) as maximum instantaneous rate of rise.

One way of defining a transient of the displaced-cosine form, as produced by circuits having lumped parameters, has been proposed by Ferguson *et al.* (1968), and involves the second derivative of voltage with respect to time, i.e. $d^2 V/dt^2$, but this has the disadvantage that it cannot be generally applied to all types of circuits, since the second derivative of a voltage rising linearly with time is zero.

A more general method of defining the severity of transient recovery voltage is the 'four parameter method' currently embodied in IEC document 56. This allows a two-part rate of rise to be defined in terms of dV/dt over specified periods of time, see Fig. 3.9.

Some other aspects of circuit breaker transient recovery voltage have been published by Beehler *et al.* (1965) and Buettner *et al.* (1969).

3.3.2 Short-line faults

The maximum fault current, that a circuit breaker may be called on to interrupt, is that arising with a fault immediately at its terminals. Although this is the most onerous from a short-circuit current viewpoint, it is not necessarily so in respect of the transient recovery voltage.

One of the most frequent system-fault conditions is that of an overhead line fault, remote to the circuit breaker. Under these circumstances, a double-frequency

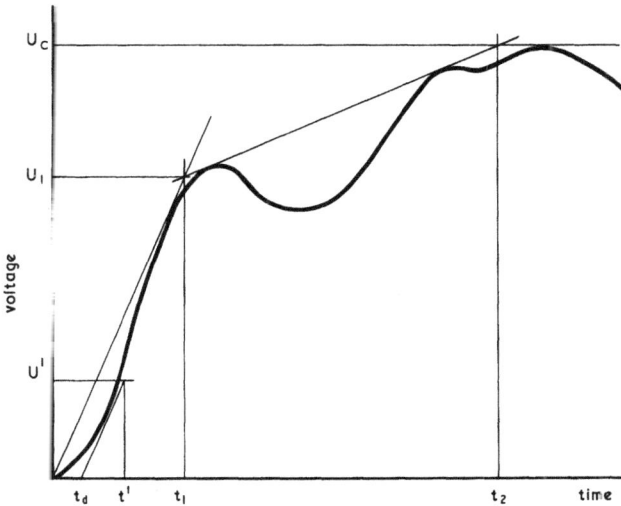

Fig. 3.9 Four parameter method of defining rate of rise of recovery voltage

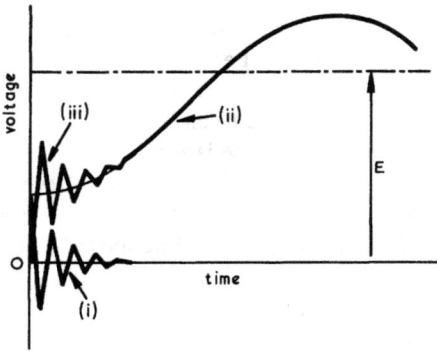

(i)   voltage at line terminal of circuit-breaker
(ii)   voltage at source terminal of circuit-breaker
(iii)   voltage across circuit breaker

Fig. 3.10 Voltage waveforms associated with a short-line fault

recovery voltage transient is applied across the circuit breaker, comprising of the vector difference of source and line-side transient recovery voltages. In the particular case of the fault being some few kilometres distant from the circuit breaker, an extremely onerous fault condition results. The voltages associated with this type of fault, commonly known as a short-line or kilometric fault, are illustrated in Fig. 3.10.

At zero current, the voltage at the circuit breaker line-side terminal is equal to the voltage drop along the faulted line section. On interruption of current, the line remains charged at this voltage. The stored energy of the line is then dissipated by means of a voltage wave traversing the line length, resulting in a high-frequency voltage transient of sawtooth waveform impressed on the circuit breaker as shown in Fig. 3.10(i).

The peak voltage and frequency of oscillation of this line-side transient are a function of the distance of the fault from the terminal of the circuit breaker. Although the time to peak of the line-side transient may be short, the source-side contribution during this period, as shown in Fig. 3.10(ii), generally, cannot be ignored. The circuit breaker is subjected to a total transient recovery voltage comprising of the vector difference of source and line-side transients [Fig. 3.10(iii)].

The rate of rise of recovery voltage is considerably greater for the 'short-line' fault condition than for the terminal-fault case, although the peak transient voltage is comparatively low (Fig. 3.10). However, considerably more energy is injected into the circuit breaker interruptor during the post-arc current period from the line-side transient, rather than from the source transient, and it is this factor which accounts for the increased severity of duty compared with the terminal-fault condition.

The rate of rise of the line-side transient is equal to the product of the natural impedance of the line and the rate of change of current at the instant of current interruption, that is:

$$\frac{dV}{dt} = Z_n \frac{di}{dt} \qquad (3.18)$$

The period of the sawtooth waveform is equal to four times the reflection time from the circuit breaker to the fault point. The severity of the recovery voltage depends on the fault position as this will determine both the magnitude of the fault current and hence di/dt and the peak value of the sawtooth waveform, since this is proportional to the reflection time from the fault.

Determination of natural impedance
The voltages present in a multiconductor system are given by Hawley (1945) as follows:

$$e_1 = Z_{11}i_1 + Z_{12}i_2 + Z_{13}i_3 \quad .. \quad .. \quad .. \quad + Z_{1n}i_n$$

$$e_2 = Z_{21}i_1 + Z_{22}i_2 + Z_{23}i_3 \quad .. \quad .. \quad .. \quad Z_{2n}i_n$$

$$e_n = Z_{n1}i_1 + Z_{n2}i_2 + Z_{n3}i_3 \quad .. \quad .. \quad .. \quad Z_{nn}i_n$$

where e_n = voltage at any point on conductor n and i_n = current at any point in conductor n.

The above equations may be simplified in the case of a single-circuit three-phase line with an earth wire as follows:

$$e_1 = Z_d i_1 + Z_m i_2 + Z_m i_3 + Z_{mg} i_4 \qquad (3.19)$$

$$e_2 = Z_d i_2 + Z_m i_1 + Z_m i_3 + Z_{mg} i_4 \qquad (3.20)$$

$$e_3 = Z_d i_3 + Z_m i_2 + Z_m i_1 + Z_{mg} i_4 \qquad (3.21)$$

where $e_4 = Z_g i_4 + Z_{mg}(i_1 + i_2 + i_g)$

$$Z_d = 60 \log_e \frac{2h}{a}$$

$$Z_g = 60 \log_e \frac{2h'}{b}$$

$$Z_m = 60 \log_e \frac{2h}{d}$$

$$Z_{mg} = 60 \log_e \frac{h + h'}{d'}$$

The appropriate natural impedance is given by e_1/i_1. The values assigned to e_2, e_3, i_2, and i_4 are dependent on whether or not conductors 2 and 3 are energised. With bundle conductor lines the natural self impedance is given by

$$Z_d = 60 \log_e \frac{2h}{n^{rs}(n-1)}$$

Solution of eqs. 3.19, 3.20 and 3.21 for the case of the first, second and third poles to clear with a 3-phase fault, yields the following:

first pole to clear

$$Z_n = Z_d - \frac{2Z_m{}^2 + \dfrac{Z_{mg}{}^2}{Zg}(Z_d - 3Z_m)}{Z_m + Z_d - \dfrac{2Z_{mg}{}^2}{Zg}} \tag{3.22}$$

second pole to clear

$$Z_n = Z_d - \frac{Z_m{}^2 + \dfrac{Z_{mg}{}^2}{Zg}(Z_d - 2Z_m)}{Z_d - \dfrac{Z_{mg}{}^2}{Zg}} \tag{3.23}$$

third pole to clear

$$Z_n = Z_d - \frac{Z_{mg}{}^2}{Zg} \tag{3.24}$$

In the event of two poles clearing simultaneously

$$Z_n = Z_d - Z_m \tag{3.25}$$

When overhead earth wires are not provided the term Z_{mg}^2/Zg can be neglected.

In utilising eqs. 3.22, 3.23, 3.24 and 3.25, it should be noted that the results are not precise, but are sufficiently accurate for practical purposes. In the case of a double-circuit line with earth wires, or when checking actual system measurements, a more rigorous solution of the general equations is required as reported by Bolton *et al.* (1964).

Factors influencing natural impedance
(a) Line configuration
As shown previously, major factors influencing the natural impedance of an overhead line are the conductor spacing, height of conductor above ground, and the number of conductors per phase, thus each particular voltage rating has an inherent natural impedance.

The actual value of the natural impedance is a function of the particular pole attempting to interrupt the fault and the state of adjacent poles at the same instant, i.e. whether de-energised or carrying current. Calculation and system measurements have shown that the highest natural impedance is encountered by the last pole to clear any particular fault condition, whether the line is of single or double-circuit configuration. The value of the natural impedance is thus quite independent of the type of fault. The difference for first and last poles to clear a 3-phase fault has been shown by Hawley (1945) to be of the order of 10%.

(b) Earth resistivity

Practical conditions dictate that the effect of a significant earth-return impedance must be considered in the determination of the natural impedance. A fault close to the circuit-breaker terminal would result in a very high-frequency transient being produced. In the presence of high earth resistivity, current penetration of the earth during the initial portion of this transient is only to a shallow depth, resulting in the effective line inductance, and hence reactance, being lower than that prevailing under power-frequency conditions. The depth of current penetration increases with time, resulting in an increase of reactance, but still not to the extent of approaching the power-frequency value. The effect of this phenomena on the wave form of the transient has been measured by Bolton *et al.* (1964).

Under conditions of high resistivity, therefore, the effective natural impedance during the initial portion of the line transient is equal to that calculated assuming zero earth resistivity, but increases with time as greater earth penetration is achieved. Calculation assuming a homogeneous earth have shown the increase to be of the order of 3% for resistivities of up to 80 Ωm, to as much as 12% at 1000 Ωm. In the case of earth resistivity being low, say, less than 30 Ωm, the impedance of the earth return path will be negligible.

The presence of an earth wire presents an additional earth-current return path, which may tend to counteract the effect of high-ground resistivity, thereby producing a reduction in natural impedance compared with the case where no earth wire exists. It should be noted that this effect depends on the earth-wire material. For example, on a typical 330 kV overhead line, the natural zero-sequence impedance without an earth wire is 746 Ω, with a steel earth wire, this is reduced to 660 Ω, whilst with an a.c.s.r. earth wire, the value falls to 552 Ω.

Relative severity of 3-phase and single-phase earth-faults

In the event of a 3-phase insulated terminal fault occurring, the first phase to clear is subject to a voltage of 1·5 times normal phase to earth voltage (Section 3.2.1). It should be noted that this voltage distribution is phase to neutral voltage above ground on the source side and 0·5 times phase to neutral voltage below ground on the line side, this can improve the voltage distribution across a circuit breaker so that the voltage grading is assisted. The overall result is not, therefore, as onerous as might be expected. Of course, under conditions where near perfect voltage grading exists, this distribution will not make any appreciable difference. These same voltage conditions are present should be fault be a short distance from the circuit-breaker terminals and it can be postulated that tests should be made with the source side transient oscillating about this voltage. The probability of an insulated 3-phase fault occurring is, however, extremely low, and the probability of a short-line fault under such circumstances is even more remote.

In heavily interconnected power systems, the zero-sequence source impedance is frequently lower than the positive-sequence source impedance (Section 3.2.2), resulting in the fault current under the single-phase condition being greater than that of the 3-phase condition, typically by some 20–30%. The severity of the short-line fault duty is, therefore, correspondingly increased. This effect is illustrated in Fig. 3.11.

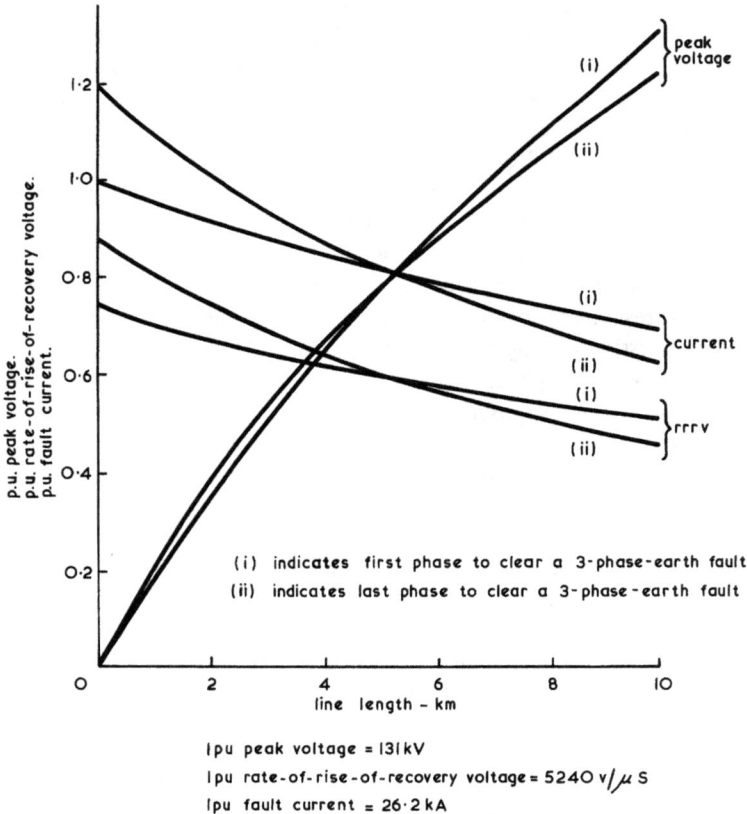

1pu peak voltage = 131kV
1pu rate-of-rise-of-recovery voltage = 5240 v/μ S
1pu fault current = 26·2 kA

Fig. 3.11 Severity of short-line fault when source zero-sequence impedance is smaller than source positive-sequence impedance

The above factors, together with the increased natural impedance under single-phase conditions, have led to an international agreement that short-line fault tests need only be made for the single-phase case.

Effect of equipment at the circuit breaker terminal
A practical installation may well have a considerable capacitive load between the circuit-breaker terminals and the line-entry position, e.g. instrument transformers, line traps etc. The presence of these items has a significant effect upon the line-side transient, and has been investigated by Mazza *et al.* (1972). They demonstrated that the added capacitance of a capacitor voltage transformer significantly reduces the line transient frequency, thus easing the interruption duty imposed on the circuit breaker. The presence of a line trap, however, introduces a high-frequency component into the line transient, the value being dependent on the line-trap inductance.

The significance of this additional transient frequency is a function of the type of circuit breaker, and may present a severe duty to those designs sensitive to initial the rate of rise of recovery voltage.

The beneficial effect afforded by a capacitive load at the circuit-breaker terminals can be utilised in place of damping resistors to improve the short-line fault performance of a circuit breaker. An instance of this has been reported by Berkebile *et al.* (1971). This factor is particularly useful when attempting to uprate existing equipment. A great deal of information has been published on the short-line fault, and mention may be made of the contributions by Pouard (1958), Baltensperger *et al.* (1960*b*), Petitpierre (1960), Baltensperger (1962), Young *et al.* (1963), Eidinger *et al.* (1964) and Kummerow (1964).

3.4 Load and system switching

3.4.1 Effect of load power factor
It has already been pointed out that recovery voltage is one of the principle factors determining the ability of a circuit breaker to clear a circuit successfully. The power factor of the circuit being switched is an important factor in determining the transient component of this recovery voltage and in many of the cases considered in this chapter, e.g. short-circuit switching, reactor switching and capacitor switching, the power factor rarely exceeds a value of 0·1. In the case of switching normal loads, however, the power factor is likely to be nearer unity. Flugum *et al.* (1965) considered the various basic configurations of load circuits and compared two simple ideal circuits, one containing resistance only and the other containing inductance only to illustrate the general effect of the power factor on transient recovery voltage.

Fig 3.12*a* shows a purely resistive circuit where the current and voltage are in phase. At the first current zero after the circuit breaker opens, the voltage is also

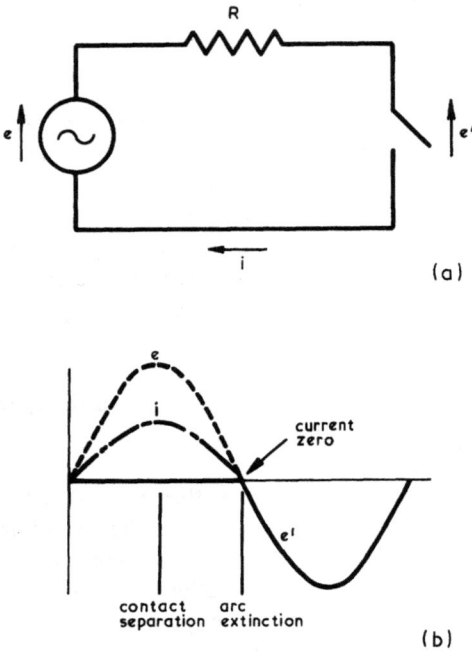

Fig. 3.12 Recovery voltage in a resistive circuit

zero, the arc is extinguished and the recovery voltage e' rises across the circuit breaker at normal power-frequency rate as shown in Fig. 3.12b.

With a purely inductive circuit, however, as shown in Fig. 3.13a, the following conditions hold good:

$$e = E \sin w_s t \tag{3.26}$$

$$i = \frac{-E}{w_s L} \cos w_s t \tag{3.27}$$

at an instant of arc extinction, i.e. at $t = 3\pi/2$, the recovery voltage across the circuit breaker is equal to $-E$. Fig. 3.13b shows that the voltage tries to recover instantaneously to its maximum value, thus imposing an extremely steep recovery voltage wavefront across the circuit breaker.

The recovery voltage characteristic of a purely capacitive circuit (Fig. 3.14a), is also of interest, because it too has a power factor equal to zero. This may be

(a)

(b)

contact
separation

arc
extinction

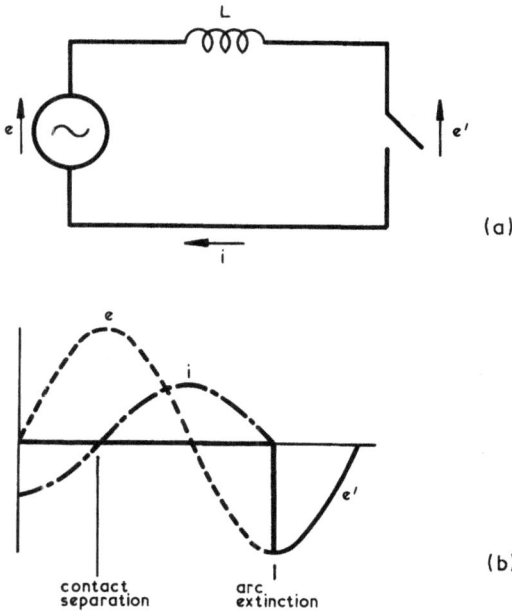

Fig. 3.13 Recovery voltage in an inductive circuit

analysed as follows:

at the instant of arc extinction, i.e. at $t = 3\pi/2$,

$$e_c = -E$$

and

$$e = E \cos w_s t$$

since

$$e' = e - e_c$$

$$= E(1 - \cos w_s t) \tag{3.28}$$

i.e. at $t = 0$, $e' = 0$

The resultant recovery voltage is a displaced cosine wave, as shown in Fig. 3.14*b*. At the instant of current zero, the voltage across the circuit breaker is zero. After one half-cycle, the voltage across the contacts rises to a maximum of $2E$ and if the dielectric strength of the interruptor gap is not adequate, a restrike may occur.

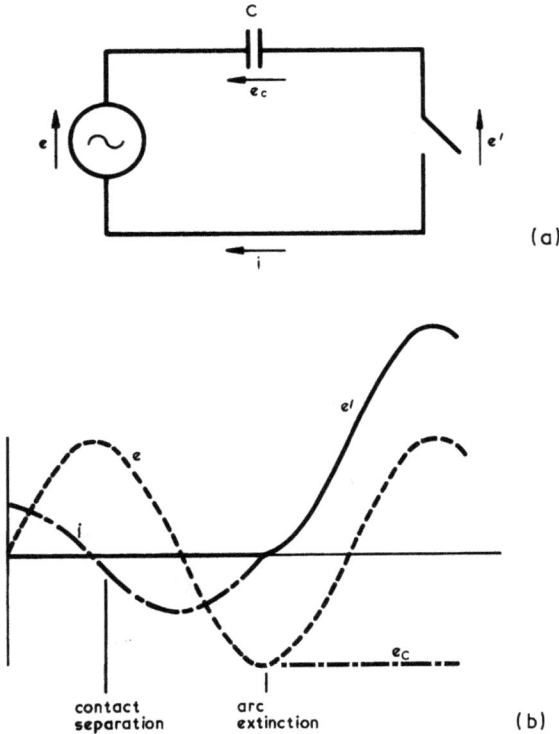

Fig. 3.14 Recovery voltage in a capacitive circuit

Having established that the power factor influences interruption severity profoundly, it can also be shown that different arrangements of circuit parameters can also effect interruption severity, even though the applied voltage, current and power factor may be the same. The use of the power factor as a single criterion of severity is, therefore invalid.

The impedance of the load circuit being switched can be a series combination of resistance and reactance, a parallel combination of resistance and reactance or a series-parallel combination of resistance and reactance. In the series circuit of Fig. 3.13a, for example, at the instant of current zero, the voltage across the circuit breaker tries to change instantaneously from the arc voltage to the peak value of the system voltage. In practice, due to the interchange of energy between the inductance and stray capacitance in the circuit, the rate of voltage rise across the contacts will be reduced, but the voltage will overshoot and oscillate about normal

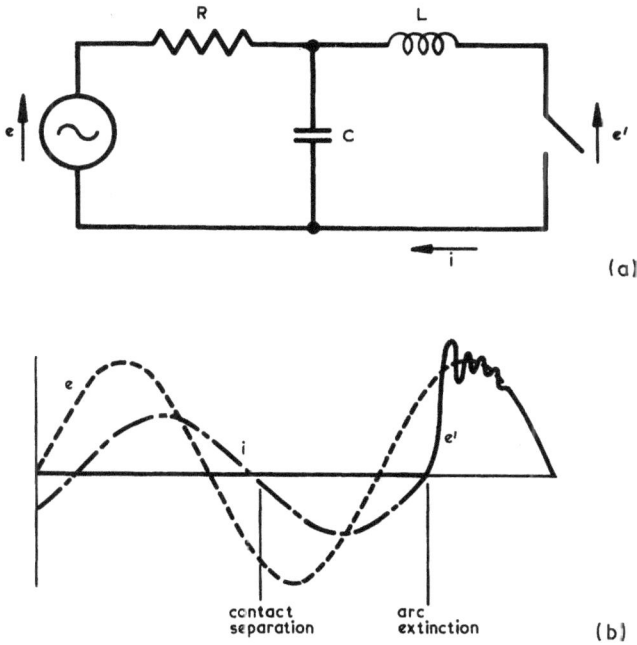

Fig. 3.15 Recovery voltage in an RLC circuit

voltage. These conditions are illustrated in Fig. 3.15, although a more rigorous analysis of this case is given in Section 3.2.4.

If the circuit elements are connected in parallel, as shown in Fig. 3.16a, different conditions prevail, even though the total current and power factor may be the same. At the instant of total current zero in the circuit breaker a circulating current flows between the inductive and resistive branches, causing the voltage e_L across the circuit elements to be equal to the source voltage e. Consequently, at current zero, the voltage across the circuit breaker contacts is zero. The voltage e_L across the circuit elements then decays in a manner proportion to $\exp(-Rt/L)$.

The resultant recovery voltage rises from zero to the system voltage (Fig. 3.16b), but it has a much slower rate of rise than that of the series circuit shown in Fig. 3.15a.

The most probable load circuit in practice is, however, one consisting of both series and parallel components. The recovery voltage, and hence the resulting interruption severity, for a given power-factor, depends in this case on the proportion of series and parallel elements. If the series impedance component is

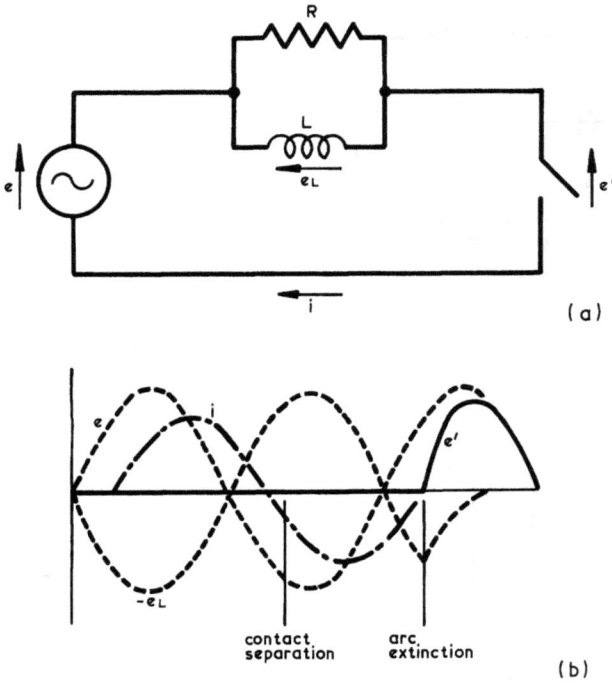

Fig. 3.16 Recovery voltage in a parallel RL circuit

larger than the parallel component, the recovery voltage wavefront will be steep and interruption will be more difficult. Conversely, if most of the impedance is contained in the parallel branches, the recovery voltage will approximate to that shown in Fig. 3.16b and the circuit will be cleared more easily. Boehne (1963) showed that a compound series-parallel circuit is the exact equivalent of the great majority of load circuits commonly encountered.

The difference in the recovery voltage for the series and parallel circuits at a given power factor can be further illustrated by considering the fundamental equation of the voltage across the circuit breaker. This consists of power and natural or transient frequency components. For example, in the series circuit of Fig. 3.15a the recovery voltage is

$$e' = E \sin(w_s t + \phi) + E \sin \phi (1 - \cos w_n t) \qquad (3.29)$$

where

$$\phi = \text{initial power factor} = \tan^{-1} \frac{w_s L}{R}$$

The rate of rise of recovery voltage at any time is obtained by differentiation of eq. 3.29, i.e.

$$\frac{de'}{dt} = w_s E \cos(w_s t + \phi) + w_n E \sin \phi \sin w_n t \tag{3.30}$$

Graphically (Fig. 3.16*b*), the recovery voltage of the parallel circuit shown is the sum of the power frequency voltage and the voltage across the inductance, therefore,

$$e' = E \sin(w_s t + \phi) - E \sin \phi \exp(-Rt/L)$$

where, in addition to the other terms,

$$L = \frac{X}{w_n}$$

hence

$$\frac{de'}{dt} = w_s E \cos(w_s t + \phi) + \frac{w_n R}{X} E \sin \phi \exp\left(\frac{w_n R t}{X}\right) \tag{3.32}$$

For purposes of comparison, the recovery voltage of the capacitive circuit (Fig. 3.14*a*) has a sinusoidal wave shape, hence the differential eq. 3.28 with respect to time is

$$\frac{de'}{dt} = w_s E \sin w_s t \tag{3.33}$$

The rate of rise of recovery voltage may be evaluated for the three circuits by means of eqns. 3.30, 3.32 and 3.33, respectively, using the same relative value for w_s and a power factor of $0\cdot7$ at an arbitrarily chosen time of 200 μs after current interruption. The rate of rise of recovery voltage for the series circuit is found to be $14\cdot5E$, for the parallel circuit $1\cdot2E$, and for the capacitive circuit $0\cdot11E$. Using the rate of rise of recovery voltage as a criterion, in this example, the series circuit is about 12 times more severe than the parallel circuit with the same current, voltage and power factor. It should be noted that the ratio of 12 changes as the transient dies out, and of course depends on the power factor.

3.4.2 Out-of-phase switching

An important circuit-breaker requirement to be considered in relation to interconnectors between generating stations or subsystems is the ability of the circuit breaker to open satisfactorily under asynchronous conditions. In this case the recovery voltage across an opening circuit breaker may be much higher than under short-circuit conditions and the duty may be more severe in some respects than the short-circuit duty.

The phenomenon has been covered by Lackey (1951*a*), and Leeds *et al.* (1952).

They showed that the magnitude of the recovery voltage depends on the phase angle between the internal voltages of the two sources at the instant of opening, and also on the method of earthing the power system neutral (Section 3.2.3). With modern high-speed protection, tripping normally takes place while the displacement angle between the two sources is still small but with slower forms of protection, e.g. some types of back-up protection, the angle at which tripping takes place may be large. When considering autoreclosing circuit breakers, the instance of a second fault clearance, after an unsuccessful reclosure, may occur when the displacement between the two voltages is considerable. It follows, therefore, that if no special provision is made to relate the opening with the phase displacement, it must be assumed that a circuit breaker may open at any phase angle.

The worst condition in a solidly-earthed system occurs when opening with the two sources 180° out of phase (Fig. 3.17a). The recovery voltage then has a maximum value of about twice the phase to neutral voltage. The Ferranti effect can increase this value if the circuit breaker is located towards the middle of a long transmission system. For example, the asynchronous switching duty required of circuit breakers located at an intermediate substation in a 1000 km 500 kV transmission circuit, is 2·4 times the phase-neutral voltage.

Fig. 3.17b shows the conditions in an insulated neutral system with the maximum recovery voltage about three times the phase-to-neutral voltage.

In Fig. 3.17c, a Petersen coil earthed system is shown with two earth-faults, one in phase A and the other in phase B. The associated vector diagram shows that, under these circumstances, the voltage across the circuit breaker in phase C can reach 3·4 times the phase-to-neutral voltage.

The maximum current with which a circuit breaker has to deal under asynchronous conditions is

$$i = \frac{e + e'}{Z} \tag{3.34}$$

where e and e' are the phase-to-neutral voltages of the two sources, and Z is the total impedance of the sources and the interconnecting line. This value of current is generally considerably less than the maximum short-circuit fault current in the circuit. Fig. 3.18 shows the relationship between the asynchronous current and the ratio of the source impedances Zs and $Z's$ on each side of the circuit breaker. When this ratio is unity, the current is 50% of the short-circuit fault current and when the ratio is 2, for example, the current is approximately 44% of the short-circuit fault current. It is clear that all conditions could be met by arranging for circuit breakers exposed to the risk of operating under asynchronous conditions to be capable of interrupting 50% of rated short-circuit fault current at say 2·4 times normal voltage. This requirement, however, may be unnecessarily severe. IEC Document 56 calls for an asynchronous switching capability of interrupting 25% rated short-circuit current at 2·0 times normal voltage, although some supply authorities call for the asynchronous voltage to be 2·6 times the normal.

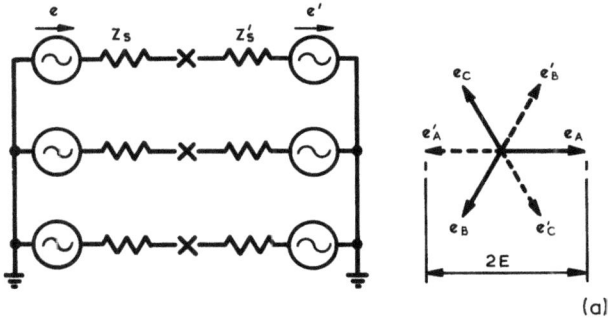

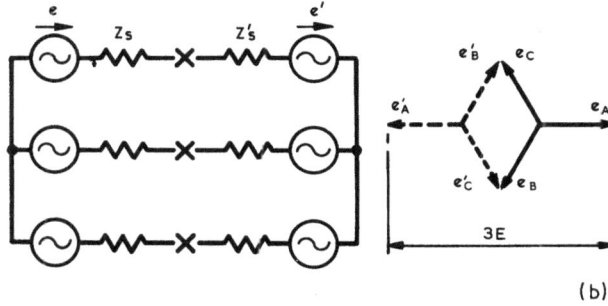

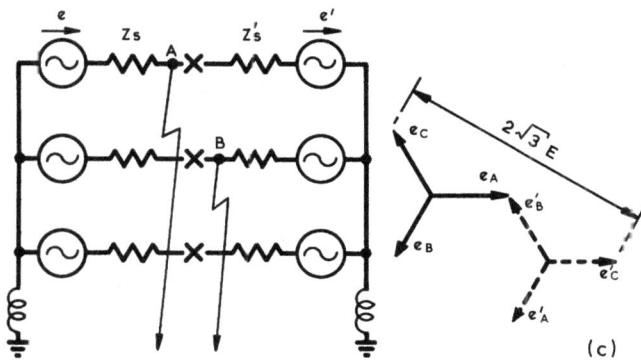

Fig. 3.17 Voltages across a circuit breaker under asynchronous conditions

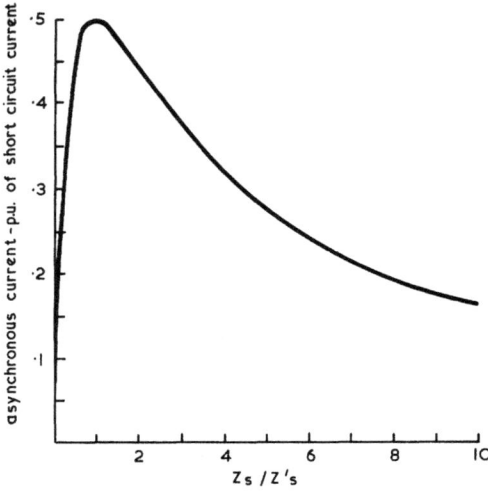

Fig. 3.18 Relationship between synchronising current and ratio of source impedances during asynchronous conditions

A factor that eases asynchronous switching conditions is the severity of the transient recovery voltage, usually small compared with that arising during short-circuit switching. Referring to Fig. 3.19, and considering a circuit breaker at one end of an interconnector, if v is the voltage at busbar M, the fall in voltage at the busbar due to the flow of synchronising current is $e - v$. This is not large if the source impedance is small in comparison to the line impedance Z_L. The amplitude of the transient recovery voltage is, therefore, small as shown by curve (i) of Fig. 3.19b. On the line side of the circuit breaker the voltage has to change at the time of clearance from the busbar voltage v to the open-circuit voltage of the remote source e', which is a change approaching twice the normal system voltage. The transient recovery voltage on the line side of the circuit breaker is, however, heavily damped by the line, as shown by Fig. 3.19b, (ii) so that the amplitude is small. If the circuit breaker interrupting the interconnector is located at a substation situated between the two sources, the conditions may be easier still since there is then heavy damping on both sides of the circuit breaker.

In general, therefore, the severity of the transient recovery voltage when switching takes place under asynchronous conditions is much smaller than that associated with many short-circuit conditions.

3.4.3 Single-phase switching

The occurrence of high induced voltages at the terminals of an unearthed unloaded 3-phase transformer when only one or two phases are energised, Figs. 3.20a and

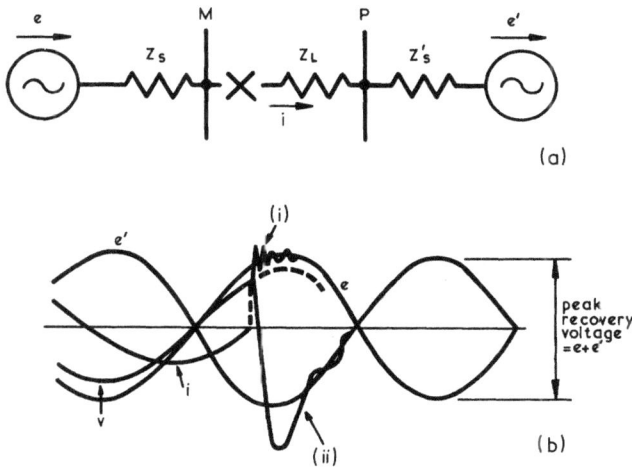

Fig. 3.19 Recovery voltage oscillations during asynchronous switching

3.20*b* is a well known phenomena, commonly referred to as 'neutral inversion', which has been described by many authors including Clarke *et al.* (1941), Petersen (1951), Rittenhouse *et al.* (1963) b and Hopkinson (1965). These high voltages may occur after the blowing of a fuse or the nonsimultaneous opening or closing of circuit-breaker contacts. The phenomena may be illustrated by the following example, which is typical of many that can be envisaged.

When one phase of a 3-phase circuit supplying an unearthed unloaded transformer bank is open, there is a path for currents from the two closed phases through the parallel combination of the magnetising impedance of the transformer, and the capacitance of the two closed phases to the open phase, and thence to earth through the capacitance to earth of the open phase. There is a similar path with two phases open. When the 3-phase system is earthed, the path is completed through the system neutrals. If the system is unearthed, however, the path is through the capacitance to earth on the system side of the open phase. Under this latter condition, resonance can occur and high voltage to earth at the open transformer terminal, or terminals, may result depending upon the magnetising impedance of the transformer and the capacitive reactance of the line. The capacitance of the transformer bushings and busbar structures, with no intervening length of line, may be sufficient to reduce this high voltage on small transformers such as those used for instruments or protection.

For either case, the equivalent circuit is shown in Fig. 3.20*c*. If X_L is the magnetising reactance of the transformer and X_c is the capacitive reactance of the

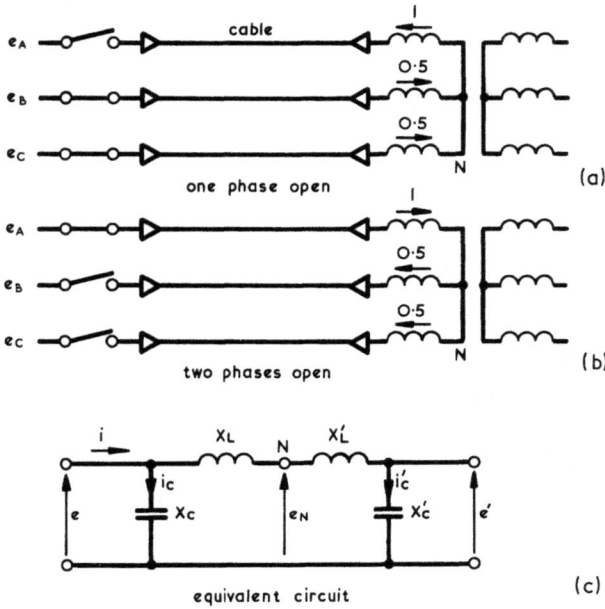

Fig. 3.20 Conditions for neutral inversion and equivalent circuit of ungrounded transformer

line or cable, for one phase open:

$$X_L' = 2X_L \tag{3.35}$$

$$X_c' = 2X_c \tag{3.36}$$

for two phases open:

$$X_L' = 0\cdot5X_L \tag{3.37}$$

$$X_c' = 0\cdot5X_c \tag{3.38}$$

The equivalent circuit is clearly similar to that of an inductive transmission line feeding a capacitive load and it is well known that this combination results in a rising voltage along the inductive line (Ferranti effect). It follows directly, therefore, that the voltage at the neutral of the transformer and at the transformer terminals connected to the open cables will be higher than the supply voltage.

The relationships of the vector diagrams shown in Fig. 3.21 are solved to yield:

$$e_N = e\left(\frac{X_c' - X_L'}{X_c' - X_L - X_L'}\right) \tag{3.39}$$

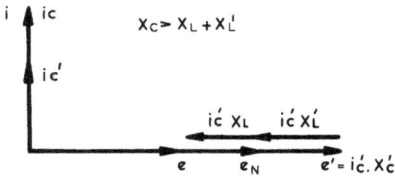

Fig. 3.21 Vector diagram of voltages and currents associated with neutral inversion

$$e' = e\left(\frac{X_c{}'}{X_c{}' - X_L - X_L{}'}\right) \tag{3.40}$$

Therefore, using eqns. 3.35 and 3.36, with one phase open, and with X_c greater than $1 \cdot 5 X_L$:

$$e_N = \frac{e}{2}\left(\frac{X_c - X_L}{X_c - 1 \cdot 5 X_L}\right) \tag{3.41}$$

and

$$e' = \frac{e}{2}\left(\frac{X_c}{X_c - 1 \cdot 5 X_L}\right) \tag{3.42}$$

and using eqns. 3.37 and 3.38, with two phases open:

$$e_N = e\left(\frac{X_c - X_L}{X_c - 3 X_L}\right) \tag{3.43}$$

and

$$e' = e\left(\frac{X_c}{X_c - 3 X_L}\right) \tag{3.44}$$

where e is the system phase to neutral voltage.

The infinite value of voltage which is indicated for exact equality in the denominator of eqns. 3.41, 3.42, 3.43 and 3.44 would not of course occur in practice, since the voltage rise would be limited by saturation and resistive effects, although substantial overvoltages can, nevertheless, occur.

3.4.4 Closing and reclosing on long transmission lines
Line energisation overvoltages
As well as overvoltages that may arise due to restrikes occurring at line de-energisation, severe overvoltages can also result from the energisation of

transmission lines. Even more severe overvoltages can occur as a consequence of line reclosing after a momentary interruption, as has been shown by Johnson *et al.* (1964), Hedman *et al.* (1964), Petitpierre *et al.* (1964), Paris (1968) and Greenwood (1971). This is a very important aspect of e.h.v. and u.h.v. system design, as the overvoltages produced by this type of switching may be those which determine the basic insulation level of the power-system equipment.

Basically, the phenomenon of line energisation can be considered as follows. As the line-charging circuit breaker closes, but immediately prior to the circuit being completed, a voltage exists across its contacts. If the line is initially de-energised, this voltage could lie anywhere between zero and the peak system voltage. At the moment that the contacts make or are joined by a prestriking discharge, this voltage disappears, only to reappear distributed around the circuit. This distribution will be in accordance with the impedance of the various parts of the circuit. Looking back from the circuit breaker to the source there is the source impedance paralleled by the natural impedance of other lines connected at that time. Let this combined impedance be Z_x. Looking from the circuit breaker into the line, it is the natural impedance of the line which is apparent. Let this natural impedance be Z_n.

If the instantaneous voltage that appears after the line is energised is V, a voltage V_1 will appear momentarily on the line, where

$$V_1 = \left(\frac{Z_n}{Z_x + Z_n} \right) V \qquad (3.45)$$

and a voltage V_s will appear across the source where

$$V_s = \left(\frac{Z_x}{Z_x + Z_n} \right) V \qquad (3.46)$$

If $Z_n \gg Z_x$, as in a heavily interconnected system, most of the voltage will appear across the line. This voltage will travel down the line as a wave, be reflected from its remote end and return to the source. If the far end of the line is open circuited, the reflected wave will add to the incident wave and a voltage approaching 2 p.u. will be impressed on the line. The propagation of such surges has been described by McElroy *et al.* (1963). The de-energisation of a healthy transmission line or the clearance of an asymmetrical fault can result in the healthy phase or phases being left charged to a voltage that can be of the order of the peak system voltage. The subsequent restoration of the supply to a line in this charged state, if the polarities are in opposition, can theoretically increase the travelling wave voltages already described to approximately 3·0 p.u., although in practice, resistive and corona losses tend to limit the magnitude of the overvoltage. The basic considerations involved are relatively simple but the number of variables involved makes analysis difficult without the assistance of a computer. Miller *et al.* (1964), Bickford *et al.* (1967), Lauber (1968) and Clerici *et al.* (1970)a, have published the results of their computer assisted investigations.

Generally, as already stated, the value of the transient voltage depends on the point on the supply-voltage wave at which reclosure is effected and on the trapped charge voltage. The maximum transient voltage on the line may be expected to occur when the line is energised at a peak of the supply-voltage wave of opposite polarity to the trapped charge voltage on the line. Other factors which affect the magnitude of the overvoltage include the nature and impedance of the source, the amount of reactive compensation connected to the line, the line losses and, due to mutual effects between phases, nonsimultaneous closure of the three phases of the circuit breaker energising the line. With air-blast circuit breakers having a rotating-arm type of isolator which is used also as a closing switch, the first phase to close usually does so at, or about, peak voltage, owing to sparkover occurring as the moving contacts approach the fixed contacts. Thus, even if the three phases can be adjusted to make metallic contact simultaneously one phase is likely to close before the other two because of sparkover. With the pressurised-head type of air-blast circuit-breaker where the closing operation is performed relatively quickly under pressure, sparkover is not so likely to occur and near simultaneous closure of the three phases should be attainable.

When lines or individual phases are disconnected from the supply, they remain charged for a finite period. The line will discharge eventually through leakage paths across insulators for example, but the rate at which discharge occurs is governed principally by the prevailing climatic conditions. The time constant of the discharge is usually in the range 20–60 s, but under extremely dry conditions the time constant may be as much as 3 min. A very real possibility exists, therefore, even in the case of manual reclosure, of closing onto a line with a large trapped-charge voltage since the line may remain almost at its initial trapped-charge voltage for many seconds after current interruption, the discharge time constant of the line being typically longer than the dead time of high-speed auto reclosure schemes.

Very often factors other than those associated with the line itself are present which tend to accelerate the discharge of the line. Opening resistors may be fitted to the circuit-breaker shunt reactors or electromagnetic voltage transformers may be directly connected to the line, all of which discharge the line to some extent before the reclosing operation takes place.

The effect of a circuit-breaker opening resistor in reducing the discharge time of the line and the trapped charge voltage on the line depends, of course, on the value of the resistor, the length of line and the time the resistor is in circuit. The resistance and the time it is in circuit are functions of the circuit breaker design. Opening resistors used are generally of the order of thousands of ohms, the contacts of the resistor break may separate at times of the order of 30–60 ms after the main break. The resistor may, however, be in circuit for times longer than this owing to arcing at the contacts. The time constant of the discharge may be calculated from the resistance and the shunt capacitance of the line.

When the line is compensated by shunt reactors, the line will discharge through the reactors in an oscillatory manner as shown by Clerici *et al.* (1970*b*). The frequency of the oscillation is determined by the reactor inductance and the line capacitance and the oscillation will decay at a rate determined by the losses of the line and the reactor. The frequency of oscillation is very often of the same order as the supply frequency and, in general, a difference in the frequencies of the voltages on either side of the circuit breaker will exist. There is, therefore, always a possibility of reclosing the supply onto the line in antiphase to the instantaneous trapped charge voltage.

As stated earlier, electromagnetic voltage transformers are also effective in discharging the trapped charge of a line. It has been reported by Cahill (1964) that, on a line equipped with this type of instrument transformer, the trapped charge voltage is negligible after 0·4s. Some other aspects of this system switching phenomena have been published by Althammer *et al.* (1964), Glavitsch (1966) and Glavitsch *et al.* (1968).

Reduction of energisation transients
Despite the effects of trapped-charge voltage decay and resistive and corona losses, the energisation transients may still exceed the insulation level of the system. These transient-voltage levels are typically in the range 2·0 to 2·5 p.u. of the peak phase-to-neutral voltage of the system and where energisation transient voltages do not fall inherently below the economical insulation level of the system, they must be reduced by other means.

Although over the past few years surge diverters have been developed which are capable of dealing with the energies of switching surge waves having long durations, their protective capacity is likely to be impaired should they be required to operate with every switching operation.

Since the switching surges are dependent on the voltages across the circuit-breaker contacts at the instant of closing and are reduced considerably if this voltage is zero another possible method exists to control switching surge overvoltages by so-called 'synchronous switching', see Section 9.4.2. The variation of the maximum switching surge at the receiving end of a line with the instant of closing is indicated in Fig. 3.22. The possibility exists, therefore, with the pressurised-head type of air-blast circuit breaker, of closing the circuit breaker at the instant at which the voltage is zero. This technique has been described by Thoren (1971) and the results of tests using this method are described by Maury (1966).

The method usually adopted to reduce line energisation overvoltages, however, involves the use of preclosing resistors, as shown in Fig. 3.23.

Initially interrupters B are open and the line is energised through the resistor R by closure of the interrupters A. After a short time, which must be in excess of twice the transit time of the line, interrupters B close thus short-circuiting the resistor. The line is thereby energised in two stages and two switching surges are produced; one owing to the initial energisation through the resistor and the second owing to the short circuiting of the resistor. Both of these surges have a reduced

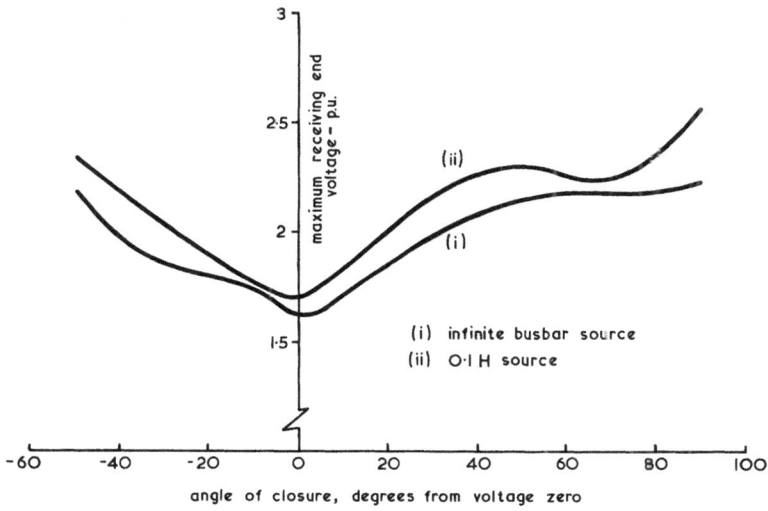

Fig. 3.22 Effect of angle of closure on maximum receiving-end voltage of line

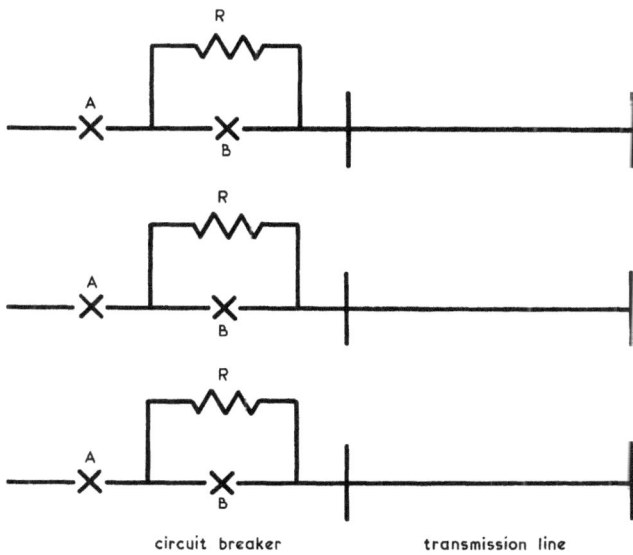

Fig. 3.23 Arrangement for energising a line using switching resistors

amplitude when compared with the surges that would result if preclosing resistors were not employed.

3.5 Reactive current switching

3.5.1 Current chopping with small reactive currents
The phenomena of current chopping
The analyses of the previous Sections have been based on the assumption that current would be extinguished at a normal current zero; i.e. that the current would proceed to zero naturally, and the arc would not restrike.

Since the arc in an interrupter is generally subjected to intense extinguishing forces, it may be extinguished prior to a natural current zero. This phenomena is known as current chopping and may give rise to very severe transient recovery voltages. The phenomena has been extensively investigated by Young (1953) and Greenwood *et al.* (1960).

Voltage produced by current chopping
Fig. 3.24 shows the equivalent circuit of an unloaded transformer or shunt reactor being switched. Before chopping, the current through the circuit breaker flows mainly through the magnetising inductance L with small components through C, the winding capacitance and R, the eddy-current loss of the core material. When the current through the circuit breaker ceases the current through the inductance cannot change abruptly and continues to flow by circulating in the capacitance C. This causes the voltage across C and L to increase.

The initial rate of rise of this voltage is given by

$$\frac{de'}{dt} = \frac{i' - i_c{}'}{C} \tag{3.47}$$

where i' is the current in the circuit breaker at the instant of current chopping and

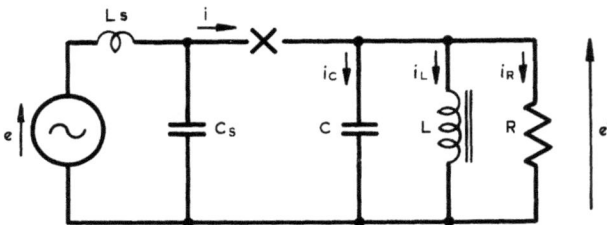

Fig. 3.24 Equivalent circuit for analysis of current chopping

i_c' is the current in the capacitance immediately before the instant of current chopping.

Since i_c' is usually very small in relation to i', eqn. (3.47) may be written to a first approximation as

$$\frac{de'}{dt} = -\frac{i'}{C} \qquad\qquad (3.48)$$

Eqn. 3.48 is often useful in determining the approximate value of the instantaneous current chopped when suitable records of current waveforms are not available. Fig. 3.25 illustrates the currents and voltages after chopping when switching an

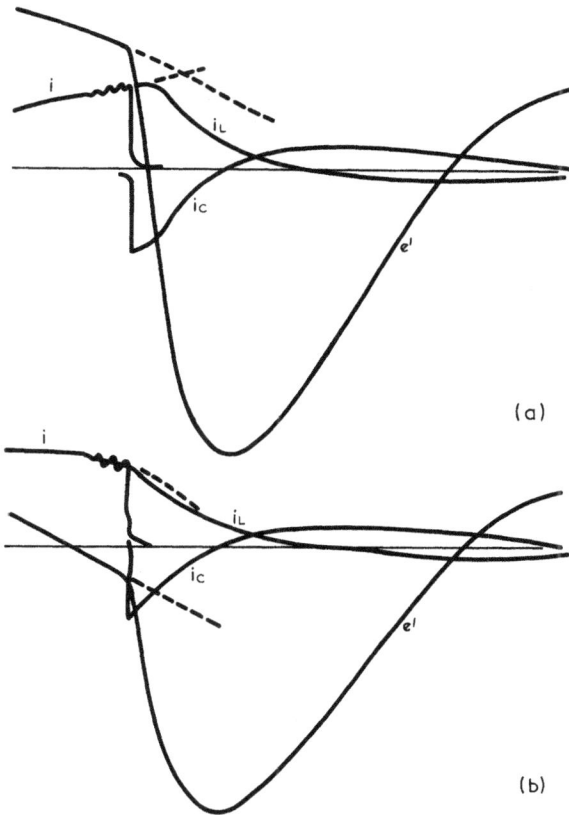

Fig. 3.25 Voltage and current waveforms associated with current chopping

unloaded transformer for the following two principal cases:

(*a*) when the current is rising from zero at the instant of chopping (Fig. 3.25a)
(*b*) when it is falling towards zero at the instant of chopping (Fig. 3.25b)

The voltage oscillates at a frequency determined by the magnetising inductance L and the winding capacitance C, and is damped partially by the effect of the eddy-current loss equivalent resistance R and partially by the hysteresis loss of the inductance L.

The nature of the phenomena is most readily understood by considering a simple case with a linear, loss-free inductance.

Case without loss
Equating the energy stored in the inductance before and after discharge into the capacitance gives

$$\tfrac{1}{2} Ce'^2 = \tfrac{1}{2} Li_L'^2 \tag{3.49}$$

where i_L' is the current through the inductance at the instant of current chopping. The voltage across the transformer or shunt reactor in this case is then given by:

$$e' = i_L' \sqrt{\left(\frac{L}{C}\right)} \sin w_n t + e'' \cos w_n t \tag{3.50}$$

where e'' is the voltage across the transformer or reactor also at the instant of current chopping. The voltage across the circuit breaker is given by

$$e' = i_L' \sqrt{\left(\frac{L}{C}\right)} \sin w_n t + e'' \quad (\cos w_n t - 1) \tag{3.51}$$

where

$$w_n = \sqrt{\left(\frac{1}{LC}\right)}$$

The peak value of the voltage across the transformer or reactor is given by

$$e' = \sqrt{\left(i'^2 \frac{L}{C} + e''^2\right)} \tag{3.52}$$

This expression has been plotted against r.m.s. current in curve (i) of Fig. 3.26 assuming negligible source impedance, constant capacitance and constant supply voltage.

For these curves it has been assumed that the circuit breaker can chop a maximum instantaneous current of 10 A. It follows that for r.m.s. currents above $10/\sqrt{2}$ A, the instantaneous current chopped is 10 A, whereas, for r.m.s. currents below this value, the current chopped is the peak value of the r.m.s. current. The maximum prospective peak voltage occurs at $10/\sqrt{2}$ A r.m.s.

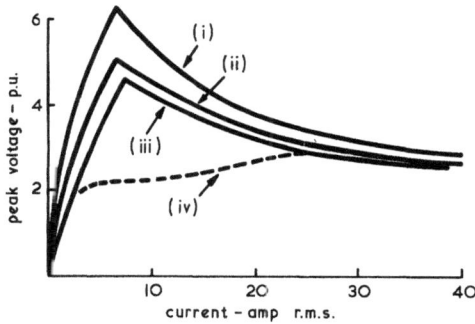

Fig. 3.26 Peak voltage across reactor as a function of RMS current

(i)   case without loss
(ii)   effect of eddy current
(iii) effect of hysteresis loss
(iv)   limitation of voltage by circuit-breaker contact gap

The effect of eddy-current loss
If the effect of eddy-current loss is now considered, the equivalent circuit can be modified by including a shunt resistance R across the inductance since this loss is independent of frequency at constant voltage. Eddy-current loss is therefore considered separately from hysteresis loss, which varies with frequency for constant voltage. The effect of a damping resistance R is to modify the expression for voltage across the transformer or reactor by a damping term so that approximately

$$e' = \exp(-xt) \left\{ i' \sqrt{\left(\frac{L}{C}\right)} \sin w_n't + e'' \cos w_n't \right\} \tag{3.53}$$

where

$$x = \frac{1}{2RC}$$

and

$$w_n' = \sqrt{(w_n^2 - x^2)}$$

This expression shows that the damping of the voltage across the transformer or reactor is dependent upon x and t. Since x is independent of the inductance L, and consequently independent of the r.m.s. current of the circuit, taking t as the time between the successive peaks,

$$t \propto \frac{1}{f} \propto \sqrt{(L)} \propto \frac{1}{I} \tag{3.54}$$

Thus the larger the current the smaller the value of xt, and the smaller the damping. This is illustrated by the relation of curve (ii) to curve (i) of Fig. 3.26.

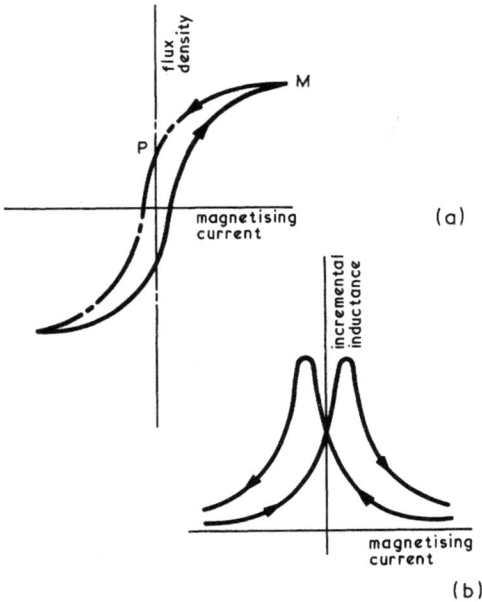

Fig. 3.27 Effect of hysteresis on inductance

The effect of hysteresis loss

The peak value of the steady-state magnetising current of a transformer or reactor is often less than the largest value of instantaneous current that can be chopped by a circuit breaker. There is, therefore, a possibility of chopping the peak value of the magnetising current. At this instant the flux in the transformer core is near maximum value and the transient voltage owing to current chopping is produced during the flux decrement; i.e. in the region of M–P in the hysteresis curve of Fig. 3.27a. In this region the incremental inductance L of the transformer (Fig. 3.27b) is considerably lower than the mean value. This results in a decrease in voltage as compared with the value calculated by the mean value of inductance. Because the major part of the voltage is proportional to $i'\sqrt{(L/C)}$ this factor is important.

Curve (iii) of Fig. 3.26 shows how the voltage is modified by hysteresis. This curve has been constructed on the assumption that an increase of r.m.s. magnetising current above 2 A is obtained by use of an airgap in the core.

For large r.m.s. magnetising currents, current chopping takes place when the flux in the core is very small and consequently hysteresis has a correspondingly small effect upon the voltage.

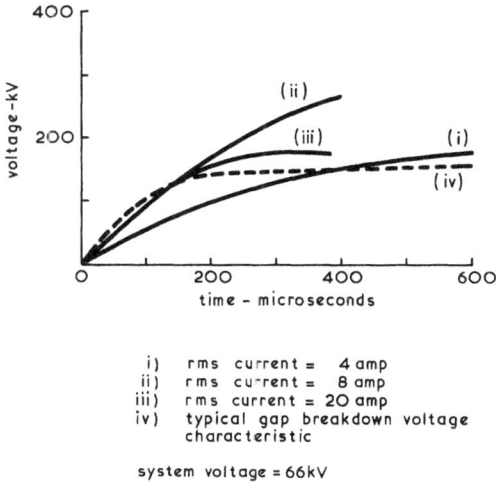

i)   rms current =   4 amp
ii)   rms current =   8 amp
iii)   rms current = 20 amp
iv)   typical gap breakdown voltage
        characteristic

system voltage = 66kV

Fig. 3.28 Voltage across circuit breaker as a result of current chopping

Limitation of voltage by circuit-breaker contact gap

Consideration has been given hitherto only to the prospective voltage. If this is very high or if it has a very high rate of rise, it may not be reached because of the limiting action of the circuit-breaker contact gap as shown in curve (iv) of Fig. 3.26. The prospective voltage across the circuit breaker, taking the damping effect of eddy current and hysteresis loss into account, has been calculated for r.m.s. currents of 4, 8 and 20 A on the basis that a maximum current of 10 A can be chopped. These prospective values have been plotted in Fig. 3.28 with a typical gap breakdown voltage characteristic also plotted.

The maximum possible voltage across the circuit breaker is given by the intersection of the prospective curves with the gap breakdown curve.

The maximum voltage across the transformer or reactor is given by the voltage across the circuit breaker added to the instantaneous value of the system recovery voltage. This prospective voltage is shown in curve (i) of Fig. 3.29.

The prospective maximum voltage has a small probability of occurrence, partly because of the wide variation of gap breakdown voltage below the maximum shown and partly because chopping occurs frequently before the full contact gap has been established, particularly in the case of small currents such as magnetising currents. Curve (ii) of Fig. 3.29 illustrates measured voltages for the same conditions as those for which curve (i) was calculated.

Mention should perhaps be made at this point that problems associated with the

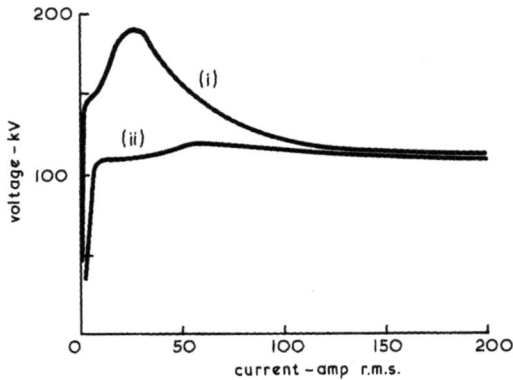

Fig. 3.29 Measured and calculated voltages across reactor due to current chopping

current chopping in vacuum interruptors when switching inductive circuits exist and for an analysis of the problem reference should be made to Greenwood *et al.* (1962).

Comparison of voltages under 3-phase and single-phase conditions
Consideration has so far been concentrated on current chopping in single-phase circuits, but when considering current chopping under 3-phase conditions, the following main factors must be taken into account:

(*a*) The influence of the system neutral earthing, and of the transformer or reactor on the instantaneous value of the recovery voltage.
(*b*) The influence of the type of magnetic circuit on the value of the instantaneous recovery voltage, and on the effective inductance of the winding; i.e. whether 3-limbed, 5-limbed or separate cores per phase are employed.
(*c*) The influence of the winding connections; i.e. star or delta on the instantaneous recovery voltage, and on the effective inductance.

The major factors that may affect the transient voltages due to chopping are power frequency recovery voltage, prospective transient voltages per ampere chopped, and the maximum current that the circuit breaker will chop. Table 3.1, published by Young (1953), shows the values of the instantaneous power-frequency recovery voltage, and the major component of the prospective current-chopping voltage across the circuit breaker for 3-phase conditions, and for various arrangement of cores and windings.

The value of the power frequency recovery voltage when one phase clears varies from zero to 1·5 times the phase-to-neutral voltage, and the chopping component

Table 3.1 Recovery voltage and prospective voltage due to current chopping for some 3-phase transformer arrangements

unit being switched	type of core	recovery voltage p.u.			chopping component of transient voltage *		
		a	b	c	a	b	c
(diagram)	(core)†	1	1	1	$i'\sqrt{\left(\frac{L}{C}\right)}$	$i'\sqrt{\left(\frac{L}{C}\right)}$	$i'\sqrt{\left(\frac{L}{C}\right)}$
(diagram)	(core)	1·5	0·87	0·87	$1\cdot5\,i'\sqrt{\left(\frac{L}{C}\right)}$	$i'\sqrt{\left(\frac{L}{C}\right)}$	$i'\sqrt{\left(\frac{L}{C}\right)}$
(diagram)	(core)	0	0·87	1	0	$\frac{i'}{2}\sqrt{\left(\frac{L}{C}\right)}$	$\frac{2i'}{3}\sqrt{\left(\frac{L}{C}\right)}$
(diagram)	(core)	1·5	0·87	0·87	$1\cdot5\,i'\sqrt{\left(\frac{L}{C}\right)}$	$i'\sqrt{\left(\frac{L}{C}\right)}$	$i'\sqrt{\left(\frac{L}{C}\right)}$
(diagram)	(core)	0	0 87	1	0	$\frac{i'}{2}\sqrt{\left(\frac{L}{C}\right)}$	$\frac{2i'}{3}\sqrt{\left(\frac{L}{C}\right)}$
(diagram)	(core)	1·5	0·87	0·87	$1\cdot5\,i'\sqrt{\left(\frac{L}{C}\right)}$	$i'\sqrt{\left(\frac{L}{C}\right)}$	$i'\sqrt{\left(\frac{L}{C}\right)}$
(diagram)	(core)	0	0 87	1	0	$\frac{i'}{2}\sqrt{\left(\frac{L}{C}\right)}$	$\frac{2i'}{3}\sqrt{\left(\frac{L}{C}\right)}$
(diagram)	(core)	1·5	0·87	0·87	$1\cdot5\,i'\sqrt{\left(\frac{L}{C}\right)}$	$i'\sqrt{\left(\frac{L}{C}\right)}$	$i'\sqrt{\left(\frac{L}{C}\right)}$
(diagram)	(core)	1·5	0·87	0·87	$\frac{\sqrt{3}\,i'}{2}\sqrt{\left(\frac{L}{C}\right)}$	$\frac{\sqrt{3}\,i'}{2}\sqrt{\left(\frac{L}{C}\right)}$	$\frac{\sqrt{3}\,i'}{2}\sqrt{\left(\frac{L}{C}\right)}$

† five limbed core and one core per phase

* this is the peak value of the major component of the transient voltage due to current chopping

of the transient voltage varies from zero to 1·5 times the normal per-phase value.

The variation of power-frequency recovery voltage is of greatest importance in oil-break circuit breakers, where the transient voltage and the arc duration depend directly on it.

Consider, for example, a 3-limbed star-delta transformer with a solidly earthed neutral, switched on the star-connected side. The first phase clears almost immediately after contact separation, and the second and third phases usually clear independently several cycles later, with the largest transient recovery voltage across the third phase. When a 5-limbed star-connected shunt reactor with solidly earthed neutral is switched by an oil-break circuit breaker, the recovery voltage is the same for each phase, the three phases clear independently in random order, and there is no significant difference between the transient recovery voltages across the three phases.

With air-blast circuit breakers the more important factor is the prospective current-chopping voltage. Considering the same two examples, the first phase of the transformer clears immediately after contact separation without a significant transient recovery voltage, the second two phases usually clear within a cycle of the first, and the recovery voltages of the two phases usually interact. The interaction makes analysis difficult, but the largest transient usually occurs on the last phase to clear. When a shunt reactor is switched, each phase clears separately because each pole of the circuit breaker must wait until its associated phase current is small enough to chop. The transient voltages should show no significant difference in magnitude. Some other aspects of switching small reactive currents have been published by Baltensperger (1960).

The use of switching resistors to limit current chopping overvoltages
Switching resistors may be used to reduce the transient recovery voltage on short-circuit switching with both air-blast and oil-break circuit breakers (Section 3.2.4). When resistors are fitted to oil-break circuit-breakers they are generally designed for the lower values of current only (0–30% of rated breaking current) because it is within this current range that transient recovery voltage influences the performance.

On the other hand, for an air-blast circuit breaker the influence of the transient recovery voltage at the full rating is relatively great and if the design calls for the use of resistors it is necessary to make them of low ohmic value compared with those for an oil-break circuit breaker of the same rating. For a 132 kV circuit breaker of 2500 MVA rating, for example, suitable values of resistance would be 2000 Ω for an oil-break circuit breaker and 300 Ω for an air-blast circuit breaker.

When shunt reactors or unloaded transformers are being switched, the resistor is connected in series with the circuit at the instant of switching as shown in Fig. 3.30. The most significant effect of the resistor will be to decrease the instantaneous value of the recovery voltage at the instant of final arc extinction. It

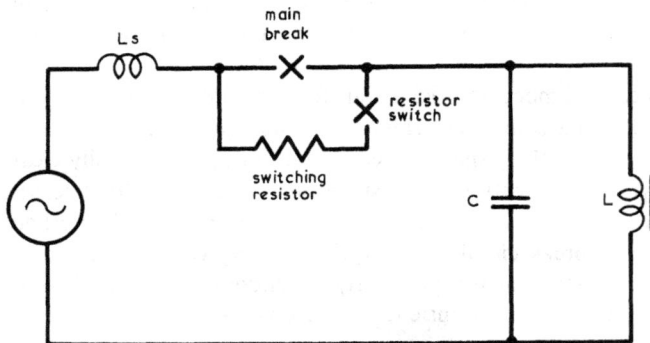

Fig. 3.30 Use of switching resistor to limit current chopping overvoltages

is of interest to consider, for example, the switching at 132 kv of a 45 MVA unloaded transformer and of a 15 MVA shunt reactor with magnetising currents of
5 A and 66 A, respectively. The transformer would have a reactance of about
15200 Ω and it is clear that the connection of a resistance of even 2000 Ω in series
with this would make little difference to the instantaneous value of recovery
voltage. The reactance of the shunt reactor would be about 1150 Ω, and the
connection of a 300 Ω resistor in series would have little effect but 2000 Ω would
reduce the instantaneous recovery voltage by approximately one half. It follows,
therefore, that when a transformer is being switched by a circuit breaker which
incorporates a switching resistor having a value suitable for reduction of transient
recovery voltages, the conditions when the resistor current is being broken are
substantially the same as when there is no resistor. The only difference of any
significance is that when a restrike occurs the discharge current from the
transformer capacitance is limited by the switching resistor. The chopping voltages
are, therefore, the same, whether or not the circuit breaker is fitted with resistors.

The instantaneous current chopped under identical loading conditions for both
shunted and unshunted breaks in oil-break circuit breakers have been measured by
Young (1953) and no significant difference has been observed in the current
chopped. With regard to the gap recovery characteristics of oil-break circuit
breakers, the envelope characteristic of the resistance break is substantially the
same as for the main break at small currents, and no difference in the voltages due
to current chopping can be expected on this account. For air-blast circuit breakers
the gap-recovery characteristics for the unshunted gap can be chosen for its special
duty. This does not, however, make it possible to use a characteristic particularly
favourable to the limitation of chopping voltages, although some small gain is
possible. The reason for this is that the gap must be designed to withstand the full
recovery voltage without damping of the system transient recovery voltage for all
conditions including, for example, asynchronous conditions.

It can be concluded that the use of switching resistors of value suitable for
control of transient recovery voltage cannot produce any significant reduction of
current-chopping voltage when transformers are switched. For shunt-reactor
switching, a significant reduction in chopping voltage is possible for the higher
reactor current with an oil-break circuit breaker but little reduction is possible with
air-blast circuit breakers.

3.5.2 Capacitance switching

In considering the *shunt-capacitor switching* of capacitive circuits, it is important
that the terms 'reignition' and 'restriking' are defined. In general, dielectric breakdown
can occur in an interruptor at any time after initial arc extinction if the conditions are

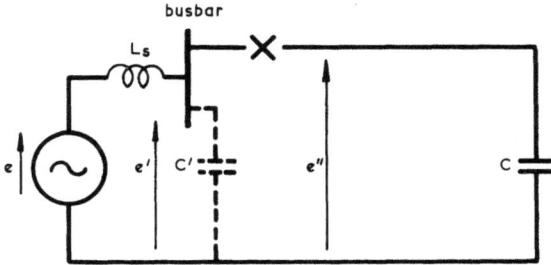

Fig. 3.31 Interruption of a capacitor bank

unfavourable. When switching capacitive circuits, this breakdown is termed 'reignition' if it occurs within 0·25 cycle of initial arc extinction, but at any time after this period the phenomena is called 'restriking'.

The phenomenon of circuit-breaker restriking associated, with the interruption of capacitive currents, has been described by Leeds *et al.* (1947) and may best be described by considering the single-phase circuit shown in Fig. 3.31 in which a lumped capacitance *C* is connected to a high-voltage busbar through a circuit breaker. The small parallel capacitance *C'* represents the capacitance of the busbar together with that of any directly connected equipment. The series inductive reactance between the generated voltage and the busbar, L_s, whilst often quite small in comparison with the capacitive reactance, is a factor which can cause reignition (as opposed to restriking) as will be shown later.

Many circuit breakers required to interrupt load current or fault current can not do so at the first current zero because of the inadequacy of the dielectric strength of the interrupter at that instant. In capacitance switching, the current involved is frequently small and, at the first current zero, the recovery voltage is relatively low due to the trapped charge on the capacitor (Section 3.2.1). There is, therefore, a high probability that the circuit breaker is capable of interrupting the capacitor current at the first current zero. If this should occur soon after the contacts have parted, a recovery voltage of $2E$ can appear across the contacts when their separation is small. There is, therefore, a substantial probability of the interrupter restriking. Should a restrike occur precisely when the voltage reaches its peak, at say P in Fig. 3.32a, the circuit behaves as though the circuit breaker were suddenly closed at that instant.

The *LC* circuit then oscillates at its natural frequency, *fn*, given by

$$f_n = \frac{1}{2\pi \sqrt{(L_sC)}} \quad \text{if } C' \ll C \tag{3.55}$$

This oscillation is shown at S in Fig. 3.32a. The voltages in the circuit following the

(a)

(b)

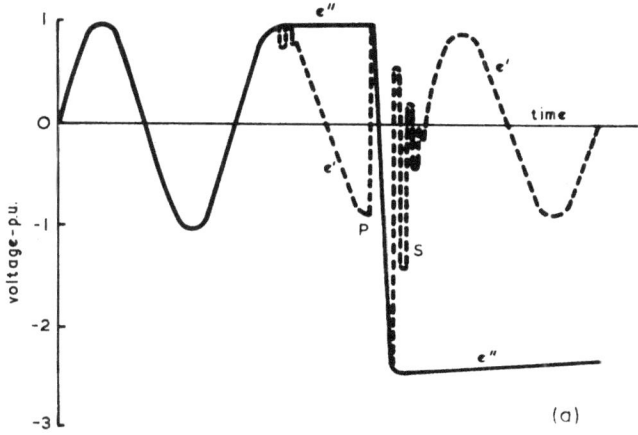

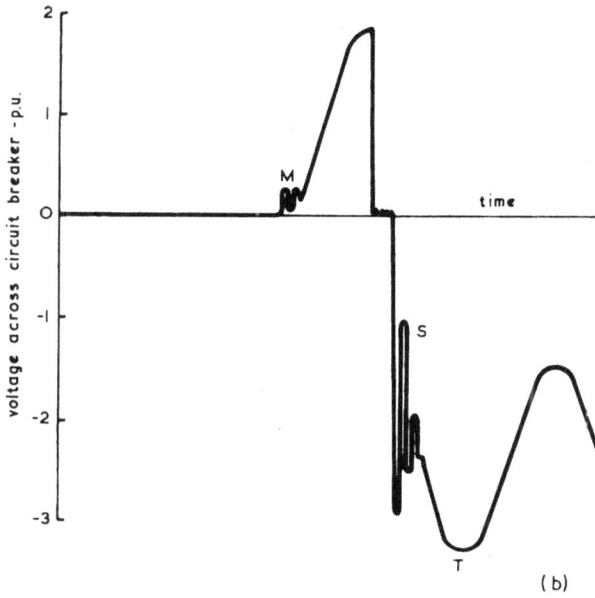

Fig. 3.32 Voltage waveforms associated with the interruption of a capacitor bank

restrike as shown in Fig. 3.32a are given by:

$$e - e'' = L_s \frac{di}{dt} \tag{3.56}$$

but

$$e'' = E' + \frac{1}{C} \int i dt \tag{3.57}$$

where E' = voltage across capacitor C at the instant of restrike.

Assuming the restrike occurs at voltage maximum, i.e. P in Fig. 3.32a, the solution of eqns. 3.56 and 3.57 result in an expression for the current through the circuit breaker after the restrike of

$$i = (E - E') \sqrt{\left(\frac{C}{L_s}\right)} \sin w_n t \tag{3.58}$$

The quantity $E-E'$ represents the voltage across the circuit breaker at the instant of restrike. The magnitude of this surge current can be quite large and quite damaging to the circuit breaker if the source impedance L_s is small. The frequency of this transient current oscillation is equal to the natural frequency of the circuit. In a practical circuit, this current oscillation would, of course, be damped in the same manner as the voltage oscillation shown in Fig. 3.32a

The capacitor voltage after restrike is given by

$$e'' = E' + \frac{1}{C} \int_0^t (E - E') \sqrt{\left(\frac{C}{L_s}\right)} \sin w_n t \, dt \tag{3.59}$$

If the voltage E' due to the trapped charge on the capacitor is equal to E and the restrike occurs at system maximum negative voltage $-E$, as shown in Fig. 3.32a, eqn. 3.59 can be written as follows:

$$e'' = -E + \frac{2E}{\sqrt{(L_s C)}} \int_0^t \sin w_n t \, dt$$

$$= E(1 - 2 \cos w_n t) \tag{3.60}$$

The maximum value of voltage across the capacitor is therefore $3E$. The trend is shown by e'' in Fig. 3.32a. It is assumed that the circuit breaker characteristics are such that the restrike current is extinguished after only one half-cycle oscillation at the natural frequency of the circuit. The result of this restrike and the resulting voltage overswing is to leave the capacitor C charged to a voltage between two and three times the peak system phase to neutral voltage.

Fig. 3.32b shows the voltage conditions across the circuit-breaker contacts as a function of time. At one half cycle after current zero, the difference of potential across the circuit-breaker contacts reaches approximately twice the peak value of the voltage due to the trapped charge on the capacitor C. It is now assumed that the dielectric strength between the circuit-breaker contacts is just insufficient to maintain this voltage, and a restrike occurs. Immediately after this restrike is extinguished, the busbar voltage swings back to the generator voltage by way of a high-frequency oscillation S determined by L_s and C. Following the transient, the voltage across the breaker builds up to a normal peak T as the generator voltage reverses polarity at the normal power-frequency rate. If a second restrike were now to follow the first at the next peak of the generator voltage, the voltage across the capacitor C could carry it to a peak value of between four and five times normal voltage. The small transient shown at M in Fig. 3.32b is due to the sudden change in voltage across L_s as the flow of capacitive current ceases at first interruption. This transient can often be neglected but when high source impedances prevail the magnitude of the transient may precipitate an immediate reignition and result in prolonged arcing.

In the simple capacitive circuit of Fig. 3.31, there would theoretically be no limit to the voltage which could be attained by repeated restriking under the most unfavourable conditions. If, however, the effects which damp the high-frequency transients during restriking are taken into account, a limiting voltage would be reached, and is a maximum regardless of the number of restrikes. This ultimate value, expressed as a multiple of the system peak voltage, depends upon the damping of the transient restrike current and on the magnitude of the arc voltage drop in the circuit breaker. An analysis of the most probable restrike conditions requires some knowledge of the rate at which the dielectric strength is built up between the contacts of a circuit breaker immediately following current zero. Although this is a function of the type of circuit breaker and the manner in which the arc is deionised and interrupted, certain conclusions may be drawn from some simple assumptions. Fig. 3.33 shows the voltage rise across the circuit-breaker

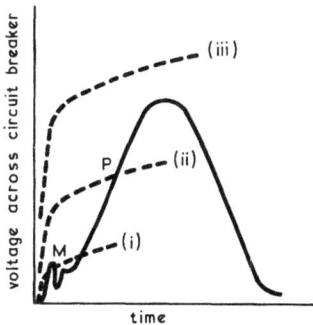

Fig. 3.33 Transient recovery voltage across circuit breaker when interrupting capacitor charging current

together with a family of curves, (i), (ii) and (iii), representing the rise of dielectric strength for given contact gaps immediately following the initial interruption of charging current. At current zero, the dielectric strength has a definite value, depending on the contact separation and the intensity of deionising activity. The initial slope of the dielectric recovery curve may be fairly steep but later the curve tends to level off at the breakdown strength of the deionised maximum contact gap.

If the interrupting effectiveness is low, as shown by curve (i), or if the current zero occurs only a small fraction of a cycle after contact parting, a reignition may occur in a very short time by way of a high-frequency voltage oscillation at the beginning of the recovery voltage wave, as at point M. With greater contact separation, restriking may occur after say one quarter of a cycle as at point P on curve (ii). With greater interrupting effectiveness or greater contact separation, as indicated by curve (iii), the interruption will be complete without any restriking. Actual circuit-breaker tests show that most restrikes occur before the voltage across the circuit breaker rises to $2E$ thus tending to avoid those conditions which lead to extremely high overvoltages. As a result the probability of producing a succession of restrikes with the voltage across the circuit breaker building up each time to the maximum peak is low.

The above analysis applies to 3-phase capacitor bank switching only if the source neutral and capacitor bank neutral are solidly earthed. Where this is not the case, i.e. when switching out an unearthed capacitor bank, the recovery voltage can be as high as $3 \cdot 0$ p.u. and restriking can further increase this figure. It can be shown that the ratio of positive sequence capacitance C_1 to zero sequence capacitance C_0 of the circuit influences the value of the recovery voltage across the first pole to clear. This is illustrated in Fig. 3.34.

When all the system capacitance is solidly earthed the ratio C_1/C_0 is equal to unity. This ratio is infinity, however, when all the system capacitance is unearthed although this condition is unlikely in practice. Further aspects of capacitive current switching have been published by Butler (1940), Van Sickle *et al.* (1951), Johnson *et al.* (1955) and Baltensperger (1960).

Transmission line switching
Transmission line switching, including overhead line and cable circuits, represents a special case of the more general problem of capacitance switching. It was stated in the previous section that the ratio C_1/C_0 influences the recovery voltage across the first pole to clear when interrupting a capacitive circuit. Since in practice overhead lines and unshielded cables usually have C_1/C_0 ratios of between $0 \cdot 6$ and $2 \cdot 0$, it can be seen from Fig. 3.34 that the recovery voltage is likely to lie between $2 \cdot 2$ and $2 \cdot 4$ times system peak voltage. Single-core cables and shielded 3-core cables have a C_1/C_0 ratio of unity, therefore, the recovery voltage across the first pole to clear will be twice system peak voltage.

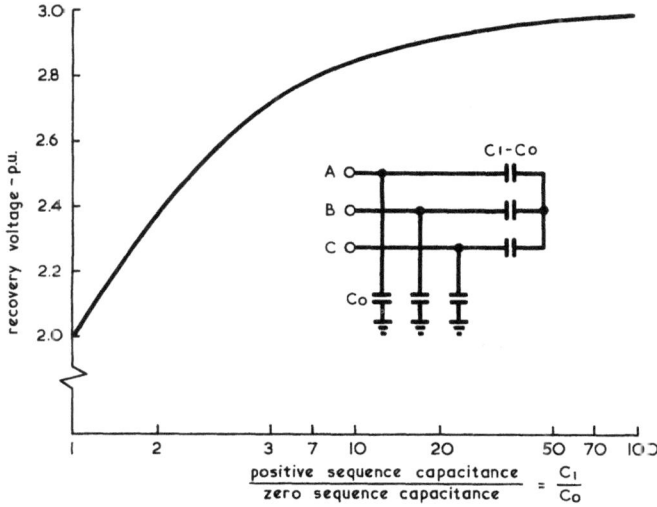

Fig. 3.34 Variation of recovery voltage in a capacitive circuit

Fig. 3.35 shows the equivalent circuit of transmission line switching and Fig. 3.36a indicates the transient voltage variations of the busbar and line potentials when the restrike occurs one half cycle after current zero. The most obvious difference between this case and the switching of a simple lumped capacitance is the more abrupt change in the voltage and current at the beginning and end of the restrike due to the characteristics of travelling waves on a transmission line. The magnitude of the transient current is equal to the potential difference across the

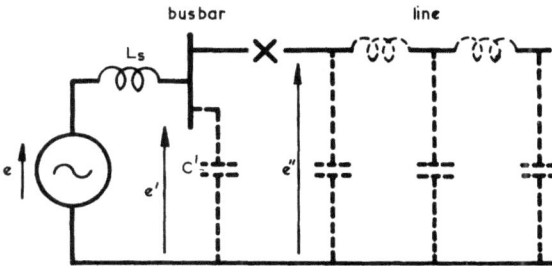

Fig. 3.35 Interruption of line-charging current

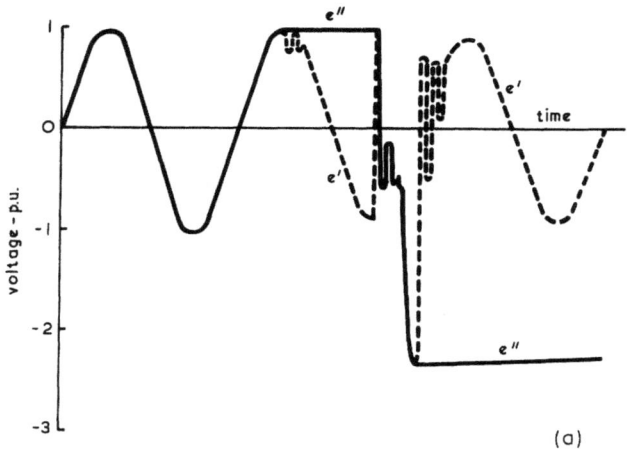

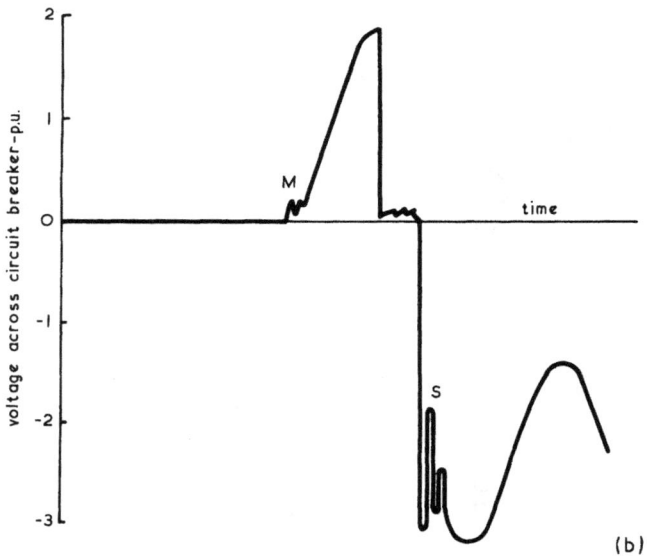

Fig. 3.36 Voltage waveforms associated with the interruption of line-charging current

circuit breaker at the instant of restrike divided by the sum of the natural impedance of the line and the source impedance.

The magnitude of this transient current is generally less than that occurring for a restrike in a circuit breaker switching a capacitor bank because of the limiting effect of the natural impedance Z_n of the line. The magnitude of the surge current is

$$i = \frac{2E}{Z_n} \left\{ 1 - \exp\left(\frac{-Z_n t}{L_s}\right) \right\} \tag{3.61}$$

The duration of the restrike current pulse is primarily a function of the line length, representing the time for a wave to travel the length of the line, reflect at the remote end and return to the circuit breaker, together with a small additional time interval for the current to fall to zero which depends on the source impedance L_s. During the time that the wave is in transit there is appreciable attenuation which reduces the amplitude of the overvoltage at the circuit-breaker when the wave returns. If L_s is large, the rate of change of voltage on the busbar during a restrike will be reduced, the equivalent time constant being approximately L_s/Z_n. Taking into consideration the busbar capacitance C' results in a superimposed oscillation on the rising surge voltage as indicated at M in Fig. 3.36b. Provided that the restrike current is extinguished after only one half cycle, the voltage oscillation is dependent on the constants of the circuit on the busbar side and is not influenced by the nature of the connected load on the other side of the circuit-breaker. Fig. 3.37 is a reproduction of the oscillographic record given by Leeds *et al.* (1947) for a field test where a charging current of 155 A associated with 290 km of line was switched with a busbar voltage of 241 kV.

A single restrike took place approximately 0·2 cycles after a current zero. The transient current was interrupted after a single pulse, the busbar voltage rising momentarily to about 1·35 times the voltage immediately after the line was disconnected, but only 1·2 times the voltage on the busbar while charging current was flowing. The maximum voltage across the circuit breaker, one half cycle after the restrike, was 2·5 times the busbar voltage ultimately attained. So far as disturbance to the system is concerned, this particular operation would not be of major importance, being comparable to the voltage fluctuations to be expected during the energising of the line.

The switching of transmission lines containing short lengths of underground cables has been covered by Eidinger *et al.* (1968), and they conclude that stresses imposed on the circuit breaker by the transient recovery voltage under these conditions are considerably reduced. The presence of an unloaded transformer at the end of a transmission line has been studied by Johnson *et al.* (1960), their conclusions were that phase-earth overvoltages in excess of 3·0 p.u. could be generated on energising such a circuit. Lightning arresters on the incoming terminals of the transformers would reduce these to a certain extent, but this would obviously greatly increase the duty required from the arrester.

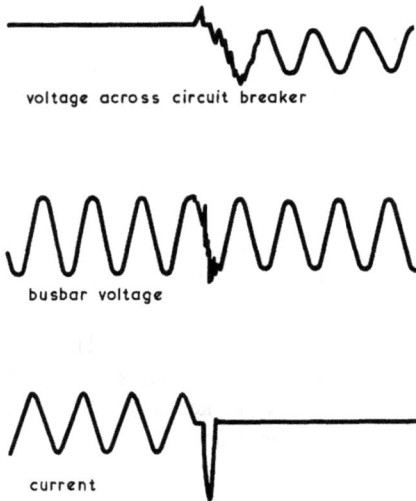

voltage across circuit breaker

busbar voltage

current

Fig. 3.37 Current and voltage waveforms with disconnecting 290 km of 230 kV line

A study of the magnitude of the phase-phase overvoltages when energising a transformer terminated line has been made by Johnson *et al.* (1962). Their main conclusions were:

(*a*) Phase-Phase transient voltages in excess of 2·6 p.u. normal phase-phase voltage may be experienced.
(*b*) This overvoltage increases with length of line.
(*c*) It appears to be independent of transformer zero-sequence impedance.

Some aspects of de-energising unloaded transmission lines have been covered by Petitpierre *et al.* (1964).

Use of resistor switching
Resistors can be of value in capacitance switching or line de-energising and the technique has been described by Van Sickle *et al.* (1954). In the case of resistors being applied, the process of disconnection of the capacitor bank or transmission line involves insertion of the resistor or sequential insertion of a number of resistors such that the actual load becomes a series combination of R and C. The system voltage is, therefore, no longer at a peak at current zero when interruption takes place and consequently the value of trapped voltage on the capacitor or line capacitance is smaller. This reduces the likelihood of a restrike. The effectiveness of the resistor depends, of course, on its value in relation to the capacitive reactance

being switched but even if a restrike does occur the amplitude of the resultant voltage transient is significantly reduced. It should be noted that when the resistor has a low ohmic value, the main interrupting duty is transferred from the main break to the resistor break.

The use of resistors in switching capacitor banks at the end of relatively long supply lines, i.e. a total circuit containing both lumped and distributed parameters has been described by Elgerd (1957).

Series-capacitor switching

The switching of series capacitors does not usually give rise to limiting conditions as far as any associated circuit breakers are concerned, particularly when incorporated in a transmission-line circuit and switched as a unit with that circuit. In the case where a series capacitor bank is individually switched, located at a substation and connected between busbars for example, the bank is invariably bypassed prior to any routine switching operation. In addition, series capacitor banks are protected from excessive voltage stresses during network faults. This protection takes the form of an automatic bypassing arrangement, either by a spark gap and a bypass circuit breaker, by a nonlinear resistor or by a combination: Even when switched under fault conditions, therefore, the series capacitor does not give rise to difficult switching problems. The principal area of interest associated with the switching of series capacitors relates to their rapid re-insertion. It is necessary to ensure that the voltage appearing across the capacitor upon re-insertion does not unduly stress the equipment, and hence require a further operation of the bypass circuit.

The subject of series capacitor voltages and re-insertion techniques are covered by AHLGREN *et al.* (1979) and ILICETO *et al.* (1976).

See also Appendix A, Virtual current chopping, p. 591.

3.6 References

AHLGREN, L., FAHLEN, J. and JOHANSSON, K.E. (1979): 'EHV series capacitors with dual gaps and nonlinear resistors provide technical and economic advantages,' IEEE A79 049-8

ALLAN, A., and AMER, D. F. (1947): 'The extinction of arcs in air-blast circuit breakers', *J. IEE*, 94, Pt. II, pp. 333–350

ALTHAMMER, P., and PETITPIERRE, R. (1964): 'Switching operations and switching surges in systems employing extremely high voltages', *Brown Boveri Rev.*, 51, pp. 33–46

BALTENSPERGER, P. (1960): 'The shape and magnitude of overvoltages when interrupting small inductive and capacitive currents in HV systems', *ibid.*, 47, pp. 195–224

BALTENSPERGER, P., and RUOSS, E. (1960): 'The short-line fault in high-voltage systems', *ibid.*, 47, pp. 329–339

BALTENSPERGER, P. (1962): 'New knowledge in the field of switching phenomena and circuit-breaker testing', *ibid.*, 49, pp. 381–397

BICKFORD, J. P., and DOEPEL, P. S. (1967): 'Calculation of switching transients with particular reference to line energisation', *Proc. IEE*, 114, pp. 465–477

BEEHLER, J. E., NAEF, O., and ZIMMERMAN, C. P. (1965): 'Proposed transient recovery voltage ratings for power circuit breakers', IEEE Trans., PAS-84, pp. 580–608

BERKEBILE, L. E., BUETTNER, D. E., and COLCLASER, R. G. (1971): 'The effect of capacitors on the short-line fault component of transient recovery voltage' *ibid.*, PAS-90, pp. 660–669

BOEHNE, E. W. (1963): 'The switching severity of industrial circuits', *ibid.*, PAS-82, pp. 255–261

BOEHNE, E. W. (1965): 'EHV surge suppression on interrupting light currents with air switches: 1 – capacitive currents', *ibid.*, PAS-84, pp. 906–923

BOLTON, E., EHRENBERG, A. C., HAMILTON, F. L., HAWKINS, A. G., MATRAVERS, F. P., and THOMAS, J. A. (1964): 'British investigations of short-line fault phenomena'. CIGRÉ Report 109 and 109a

BUETTNER, D. E., and COLCLASER, R. G. (1969): 'The travelling wave approach to transient recovery voltage', *IEEE Trans.* PAS-88, pp. 1028–1035

BUTLER, J. W. (1940): 'Analysis of factors which influence the application, operation and design of shunt capacitor equipments switched in large banks', *Trans Am. Inst. Elec. Eng.*, PAS-59, Pt. 1, pp. 795–800

CAHILL, L. (1964): 'La premiere transmission d'energie electrique a 735 kV Manicougan-Montreal', *Bull. Ass. Suisse Electr.*, 55, p. 519

CARTER, G. W. (1950): 'The simple calculation of electrical transients' (Cambridge University Press)

CLARKE, E., LIGHT, P. H., and PETERSON, H. A. (1941): 'Abnormal voltage conditions in three-phase systems produced by single-phase switching', *Trans Am. Inst. Elec. Eng.*, PAS-60, pp. 329–339

CLERICI, A., and TASCHINI, A. (1970): 'Over-voltage due to line energisation and re-energisation versus over-voltage caused by faults and fault clearing in EHV systems', *IEEE Trans.*, PAS-89, pp. 932–941

CLERICI, A., RUCKSTUHL, G., and VIAN, A. (1970): 'Influence of shunt reactors on switching surges', *ibid.*, PAS-89, pp. 1727–1736

COX, H. E., WILCOX, T. W. (1944): 'The influence of resistance switching on the design of high-voltage air-blast circuit breakers', *J. IEE*, 91, pt. II, pp. 483–509

COX, H. E., WILCOX, T. W. (1947): 'The performance of high-voltage oil circuit-breakers incorporating resistance switching', *ibid.*, 94, pt. II, pp. 351–371

EIDINGER, A., and JUSSILA, J. (1964): 'Transients during 3-phase short-line faults', *Brown Boveri Rev.*, 51, pp. 303–319

EIDINGER, A., and KOPPL, G. (1968): 'Transient recovery voltage caused by different switching operations in HV networks containing underground cables', *ibid.*, 55, pp. 175–181

ELGERD, O. I. (1957): 'Capacitor switching phenomena in networks containing long transmission lines', *Trans. Am. Inst. Electr. Eng.*, PAS-76, pp. 1157–1163

FERGUSON, J. E., and HARNER, R. H. (1968): 'Voltage acceleration – a basic measure of severity of oscillatory transient recovery voltage', *IEEE Trans.*, PAS-87, pp. 721–728

FLUGUM, R. W., and HILLESLAND, G. G. (1965): 'Load breaking on distribution circuits', *ibid.*, PAS-84, pp. 1143–1150

GENG, P., and KOPPL, G. (1966): 'Determination of the transient recovery voltage in symmetrical 3-phase systems', *Brown Boveri Rev.*, 53, pp. 311–325

GLAVITSCH, J. (1964): 'Power-frequency over-voltages in e.h.v. systems', *ibid.*, 51, pp. 21–32

GLAVITSCH, H. (1966): 'Problems associated with switching surges in e.h.v. netwoiks', *ibid.*, 53, pp. 267–277

GLAVITSCH, J., PETITPIERRE, R., and RUOSS, E. (1968): 'The influence of the circuit-breaker on switching surges in e.h.v. networks', *ibid.*, 55, pp. 167–174

GREENWOOD, A., and LEE, T. H. (1960): 'The effect of current chopping in circuit-breakers on networks and transformers', *Trans. Am. Inst. Electr. Eng.*, PAS-79, pp. 535–555

GREENWOOD, A., LEE, T. H., and POLINKO, G. (1962): 'Design of vacuum interruptors to eliminate abnormal overvoltages', *ibid.*, PAS-81, pp. 376–384

GREENWOOD, A. (1971): 'Electrical transients in power systems' (Wiley Interscience)

HAMMERLUND, P. (1946): 'Transient recovery voltage subsequent to short-circuit interruption with special reference to Swedish power systems'. *Proceedings of the Royal Swedish Academy of Engineering Sciences*, No. 189

HARLE, J. A., and WILD, R. W. (1944): 'Restriking voltage as a factor in the performance, rating and selection of circuit-breakers', *J. IEE*, 91, Pt. II, pp. 469–482

HAWLEY, W. G. (1945): 'The surge characteristics of a single-circuit 3-phase transmission line having a single earth-wire', *J. IEE*, 92, Pt. II, pp. 39–46

HEDMAN, D. E., JOHNSON, I. B., TITUS, C. H., and WILSON, D. D., (1964): 'Switching of extra-high-voltage circuits' 2 – surge reduction with circuit-breaker resistors, *IEEE Trans.*, PAS-83, pp. 1196–1205

HOPKINSON, R. H. (1965): 'Ferro-resonance during single phase switching of 3-phase distribution transformer banks', *ibid.*, PAS-84, pp. 289–293

ILICETO, F., CINIERI, E., CAZZANI, H., and SANTAGOSTINO, G. (1976): 'Transient voltages and currents in series-compensated e.h.v. lines,' *Proc. IEE*, 123, pp.811–817

JOHNSON, I. B., SCHULTZ, A. J., SCHULTZ, N. R., and SHORES, R. B. (1955): 'Some fundamentals of capacitance switching', *Am. Inst. Electr. Eng.*, PAS-74, pp. 727–736

JOHNSON, I. B., and SCHULTZ, A. J. (1960): 'Switching surges on energising a transformer-terminated line', *ibid.*, PAS-79, pp. 241–245

JOHNSON, I. B., SILVA, R. F., and WILSON, D. D. (1962): 'Phase-to-phase switching surges on line energisation', *ibid.*, PAS-81, pp. 298–302

JOHNSON, I. B., PHILLIPS, V. E., and SIMMONS, H. O. (1964): 'Switching of extra-high-voltage circuits: 1 – system requirements for circuit-breakers', IEEE Trans., PAS-83, pp. 1187–1196

KIMBARK, E. W., and LEGATE, A. C. (1968): 'Fault surge versus switching surge; a study of transient over-voltages caused by line-to-ground faults', *ibid.*, PAS-87, pp. 1762–1769

KUMMEROW, G. (1964): 'Die spannungsbeanspruchung von hochspannungs hochleistungs-schaltern beim auschalten von abstandskurzschlussen', *Siemens Zeitschrift*, 38, pp. 350–356

LACKEY, C. H. W. (1951): 'Some technical considerations relating to the design, performance and application of high-voltage switchgear', *Trans. S. Afr. Inst. Electr. Eng.*, 42, pp. 261–298

LACKEY, C. H. W. (1951a): 'Fault calculations' (Oliver & Boyd)

LAUBER, T. S. (1968): 'Determining maximum over-voltages produced by EHV circuit-breaker closure', *IEEE Trans.*, PAS-87, pp. 1026–1032

LEEDS, W. M., VAN SICKLE, R. C. (1947): 'The interruption of charging current at high voltage', *Trans. Am. Inst. Electr. Eng.*, PAS-66, pp. 373–382

LEEDS, W. M., and POVEJSIL, D. J. (1952): 'Out-of-phase switching voltages and their effect on high-voltage circuit-breaker performance', *ibid.*, PAS-71, pp 88–96

MAURY, E. (1966): 'Synchronous closing of 525 and 765 kV circuit-breakers: a means of reducing switching surges on unloaded lines'. CIGRÉ Report 143

MAZZA, G. and ROVELLI, S. (1972): 'The influence of transient recovery voltage parameters on the behaviour of high-voltage circuit-breakers'. CIGRÉ Report 13–11

McELROY, A. J., SMITH, H. M. (1963): 'Propagation of switching surge-wave-fronts on EHV transmission lines', *Trans. Am. Inst. Electr. Eng.*, **PAS-81**, pp. 983–998

MILLER, R. W., and URAM, R. (1964): 'Mathematical analysis and solution of transmission line transients', *IEEE Trans.*, **PAS-83**, pp. 1116–1137

PARIS, L. (1968): 'Basic considerations of magnitude reduction of switching surges due to line energisation', *ibid.*, **PAS-87**, pp. 295–305

PARK, R. H., and SKEATS, W. F. (1931): 'Circuit-breaker recovery voltages, magnitudes and rates-of-rise', *Trans. Am. Inst. Elec. Eng.*, **PAS-50**, pp. 204–239

PETERSON, H. A. (1951): 'Transients in power systems' (Wiley)

PETITPIERRE, R. (1960): 'Airblast circuit-breakers with relation to stresses which occur in modern networks, with particular reference to the interruption of short line faults', *CIGRÉ*, Report 115

PETITPIERRE, R., and SCHNEIDER, J. (1964): 'Transient phenomena during opening and closing operations on EHV lines to 750 kV and the special requirements resulting for the breakers'. CIGRÉ Report 113

POUARD, M. (1958): 'New concepts on rates of rise of restriking voltage at the terminals of high-voltage circuit-breakers', *Bull. Soc. Francais*, **8**, (95), pp. 748–764

RITTENHOUSE, J. W., and ZABORSKY, J. (1963): 'Some fundamental aspects of recovery voltages', *Trans. Am. Inst. Elec. Eng.*, **PAS-81**, pp. 815–822

RITTENHOUSE, J. W., and ZABORSKY, J. (1963): 'Fundamental aspects of some switching over-voltages on power systems', *ibid.*, **PAS-81**, pp. 822–830

THOREN, H. B. (1971): 'Reduction of switching over-voltages in e.h.v. and u.h.v. systems', *IEEE trans.*, **PAS-90**, pp. 1321–1326

VAN SICKLE, R. C., and ZABORSKY, J. (1951): 'Capacitor switching phenomena', *Trans. Am. Inst. Elec. Eng.*, **PAS-70**, Pt. 1, pp. 151–158

VAN SICKLE, R. C., and ZABORSKY, J. (1954): 'Capacitor switching phenomena with resistors', *ibid.*, **PAS-73**, pp. 971–977

WAGNER, C. F., and EVANS, R. D. (1933): 'Symmetrical components' (McGraw-Hill)

YOUNG, A. F. B. (1953): 'Some researches on current chopping in high-voltage circuit-breakers', *J. IEE*, **100**, pt. II, pp. 337–361

YOUNG, A. F. B., and ELLIS, N. S. (1963): 'The short-line fault', *Electrical Rev.*, **182**, pp. 271–274

3.7 General bibliography

BOLTON, E. (1971): 'Circuit-breaker performance requirements for higher voltage systems', *GEC J.* (4), pp. 142–150

CANAY, M., WERREN, L. (1969): 'Interrupting sudden asymmetric short-circuit current without zero-transition', *Brown Boveri Rev.*, **56**, pp. 484–493

GRABER, W., and GYSEL, T. (1980): 'The interruption of fault currents with delays and current zeros by HV circuit breakers,' *ibid.*, **67**, pp.237-243

KALSI, S.S., STEPHEN, D.D., and ADKINS, B. (1971): 'Calculation of system-fault currents due to induction motors,' *Proc. IEE*, **118**, pp.201-215

KOPPL, G., RUOSS, E. (1970): 'Switching over-voltages in EHV and UHV networks,' *Brown Boverie Rev.*, **57**, pp.554-561

KOPPL, G. RUOSS, E. (1970): 'Switching over-voltages in EHV and UHV networks', *ibid.*, **57**, pp. 554–561

MORTLOCK, J. R. (1956): 'AC switchgear' (Chapman & Hall Ltd.)

MORTLOCK, J. R., HUMPHREY-DAVIS, M. W. (1952): 'power system analysis' (Chapman & Hall Ltd.)

RUDENBERG, R. (1950): 'Transient performance of electric power systems' (McGraw-Hill)

(1970) 'Switching of capacitive circuits exclusive of series capacitors' *IEEE Trans.* **PAS-89,** pp. 1203–1207

(1961): 'Switching surges' *Trans. Am. Inst. Elec. Eng.,* **PAS-80,** pp. 240–261

3.8 List of symbols

a = radius of conductor

a, b, c = suffixes used to denote quantities associated with phases A, B and C

b = radius of earthwire

C = capacitance, F

d = average distance between conductors

d' = average distance between earthwire and any conductor

d = suffix used to denote natural self impedance of one phase of a multiconductor transmission line

e = instantaneous voltage

E = peak value of normal generated power frequency voltage

$f_n = \dfrac{w_n}{2\pi}$ = natural frequency

$f_s = \dfrac{w_s}{2\pi}$ = normal generated-system-power frequency

F = point of fault on diagrams

g = suffixes used to denote quantities associated with an earthwire

h = height of conductor or bundle of conductors above earth

h' = height of earthwire

i = instantaneous current

I = peak current

l = line length

L = inductance, H

m = suffix used to denote natural mutual impedance between phases of a multiconductor transmission line

mg = double suffix used to denote natural mutual impedance between phase and earthwire of a multiconductor transmission line

M, P, S, T = reference points on diagrams

n = number of individual conductors in a bundle conductor

q = charge, C

r = radius of individual conductor in a bundle conductor

R = resistance, Ω

s = average spacing of individual conductors in a bundle conductor

t = time

v, V = voltage at an assigned point in a circuit

w_n = angular velocity of natural frequency

w_s = angular velocity of normal generated system power frequency

x, y, z = suffixes used to denote amplitude etc.

$X'd$ = direct axis transient reactance

$X''d$ = direct axis subtransient reactance

X = reactance, Ω

Za, Zb, Zc = impedance of phases A, B and C, respectively

Zn = natural or surge impedance

Zs = source impedance

1, 2, 0 = suffixes used to denote positive, negative and zero phase-sequence quantities, respectively

11, 12, 13, 21, 22, etc. = double suffixes used to denote self and mutual impedances between conductors of a multiconductor transmission line

θ = general symbol used to denote phase-angle difference

ϕ = power factor

Chapter 4

Oil circuit breakers

D. F. Amer, C.Eng., F.I.E.E.

4.1 Introduction

This chapter refers to oil circuit breakers, i.e. a circuit breaker in which the contact members of the interrupting device are separated in oil.

Oil circuit breakers have been continuously developed during the 20th century and have retained a pre-eminent position, although many other forms of equipment have come into being. The main function of a circuit breaker is to isolate automatically a fault section of a power system. Viewed purely from what is likely to give reliable service, however, the following qualities are required of circuit breakers:

(*a*) To act as an insulator most of its service life
(*b*) To carry load current (not necessarily full load current) most of its service life
(*c*) To make and break load current frequently
(*d*) To carry, to break and to make short-circuit currents, normally infrequently.

In most applications, these functions can be performed equally well by oil circuit breakers as by gas-blast or vacuum circuit breakers. This, together with the fact that they are generally of lower cost than other forms of equipment, will ensure their continued use.

There are two categories considered, one in which the enclosure for the oil and the contacts is metallic and earthed, and the other in which the chamber containing the oil is insulated from earth. See Fig. 4.1. In the latter, the container itself is principally of insulation. The former will be referred to as a dead tank oil circuit breaker and the latter as a live tank oil circuit breaker. Normally live tank oil circuit breakers contain less oil than dead tank oil circuit breakers. In both types, the solid byproducts of the interrupting processes, mainly carbon, are contained in the circuit breaking tanks. Fig. 4.1*a* and 4.1*b* illustrate one phase of a live tank and a dead tank circuit breaker in their simplest forms.

More complex forms for use on extra-high-voltage systems employ several

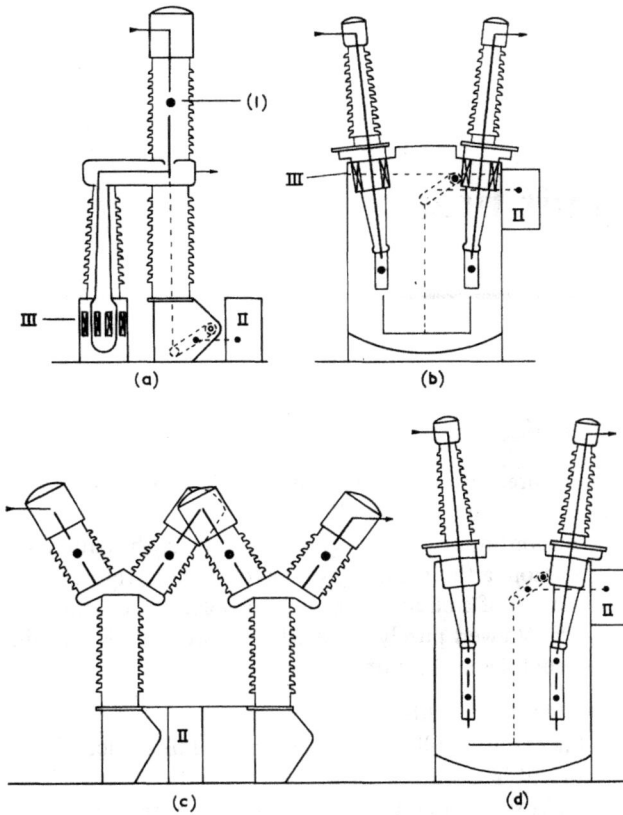

Fig. 4.1 Live tank and dead tank circuit breakers

     a Live tank single break
     b Dead tank double break
     c Live tank 4 break
     d Dead tank 4 break

     I interrupters
    II mechanism
  III current transformer

interrupters in series, and these are shown in Fig. 4.1*c* and *d*. In this type of equipment, resistors and/or capacitors are sometimes used in parallel, with the interrupters to control the voltage distribution across the interrupters during circuit breaking. These resistors must be rapidly switched out, because they are not continuously rated.

In some live tank circuit breakers, a resistor and series connected switch are arranged in parallel with each interrupter. These series connected switches are simultaneously opened following the opening of the main interrupters; thus the resistors remain in parallel with the interrupters for a predetermined time, and this time interval permits the main interrupters to break the current passing through them. Similar principles can be used for dead tank circuit breakers, but it is usual to arrange the switch, which interrupts the resistor current, in series with the main interrupters because a simpler construction is inherent in this arrangement. This series connected switch is also opened at a predetermined time following the opening of the main interrupters.

The chapter is confined to design features directly associated with the function of the circuit breaker, i.e. the interrupters, contact design, mechanisms; insulation problems, general construction and the effects of environment and features at the interface of the circuit breaker and other equipment with which it forms an entity.

Although many of the design principles are common to oil circuit breakers from 11 kV through to the highest voltage (750 kV), and indeed to other types of circuit breaker, some constructional features are determined solely by environment, and problems at the interface of application.

4.2 Interrupters: design principles

4.2.1 General

The heart of the circuit breaker is the arc-control device. The purpose of an arc-control device is to bring about a condition following the opening of the circuit-breaker contacts so that the path of the resulting arc 'loses its memory' concerning its conducting state, at the earliest current zero after contact separation. The current will then cease to flow and, provided also that the arc path and all insulation in parallel with it is able to withstand the ensuing transient and steady-state open-circuit voltages, effective circuit breaking will occur.

Arc-control devices in oil circuit breakers are designed and developed by experimental techniques in high-power testing laboratories. Literally thousands of tests are made during the development of a new device which may extend over several months or years, so that one must think in terms of spending £100 000 or more for this part of the development alone.

They can be conveniently classified into 'transverse' or 'cross-blast' interrupters and 'axial blast' interrupters, although a combination of both features is sometimes used.

Special significance is attached to arc-control units when connected in series to achieve higher-voltage ratings, and also when they operate in parallel.

4.2.2 Cross blast

A cross-blast interrupter, that is based on the early ERA baffle (Whitney *et al.*, 1930), is envisaged to work as shown below, and recent supporting photograph

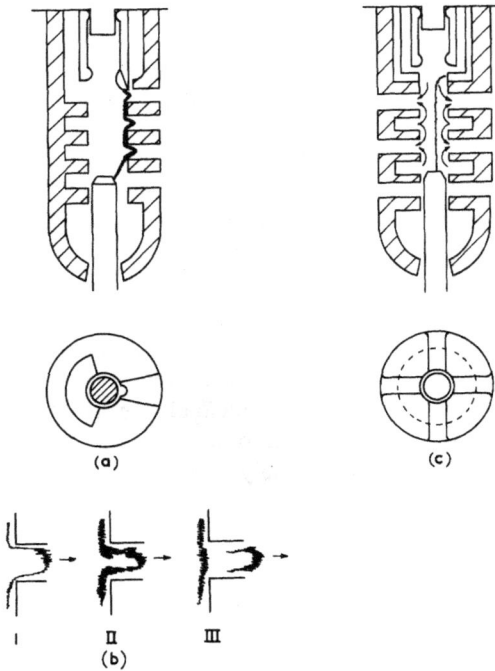

Fig. 4.2 Schematic illustration of cross-blast and axial-blast interrupters

    a Crossblast
    b arc short circuiting
    c axial blast

(Amer *et al.*, 1968) seems to confirm much of the original thinking concerning this.

Here the arc is drawn in front of a series of lateral vents. See Fig. 4.2*a*. The heat of the arc vaporises the oil and the gases formed (mainly hydrogen) increase in pressure and force the arc to bow into the vents. An increased voltage is required to maintain the longer portion of arc inside the vents. Before the arc can escape from the vents it short circuits itself at the entry to the vents (Fig. 4.2*b*).

This process continues throughout the arcing period at intervals of time of the order of tens of microseconds, though these time intervals are not constant because the relevant events both inside and outside the arc control device are changing continuously.

Ultimately, when the pressure inside the arc-control device becomes sufficiently high, and the length of the arc is also sufficiently extended at power frequency current zero, the arc is extinguished.

The arc always burns inside a bubble of gas, and this bubble extends and expands through the vents to the outside of the arc control device. The hot gases emerging from the vents are initially still ionised and it is essential to ensure by correct vent design that no breakdowns occur between the vents external to the arc-control device. This is particularly important for e.h.v. interrupters, where multiple series vent arrangements are invariably used.

4.2.3 Axial arc-control units

Many forms of this device exist, see Fig. 4.2c for a typical example. The principle is similar to the cross blast in that the arc plasma is frequently renewed, since the ionised gas, itself the carrier of the current, is being removed from the arc zone via the vents. Improved dispersion of the gas issuing from the interrupter is inherent in the design, together with a balanced reaction from the issuing gases.

One disadvantage of this form of device is that the scavenging action of the gases in the neighbourhood of the arc is determined by the size of the hole through which the contact pin passes, as well as by the vents. Better arc control can be achieved when the contact diameter is small. 20 mm or less is known to work well, and permits greater freedom in optimising venting.

Special consideration must be given to arrangements where high normal current ratings are required. For example in some designs using a large-diameter contact stem, oil injection is known to improve performance. A parallel-contact arrangement is sometimes used and it effectively separates the current carrying function from that of circuit breaking, and thus permits both functions to be optimised. With cross-blast arc-control units this is done mainly to avoid the development of another arc-control unit.

4.2.4 Factors controlling performance

4.2.4.1 Speed of break

The speed of break necessary for successful interruption is dependent on a number of factors that mainly apply to the following:

(a) the particular form of the arc control chamber itself
(b) the actual voltage rating of the equipment.

The arcing time must of necessity be kept to a minimum so that the arc energy and burning of contacts is reduced. In practice, it has been found that there is a minimum arcing time during short circuit interruption that is extremely difficult to reduce, this is approximately 0.02 of a second, and it is usual to ensure that this minimum is achieved at the full short-circuit rating of the device, if not at reduced levels.

The minimum arc gap on the other hand is a function of the voltage, and generally takes the form of $G = kV^n$ where n is of the order of 0.33, and k is a parameter dependent on the form of the arc-control device, the circuit breaker as a whole and the current being interrupted. Thus, if the minimum arcing time is to be achieved and the minimum arc gap is known for the particular form of device, the speed of break can be determined, which will give the minimum arc energy.

The optimum speed of break is however dependent upon functions other than short-circuit performance; for instance the interruption of capacitive loads, where an increased speed of break may be desirable. This unfortunately would result in increased arc energy under high-power interrupting conditions. In order to overcome this, means such as oil injection are incorporated, which assist the breaking of capacitive currents and thus avoid the need for excessive speeds of break. In many modern designs, oil is injected only during the interruption of small values of current, and thus obviates the need for powerful pumping against the high gas pressure generated when interrupting large short circuit currents.

In addition to these considerations, the speed of break is, in some constructions, affected by the interrupting duty. The pressure (Section 4.2.4.3) in the arc-control device can act to increase the speed, as can the electromechanical forces on conductors, particularly with dead tank circuit breakers, because of the conductor arrangement (Section 4.3.3).

It is normal to endeavour to retain the speed within a narrow range whatever the duty. Oil-immersed speed-control devices are therefore frequently used to combat the above effects. These are designed to produce counter forces which prevent the acceleration of the moving parts.

Typical average speeds of break of commercial circuit breakers (measured over one cycle after contact separation) are shown in Table 4.1.

Table 4.1   Speeds of break

11 kV circuit breakers	2-3 m/s
33 kV circuit breakers	4-5 m/s
132 kV circuit breakers	7-10 m/s

4.2.4.2 Speeds of make

The optimum speeds of make can be extremely significant – particularly with heavy-duty small dimensioned equipment. For example, it may be desirable to reach the fully engaged position before the peak making current is attained, thus reducing the effect of magnetic blowoff and pinch effects, consequently permitting the use of lighter mechanisms. With smaller circuit breakers, (of say 400 A, at 3·3 kV to 11 kV) it may be necessary to pay special attention to the kinetic energy of the moving parts at the point of contact make, to overcome the effects of electromagnetic blowoff (Section 4.3.3.1).

The problem then arises of absorbing the energy of a high speed contact during no-load operations.

The electromagnetic forces resisting closure have a relatively lower retarding effect in circuit breakers of larger normal current carrying capacity, because of the kinetic energy of the increased moving mass.

In general, for high-voltage and extra-high-voltage circuit breakers it can be stated that speed of make and the geometry of the moving and fixed contacts should be chosen so that the closing operation is 'not hesitant', and yet free from contact bounce, so that arcing is minimised; this obviates the generation of high gas pressures and contact burning, both of which tend to prevent contact engagement. Typical average speeds of make in commercial circuit breakers (measured over the half cycle before the contacts touch) are as shown in Table 4.2.

Table 4.2   Speeds of make

11 kV circuit breakers	3-4 m/s
33 kV circuit breakers	5-6 m/s
132 kV circuit breakers	9-10 m/s

4.2.4.3 Gas generated during circuit breaking

It has been found that the amount of gas generated is a function of the arc energy. This has an important bearing on the size of vents in an arc-control device, and also on the whole structure of the circuit breaker, although each arc-control device and circuit-breaking structure must be considered as an entity.

Some information concerning this should therefore be of interest, and the following data (Bruce *et al.*, 1933) (Vogelsanger, 1946) has been found to be of value during the development of arc-control means.

The gas generated should be the minimum consistent with successful operation, and therefore the arc voltage should be as small as possible. In commercial circuit breakers operating in ambient conditions of normal temperature and pressure, it has been found that the mean arc voltage can be kept within a value in the order of 10 V/mm.

The volume of gas generated and measured at normal temperature and pressure is in the order of 70 cm^3 kW s of arc energy. Other measurements and calculations show that at the 'mean hot temperature' during circuit breaking, 700 cm^3 of gas are generated per kilowatt second.

The composition of the gas has been found to be as follows:

(*a*) Hydrogen   66%
(*b*) Acetylene   17%
(*c*) Methane   9%
(*b*) Other   8%

The 'effective vent area' of an arc-control device varies in a cyclic manner, (Amer *et al.*, 1968) (Lawson, 1972) because the arc tends to 'choke' the vents, in a way which is dependent on the instantaneous current. Thus the pressure in the arc-control device varies in a similar way (Fig. 4.3).

(a)

(b)

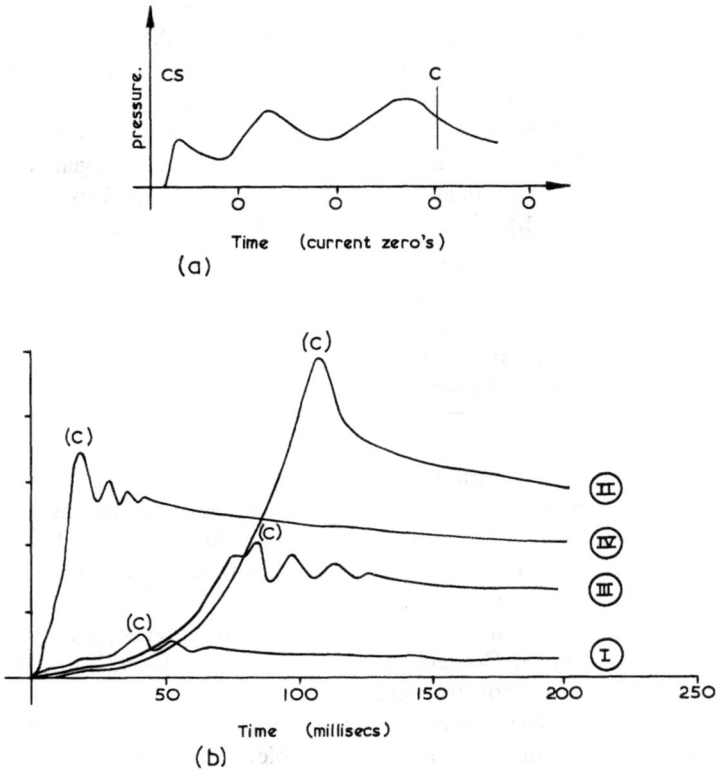

Fig. 4.3 Pressure characteristics in interrupters and circuit-breaker tanks

a inside the interrupter
b outside the interrupter
c contact separation
d clearance

I   5% rated short-circuit current
II   10% rated short-circuit current
III 30% rated short-circuit current
IV 100% asymmetrical

Peak pressures measured in arcing chambers are generally less than 70 kg/cm^2.

The physical vent areas of typical arc control means are shown in Table 4.3.

The gas generated rapidly displaces the oil outside the arc control device, and the possible damaging effects of oil movement and hot gas must be given consideration in the overall design. The pressure characteristic in the oil outside the arc-control device consists of an initial pressure pulse followed by a steady decaying pressure.

Table 4.3 Typical vent areas

Voltage kV	Number of vents	Breaking capacity MVA	Area of vents mm^2
11	2	500	370
33	4	1500	900
66	4	2500	1200
132	6	5000	2000

The latter is governed mainly by the gas cushion above the oil level and by the exhaust system.

This information, together with information concerning critical arc gaps and arcing times, forms the basis on which judgment can be exercised during the development of arc-control means.

4.2.4.4 Typical interrupter characteristics

The short-circuit-performance limits reached by interrupters at the present time are approximately as given in Table 4.4.

These have been obtained by optimising the speed of break by improving interrupter structure, and by improving the circumambiency in which the interrupter is used.

Where a combination of high voltages and higher currents are needed, it is usual to connect lower voltage units in series.

Individual interrupters have a current interrupting ability inversely proportional to some power of the voltage; this is fortunate because it matches the service requirements. It has been the practice to aim at achieving approximately 25% of the full short-circuit rating at double the rated phase to neutral voltage, although in some designs this has been as high as 2·5 times the rated phase-to-neutral voltage. This corresponds to the asynchronous levels of typical systems.

Table 4.4

kV	r.m.s. symmetrical current	r.m.s. asymmetrical level kA
20	70	80
60	60	67
132	21	23

*50% d.c. component

A desirable feature is to obtain a restrike free performance when switching capacitive loads. This ensures that overvoltages are not produced on the system. It also ensures that the arc-control device itself is not subject to the high impulse forces caused by a release of energy from the capacitors. These forces have been known to cause cracks in the interrupter plates or containers. In addition, the absence of restriking eliminates high-frequency oscillatory discharges which may superficially puncture the surface of the insulating material of the arc-control device — a feature of high-frequency oscillatory discharge.

In the present state of the art, it has been found that it is almost impossible to achieve restrike free performances where the voltage exceeds 60 kV per interrupter. The limiting capacitive switching performance can only be determined for each construction.

No doubt with adequate development restrike free performance could be achieved on e.h.v. single interrupters, but where such performance is needed at voltages in excess of 66 kV, it is usual to employ a larger number of interrupter units connected in series, or to use resistors connected in parallel with the interrupters. Oil injection improves the contact between the arc and the oil itself, thus producing more effective deionising action. In addition, it tends to envelop the contacts in oil, or at least a gas of higher pressure, thereby improving the dielectric strength, and thus tending to overcome restriking.

A resistor of suitable choice (equal approximately to the surge impedance of a line or cable) will permit the capacitance load to discharge its stored energy and thus ease the voltage impressed across the interrupter. The energy is dissipated in the resistor, which must therefore be capable of absorbing this energy without overheating.

During recent years much attention has been drawn to the performance of oil circuit breakers in interrupting short line faults. While good performance is dependent on the form of the device, recent tests, within the contributor's experience, have not shown that this presents any major problems in oil circuit breakers.

In the same way, the sensitivity of oil circuit breakers to high rates of rise of recovery voltage has also been questioned. Here, actual tests have shown that the constructions tested were relatively insensitive to changes in the rate of rise of recovery voltage. This may not, however, apply to all equipment.

The foregoing generalisations should be treated with caution. Each device has its own particular characteristics.

4.2.5 Interrupters in series
To enable arc-control units to be used in series for higher voltages, it is preferable that the performance of each unit is consistent, and that deionisation occurs in a synchronous manner. This cannot always be assured.

In circumstances where this does not apply, the co-ordination of the dynamic states of all the insulation in the circuit breaker becomes of special significance. The need is to ensure that reignitions, if and when they occur, do so in the right place, i.e. in the interrupter.

It is well known that the behaviour of oil interrupters is dependent on the current broken and the arc energy. This can be seen from any 3 phase short-circuit test by examining the last two phases to clear the short circuit, where the time period of one current loop shortens and the other lengthens. Here it is often observed that only one phase clears, and that the arc path in the other remains conducting for a longer period. The first condition, therefore, in multibreak oil circuit breakers is to ensure that the contacts separate at the same time and at the same speed.

It is also known that during repeat test duties at the same short-circuit level, and with the same point of wave for contact separation, the arcing time of a single interrupter may vary. Nevertheless successful interruption ensues. It is, however. preferable that this variation should not occur in interrupters that are to be used in series. Variation in arcing time clearly indicates that an arc-control device is operating in an inconsistent manner; either the deionisation is not occurring consistently or else the dielectric recovery characteristic is not consistent. In practice, it is possible to ensure consistent operation in single interrupters over a wide range of currents interrupted (which is fortunate, as it forms the basis of unit testing), but at low currents (less than 500 A) the arc is unstable over a large part of the arcing time, and in consequence the conduction and dielectric strength of the arc path varies also in a random manner (which again is fortunate because at these current levels full voltage tests can be done).

The effect of this on the design of multibreak equipment is that the assumption must be made that discharge may occur in one arc-control unit and not in others. Total reignition may then follow in time spaced steps (cascading). Consequently, it is vital to ensure that all insulation on any component that is in parallel with the arc path has a higher level than the arc path itself, so that any subsequent breakdown occurs in an arc-control device. This problem is not so important in oil immersed multibreak circuit breakers, but is of greater significance in live tank circuit breakers where the external insulation (normally in air) is subject to atmospheric pollution, rainfall, fog, etc., all of which are known to affect adversely the external switching surge level of equipment. It is therefore essential to use higher external insulation levels for each circuit-breaking chamber than would otherwise be necessary. This also has a bearing on the design of resistors and capacitors, each of which must be designed on the assumption that some may be short circuited.

The overall co-ordination of these factors can, fortunately, be checked in a testing station by full voltage testing for practically all circuit breakers required at the present time.

4.2.6 Features of construction
The form of an arc-control device should be such that, during circuit breaking, ionised gas is generated and removed from the arc at an optimum rate, i.e. with the

minimum arc energy necessary to bring about the self destruction of the arc. In other words, by eliminating from the arc path its 'memory' concerning its previous conducting state, the arc will then cease to exist at the appropriate current zero. Secondly, the device must be capable of establishing an adequate dielectric strength to withstand the subsequent voltage between the terminals. This will prevent reignition of the arc.

Typical forms of construction are shown in Fig. 4.4. These are single interrupters, but they can be used in multibreak constructions. Parameters concerning them are also given in Fig. 4.4.

The geometric forms of these interrupters must be designed in the light of the following considerations:

(*a*) The size of the vents must be adequate to cope with the heaviest of short circuits.

(*b*) At lower short-circuit levels, and other light duties when smaller vents would permit shorter arcing times to be obtained, it is sometimes desirable to use larger vents and accept the longer arc durations.

The conflicting requirements of (*a*) and (*b*) can be solved by the injection of oil. Earlier forms of oil injection were mainly based on pumping oil into the arc-control device through the bore of the moving contact, and are still used. More recent designs however confine the oil injection to the interruption of the lower-current spectrum, as the self generation of gas is more than adequate to extinguish the arc at high short-circuit levels. This form of oil injection can be achieved by the use of spring-loaded pistons and bypass valves, or by the application of gas pressure on the exhaust side of the interrupter (Experience has shown that a pressure of 2 kg/cm^2 works excellently in some forms of construction.) Gas pressure at this level becomes insignificant at higher levels of short-circuit interruption, where the self-generated gas may reach 70 kg/cm^2.

The materials used in the main structure of an arc-control device are insulation capable of withstanding a wide variety of switching transients and 50 cycle voltages, as discussed in Chapter 3. The insulation must also be of consistent mechanical strength suitable for withstanding high pressure impulses.

Fibre-glass structures are used for pressure containers or as tension members.

Suitable materials for vent packs are fibre, permali (impregnated wood), polycarbonates, and, for lighter duties, polyamide (nylon). These also have to withstand high impulse pressures and must be resistant to erosion by the arc. The vent packs in some constructions are arranged to withstand only hoop stresses while longitudinal stresses are taken on tie bolts.

In other constructions, the vent packs are held together by springs so that free movement (a frangible effect) of the individual plates can take place if the pressures are excessive. This type of construction has the advantage that the springs retain the vent packs under virtually consistent compressive load, thus reducing the need for periodic tightening of the vent packs during, for example maintenance periods.

The thermal capacity and the design voltage level of shunt impedances is also of

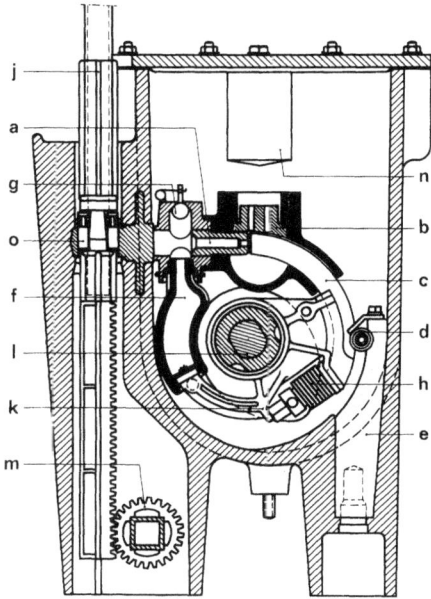

Fig. 4.4A Laterial vented interrupter
Arcuate moving contact and oil
injection feature [Hazemeyer,
Holland]

a   fixed contact
b   arc control device
c   arcuate contact pin
d   roller contact
e   connection at cable side
f   injector
g   valve
h   compression spring
j   busbar isolator
k   insulation holder
l   insulation shaft
m   isolator drive
n   gas spreader

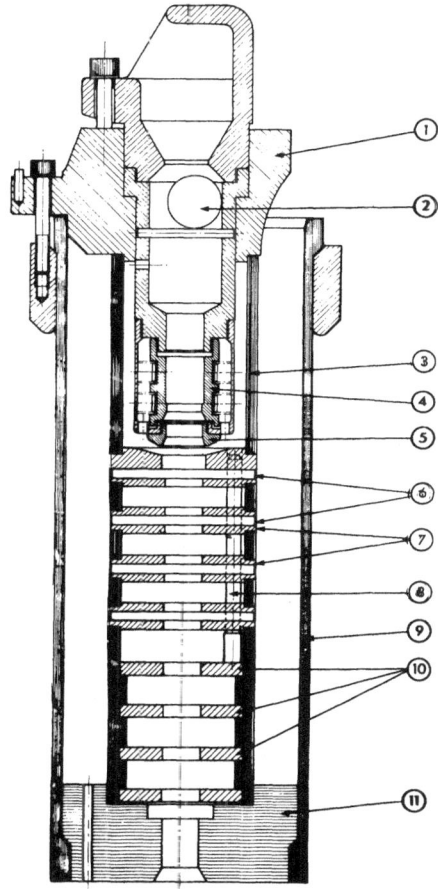

Fig. 4.4B Axial arc-control unit [Laur
Knudsen Ltd., Denmark]

1   contact holder
2   ball valve
3   spacing tubes
4   upper fixed contact
5   arcing ring
6   vents
7   vent plates
8   locating rods
9   fibre glass tube
10 spacer rings
11 throat washer

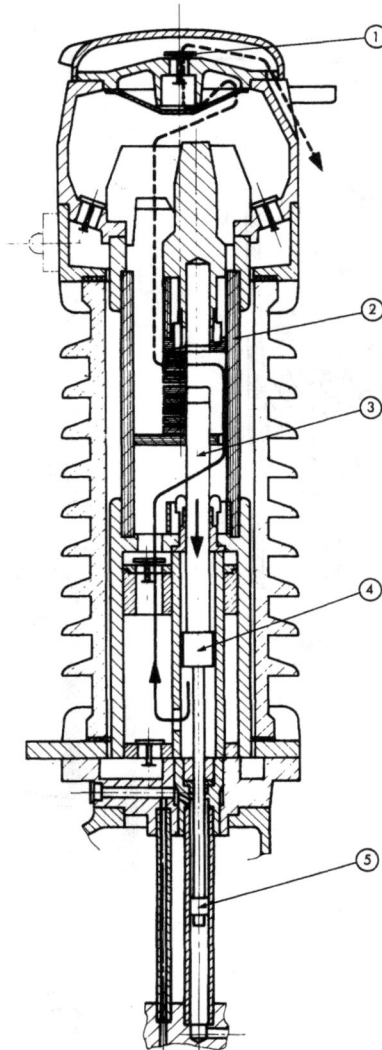

Fig. 4.4C Lateral vented interrupter showing oil injection feature [Brown Boveri & Co. Ltd., Switzerland]

1 exhaust valve
2 arc control chamber
3 moving contact
4 oil injection piston
5 drive piston

importance (Section 4.2.5). Resistors designed on an arbitrary basis so that they have to carry current associated with either twice the phase to neutral voltage for one cycle, or in live tank construction the 50 cycle flashover voltage of the live tank interrupter for one half cycle, which ever is the lower, have been found to cover all conditions likely to arise in service.

Noninductive wire-wound resistors of high or low ohmic value are frequently used in oil circuit breakers.

Ceramic resistors are also used, but these are usually confined to high-ohmic applications, because materials have not been developed for use in oil to the same extent to which they have in gas.

Capacitors used for voltage-grading purposes have taken many forms, but, more recently, ceramic capacitors are widely used because of their extreme flexibility of application and compactness arising from high permittivity of the ceramic mix. Here again the assumption must be made that a capacitor across one unit may be subject to a higher voltage, just as in the case of a resistor.

4.3 Contacts and conductors – design principles

4.3.1 General

This Section is concerned with the design of a current-carrying system that has to withstand the effects of short-circuit currents and/or the normal current.

The dimensions of the conductor system and the contacts can be readily determined in the light of their ability to carry the short-circuit current, i.e. in respect of temperature rise or contact welding, since the heating of each component is relatively unaffected by the others in the time scale of a short circuit. the electromechanical forces on the conductor system can also be determined with reasonable accuracy from mathematical considerations.

The design of the conductor system to meet the normal current is, however, markedly different in that the temperatures of specific components are dependent on the ambient temperatures derived from the complex flow of heat from the remainder of the conductor system.

The contacts of the arc-control device have also to withstand the effects of arcing – particularly during the interruption of the heaviest short-circuit currents – while still retaining their normal current carrying function.

Other components that do not normally carry current may have to accept short-circuit current during the occurrence of a fault. Such components include earthing and other metallic components subject to connection by flashover of insulation (Section 4.5.4.2).

4.3.2 Normal current carrying

A uniform conductor of cross-sectional area A, carrying a current I, will rise to uniform temperature T above ambient if dissipation by convection and radiation is also uniform along its length. This bears little relationship to commercial circuit breakers.

Here, the following apply:

(a) intense local sources of heat generation (the contacts and joints etc)
(b) distributed sources of heat generation e.g. the main conductors.

These have to be considered together with:

(c) closely thermally connected components (e.g., adjacent metal work) that act as a sink from which little heat can be extracted owing to its enclosure in poor thermal conductors, such as electrical insulation or stationary fluids (see Table 4.5)
(d) closely thermally connected components (i.e. adjacent metalwork) from which considerable heat can be extracted by conduction through solids or by convection through liquids or gases.

Table 4.5   Thermal conductivity of typical materials

Bakelised paper	0·002 W/mk
Porcelain	0·01 W/mk
Oil (Stationary)	0·0015 W/mk
Copper	3·8 W/mk
Cast Iron	0·63 W/mk
Aluminium	2·0 W/mk

The heat flow determines the temperature gradient and the temperature rise at various locations, but all heat must ultimately be wasted to the atmosphere mainly by convection. The results of two constructions can be seen from Fig. 4.5a and b.

The total heat generated is dependent on the millivoltage drop across the circuit breaker, and on the current being carried.

The most intense heat generating sources are in the contacts, particularly near the fixed and moving (or articulated) contact points, where the cross sectional area of the current carrying path is small, thus producing considerable constriction resistance (Holm, 1967). This heat is conducted from the contacts mainly via metal parts and subsequently by convection to the oil surface.

The heat generated can be reduced by the following:

(a) using more contacts in parallel
(b) using the highest permissible contact loads
(c) eliminating joints
(d) using conductors of effectively larger section or better electrical conductivity.

The temperature can be reduced by enhancing the areas from which heat can be extracted. The main area from which heat is extracted is in the metal parts at the

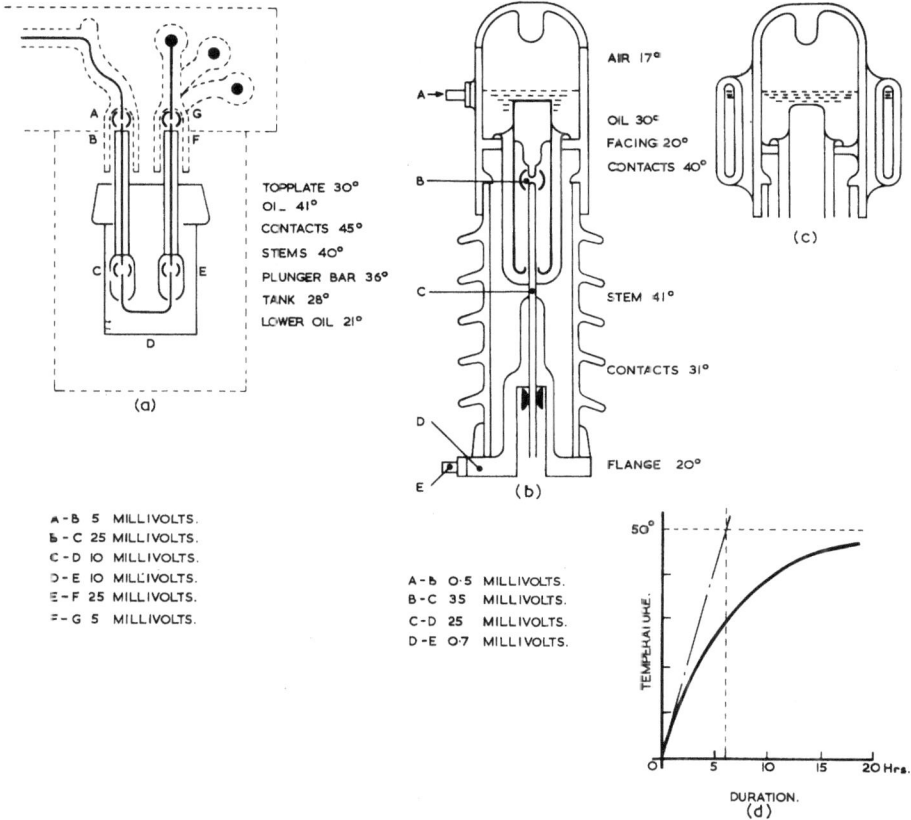

Fig. 4.5 Distribution of temperature in typical circuit-breaker constructions (degrees Kelvin above ambient)

a metal enclosed construction
b live tank construction
c live tank construction with increased cooling surface
d typical temperature versus duration

top of the circuit breaker. A typical solution of the latter is shown in Fig. 4.5*c*, which effectively raises the rating of a live-tank circuit breaker from 1200 A to 2000 A by extracting heat from the upper layers of the oil.

The permitted temperature rise is determined by the temperature limits of the insulation both solid and fluid (Section on specification); for the former because there is an upper temperature limit when the insulation deteriorates due to excessive dielectric losses brought about by thermal instability, (approximately

90 K). The permitted temperature of the top oil in circuit breakers derives from work on the oxidisation and sludging of oil for transformers and has been agreed (on an arbitrary basis) for oil circuit breakers to be 80 K (IEC specification).

Other sources of heat are created by magnetic induction in surrounding iron work. Theoretically the power dissipated in 'iron losses' is proportional to the square of the current, and inversely proportional to the length of the magnetic path.

Magnetising heat can be virtually eliminated by using nonmagnetic materials; or minimised by increasing the length of the path of magnetic induction; or by the use of a contrawinding, or by only using magnetic material outside a 3-phase group of conductors.

All designs require detail consideration, but it has been found in practice, and demonstrated in laboratories, that the size of a hole in magnetic material (say a flange in free atmosphere) must be at least 25 mm in diameter for every hundred amperes carried by the conductor passing through it, to limit the temperature rise in the flange to 30 K.

Heat arising from circulating currents caused by induced effects can be minimised by using high-resistance materials such as nonmagnetic spheroidal graphite iron.

It should be noted that the normal current rating of equipment is judged in accordance with set standards. The actual ambient conditions in service vary widely, and it is therefore the duty of the user to determine the safe current-carrying capacity of the equipment if the ambient conditions exceed those stated in the appropriate specifications. As a guide, the temperature rise is approximately proportional to the square of the current.

Overload ratings are not recognised in switchgear specifications because of the short thermal time constant of circuit breakers and fear of damage to the insulation. Nevertheless the temperature versus time characteristic of a typical construction is worthy of note, see Fig. 4.5d.

4.3.2.1 Skin and proximity effects

The effective resistance of a conductor is greater when carrying alternating current than direct current, because the self inductance of the inner portion is greater than that of the outer. This is known as the skin effect. The effective penetration depth of 50 cycle alternating current is one centimetre in a circular sectioned conductor, see Fig. 4.6.

The distribution of current (Copper Development Association 1936) and 'switchgear stages' (Clothier, 1933) in a conductor is also adversely affected by the priximity of other conductors carrying currents, and thus interaction between phases must be taken into account.

4.3.3 Short-circuit current carrying

4.3.3.1 Electromechanical forces

A conductor carrying current is subject to mechanical forces when it is placed in a magnetic field. The calculation of such forces is usually based on the Biot—Savart

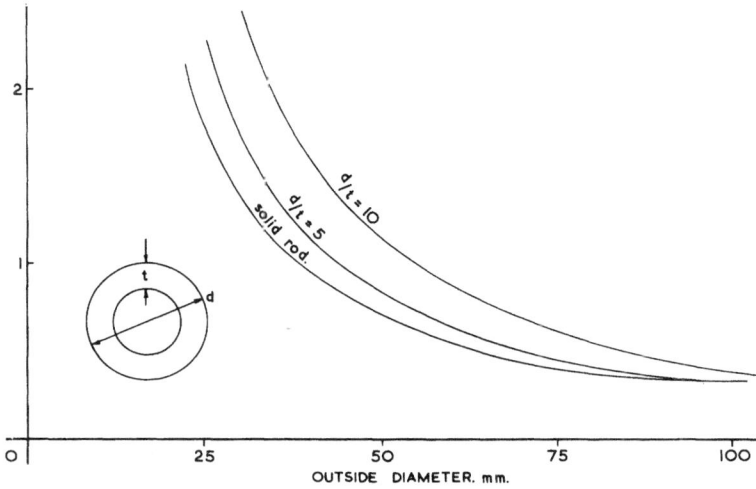

Fig. 4.6 Effective resistance of conductors

laws and on the work of Laplace; namely the force on a current carrying element $d\ell$ is proportional to the product of the current in the element and the strength of the magnetic field derived from the other current-carrying parts.

The distribution of this force in typical current carrying component of circuit breakers is shown in Fig. 4.7a, b and c together with the equations to determine the actual forces. Such calculations are near approximations, because the actual distribution of the current is assumed to be disposed symmetrically in the conductors.

A notable effect is that of electromechanical forces between phases, where it will be seen from Fig. 4.7d (V. Zajic, 1957) that the forces on the centre phase vary from positive to negative, and those on the outer phases remain biased in one direction. This has a bearing on the design of contacts, since there could be a considerable alternating load on the contacts of the centre phase, with the possibility that contacts might not, in some constructions, remain engaged with the resultant deterioration of contacts by commutation burning.

Unbalanced forces, may arise from single phase or 2-phase faults, must also be given consideration during design.

4.3.3.2 Carrying of short-time currents
Switchgear is normally rated to carry a specified peak 50~ current, and an average r.m.s. current for a prescribed time. Normally this is specified as the rated

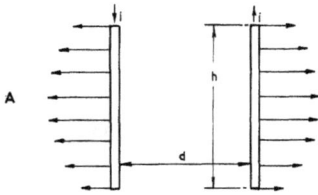

A   TOTAL FORCE $= i^2 \times 10^{-8} \sqrt{\left(\frac{h}{d}\right)^2 + 1} - 1$

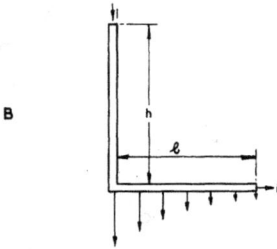

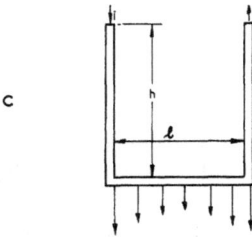

B   TOTAL FORCE $= i^2 \times 10^{-8} \log_e \frac{a}{\ell}\left[\frac{\sqrt{\left(\frac{\ell}{h}\right)^2 + 1} - 1}{\sqrt{\left(\frac{a}{h}\right)^2 + 1} - 1}\right]$

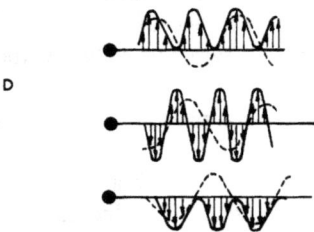

a = radius of conductor

C   TOTAL FORCE $= 2i^2 \times 10^{-8} \log_e \frac{a}{\ell}\left[\frac{\sqrt{\left(\frac{\ell}{h}\right)^2 + 1} - 1}{\sqrt{\left(\frac{a}{h}\right)^2 + 1} - 1}\right]$

D

Fig. 4.7 Electromagnetic forces on conductors

　　　a parallel conductors
　　　b conductors at right angles
　　　c conductor at right angles to two parallel conductors
　　　d 3-phase parallel conductors

Note: unidirectional forces on outer conductors, and positive and negative forces on the inner conductor

short-circuit current for 0·5 s, 10 s or 3 s, and the peak level corresponding to the peak making current.

The temperature of a conductor carrying the current for a short time can be derived on the basis that there is no heat lost by radiation, convection or conduction. The temperature rise per unit length in these circumstances is given by the following formula:

$$T = \frac{I^2}{A^2} \, t = \text{rise in temperature, K}$$

I = current, A

A = cross-sectional area of the conductor, cm²

t = time, s

ρ = resistance, Ω/cm^3

The ability of contacts to carry short-time current and resist welding is dependent on the type of material of which they are made, on the forces between the contacts during the passage of current, and on the number of contacts used. A typical result is given for hemispherical-ended silver-plated copper contacts (Fig. 4.8).

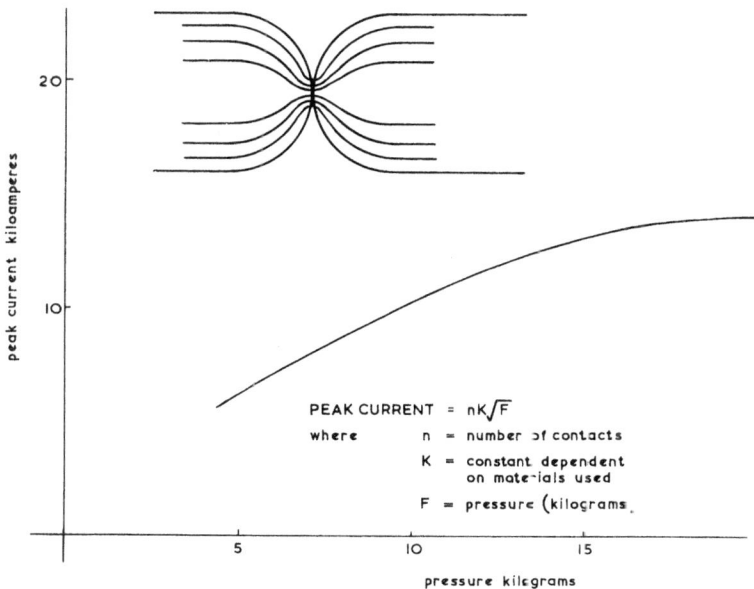

Fig. 4.8 Peak current carrying capacity versus contact pressure (silver plated copper contacts)

Table 4.6   Electrical and thermal conductivity of materials

	Specific Gravity	Electrical Conductivity	Thermal Conductivity
Copper	8·89	97% of silver	0·92
Aluminium	2·7	60% of silver	0·5

4.3.4 Features of construction

Contact and conductors are normally of copper, aluminium or other nonferrous materials. Some physical properties of these materials are shown in Table 4.6.

An important principle is that for the same physical dimensions and the same overall temperature rise, the current-carrying capacity is approximately proportional to the square root of the conductivity of the material used.

It is also customary to either tin or silver plate components within oil circuit breakers, otherwise the surface resistance of these components become unacceptably high. Blackening of the copper from oxidisation and deterioration of the contacts ensue, particularly when frequent load-breaking operations occur. In addition, plating has the advantage – where there are fixed joints – of providing a film which cannot readily deteriorate.

Sintered materials are also useful particularly for arcing contacts where combinations of either silver and tungsten, or copper and tungsten, are frequently used.

Springs which may carry current should be nonmagnetic, and should not become ductile due to high temperature. High dural bronze, beryllium copper, and stainless steel are suitable in certain applications.

In other cases, it may be essential to ensure that current is not passed through such items as the contact springs, articulated joints, or rubbing surfaces. In these circumstances, it is usual to insulate at least one of the components.

To prevent deterioration at the normal current carrying points, burning during short-circuit interruption should be confined to areas away from the normal contact position. To ensure this, the change in contact constriction resistance and electromechanical blow-off effect must be minimised up to the point of disengagement. One way to ensure this with a multifinger and poker construction (Fig. 4.9a) is to arrange the contact fingers so that

(a) in the fully engaged position, all (or most) of the contacts mate on normal-current-carrying faces, of, for example, silver
(b) during opening, all (or most) of the contacts remain engaged, on arc-resistant material of the poker
(c) final contact separation occurs at the determined arcing area, between arc-resistant material of the finger contacts and poker, or in a position where the arc can be transferred rapidly to arc-resistant material.

An alternative rectangular contact arrangement is shown in Fig. 4.9b.

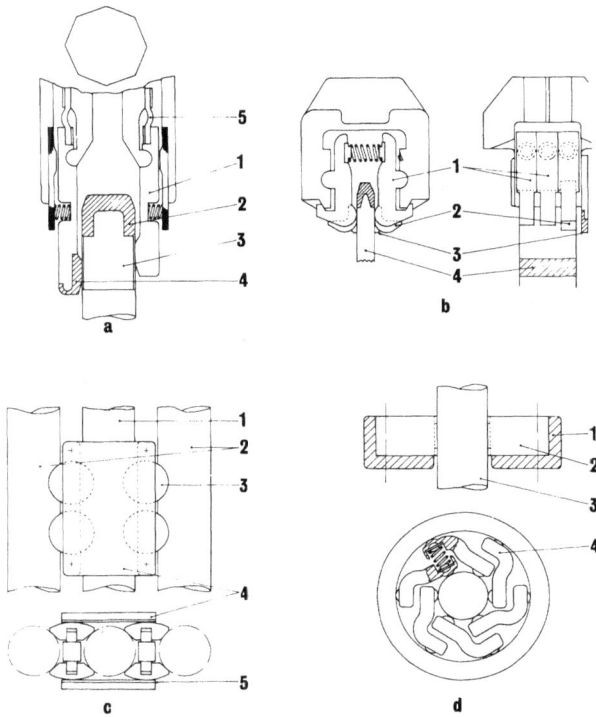

Fig. 4.9 Typical contact arrangements

a Octagonal cluster arrangement

  1 silver plated finger contacts
  2 tungsten-copper arcing tip
  3 silver faced rod contact
  4 tungsten copper arcing tip
  5 laminated connector

b Cluster with rectangular contact

  1 silver plated finger contact
  2 tungsten-copper arcing tip
  3 tungsten-copper arcer
  4 silver plated contact blade

c Pass-through contact

  1 silver plated moving contact
  2 silver plated fixed contact and guide
  3 roller contacts (silver plated)
  4 carrier

d Pass-through contact
  1 fixed carrier (silver plated)
  2 contact cluster
  3 silver plated moving contact
  4 silver plated finger contact

With the normal cluster-type finger contact used in an arc-control device it is sometimes preferable to use a flexible connection at the end of the cluster. This avoids the tendency to burn due to blow-off action in the transfer period during the rupturing process. It has the added advantage that the millivolt drop is reduced and that heat can be conducted metallically from the fingers (See Fig. 4.9*b* and *c* for typical arrangements).

Some of the foregoing remarks apply to 'sliding' contacts (Fig. 4.9*d*) that are commonly used in live-tank circuit breakers. Generally these contacts have to be of a 'low friction' type, and sometimes a roller type of contact is used (see Fig. 4.9*c*). If these are silver plated, it is preferable that one component should use a harder silver than the other.

4.4 Mechanisms — design principles

4.4.1 General

There are four different mechanisms in common use in oil circuit breakers:

(*a*) manually closed and spring opened mechanisms — closing being dependent upon the force used by the operator
(*b*) independent manually operated closing and spring opened mechanisms — closing is independent of the force used by the operator
(*c*) power closed from external source and spring opened
(*d*) power closed and power opened from an external source.

Type (*a*) is confined by specification to applications where short-circuit interruption of the circuit breaker is not more than 6 kA (Reference IEC, 1959) because of danger to the operator inherent in a hesitant closing operation on short circuit, and also because the electromechanical (and other) forces acting against closure may be transmitted to the operator (new IEC specifications no longer recognise this form of circuit breaker closing mechanism).

All mechanisms have, however, one common feature, namely that they must permit safe opening of the circuit breaker following the application of a tripping impulse, even if this impulse occurs during the closing operation.

Many modern mechanisms of the types (*a*), (*b*) and (*c*) are made 'mechanically trip free', i.e. they consist basically of two parts (Fig. 4.10) as follows:

(*a*) closing elements and a holding latch or toggle
(*b*) a collapsible yet resettable linkage associated with a tripping mechanism and latch.

A holding latch or toggle permits the removal of the source of the closing energy following closure of the circuit breaker.

It is however customary to arrange the mechanism design and the controls, so that if the closing movement of the contact is started, normal closing will be

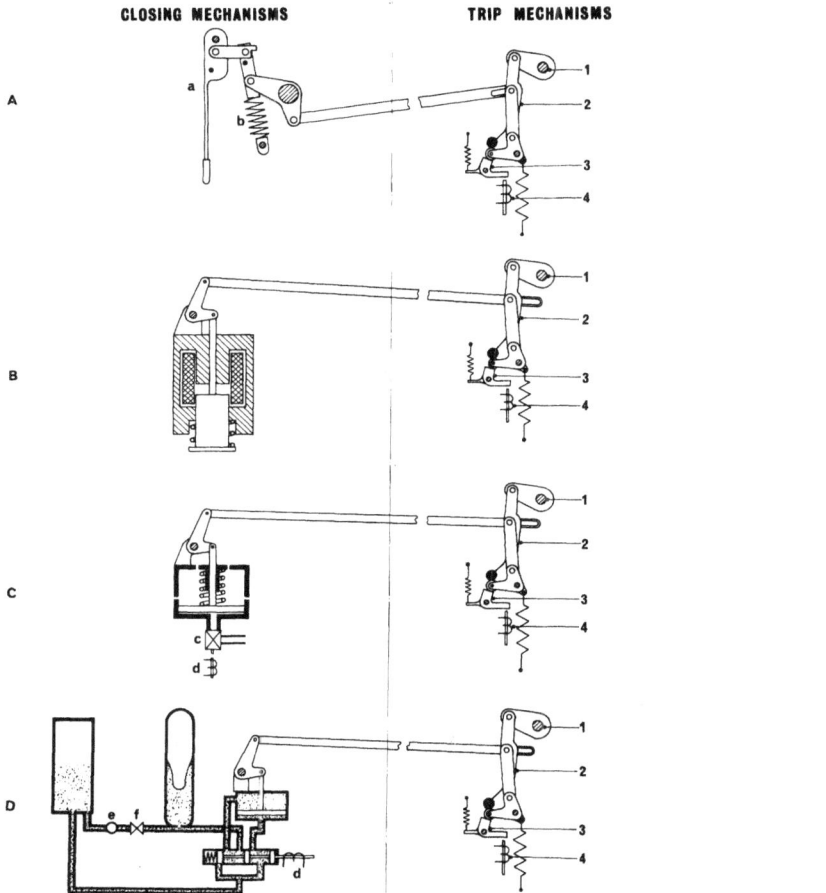

Fig. 4.10 Schematic arrangement of mechanism

A independent manual closing mechanism
B solenoid closing mechanism
C pneumatic closing mechanism
D hydraulic accumulator closing

1 circuit-breaker shaft
2 toggle
3 trip latch
4 trip actuator

a closing handle
b closing spring
c closing valve
d closing actuator
e motor
f nonreturn valve

completed before the tripping function will take place. The reason for this is that many designs of circuit breaker would give an adverse interrupting performance, if the contacts did not start from full engagement.

This has brought into question the need for mechanically 'trip-free' mechanisms with their associated complexity, particularly where automatic reclosing circuit breakers are required. Clearly the elimination of resetting features would result in simplified mechanisms, yet the predominance of the tripping operation over the closing force at the correct time could be retained.

Mechanisms of the type (*d*) are based on the use of an external compressed air supply for opening and closing, or on the use of a hydraulic accumulator for this purpose.

Many systems need high-speed circuit breakers. This necessitates mechanisms which permit opening times of 1½ cycles, and reclosing times of 15 cycles.

Circuit breaker mechanisms use extremely high bearing loads. This, together with the very high rates of acceleration and deceleration of the components make the design of circuit breaker mechanisms unique.

4.4.2 Energy levels
The total energy required to open the circuit breaker is the sum of:

(*a*) the kinetic energy required to accelerate the masses of the moving parts of the circuit breaker, and the associated coupled side of the mechanism itself
(*b*) the energy to overcome contact friction
(*c*) the energy to overcome mechanism friction
(*d*) the energy to overcome oil drag
(*e*) the energy associated with 'oil injection' when this is used.

The closing-energy requirements are the sum of the energies required to:

(*a*) accelerate the masses of the moving parts of the circuit breaker, and all the moving parts of the closing and opening mechanism
(*b*) overcome contact friction
(*c*) overcome oil drag
(*d*) overcome electromagnetic blow-off effect
(*e*) charge the opening springs when these are fitted
(*f*) overcome friction losses in the mechanism.

Attention must also be given to forces arising from electromagnetic effects on the contacts, and from pressure in the arc-control device during circuit breaking and making.

All of these energies have to be computed, and consideration given to the form of the design to ensure that the moving contact functions within the correct speed range (Fig. 4.11). Experience indicates that the opening energy should be disposed preferably so that the contacts can attain the critical arc-extinction length in approximately 0·02 s.

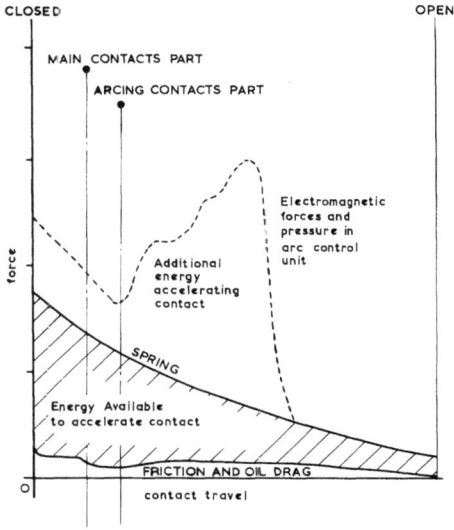

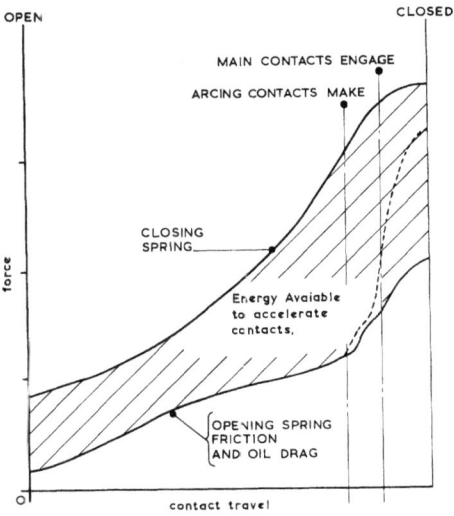

Fig. 4.11 Static forces versus travel diagrams

$$F = \mu N \tag{4.1}$$

If there is no external torque, then there is no friction force, and the shaft will occupy the lowest position making contact at point B. When an external torque is applied in the direction shown in Fig. 4.12b, the shaft first rolls along the surface of the bearing to a position (point A) where the angle of the tangent to the point of contact is equal to the friction angle. At this point slip occurs. During rotation of the shaft the position of point A remains stationary.

$$M_f = Fr = \mu N r \tag{4.2}$$

Referring to figure b and substituting eqn. 4.1

$$R = \sqrt{N^2 + F^2} = \sqrt{N^2 + \mu^2 N^2} = N\sqrt{1 + \mu^2}$$

or

$$N = \frac{R}{\sqrt{1 + \mu^2}} = \frac{Q}{\sqrt{1 + \mu^2}}$$

Substituting in eqn. 4.2

$$M_f = \frac{\mu r Q}{\sqrt{1 + \mu^2}}$$

Since R and Q form a couple with arm h

$$M_f = Qh = \frac{\mu r Q}{\sqrt{1 + \mu^2}}$$

and

$$h = \frac{\mu r}{\sqrt{1 + \mu^2}} \approx \frac{\mu r}{1 + \mu^2/2}$$

and $\mu^2/2$ can be neglected compared with unity, then

$$h = \mu r$$

where h may be described as the radius of the friction circle. The notion of the radius of the friction circle permits a simple graphical representation of forces acting in a mechanism.

If we consider a simple 2-pin-connection link and ignore friction, the line of action of the force, transmitted through the link coincides with the geometrical centreline of the link. If however friction is considered, the line of force will pass through the friction circles, but its precise direction is indeterminate. In the extreme, it can be tangential to the friction circles (see Figure 4.12a). The direction of the force can vary within the shaded area, and consequently the trip-force will vary accordingly. To ensure that there is a positive load on the trip latch, the shaded area must not overlap the friction circle of the bell-crank pivot bearing.

The necessary closing velocity must be attained before contact engagement, and thereafter the total closing energy must be adequate to overcome the additional electromechanical forces and associated friction. In some mechanisms, it is arranged that the static closing force is higher than the total of the resisting forces during the travel of the contact into full engagement. Typical closing and opening energy levels are given in Table 4.7.

The final derivation of the mechanical linkage system may be required to perform repeated opening and closing operations at very high speed.

Table 4.7   Typical opening and closing energy level

Type of equipment	Closing energy* mkg	Opening energy mkg
11 kV 250 MVA 400 A CB	8·25	2·17
33 kV 1000 MVA 1200 A CB	31·0	3·8
66 kV 2500 MVA 1200 A CB	98·0	28·8
132 kV 5000 MVA 1200 A CB	520	243

*This includes the energy necessary to charge the opening springs

4.4.3 Trip-mechanism chain

A simple kinematic chain is shown in Fig. 4.12a. The importance of minimising friction and ensuring that a positive force is available to permit the linkage to collapse on application of the tripping impulse will be understood from the following analysis. For this purpose the following conditions, which approximate very closely to reality, are assumed to apply:

(a) There is a small diametrical clearance between shafts and the bearings.
(b) The reaction force acting on a shaft is uniformly distributed along the line of contact, i.e. elastic deformations are neglected.
(c) The resultant of the uniformly distributed forces acts at some point A (Fig. 4.12b) on the line of contact.

Notation
  R = resultant force
  F = friction force (tangential component of the resultant force)
  N = normal component of the resultant force
  μ = coefficient of friction

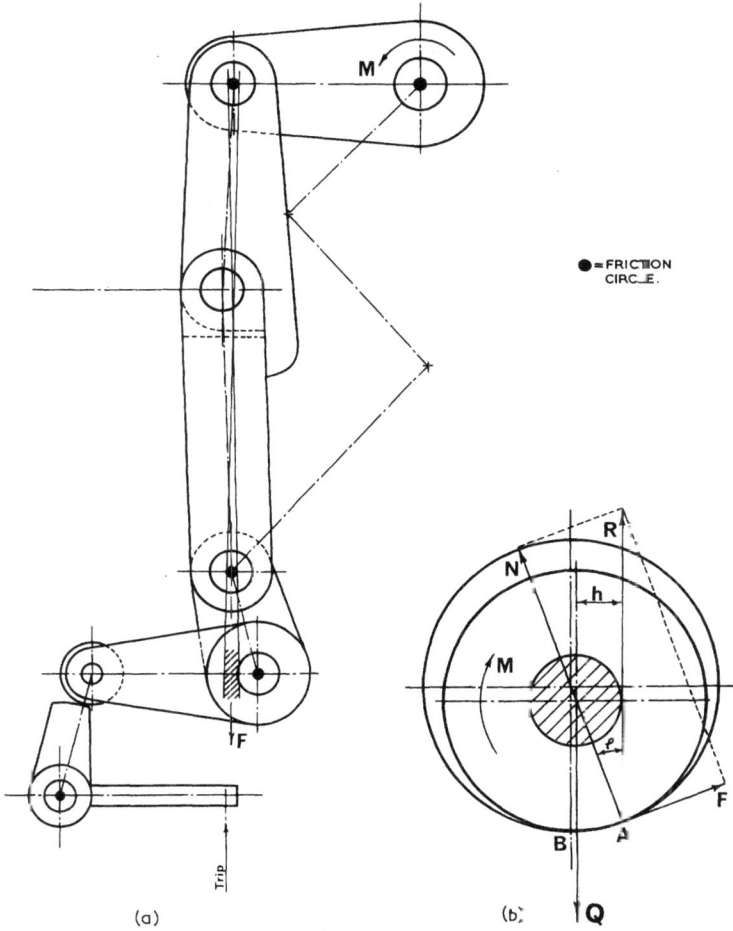

**= FRICTION
CIRCLE.

Fig. 4.12 Schematic arrangement of trip mechanism chain

Q = external load on shaft
M = external torque on shaft
M_f = torque of friction on shaft
r = radius of shaft

Resolving the resultant force into normal and tangential components, the magnitude of the friction force is proportional to the normal component.

Table 4.8   Static friction coefficient

Brass	0·34 to 0·18
Phosphor bronze	0·18
Dyn metal	0·18
Oilite bronze	0·14
Nylon	0·10
Teflon	0·05
Ball bearing	0·01

The static friction coefficients of nonlubricated bearings, utilising stainless steel pins are shown in Table 4.8.

4.4.4 Latches

The trip latch is one of the most important features of the mechanism. The force and energy required to trip the latch are mainly dependent on the number of reduction linkages involved in the mechanism chain and on the type of latch used.

The energy available to trip the latch is determined by the overall features of the protective system. This energy may derive from the output from a set of current transformers (say 10 VA) or from an external energy source, e.g. a battery connected via tripping relay contacts to an electromagnetic actuator. In the latter case, the current from the battery must be limited to values less than 30 A, i.e. the making current of the relay contacts. Some specifications require that the current from the battery does not exceed five amperes in order to minimise the voltage drop in the leads between the actuator and the battery.

During the circuit breaking operation the loads on the latch vary. These loads are as follows:

(a) shock loading (snatch)
(b) static load in the closed position of the circuit breaker
(c) dynamic variation in load during closing.

The form of the latch and the engaging materials must be chosen to meet these three conditions.

A fourth condition is important, namely, that as the latch disengages, the bearing load intensity becomes very large. The disengagement should therefore be rapid so that the time of the applied force is extremely small. Commonly used trip latches are shown in Fig. 4.13.

(a) Fig. 4.13a shows a D shaft. The force on the D shaft does not usually exceed 70 kgm. The D shaft is to be turned against the 'stiction' and the sliding friction associated with this force.

(b) Fig. 4.13b shows a roller latch that can be loaded to about 140 kgm. To trip the latch, the cam has to be turned against the rolling friction associated with this load.

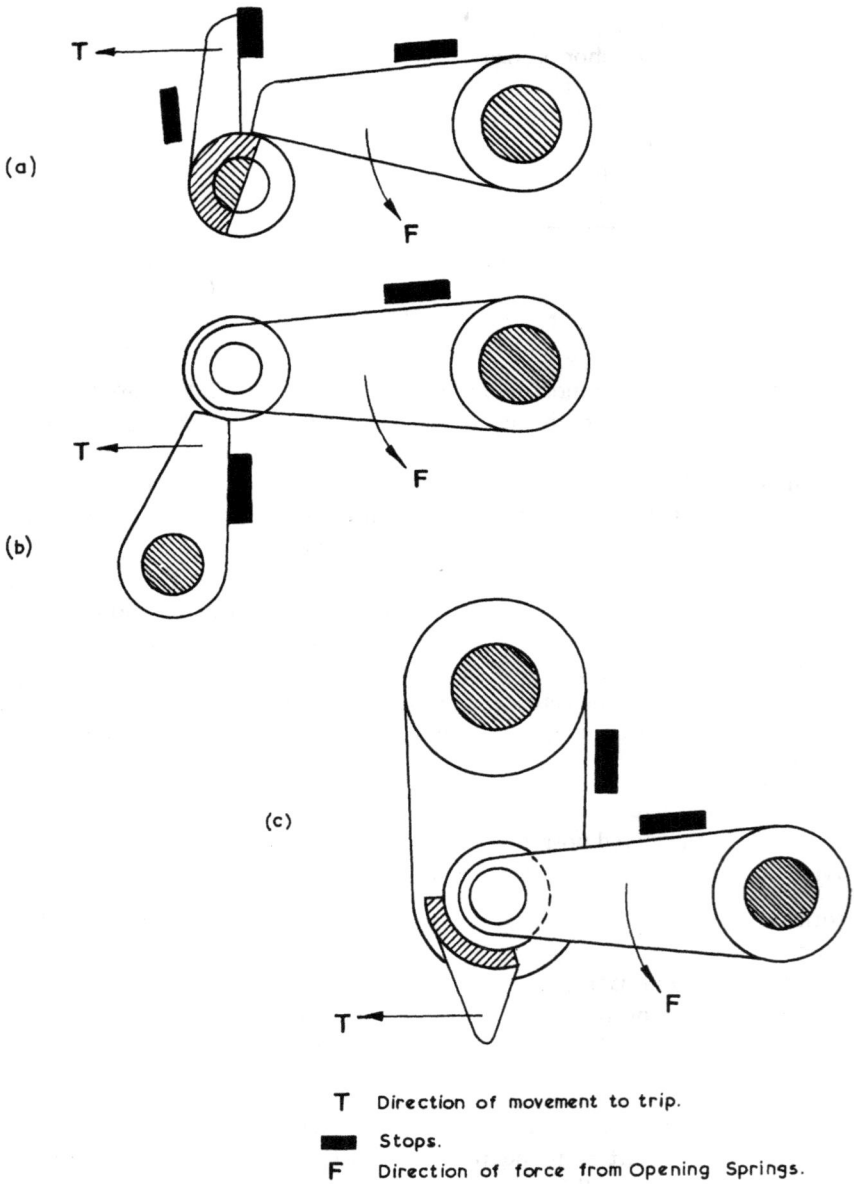

(a)

(b)

(c)

T   Direction of movement to trip.

■■ Stops.

F   Direction of force from Opening Springs.

Fig. 4.13 Latching arrangements

It should be noted that in both cases, the force is concentrated along the 'line' of contact. The 'line' is in fact an oblong area that is dependent on the elastic deformation of the metal used.

(c) Fig. 4.13c shows a latch that is known as a cradle latch; this endeavours to increase the area of the bearing face and thus minimise the depth of elastic deformation at the contacting 'line', and at the same time it reduces the angular movement of the part to be moved. This arrangement allows an extremely large increase in the load on the latch (about 450 kgm) without increasing the tripping force or energy, and it thereby permits a reduced number of parts in the mechanism chain. The 'hanger' assists the resetting, by allowing an easier access to the cradle.

It is customary to use hardened components for the engaging areas. Typical materials and levels of hardness are shown in the Table 4.9.

Table 4.9 **Typical materials for latches**

tripping D shaft or cam	latch or roller
colmonoy 6 hardness 600 h.v.	s.g. iron hardness 600 h.v.*
nitrided steel EN40B hardness 800 h.v.*	nitrided steel EN40B hardness 800 h.v.*
stellite grade I hardness 700 h.v.	needle roller hardness 750 h.v.*
mild steel EN32A	mild steel EN32A
case hardened 700 h.v.*	case hardened 700 h.v.

*h.v. = Vickers Hardness Test BS427

4.4.5 Electromagnetic actuators
4.4.5.1 General
These are devices that release the tripping latches or initiate closing of circuit breakers.

Before describing the principles of tripping actuators, it should be mentioned that an automatic tripping signal is normally received from sensing means which determine the faulty section of the system that is to be isolated. These sensing means are known as protective systems.

All protective systems are based on one or more of the following measurement techniques:

(a) magnitude of current
(b) comparison (e.g. current entering and leaving a circuit)
(c) ratio measurement (e.g. changes in the reactance of a high-voltage circuit).

As the main conductors are at high voltages, the tripping signal can be sensed by series connected devices in the main circuit, or by shunt connected devices, i.e. current transformers.

In both cases, the signal has to be transmitted from live components to the tripping mechanism via an insulated link or coupling.

4.4.5.2 Series connected tripping devices

This is a solenoid with its coil connected in series with the main conductor. The solenoid plunger operates on a tripping latch of the circuit breaker via an insulated link, where the plunger is also at high potential.

It is normally used in conjunction with a time-delay device, e.g. a hydraulic timer or mechanical escapement (a clock).

Series connected devices are commonly used on circuit breakers, see Fig. 4.14a, up to 11 kV; they simply detect overload current, and act rapidly if a short circuit occurs, and more slowly for overload conditions.

The timer is set to prevent operation under transient overload, e.g. starting current of a machine (a motor for example). These devices are integral with the circuit breaker and have an important influence on its design.

In particular, the short circuit through current time does not need to exceed that which the coil permits in its actuation of the circuit breaker. Reduced copper sections can therefore be used.

Sometimes a saturable reactor is connected in parallel with the series coil. The impedance of the reactor is high under normal load conditions and thus the major part of the current will flow through the series coil. With high fault current the reactor becomes saturated, its relative impedance falls, and it accepts a higher proportion of fault current.

4.4.5.3 Shunt-connected tripping devices

Two shunt tripping devices are indicated in Fig. 4.14b. One is a simple solenoid utilising an iron core. The tripping device can be operated directly by the output from current transformer secondaries, or indirectly through a relay utilising a secondary source of power such as battery.

When the power is taken direct from the output of the current transformers, the solenoid is designed to operate with approximately 15 VA. When a relay and battery are used a tripping power of approximately 300 W is normal.

In some cases, however, where high-speed operation is required — in the order of 5 ms — it may be as high as 2000 W. In general, the level of the current taken by these actuators is limited by the making current of the relays used — normally not more than 30 A.

The design of high-speed magnets is extremely complex, necessitating techniques of step-by-step integration (Roters, 1970).

Fig. 4.14c shows the design of a high-speed tripping unit utilising a fluid amplifier and a limiting resistor, the output from which operates directly on the

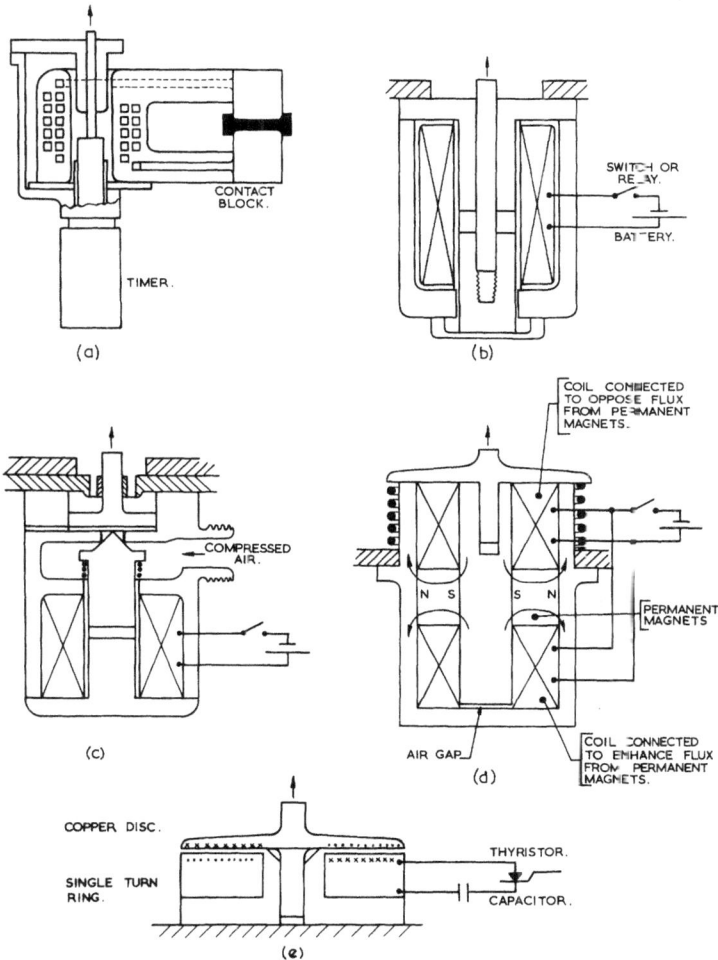

Fig. 4.14 Electromagnetic actuators

tripping latch. To prevent the current from rising to a value which is too high a switch is operated by the trip latch that inserts a resistance in the control circuit (not shown in the figure), thereby permitting the use of a coil of extremely low inductance and consequently a high speed build up of current. This ensures that the force necessary to 'crack' the valve is obtained rapidly. High-speed movement of the valve is assured by reducing the mass of the valve itself. Overall tripping times of approximately three milliseconds are possible with this form of device.

A more complex form of magnetic actuator a — flux shifting device — is shown in Fig. 4.14*d*. This device utilises a permanent magnet and a dual magnetic path, and has a spring loaded 'keep' that is held by the magnet in the path of least reluctance (Brat Kowsk *et al.*, 1968). By energising coils, which oppose the flux of the main magnetic path, it is possible to divert rapidly the magnetic flux into the other path. The 'keep' is then moved rapidly by the spring, and this releases the trip latch. This type of device is normally operated from a battery or from condensers, and is capable of high speeds of operation down to approximately two milliseconds. Such devices are arranged to be reset during the opening operation of the circuit breaker.

A feature of all flux shifting devices is that the permanent magnets used must only supply a 'force'. The magnetic energy 'lost' during operation must be minimal. In the device shown in Fig. 4.14*d* a second coil is fitted which gives a 'boost' to the permanent magnet each time the device operates. This offsets the small loss of magnetic energy, thus avoiding the demagnetisation that can occur from frequent operation.

Another device that is coming into use for rapid circuit-breaker tripping is based on the repulsion forces produced in a metal disc which is subject to a transient magnetic field. The basic circuit and form of device is shown in Fig. 4.14*e* (S. Basu *et al.*, 1969 and 1971).

The exact calculation of these designs is difficult and in consequence the hardware cannot easily be optimised.

The energy imparted to the tripping disc is a function of that given up by the capacitors in $(c^2 e/2)$, and the force driving the disc is a function of the discharge current and of the change in equivalent inductance at various spacings of the disc. These devices are capable of high speeds of operation down to parts of a millisecond.

4.4.6 Features of construction

Mechanisms consist basically of a rigid frame attached to which is a closing unit, a linkage associated with tripping, and a resettable linkage which couples these together. Tripping springs may also be included, but some circuit breakers are fitted with a 'tail spring' mounted in the circuit breaker tank. Secondary control switches, latch checking switches, and suitable means for attaching a slow opening and slow closing device, are also included.

A vital feature is the correct setting of the stops in the mechanism, and the correct setting of the latches. These are normally interdependent, and must permit consistent positioning when a mechanism is at rest (Fig. 4.15).

A trip latch, for example, is not normally loaded until the closing operation is started. Thus the mechanism moves freely for a short distance until the latch takes up the load necessary to permit the closing force to be transferred to the moving parts of the circuit breaker. It is essential to ensure that this initial 'snatch' loading can be sustained without damaging the surface of the latch. For this reason it is usual to use only small clearances between, e.g. the rollers and the latch, in the order of 0:5 mm.

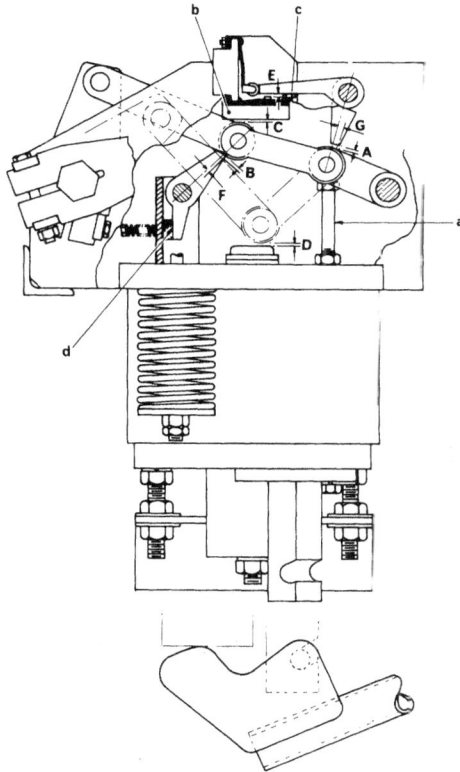

Fig. 4.15 Typical mechanism and table of settings

	Position	Dimensions*	Measurement	C.B. position	Closing drive position
A	Trip latch	0·2—0·5	clearance	open	open
B	Closing latch	0·6—0·75	clearance	closed	closed
C	Closing toggle	2·0—2·5	clearance	closed	open
D	Closing ram	2·5—3·0	clearance	open	open
E	Trip tappit	3·0—4·0	clearance	—	—
F and G	Latches	3·0—4·0	engagement	closed	open

*The smaller dimensions are normal for high-speed reclosing circuit breakers and are closely toleranced

'Overshoot' during closing is limited for similar reasons, although the 'settling back' of the mechanism on to the closing latch or toggle is normally not as severe as that at the trip latch during the closing operation.

In recent years, greater use has been made of dry bearings, e.g. p.t.f.e. impregnated sintered bearings and extremely high dynamic loading of up to 300 kg/s cm^2 can be applied successfully.

Advances in ductile castings, among which spheroidal-graphite iron may be mentioned, are re-establishing castings as suitable for high-power mechanisms (note: flame cut components have been widely used).

Mechanism components attain high velocities so that the momentum of these components is larger in relation to the distance in which they have to be brought to rest. Many of these become uncoupled in some way during the operation. Rates of deceleration as high as 50 g are not uncommon, and buffers or dashpots must be used to bring any decoupled component or group of components to rest.

4.4.7 Pneumatic-hydraulic mechanisms

Pneumatic-hydraulic mechanisms utilise the stored energy of compressed gas, the gas (nitrogen) being compressed hydraulically within an accumulator to a pressure of approximately 300 kg/cm^2. These accumulators are applied sometimes as a source of closing power only (Perri *et al.* 1954) and in other applications for opening and closing circuit breakers (Widmer, 1968).

In both arrangements, an electrically operated oil pump, controlled from a pressure switch maintains the pressure in the accumulator.

When this system is used only for closing a circuit breaker it can be applied to a normal mechanically-trip-free mechanism. Here the closing control valve permits pressure to be applied to an hydraulic ram which operates the closing linkage, and thus the stored energy is available for closing the circuit breaker (Fig. 4.10).

Fig. 4.16 shows how the stored energy of the accumulator is applied for both opening and closing a circuit breaker. In the particular example shown, the energy from the hydraulic accumulator is applied directly to the moving contact system for both opening and closing the circuit breaker. This is done via servo-operated valves that are responsive to the closing and opening command valves. The servo-mechanism itself is driven by power from the hydraulic accumulator (Widmer, 1968).

This type of mechanism is suitable for modular construction and has come into greater use with multibreak small oil volume equipment.

4.5 Insulation – design principles

4.5.1 General

Two basic insulation problems arise in the design of oil circuit breakers. One is concerned with the apparatus in its function as an insulator of the live components

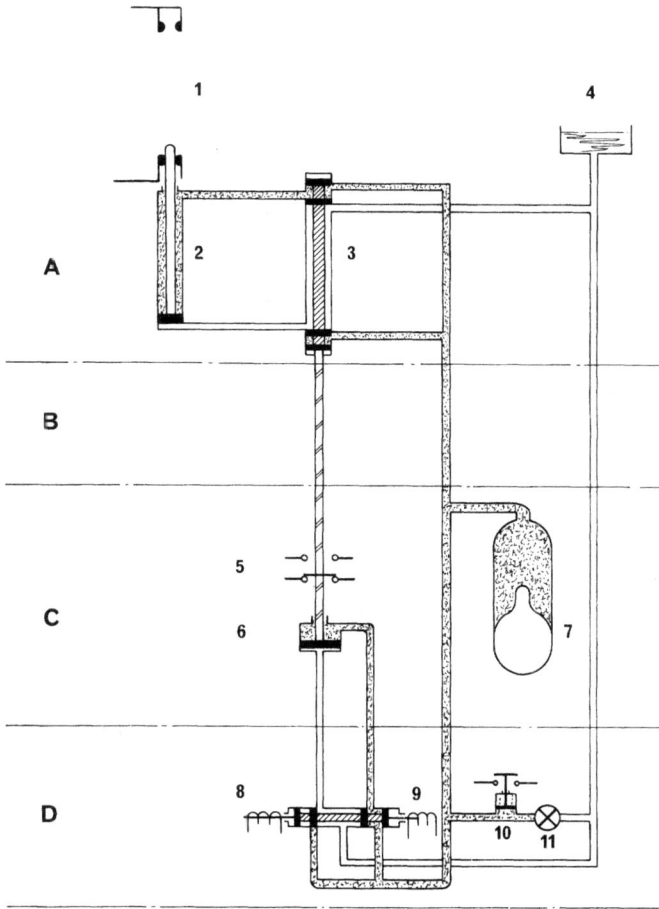

Fig. 4.16 Pneumohydraulic mechanism

    A hydraulic drive in c.b. head
    B transmission from c.b. base
    C power unit and control valve drive
    D central servounit

    1 c.b. contacts
    2 contact drive
    3 control valve
    4 low pressure oil source
    5 electrical indication
    6 control valve drive

    7 pneumohydraulic accumulator
    8 closing coil
    9 trip coil
    10 pressure control
    11 motor driven pump

when they are stationary, and the second arises during the circuit breaking operation when the co-ordination of the insulation may be upset by the disturbance in the circuit breaking compartments.

The insulators are also subject to mechanical forces. In dead tank circuit breakers, the main supporting insulators are the bushings; the guides for the main moving contacts and the insulation drive members of the moving contacts.

In live tank circuit breakers, bushings are not normally used, so that the main components are the insulators constituting the tanks; the insulation of the support columns; the contact drive rods; and the insulation of capacitor or resistor housings. In effect, a bushing is needed only in the current transformers which is normally a separate free standing unit.

The design features of insulation co-ordination, the effect of contamination, and constructional features are discussed in this Section.

4.5.2 Co-ordination of insulation

In Section 4.2.5 reference was made to the subject of insulation co-ordination during circuit breaking. Further reference is now made to this, and to insulation co-ordination related to externally caused overvoltages, such as lightning surges, switching surges – chopped and unchopped – and 50 cycle overvoltages (Fig. 4.5).

The effect of these voltages on equipment design should be considered with the circuit breaker both open and closed. Normally external flashover to earth must constitute the lowest insulation level. The puncture level of insulation should always be higher.

It is relatively easy to ensure this, in both live tank and dead tank equipment, when such equipment is in a clean and pristine condition.

It is much more difficult to obtain this under polluted conditions, and particularly with live tank circuit breakers. Here high leakage currents may occur across the support insulators (in particular the first supporting post) and in consequence there will be a tendency to impress the peak system 50 cycle voltage or surge voltages on to one unit of the circuit breaker, if the circuit breaker is in the open position.

The fail safe solution to this problem lies in ensuring that flashover will always occur before the internal breakdown of any component. The problem is basic to insulation which is connected in series, and where breakdown may occur in cascade.

This in turn must be related to the level of voltage causing reignitions during circuit breaking, where it is necessary to ensure that such reignitions occur along the arc path.

The overall conditions to be met therefore are as follows:

(a) The 'weakest' link in the co-ordination should be externally from the live terminal of the circuit breaker to earth in air, in both the open and closed positions. Note: this should be arranged to carry any arcing away from the actual supporting posts.

(*b*) Reignitions, if they occur during circuit breaking, must occur along the arc path. The insulation level of the arc path must therefore be lower than (*a*).

(*c*) The internal breakdown level must be higher than both (*a*) and (*b*) when the circuit breaker is in the steady-state open or closed position.

(*b*) and (*c*) tend to be incompatible, but in practice it has been found that the internal insulation level (along the arc path) is normally lower during circuit breaking.

Insulation co-ordination 'to earth' has also special significance in dead tank circuit breakers during circuit breaking. Here restriking voltage transients produce stresses between the live terminals and the earthed tank, and gas issuing from the interrupters must be prevented from causing a weak insulating channel to the tank wall.

To this end, the tank is frequently lined with a barrier of insulating material. If metallic scantlings are used for anchoring the barrier material, care must be taken to ensure that these scantlings are not near the blast from the arc control unit, thus constituting a weak link from live parts to earth.

It is customary to prove a design by interrupting the full short-circuit current, and with the full voltage between the tank and live components. In e.h.v. circuit breakers this is done by synthetic testing techniques.

4.5.3 Effects of carbon and water

In dead tank circuit breakers, in both the closed and open positions, the insulation components are stressed continuously; consequently carbon is drawn on to insulators in highly stressed areas, and mainly where the specific inductive capacity of any insulation surface is greater than that of the oil (see chapter 12, Section 12.4.1.1). It is important to ensure that the surface gradients in these areas are sufficiently low to ensure reliability. Experience suggests that the carbon itself is not particularly destructive, but if, in addition, moisture is present, serious deterioration of the insulation will result. In e.h.v. constructions a design stress at the ends of the bushings of not more than $0 \cdot 15$ kV/mm at phase to neutral voltage is found beneficial, and is achieved by electrically shielding or by the special arrangement of the internal layers in bushings of the condenser type. Special care is required in the design of the insulation near the areas carrying the moving contact system, especially where a single insulator carries the 3-phase moving contacts since the voltage between contacts is line voltage. It is also important to ensure an adequate clearance between the moving contact and the bottom of the tank in the open position, since contamination can be drawn in the wake of the moving contact during the closing movement that may reduce the insulation level. These adverse effects have not the same importance in small oil volume equipments in those designs where the insulation oil, which is continuously stressed between live components and earth, is separate from that used for circuit breaking.

4.5.4 Constructional features
4.5.4.1 Materials
Insulating materials in oil circuit breakers include porcelain, bakelised paper, glass fibre, oil impregnated paper, resin impregnated wood, and epoxy resin.

In oil circuit breakers, porcelain is mainly used as a weather shield. It is rarely subject to high electric puncture (i.e. radial) stresses.

The latter remark also applies to glass fibre and impregnated wood. Their insulation strength along the lamina or fibres has to be at least as good as the interface between the insulator and oil or air. This would appear to be easy to achieve in any of these materials, but in fact good quality components made under controlled conditions of processing are essential to eliminate weaknesses which can arise from lack of bonding, gas occlusion, or moisture absorption. Components made for mechanical applications may be unsuitable.

On the other hand, bakelised paper and oil impregnated paper are subject to extremely high radial stresses. These materials are generally made under very stringent quality control processes. The importance of good quality components cannot be overstressed.

4.5.4.2 Bushings
Bushing constructions include simple bakelised paper bushings suitable for indoor use, or more complex constructions with weather housings which are normally oil filled or compound filled for outdoor use.

Most of the bushings used in modern oil circuit breakers are of the condenser type (Nagel, 1906) and consist of a series of insulated concentric metallic cylinders which have a progressively increasing length from the earthed end of the bushing to the live central rod. It is normal to arrange that the voltage between these cylinders is proportional to the distance between the ends of the cylinders. Thus, a fairly uniform electric stress will exist at the surface of the bushing. In bakelised paper, these metallic layers are inserted during the winding of the paper core. This equally applies to oil impregnated paper bushings, but in other constructions rigid metallic cylinders are separated by suitable spacers.

It is essential that the first and last layers of condenser bushings have adequate current carrying capacity to deal with the capacitance current arising from high de/dt transients, i.e. lightning surges and chopped waves.

Condenser type bushings are sometimes used as a means of obtaining a voltage reference by measuring the voltage across an additional section at the earthed end of the bushing (Fig. 4.17).

Generally, circuit breaker bushings should be as small in diameter as possible since the current transformers which are normally mounted around the bushings can be smaller and consequently less costly.

For bushings up to 66 kV relatively simple constructions can be used, but for voltages in excess of this a great deal of consideration must be given not only to the

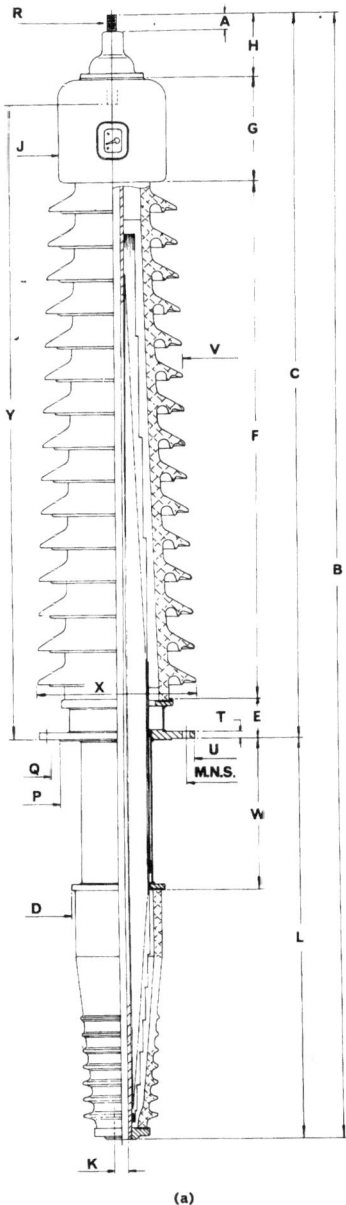

(a)

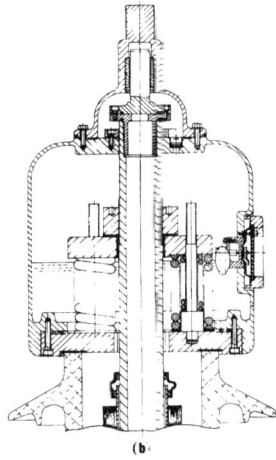

(b)

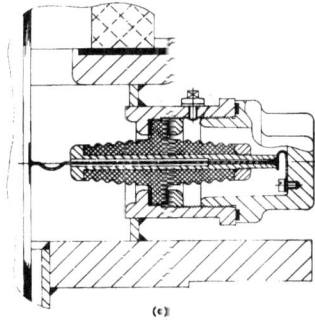

(c)

Fig. 4.17 Typical bushing construction and table
of dimensions [The Bushing Co. Ltd.,
England]

a   cross-section of bushing
b   arrangement of head
c   arrangement of tapping terminal
d   table of dimensions

	132-145 kV	220-145 kV	300-345 kV
A	64	64	75
B	3030	4400	6150
C	1850	2850	4100
D	250	355	445
E	155	155	250
F	1140	2040	3185
G	305	405	405
H	250	250	260
J	292	405	470
K	44·5	50	50
L	1180	1550	2050
M	8	12	12
N	29·2	29·2	29·2
P	314	435	535
Q	364	495	585
R	50	50	50
S	400	532	635
T	25	25	32
U	458	585	710
V	10	17	28
W	585	585	635
X	384	560	675
Y	1450	2490	3690

Typical dimensions (of Fig. 4.17a) in millimetres
V = number of sheds
M = number of holes

electrical stresses but to the severe mechanical stresses arising from short circuit operation, and also to the environmental conditions to which such bushings will be subjected.

A typical e.h.v. bushing is shown in Fig. 4.17. The construction is of special interest. It is an oil-impregnated paper bushing in which the whole assembly is clamped by the central conductor via the nest of springs. The nest of springs are precompressed as a unit before assembly in the bushing (to approximately 4000 kg). On assembly, it is arranged that the load derived from the springs tensions the main conductor, which in turn causes the porcelain insulators and the central chamber to be held in compression.

A feature of note is that in this type of design, jointing washers at the porcelain interface should be uniformly loaded if the same material is used at both ends of the porcelain insulators. Alternatively, different materials should be used at the smaller diameter ends of these insulators where a smaller area may exist in order to ensure the correct degree of compression.

A further point is that organic material, as is used for joints between porcelain and metallic components, should be effectively short circuited, or protected. This is to prevent the destruction of the joints by leakage currents over the insulator surface arising from pollution or condensation.

It should also be noted that if flashover occurs in service, short-circuit currents may flow through metallic joints that do not normally carry current. The design of metal components and fastenings must be such that failure of the bushing will not arise from a lack of current carrying capacity by these joints. In some cases, it may be desirable to bond an arcing ring to the main terminal.

4.5.4.3 Stand-off insulation

The insulation of live tank circuit breakers presents an interesting case involving advanced electrode techniques. In these designs, it is possible to structure the equipment to enhance the flashover level.

The choice of stand-off insulation involves length and diameter, and the disposition of the internal metal work. From Fig. 4.18 it will be seen that good proportioning has an important influence on the insulation level (Kappelar, 1945).

It is also worthy of note that where equipment is mounted in a manner which causes the earthed plane to be brought near to the live components, the flashover level is reduced.

4.6 Construction of circuit breakers

4.6.1 General

Brief descriptions of modern equipment are included in this section. For this purpose it is convenient to divide the classes of equipment as follows:

(*a*) open type (dead) tank
(*b*) open type (live) tank
(*c*) metalclad dead tank and live tank
(*d*) insulation enclosed switchgear

Also, comment is made on other features that are common in principle to all oil circuit breakers, namely, means for slow opening and slow closing for maintenance purposes, and secondary control features; e.g. auxiliary circuitry, interlocking, latch checking switches, and housing of current transformers.

4.6.2 Open-type dead tank circuit breakers
4.6.2.1 General

These equipments are made for voltages from 11 to 220 kV on a wide scale by many manufacturers at the present time, and at least two manufacturers have supplied equipment for up to 330 kV.

Table 4.10 gives information concerning the ratings of available equipment and other salient features.

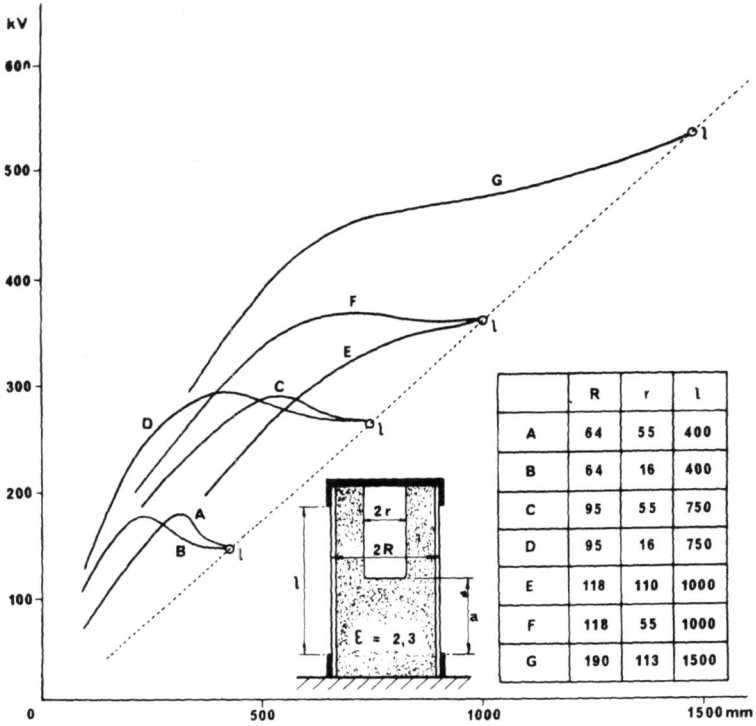

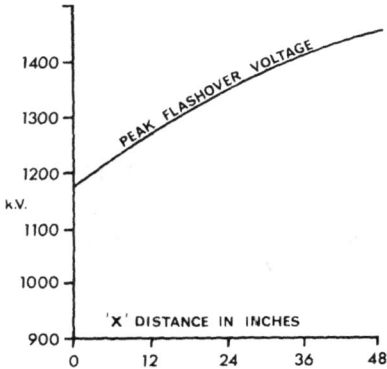

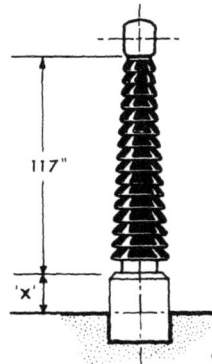

Fig. 4.18 Effect of electrode configuration flashover

    a oil filled supporting posts
    b flashover as a function of the height of the mounting plinth

Table 4.10   Equipment ratings

Voltage	Breaking capacity GVA	Normal current A	Number of interrupters per phase	Approximate weight without oil kg
11 kV	0·5	up to 2500	1 or 2	450/700
33/34·5 kV	1·5	up to 2500	1 or 2	900-220
69 kV	3·5	up to 2000	2	2800-3200
132/161 kV	15	up to 4000	2 or 4*	12000-15000
220 kV	20	up to 3000	4*	18000-20000
275/330/ 345 kV	15	up to 2000	+ to 8*	45000

*circuit breakers with 3-cycle total break time

4.6.2.2 69 kV dead-tank circuit breakers

Fig. 4.19 illustrates a typical 69 kV circuit breaker of 2500 MVA breaking capacity, which includes the three phases in one tank. The essential elements are the top plate with its six entry bushings, the fixed and movable contact system, and mechanism. The whole circuit breaker is mounted on a frame and it is normal to lower the tank to ground level to give access to the contacts and current transformers.

The moving contact of each phase of the circuit breaker is assembled on a lift rod of insulating material. The metal frame and insulation lift-rod guiding system together with overtravel stops, confine the vertical movement of the contacts. Arc-control units having transverse outlets are fitted.

Mechanisms for this type of equipment range from solenoid to hydraulic types (Section 4.4).

Current transformers are housed over the bushings. The fixings of the current transformers are mounted in such a manner that they do not form a complete electrical loop around the transformer cores. The leads from the current transformers are brought out of the top plate through compressible gas-tight seals.

It should be noted, however, that in this type of equipment where the terminations of current transformers are brought to open terminals on the inside of the top plate, they must be covered in such a manner as to ensure that a secondary explosion of gases from the circuit breaker operation does not accidently occur. When installed, it is essential to ensure that the top plate of the circuit breaker is connected to earth through copper work of appropriate section to carry the short-circuit current and also to maintain the resistance (and consequently the voltage drop) of the copper work at a level which ensures the safety of operating personnel.

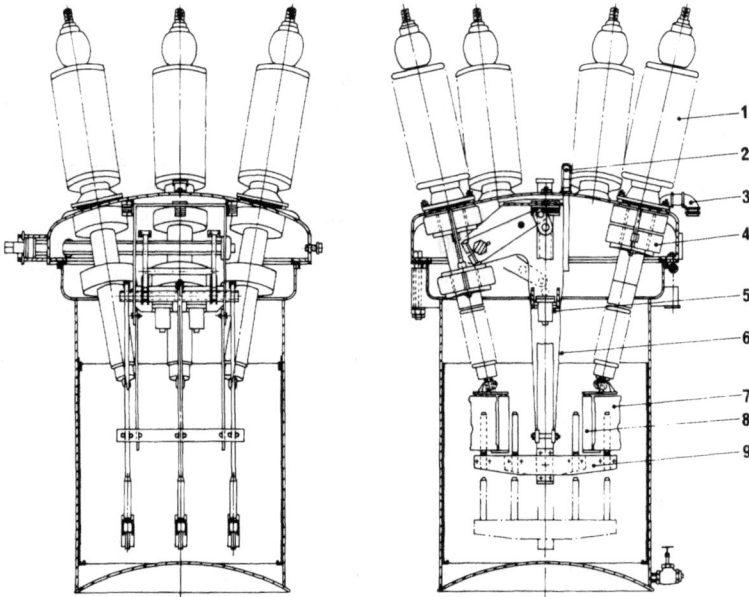

Fig. 4.19 Dead tank c.b., 3-phases in one tank [Allis Chalmers Ltd., USA]

1 bushing	6 plunger guide
2 oil level indicator	7 arc control device
3 vent	8 resistor
4 current transformer	9 plunger bar
5 dashpot	

4.6.2.3 161/220 kV dead-tank circuit breakers

This type of circuit breaker is similar in principle except that it is normal practice to house the live components of each phase in a separate tank, and also to mount the tanks on the ground. Access to the contacts and other components is gained through a manhole after emptying the oil from the circuit breaker and after precautions have been taken to ensure that the tank is free of oil vapour. Fig. 4.20 shows a typical construction.

The tank and top plate are of integral construction and the bushings which are of the paper oil condenser type are complete with porcelain housing at each end. In order to obtain a heavy interrupting duty – 220 kV 20 GVA – four arc-control devices of the transverse vented type are used in series.

During 'fault operation' with this form of circuit breaker, considerable shock is transmitted from the source of the gas bubble to the ground foundations, and in

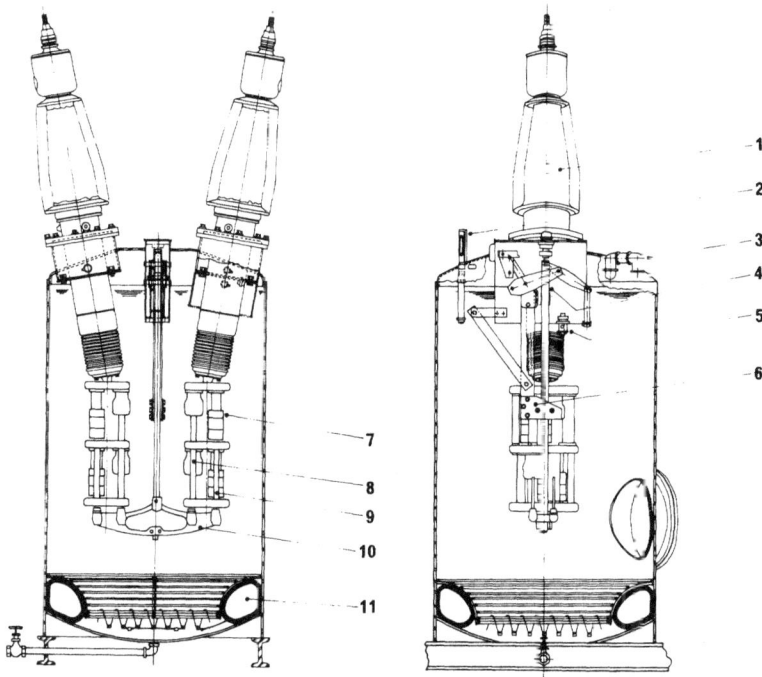

Fig. 4.20 Dead tank circuit breaker 161/220 kV
Single phase in separate tank [Allis Chalmers Ltd., USA]

1 bushing	7 arc control unit
2 oil level indicator	8 parallel contact
3 vent	9 resistor
4 linear linkage	10 plunger bar
5 dashpot	11 impulse cushion
6 guide block	

consequence the foundations must be robust and heavy and the anchorage of the circuit breaker very firm. If due cognisance of this were not taken, it is possible that damage to the bushings would arise from the sudden movement of the whole circuit breaker during the interruption of heavy short circuits.

To minimise this shock some designs have an 'impulse cushion' placed at the bottom of the circuit breaker tank (Fig. 4.20). The cushion is filled with nitrogen to approximately half an atmosphere.

Mechanisms for this type of equipment are normally confined to pneumatic or pneumo-hydraulic closing and spring opening.

4.6.2.4 300 – 380 kV dead-tank circuit breakers

Designs of higher voltage dead-tank equipment are available for use in 380 kV systems and are actually used in 300 and 330 kV systems. A typical circuit breaker is illustrated in Fig. 4.21a and b, and is rated from 7·5 to 15 GVA. Each phase is housed in a lenticular shaped tank and the entry bushings are of the barrier type. Four or six breaks are used, depending on the short circuit rating required, each of which is shunted by a wire wound resistor.

In the four break arrangement, (Fig. 4.21b) the outer moving contacts are mounted on a normal plunger bar and the inner moving contacts which are spring opened are part of the interrupter assembly. These inner moving contacts were pushed into the closed position by striker rods of insulation that are carried on the plunger bar, thus all main breaks are opened and closed simultaneously. The current flowing through the resistor after the main arc is extinguished, is interrupted on the two outer breaks.

The circuit breaker is closed pneumatically, and the opening is effected by springs of the torsion-bar type. These circuit breakers are suitable for high-speed reclosing operations. If single phase autoreclosing is required, separate mechanisms are fitted to each phase of the circuit breaker.

Fig. 4.21 Dead tank circuit breaker 300/330 kV [General Electric Co., England]

    a 3-phase circuit breaker in situ
    b section through arc control unit

Table 4.11   Construction ratings

Voltage kV	Breaking capacity GVA	Normal current A	Number of interrupters per phase	Approximate weight (kg) without oil and CTs
33/34·5	1·5	2500	1	600
69	2·5	2500	1	1000
132	5·0	2500	1 or 2	2000
200	10·0	2500	4 or 6	4000-5000
330	15·0	2500	6 or 8	5000-6500
525	40·0	2500	8 or 10	7000-10000
765	55·0	2500	10	11000

4.6.3 Open-type live-tank equipment

Circuit breakers of this form include single break and multibreak equipment. Table 4.11 shows the range of ratings that can be covered by these constructions.

Single break equipment are generally available for ratings of up to 170 kV 5000 MVA, and are extremely widely used. They are not normally restrike free. Where a restrike free performance is required — for switching capacitor banks or long cables — multi-break constructions are used, or special features are included such as oil injection.

For voltages above 170 kV it is normal to use multibreak constructions.

4.6.3.1 Single-break type

These are simple, inexpensive, of low weight, and easy to maintain. Thus, they are suitable for shipment to, and for use in, remote areas.

Each phase consists of a circuit-breaking unit, a support column and crank case (Fig. 4.22). The phases together with the operating mechanism are mounted on a frame. The mechanism normally takes the form of a motor-wound spring-closing unit, that charges the opening spring during the closing movement via a trip-free linkage. Other forms of mechanisms include pneumatic and hydraulic closing.

Closing mechanisms of the spring type can be charged by hand, and consequently circuit breakers utilising these mechanisms can be put into commission without the need for a local electrical supply. The circuit-breaking oil is normally confined to the upper circuit-breaking chamber, and because of this maintenance is only necessary if the circuit breaker is to remain open for a considerable period, i.e. when carbon line up could take place.

Access to the contacts and the arc-control device, is via the head of the circuit breaker. Where necessary a small lifting device can be used for removing the arc control unit. Special tools are normally provided for removing the movable contact tip.

Fig. 4.22 Single break live tank circuit breaker [A. Reyrolle & Co. Ltd., England]

    a 3-phase circuit breaker
    b cross-section through a single phase

1 vent valve	6 separating piston
2 terminal pad	7 terminal pad
3 oil level indicator	8 upper drain valve
4 moving contact	9 lower drain valve
5 lower fixed contact	

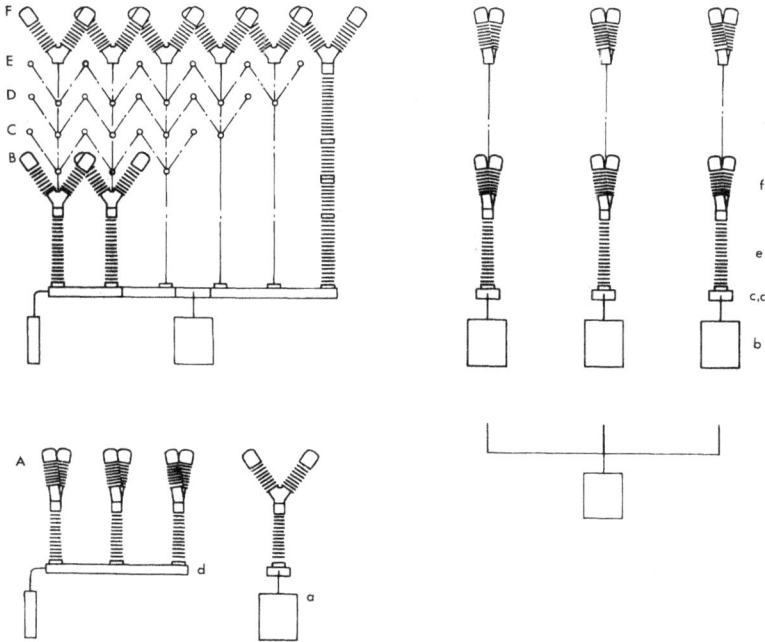

Fig. 4.23A Multibreak live tank circuit breakers [Brown **Boveri** & Co. Ltd., Switzerland]

A	72·5—123 kV	a	3-phase control unit
B	145—245 kV	b	single-phase control unit
C	245—362 kV	c	two column pole base
D	300—420 kV	d	three column pole base
E	420—525 kV	e	supporting columns
F	525—765 kV	f	breaking units

4.6.3.2 Multibreak type

The basic concept of a modular range of e.h.v. circuit breakers, that has been brought to practical fruition, is shown in Fig. 4.23. Here the entire voltage range of equipment from 72·5 kV to 765 kV is built up from a small number of construction units consisting of the following:

(*a*) control cabinet
(*b*) support columns
(*c*) interrupting chambers

Auxiliary interrupting chambers incorporating resistors similar to that illustrated in

Fig. 4.23B Experimental pole of a 12 break circuit breaker rated at 1050 kV [Brown Boveri Co. Ltd., Switzerland]

Fig. 4.4 are independent construction units (not always supplied). They can be added in parallel with the main interrupting chambers and they operate in timed sequence after the main contacts open and before the main contacts close.

Fig. 4.23*b* shows an experimental pole of a 1050 kV circuit breaker.

The mechanisms used are similar to that described in Section 4.4.6.2 where a hydraulic power source is connected to the actuators of each moving contact and to the hydraulic servo system. The servo system, and, consequently, the moving contacts respond to the operation of the appropriate electromechanical actuators through the trip or close command valves. Multibreak live tank circuit breakers are in service on systems up to 750 kV.

4.6.4 Metalclad circuit breakers

4.6.4.1 General

It has been the practice to differentiate between metalclad and metal encased switchgear and these have recently been defined by the IEC.

The only significance of this in the present context is that, in the UK, it has led to the use of dead tank circuit breakers in such equipment, while, in other European countries, live tank equipment has been developed.

Generally such circuit breakers are removable and as such constitute an isolating feature.

Equipment of this form is commonly used in systems from 415 V to 33 kV although this principle has been applied in 110 kV systems.

A significant feature of recent developments in this field is a high degree of compactness that has been achieved together with a considerable reduction in overall weight.

Two equipments are briefly described here, one of which is designed to the British Electricity Boards Specification S2, and the other is typical of that used in European countries other than the UK.

4.6.4.2 Metalclad dead-tank type

The significant features of this equipment are: that the bushings of each phase are mounted vertically; it is usual to mount the circuit breaker on a carriage which includes lifting means, and current transformer accommodation is not normally provided around the circuit breaker bushing, but are housed in the fixed portion of the board (Fig. 4.24).

Compactness is assisted by using bushings that are as short as possible externally, and which dielectrically match the orifice design on the fixed portion of the equipment so that the isolating distance (i.e. the vertical movement of the circuit breaker) is reduced to a minimum.

Most modern circuit breakers use condenser bushings that ensure a high impulse level with a comparatively short bushing.

It is also preferable that the interruption of the capacitive current of the bushings during the isolating movement of the circuit breaker is achieved in a short distance, e.g. 1 cm in an 11 kV equipment and 3 cm in 33 kV equipment. Thus the insulation level is not substantially changed during the isolating operation.

The circuit breakers are normally supplied with solenoid mechanisms, motor-wound spring mechanisms, or mechanisms utilising spring assistance so that the closing speed is independent of the operator.

It is also usual to provide a venting channel through which any gases arising from circuit breaking are ducted to the rear of the switchboard or even outside the building. This avoids possible accumulations of gases in the orifice insulators of the enclosure, which could be ignited during the isolating movement.

In this form of circuit breaker, means are incorporated to ensure that it can not be isolated until it is open, nor placed in a service position if the tank has been removed.

In other respects, the circuit breaker is similar in principle to any dead tank oil circuit breaker.

4.6.4.3 Metalclad live tank type

A typical live tank circuit breaker housed in a metal enclosure is shown in Fig. 4.25*b*. The circuit breaker is mounted on a carriage that can be horizontally withdrawn to an isolated position within the metal housing. In modern circuit breakers, only the circuit-breaking chamber includes oil (Fig. 4.25*b*); thus there is no carbonised oil path to earth metal that can be under continuous electrical stresses. The circuit breaker itself is normally closed and no stress exists across the circuit breaking chamber in its service position except for brief periods when the circuit breaker may be open.

The design of the isolating connector shown, effectively eliminates the need for a bushing type insulator on the circuit breaker. It may be of interest, however, that with this type of equipment current transformers, voltage transformers and other secondary equipment are often mounted on the truck. This has the advantage that more equipment associated with the automatic operation of the circuit breaker can be inspected without shutting down the whole of the switchboard.

4.6.5 Insulation clad circuit breakers

A new approach to producing compact circuit breaking switchboards for use indoors or in enclosures is being developed in Holland. It is based on the principle that the equipment can be safely touched although it is not metal encased. It consists of a live tank equipment, the whole of which is encased in insulation and an additional outer enclosure of insulating material that prevents personnel from touching the primary equipment; it also ensures that there is a minimum air clearance between primary equipment and personnel.

At the moment this type of equipment, does not meet British codes of practice. Nevertheless, it is an interesting development that cannot be said to involve greater risk than more orthodox equipment. It is likely to find growing application in those areas where specifications can be modified to accommodate new systems.

4.6.6 Features at the interface of application
4.6.6.1 General

Before concluding this chapter, some comment seems appropriate concerning secondary features that occur at the interface of application, and which affect the circuit breaker design.

4.6.6.2 Secondary control features

In the past, it has become the practice to festoon circuit breakers with a large variety of secondary control features, with the consequent increase in delivery time

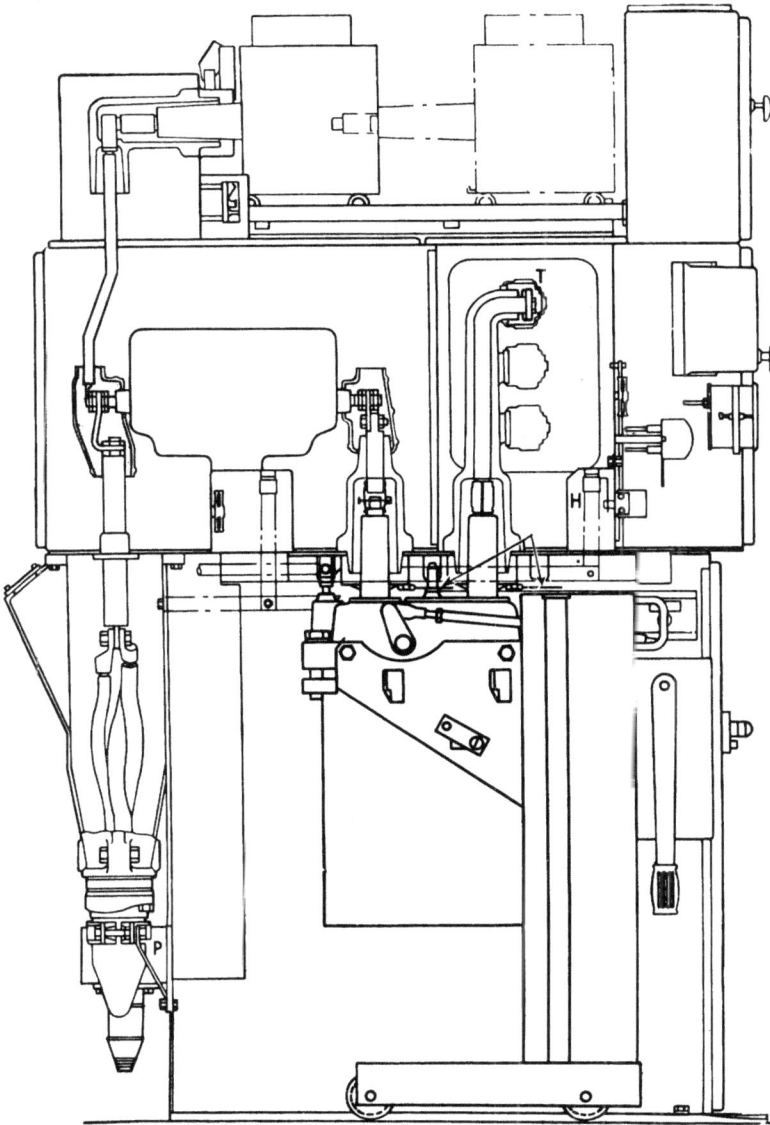

a complete panel

Fig. 4.24 Metalclad dead tank circuit breaker [A. Reyrolle & Cc. Ltd.]

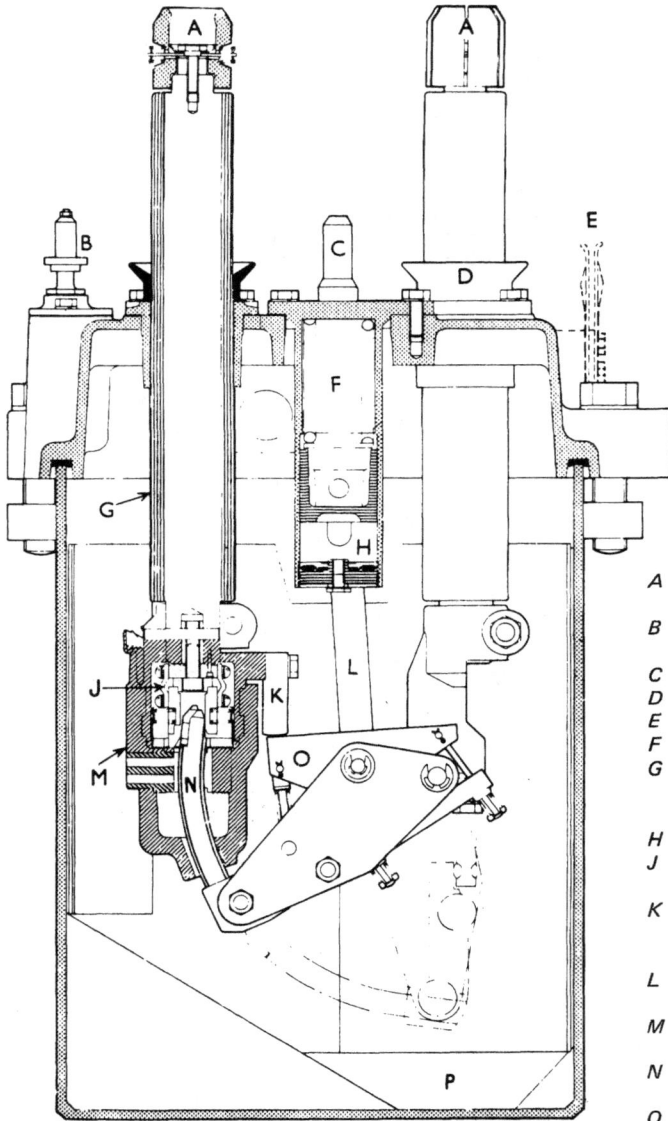

b cross-section of circuit breaker

A self-aligning socket contact
B combined guide pin and gas vent
C guide pin
D orifice sealing washer
E top-plate earthing
F kick off spring
G condenser type bakelised paper insulation
H dashpot mechanism
J self aligning fixed contacts
K external fixed load-carrying butt contacts
L moulded glass fibre operating rod
M moulded-nylon tubulator
N curved moving contact
O external moving load carrying contact
P interphase barrier

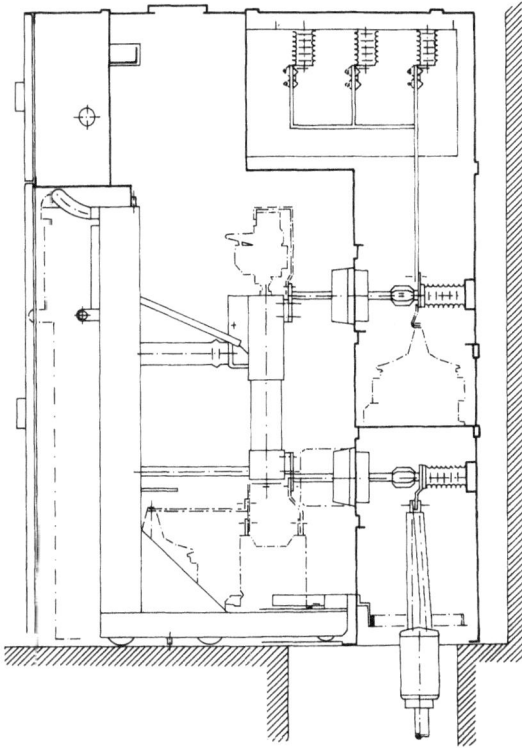

Fig. 4.25 Metalclad live tank circuit breaker [Law Knudsen Ltd., Denmark]

a complete cubicle

and cost of the average equipment. A great deal of effort in recent years has been put into the determination of standard secondary control and protection circuitry. In England, this is covered by British Electricity Board Engineering Recommendations S15. As far as design of the hardware is concerned endeavours have been made to exclude from the basic circuit breaker items that are not absolutely common for all applications and to group geographically, as distinct additions, those parts necessary for a particular application.

4.6.6.3 Foundations
Foundations have a bearing on the design of circuit breakers. Where the components of a circuit breaker are not effectively tied by a rigid structure (e.g. a steel bed plate) interphase couplings and couplings between components on one phase must be made flexible, to allow for possible variation in alignment with the passage of time.

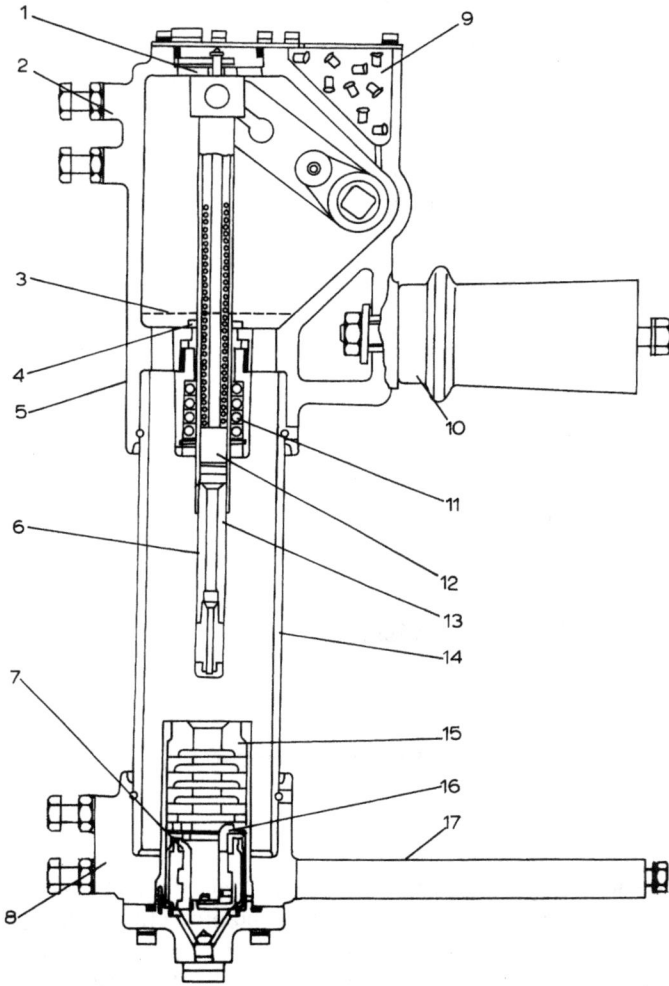

b circuit-breaker assembly

1 upper stop
2 upper terminal
3 oil level
4 inner stop
5 crank case
6 moving contact
7 main fixed contact
8 lower terminal
9 oil separator

10 epoxy resin insulator
11 sleeve contact
12 oil injection piston
13 contact pin
14 insulation tube
15 arc extinguishing chamber
16 arc contact
17 insulator

Frames, foundations and fixing bolts for bulk-oil circuit-breaking equipment are designed on the basis that the effective load of the equipment is at least equivalent to a live load, i.e. the downward thrust is at least double the weight of the equipment. Short circuit tests made on a number of circuit breakers, each of which was mounted on beams, have shown that the actual deflection of the beams was two or three times that produced by the static load alone.

4.6.6.4 Low temperatures
Heaters are frequently mounted in mechanism housings to prevent condensation or excessive viscosity of oils. These are generally in the order of 100 to 200 W.

Larger heaters may also be placed inside, or bolted to, appropriate parts of the circuit breaker to ensure that circuit-breaking oil does not become too viscous and thus interfere with the speed of operation. This applied to places such as Canada where operation in temperatures down to 230 K is sometimes required. Under such conditions heaters of several kilowatt capacity are used. In some cases, it may be desirable to use an exceptionally low viscosity oil, 'Mynas 65' is suitable.

4.6.6.5 Earthquake conditions
The ability of substation equipment to withstand seismic effects is dependent on the strength, elasticity and damping of its elementary parts. The increasing use of e.h.v. circuit breakers of the live tank type is focusing attention to this because of the slenderness of the insulating columns in relation to the mass they support in this type of circuit breaker.

It would seem desirable that the power connections to the circuit breaker should not add to the problem (associated with solutions) and that such connections should be supported separately and flexibly coupled to the circuit breaker terminals (Novoa, 1970).

4.6.6.6 Slow opening and closing
For maintenance purposes (the setting of contacts) slow opening and slow closing by means of hydraulic rams is sometimes used.

These operations should never occur with a circuit breaker alive. It is definitely dangerous to do this even though an open isolator is preventing the establishment of any 'through circuit'. It is even more dangerous to do this with circuit breakers fitted with resistors, because the resistors would carry charging current during the operation and may be overheated.

4.7 E.H.V. oil circuit breakers of the future

The trend in new switchgear design is toward small space factor equipment. This derives from the need to use much higher voltages near to load centres — cities — where space is usually at a premium.

(A) METALCLAD ARRANGEMENT (FEEDER PANEL)

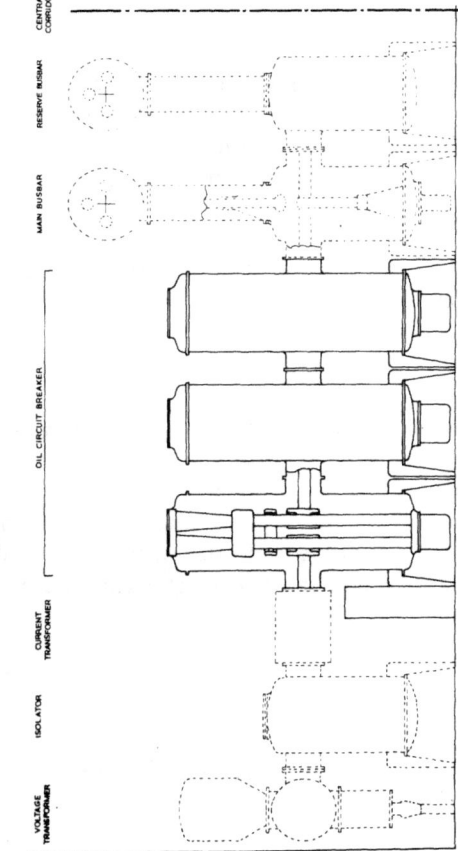

(B) OPEN TERMINAL ARRANGEMENT

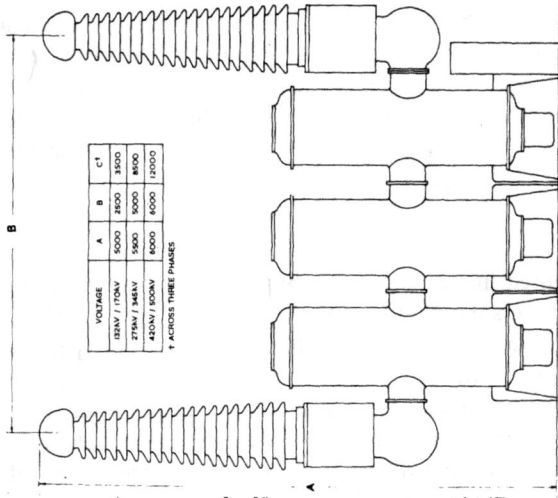

(C) SCHEMATIC ARRANGEMENT

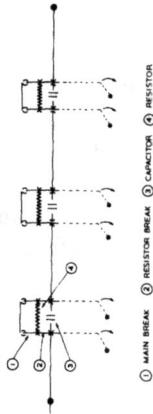

VOLTAGE	A	B	C†
132kV / 170kV	5000	2500	3500
275kV / 345kV	7500	5000	8500
420kV / 500kV	8000	6000	12000

† ACROSS THREE PHASES

① MAIN BREAK ② RESISTOR BREAK ③ CAPACITOR ④ RESISTOR

Fig. 4.26 e.h.v. oil circuit breakers for the future [A. Reyrolle & Co. Ltd., England]

For very high voltages, metalcladding is the obvious solution, and it would seem desirable that basic development for this purpose should be suitable for normal application. With this objective in mind Fig. 4.26 shows a concept of an oil-circuit-breaker range suitable for metalcladding or open terminal arrangements both for indoor or outdoor applications. The form illustrated is for a 500 kV system. It is a live tank circuit breaker enclosed within a dead-tank SF_6 gas filled housing. Similar equipment has been used in 33 kV and 110 kV systems. Buildings and foundations for housing such equipment should be relatively inexpensive, as all equipment is ground mounted.

The construction is semimodular, and is based on a 2-break dead-tank module having a rating of 150 kV, 2000 normal current-carrying capacity and 40 kA symmetrical breaking capacity. Low ohmic resistors are incorporated to ensure uniform voltage sharing during fault interruption, and ceramic capacitors ensure adequate voltage distribution when the circuit breaker is open.

By utilising the presence of gas at three bars together with low ohmic resistors, a consistent circuit-breaking performance should be ensured. A 2-cycle total break time can be confidently predicted, as well as a restrike-free capacitor-switching performance. Noise level can be expected to be low.

It is possible that there are much better ways of achieving the need (mentioned above), or that other means are preferred. Time alone will determine this.

4.8 Acknowledgments

The Author wishes to thank the Directors of A. Reyrolle & Co. Ltd. for permission to undertake the work, and, in addition, to the references quoted, acknowledges the help and the information freely given by the following organisations: Allis Chalmers, USA; Brown Boveri Ltd, Switzerland; GE, USA; GEC, England; Hazemeyer, Holland; McGraw-Edison, USA; Laur Knudsen, Denmark; ITE Co, USA and Westinghouse, USA.

4.9 References

AMER, D. F., and YOUNG, W. B. (1958): 'Researches into the Performance of High-Voltage high-breaking capacity oil Circuit-Breaking Devices with lateral venting', CIGRÉ Paper 123

BASU, S., and SRIVASTARA, K. D. (August 1969): 'Electromagnetic forces of a metal disc on an alternating magnetic field, *IEEE Trans.*, **PAS-88**, pp. 1281-1285

BASU, S., and SRIVASTARA, K. D. (May/June 1972): 'Analysis of a fast-acting circuit breaker mechanism', Pt. 1, Electrical Aspects', *ibid.*, **PAS-91**, pp. 1197-1203

BRATKOWSKI, W. V., FISCHER, W. H., and RADUS, R. J. (1966): 'High speed trip mechanisms for e.h.v. power circuit breakers', *Westinghouse Eng.*, pp. 55-57

BRUCE, G. E. R. and WHITNEY, W. B. (1933) (1951): BEAIRA Report G/XT35

CLOTHIER, W. H. (1933): 'Switchgear Stages', Contribution by Mr W. Wilson, pp. 341-342

COPPER DEVELOPMENT ASSOCIATION (1936): 'Copper for busbars', chap. 3, pp. 48-51

HOLM, R. (1967): 'Electric Contacts', p. 2

IEC specification 56-3 (1959): chap. 2, pt. 2, p. 17

KAPPELAR, H. (1945): 'brush discharges with advanced electrodes', *Micafil News*, p. 10

LAWSON, M. T. W. (1972): 'Calculations of gas pressures in oil-circuit-breaker-arc-control pots', *Proc. IEE*, 1972, **119**, pp. 74-75

NAGEL (1906): 'Electrische Kraftbetriebe und Bahnen'

PERRY, E. R., and MORELLI, N. W. (1954): 'A new hydraulic mechanism for power circuit breakers', *IEEE Trans.*, **PAS-73**, pp. 328-333

ROTERS, H. C. (1970): 'Electro magnetic devices', pp. 399-413

VOGELSANGER, E. (1946): 'Researches on arc quenching in low oil content circuit breakers', CIGRÉ Paper 121

WHITNEY and WEDMORE (1930): British Patent 366998

WIDMER, H. (February 1968): 'Hydraulic operating mechanism of the h.v. circuit breaker line F., *Bull. Oerlikon*, p. 7

ZAJIC, V. (1957): 'High-voltage circuit breakers', pp. 158-160

Chapter 5

Air-break circuit breakers

J. S. Morton, B.Sc., Sen.Mem.I.E.E.E., C.Eng., F.I.E.E.

5.1 Air-break circuit breaking

5.1.1 Principles of arc extinction

A circuit is interrupted in an air-break circuit breaker by parting a set of circuit-breaking contacts, and then controlling the resulting arc, so that it is established inside an arc chute. Here the resistance of the arc is increased, and, in turn, the current is reduced so that the arc cannot be maintained by the circuit voltage, and the circuit is cleared. The resistance of the arc, and hence the arc voltage, can be made to increase in various ways dependent on

(a) increasing the length of the arc
(b) cooling the arc
(c) splitting the arc into a number of series arcs.

Most arc chutes use at least two of these principles and apply them in various ways.
There are basically two types of arc chute, as follows:

(a) the insulated plate type that relies on stretching and cooling by means of refractory plates of special shape
(b) the bare-metal-plate type or cold cathode that splits the arc into a number of series arcs, and also cools by conduction into the plates.

In all arc chutes, the controlling forces that direct the arc into the arc chute are provided by the natural electromagnetic and thermal forces of the arc, supplemented by a strong magnetic field that is generally provided by (a) an external iron circuit around the arc, usually energised by one or more turns of a coil carrying the current being interrupted, or (b) an internal iron circuit in the form of special shaped steel plates that are energised by the arc itself.

5.1.2 Power-balance theory of arcing

When a steady arc carries a constant current, the power loss from the arc column is balanced by the power input to the arc, because the arc is in a state of thermal equilibrium. If the current decreases very rapidly, the power loss from the arc column also decreases, but, because the arc has a thermal capacity, its power loss will always lag that for the steady-state current. Thus, at any instant, the power input to the arc plus the power stored in the arc trying to restore thermal equilibrium is equal to the power loss from the arc (Young, 1953).

5.1.3 D.C. circuit breaking

When interrupting a d.c. arc, the increasing resistance of the arc automatically increases the impedance of the circuit, and hence reduces the current being interrupted. With the resistance of the arc increasing, the voltage drop across the arc also increases. This process continues until the arc voltage has become greater than the applied voltage of the system, and then the arc must extinguish, as it can no longer be maintained by the system voltage. At the instant immediately before arc extinction and at the nominal current zero at which arc extinction occurs, the arc power loss always exceeds the power input to the residual arc path, because the latter decreases as the arc voltage falls to the system voltage. When the current reaches its nominal current zero, the residual arc path still has a low resistance, which then starts to increase very rapidly after arc extinction. The nominal current zero has, in fact, a finite value, which is too small to be measured on a normal oscillogram. This finite value of current then decreases to zero, and is inversely proportional to the increasing residual arc resistance (Fig. 5.1A).

5.1.4 A.C. circuit breaking

When interrupting an a.c. arc, the interruption is similar to that of a d.c. arc, but is greatly assisted because the current passes through a zero automatically. If, at the

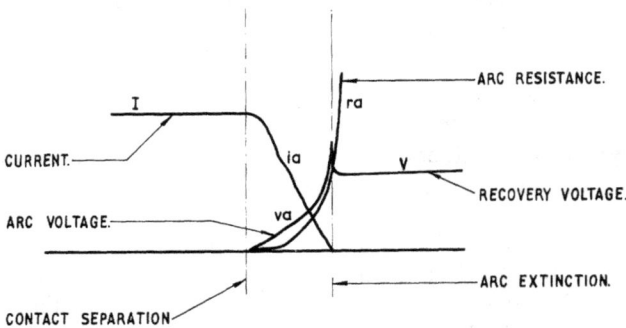

Fig. 5.1A d.c. arc extinction

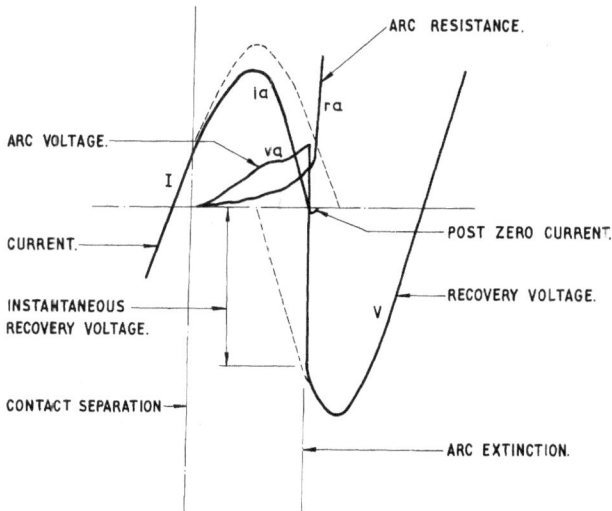

Fig. 5.1B a.c. arc extinction

instant of a current zero and during the short time following (less than 1 ms), the arc power loss always exceeds the input power to the residual arc path, arc extinction is achieved, even though some postzero current has flowed (Fig. 5.1B).

Unlike d.c. circuit breaking, with a.c. circuit breaking the input power to the residual arc path at the instant of current zero is zero, because the restriking voltage is zero. But, after current zero with the rapidly increasing restriking voltage rising to the instantaneous recovery voltage, the input power increases, and, if it can exceed the arc power loss from the residual arc path, a reignition occurs and arcing continues for another loop.

Because there is always a post-zero current flow and a fairly low resistance exists across the circuit breaker at the instant of arc extinction, high-frequency restriking voltage transients are usually damped and no excessive overvoltages are produced. An air-break circuit breaker has an inherent 'resistance – switching' arc-extinction mechanism, and, in general, does not produce switching overvoltages.

The increase in arc resistance alters the circuit phase angle, making it more resistive, and, because of the current-zero phenomenon, the arc voltage does not have to rise to that of the system instantaneous recovery voltage. When interrupting symmetrical short-circuit currents, the instantaneous recovery voltage as a ratio of the peak recovery voltage is dependent on how the circuit phase angle has been modified by the insertion of the arc resistance into the circuit. This ratio is at its maximum near the arc-chutes limit of performance, because the effect on the circuit of the arc resistance is a minimum.

5.2 Arc control

5.2.1 Contacts and arc initiation

Nearly all air-break circuit breakers are of the single-arc-chute/single-break type, fitted with a hinged arm for its moving contacts.

The contact arrangement illustrated in Fig. 5.2A is typical of that used in high-voltage air-break circuit breakers, comprising a set of main current-carrying contacts, a set of intermediate contacts and a set of arc-initiating contacts. At the upper end of the hinged arm are the main arcing contacts, so arranged that, during a break operation, the main contacts separate first and the interrupting current is diverted to the intermediate contacts and then to the arcing contacts that separate last. The intermediate contacts are situated just below the arc chute, and, when they separate, the interrupting current is diverted to the arcing contact at the upper end of the arm inside the arc chute within the influence of its magnetic field. These contacts are carefully positioned so that, during a break operation, the impedances of the transfer current paths between the sets of contacts are kept to a minimum and the time intervals between the parting of the sets of contacts are kept to an optimum minimum consistent with arc-free transfer. The arc is then initiated inside the arc chute within the influence of the magnetic field of the iron circuit, and the

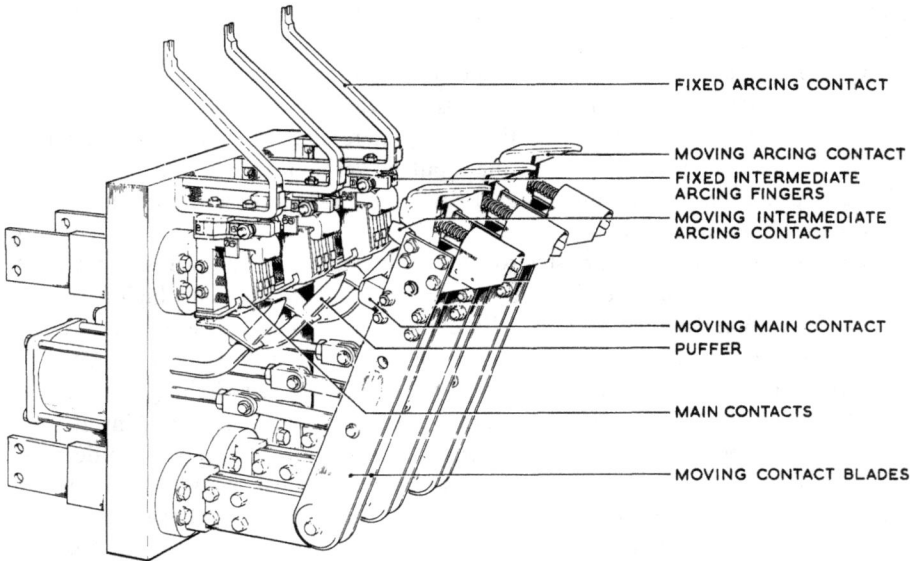

Fig. 5.2A Typical butt main contacts [Whipp and Bourne Ltd.]

ionisation due to transfer arcing below the arc chute is minimised. If this is not done, it can seriously limit the interrupting ability of the circuit breaker.

The contacts shown in Fig. 5.2A have butt-type main contacts arranged to 'blow-on' to each other, maintaining contact pressure until they separate. The intermediate contacts are the finger-and-blade type, and, because they must be capable of carrying the full short-circuit current during interruption, they are tipped with sintered copper—tungsten to limit the contact erosion and maintain the contact separation sequence. The arcing contact is a butt type arranged to 'blow off' and the tight loop of this arrangement inherently directs the arc up into the arc chute. The arcing contact is also tipped with sintered copper—tungsten.

There are many variations and combinations of the types of contacts used with air-break circuit breakers, and Figs. 5.2B and C illustrate another arrangement.

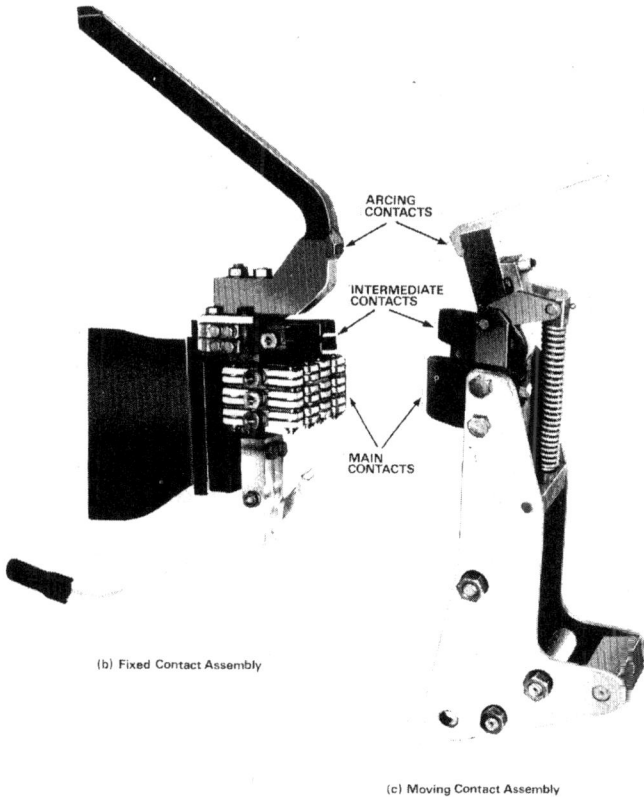

(b) Fixed Contact Assembly

(c) Moving Contact Assembly

Fig. 5.2B and C Typical finger and blade main contacts [A. Reyrolle and Company]

The main difference between this and the previously described type is that multifinger and blade main contacts are used. Care must be taken to set the angle of the blade so that, on an opening operation, the bottom fingers part first and the top fingers part last. This helps the transfer of the current from the main to the arcing contacts and minimises transfer arcing. With lower current-rated circuit breakers, the intermediate contact is often left out because the top main contact is placed just below the arcing contact.

The more common types of contacts used in air-break circuit breakers are as follows:

Main contacts: Butt, finger and blade, and bridging type.
Intermediate contacts: Butt, finger and blade, but sometimes they are not fitted at all. They greatly assist in current transfer, particularly with heavy normal current-rated contacts that cannot be situated close to the arcing contacts.
Arcing contacts: Butt and finger and blade.

5.2.2 Arc chutes

5.2.2.1 Bare-metal-plate type or cold cathode

A section through a 6·6 kV arc chute is shown in Fig. 5.3, in which the relationship between the contacts and the arc chute can be seen. The arc chute is of the cold-cathode type, comprising a number of bare-metal plates arranged at right-angles to the length of the arc chute, with spacers between the plates to allow the arc to be split up into a number of series arcs. The anode and cathode root drop of an arc on steel plates when carrying high short-circuit currents is approximately 30 V and, for N plates, the mean arc voltage in the arc chute if the arc splits completely is approximately $30 N$. This action is shown schematically in Fig. 5.4A.

The control of the arc from the point of initiation to the position of final extinction is by means of the runner system, which is designed to move the arc rapidly along the runners, and always in the correct direction. The two main runners are at their closest points at the bottom of the arc chute within the iron circuit and diverge at an angle up into the arc chute, until they become vertical and extend beyond the notch in the arc-chute plate.

At arc initiation, one arc root is on one of the main runners and the other root has to be transferred across the arc chute to the other main runner by means of the tail runner fitted to the moving arcing contact.

The arc-chute plates shown in Fig. 5.4B are shaped from a rectangular plate of steel with a notch extending from the bottom of the plate into the arc chute. The two legs of the plate extend down from the notch and are just wide enough apart to allow room for the arcing contacts and runners. To prevent the arc rooting to the legs of the plates before the arc reaches the top of the notch, an arc-resisting insulated plate is fitted across the length of the arc chute shielding the legs of the steel plates, forming a tunnel for the arcing contacts and runners. The shape of the steel plate acts as a sucker loop on the arc, drawing it up into the arc chute towards

VENT CHAMBER

HINGED FRONT
TOP COVER

INTER-PHASE
BARRIER

COLD CATHODE
ARC CHUTE

ARCING
CONTACTS

ARC CHUTE HINGE

MAIN ISOLATING
CONTACTS

INTERMEDIATE
CONTACTS

MAIN CONTACTS

MOVING
CONTACT
ASSEMBLY
KICK-OFF
SPRINGS

AIR PUFFER NOZZLE

OPERATING
MECHANISM

OPERATING SOLENOID

AUXILIARY
ISOLATING
CONTACTS

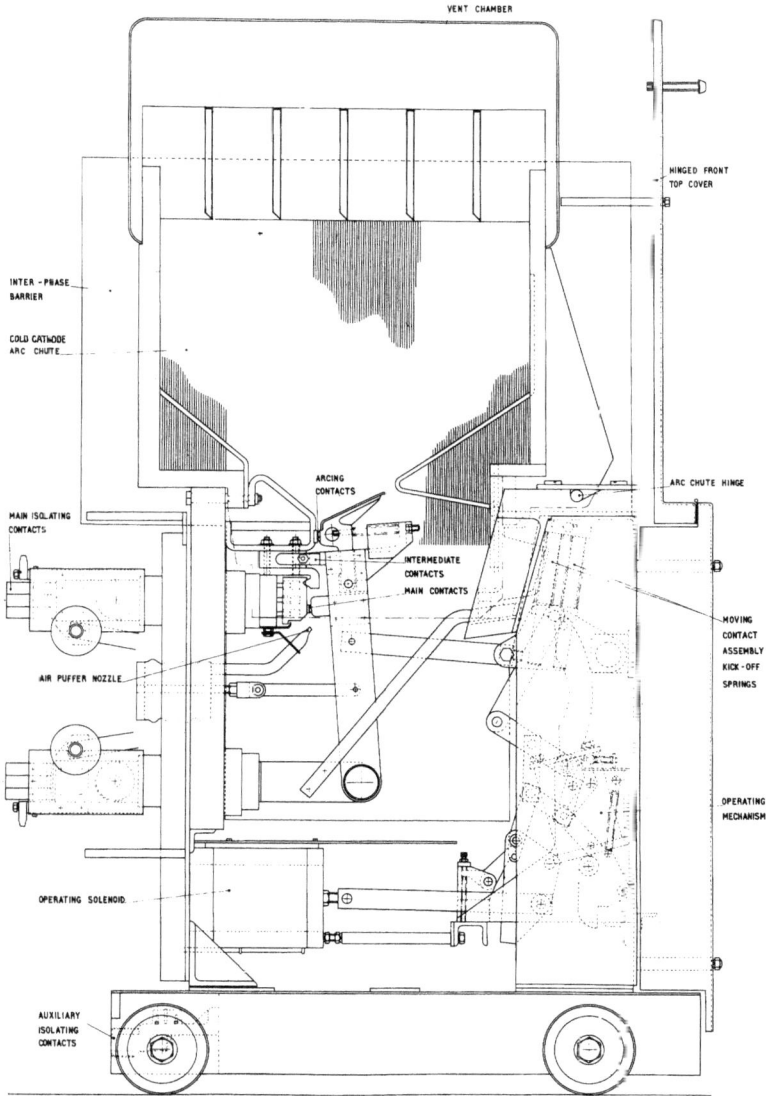

Fig. 5.3 Cold cathode arc chute and circuit breaker [Whipp and Bourne Limited]

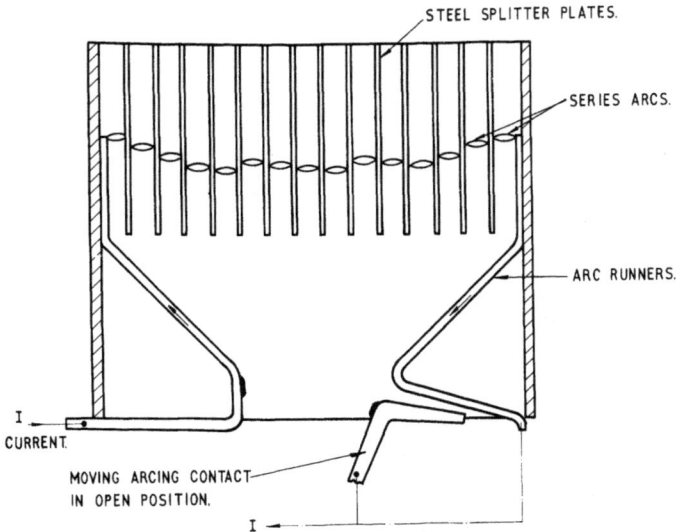

Fig. 5.4A Final arcing in cold cathode arc chute [Whipp and Bourne Limited]

the top of the notch, where further movement of the arc splits it up into a number of series arcs.

The series arcs are free to move up the steel plates, but cannot come out of the arc chute because of the insulation at the top of the plates that limits further travel of the arc roots. Exhaust gases, however, escape through the plate spacings and exhaust splitters fitted in the vent chamber prevent the escaping ionised gases flashing over the top of the arc chute. Barriers are fitted to prevent interphase and phase-to-earth flashovers, by causing the gases to travel a distance such that, when they reach the ventilator and escape into the atmosphere, they are sufficiently deionised to be harmless.

5.2.2.2 Insulated steel plate (Fay *et al.*, 1959)

A section through an 11 kV arc chute is shown in Fig. 5.5a, and shows a number of slotted plates closely spaced inside an insulating box, which is open at the top and bottom. Each plate consists of a 1/16 in thick sheet of magnetic steel moulded and completely embedded inside two 1/8 in thick mica-and-glass sheets.

The arcing contacts are arranged to initiate the arc above the bottom of the slots, so that a magnetic field is established before the arc is drawn. This, with the runner configuration, drives the arc up the chute, where it is moved into the inclined portions of the slots by the action of the steel sheets. There the arc is

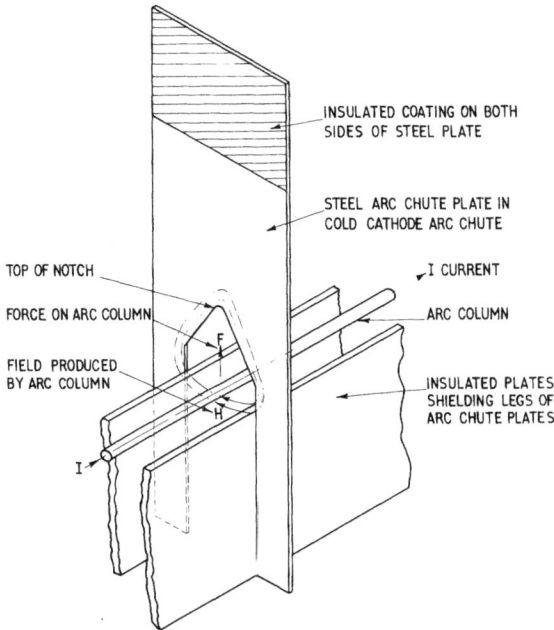

INSULATED COATING ON BOTH
SIDES OF STEEL PLATE

STEEL ARC CHUTE PLATE IN
COLD CATHODE ARC CHUTE

TOP OF NOTCH

I CURRENT

FORCE ON ARC COLUMN

ARC COLUMN

FIELD PRODUCED
BY ARC COLUMN

INSULATED PLATES
SHIELDING LEGS OF
ARC CHUTE PLATES

Fig. 5.4B Effect of sucker loop on arc column [Whipp and Bourne Limited]

lengthened by being forced to follow a compact zigzag path and is cooled by the plates and the surrounding air. At currents above the lead-current range, the arc lengthens very quickly and the arc voltage builds up with corresponding rapidity.

Arc voltage serves as a practical guide to the performance of the circuit breaker, and practice has shown that, near the limit of performance, the arc is usually extinguished if the average arc voltage reaches about 30% of the peak recovery voltage. The arc voltage modifies the waveform of the current considerably, and, except for some asymmetrical currents, it reduces the instantaneous recovery voltage, i.e. the recovery voltage at the time of arc extinction, to a value less than the peak, as shown in Fig. 5.1B.

The arc-chute plates Fig. 5.5b may be conveniently divided into three zones:

(a) The bottom zone, containing the parallel-sided portion of the slot through which the arcing contact passes and which accommodates the sloping portion of the arc runners. This is designed so that the steel is as close to the arcing contacts and runners as mechanical considerations allow.

(b) The middle zone, where the arc is elongated between the plates by means of staggered slots. The arc burns for the greater part of its life in this zone, where

CROSS - OVER

COOLER

RUNNER
POCKET

(a) **Section**

ULTIMATE
PATH OF
ARC CORE

TOP
ZONE

MIDDLE
ZONE

SLOT

BOTTOM
ZONE

MAGNETIC SHEET
EMBEDDED IN EACH
ARC CHUTE PLATE

PLATE
LEGS

(b) Exploded View

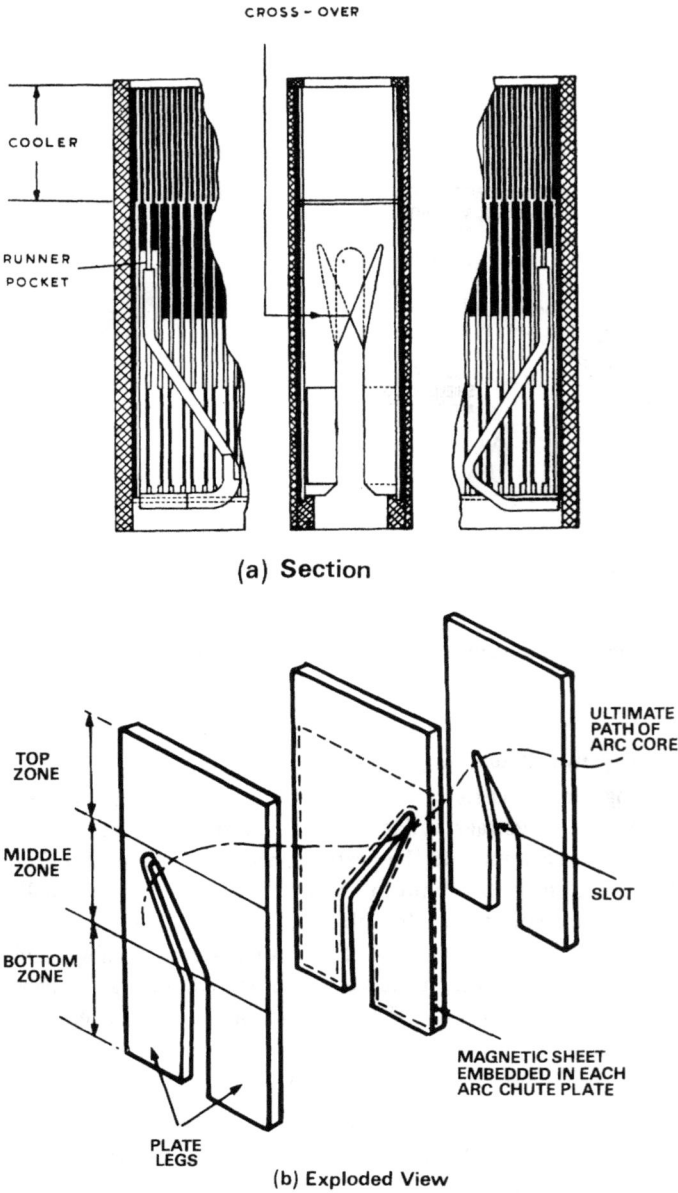

Fig. 5.5 Insulated-steel-plate arc chute [A. Reyrolle and Company]

the magnetic field is strongest and the space available for the arc to be lengthened is as great as possible. The boundary between the bottom and middle zone is the 'crossover'.

(c) The top zone, where the arc is allowed to lengthen further by looping upwards and where the gas is deionised sufficiently to prevent restrikes across the top of the chute.

5.2.2.3 Insulated plate with external iron circuit

The insulated plates used on this type of arc chute are generally similar in appearance to those of the insulated steel-plate type, but without the steel. The plates are usually made of a zircon ceramic or similar material built into an arc chute in which the external iron circuit is usually arranged to encompass the area between the arc-initiating position to the final arc-stretching position. The energising of the iron circuit is usually done by means of 'blowout' coils, which may vary in number, depending on the type of arc chute. Figs. 5.6a, b and c illustrate the more common arrangements of magnetic circuits. Fig. 5.6a uses one coil with a U shaped iron circuit, and produces the strongest field at the end fitted with the coil, the field becomes weaker towards the open end of the U. Fig. 5.6b has two coils, one at either end of the arc chute, and basically consists of two arrangements of type a in series. The advantage of this arrangement is that a strong field is produced at both ends of the arc chute with the weakest in the middle. Fig. 5.6c uses an H section of iron circuit with the coil in the middle of the arc chute, so that the strongest field is produced in the middle of the arc chute. This is basically a back to back arrangement of type (a) and can be considered as two arc chutes in series.

With all external iron circuit arc chutes using blowout coils, one of the major

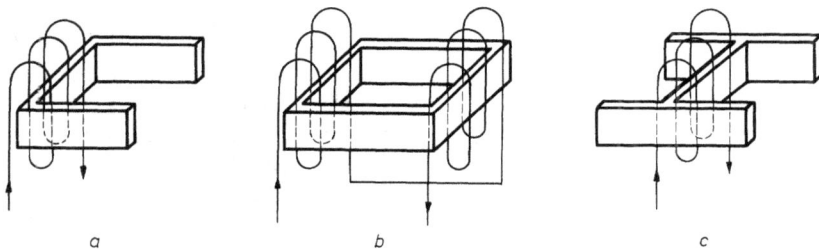

Fig. 5.6 Arrangements of external iron circuits for insulated plate arc chutes

a one coil: U-section iron
b two coils: ring-section iron
c one coil: H-section iron

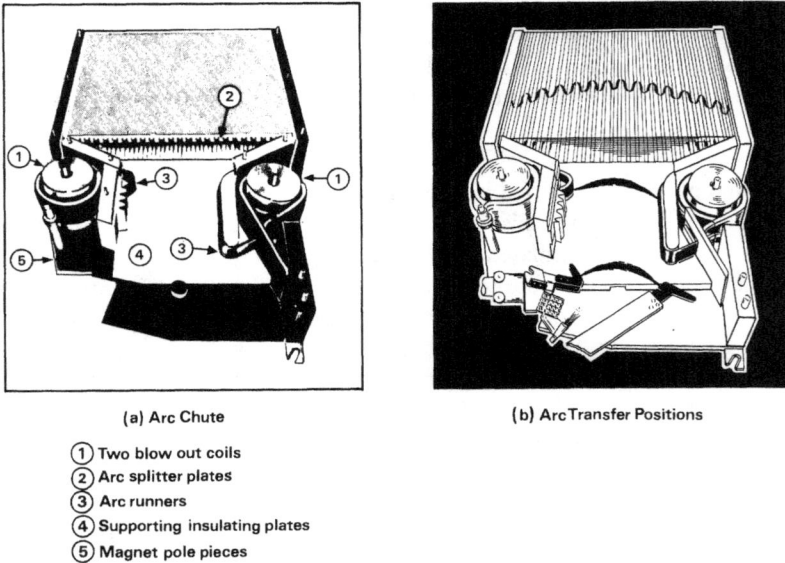

(a) Arc Chute

(b) Arc Transfer Positions

① Two blow out coils
② Arc splitter plates
③ Arc runners
④ Supporting insulating plates
⑤ Magnet pole pieces

Fig. 5.7 Coil switching [S.p.A. Costruzioni Elettromeccaniche]

problems associated with arc interruption is the 'switching-in' of the blowout coils themselves. This is done by means of the arc moving up the arc chute onto the runners connected in series with the coils. Figs. 5.7a and b illustrate this diagramatically. At the initiation of coil switching, an arc is in parallel with the coil, and to force the current to divert into the high-inductive path of the coil requires a reasonably high arc voltage across this auxiliary arc to produce a high rate of rise of coil current to bring its current up to the current being transferred. If this does not happen, the auxiliary arc continues arcing for a long period or until the next current zero at which the 'switching-in' duty is easier. The delay in 'switching-in' the coil may produce restriking across the arc chute because of the ionisation from the auxiliary arc, and results in a longer arcing period caused by the delay in the build up of the strong magnetic field. To assist the 'switching-in' process, it is more usual to fit an auxiliary or transfer arc chute across the coil to help extinguish the auxiliary arc, the size and type depending on the number of turns of the coil and iron circuit.

Fig. 5.8 shows a section through an arc chute of the type described above, which uses a blowout coil mounted in the centre of the arc chute, (Frink *et al.*, 1957). It can be seen from Fig. 5.8 that the strong magnetic field is not produced until the arc has reached the transfer arc chute, and the interruption process relies on the

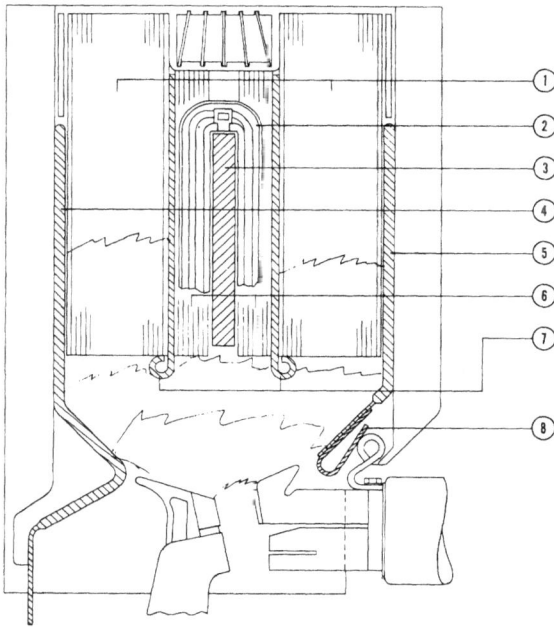

Fig. 5.8 Centre coil arc chute with external iron circuit, showing arc transfer positions [Westinghouse Electric Corporation]

1 main interrupter stacks	5 rear arc horn
2 blowout coils	6 transfer stacks
3 blowout magnet core	7 centre arc horns
4 front arc horn	8 rear arc horn disconnecting contact

inherent electromagnetic and thermal forces of the arc itself to drive it up into the arc chute until the blowout coil is switched in. A shading coil of resistance strap surrounds the magnet core and introduces a small lag in the phase of the magnetic field with respect to the current, so that appreciably higher field strengths exist at the instant immediately preceding current zero.

5.2.2.4 Insulated plate with external 'blowout' coils and no iron circuit

Arc chutes of this type are generally similar to those described in Section 5.2.3, except that the external iron circuit and its associated coils are replaced by two flat-wound coils moulded one on each of the side faces of the arc chute. Fig. 5.9a shows a 3-phase circuit breaker with its interphase barriers removed, in which the outline of the face-wound coils embedded in the moulded arc chute can clearly be seen. The two coils are connected in series in the normal manner for blowout coils,

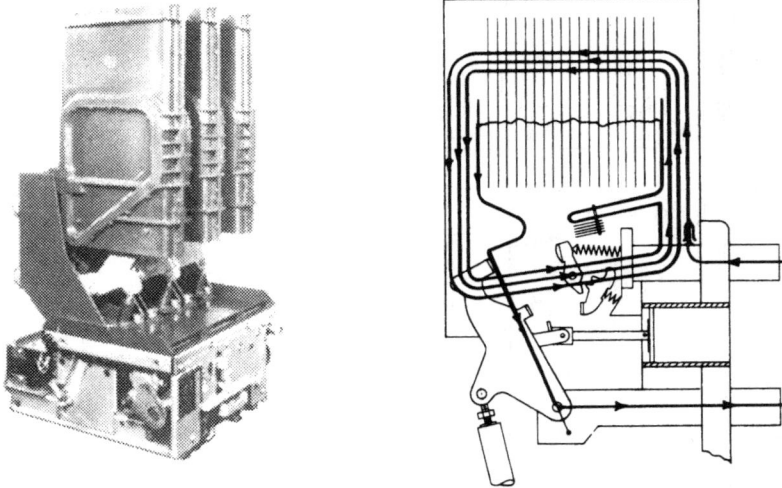

Fig. 5.9 Insulated plate with external 'blowout' coil and no iron circuit [ITE Power
Equipment]

    a Arc chutes and circuit breaker
    b Final arcing

i.e. they require switching in during arc transfer, and the area enclosed by the coils
produces a field across the middle of the arc chute similar to that produced by the
external iron-circuit arrangement of Fig. 5.6*b*. The process of arc initiation, transfer
and extinction is basically similar to the external iron-circuit arrangement, and
Fig. 5.9*b* shows schematically the current path through the arc in the final arcing
position.

    With this design of arc chute, since there is no iron circuit, the field across the
arc chute, despite the series turns, is not very strong at the lower values of current,
i.e. below 1000 A, and a large air puffer has to provide the driving force on the arc.
Also, at very high short-circuit ratings, iron shields are fitted between the phases of
the arc chutes to prevent the interaction of the field from one arc chute on an
adjacent arc chute. This interaction, if allowed to take place, could produce a field
in the wrong direction, tending to drive the arc down rather than up into the arc
chute.

5.2.2.5 Insulated plate Solenarc
The insulated plates of this type consist of specially shaped refractory plates, each
fitted with an arc-runner splitter at the bottom of the plate, so that a complete

stack of plates form a series of arcing chambers producing the same number of series arcs, each of which is allowed to stretch inside its own chamber. Depending on the performance required from the arc chute, variations in the number of arc-runner splitters may occur such that an arc-runner splitter is fitted to every second or every third plate. The intermediate plates without splitters have a special slot that allows the internally formed arc to connect between the chambers.

Figs. 5.10*a-d* show a section through a typical arc chute, and Fig. 5.10*e* illustrates the principle of the helical arcing path at final arc extinction.

The arc resistance is made to increase as a result of all the three basic principles described in Section 5.1.1, and, with the arc in the final arcing position, as shown in Fig. 5.10*e*, high arc resistances can be produced for small arc-chute sizes. These chutes are sometimes used with blowout coils and magnetic circuits, and the main technical problem with designing this type of arc chute is in establishing the arc inside the arcing chambers and keeping it inside the chambers. The voltage drop across the arc in the arcing chamber appears as a stress across the splitters, and, unless the zone around the splitter is kept reasonably free of ionised gas, restriking across the splitter will occur. To minimise this effect when interrupting the larger short-circuit currents, it is usual to fit splitters to every two or three plates.

5.2.3 Performance characteristics
5.2.3.1 Arcing time/current characteristics
The arcing time/current characteristic of a high-voltage air-break circuit breaker of the internal-iron-circuit type is shown in Fig. 5.11A. It can be seen that the longest arc durations occur in the 50-200 A range, known as the 'critical' currents. The durations would be longer if it were not for the existence of the air puffer, which is fitted below the contact assembly. The air puffer is driven by the opening mechanism, and blows a jet of air across the contact gap during a break operation and assists in moving low-current arcs up into the arc chute. At these currents, the magnetic field acting on the arc is very weak, and, without the assistance of the air puffer, very long arc durations would occur, the problem becoming more pronounced the higher the voltage.

The driving force on the arc up into the arc chute is proportional to the square of the current for currents up to about 1000 A. For higher currents, the steel plates saturate, and the driving force is only proportional to the current. Thus, the larger the driving force, the faster the arc moves up into the arc chute and the shorter becomes the arc duration, until, at very high currents, the characteristic flattens out to give a maximum arc duration of about 0·015 s. The characteristic of a high-voltage air-break circuit breaker of the external-iron-circuit type is very similar, the difference depending on the type and number of turns of the blowout coil. With a large numbers of turns, the lower-current characteristics can have shorter arc durations because of the stronger field produced at these currents. However, at the higher currents, the arc duration may be slightly longer because of the delay in switching-in the blowout coil.

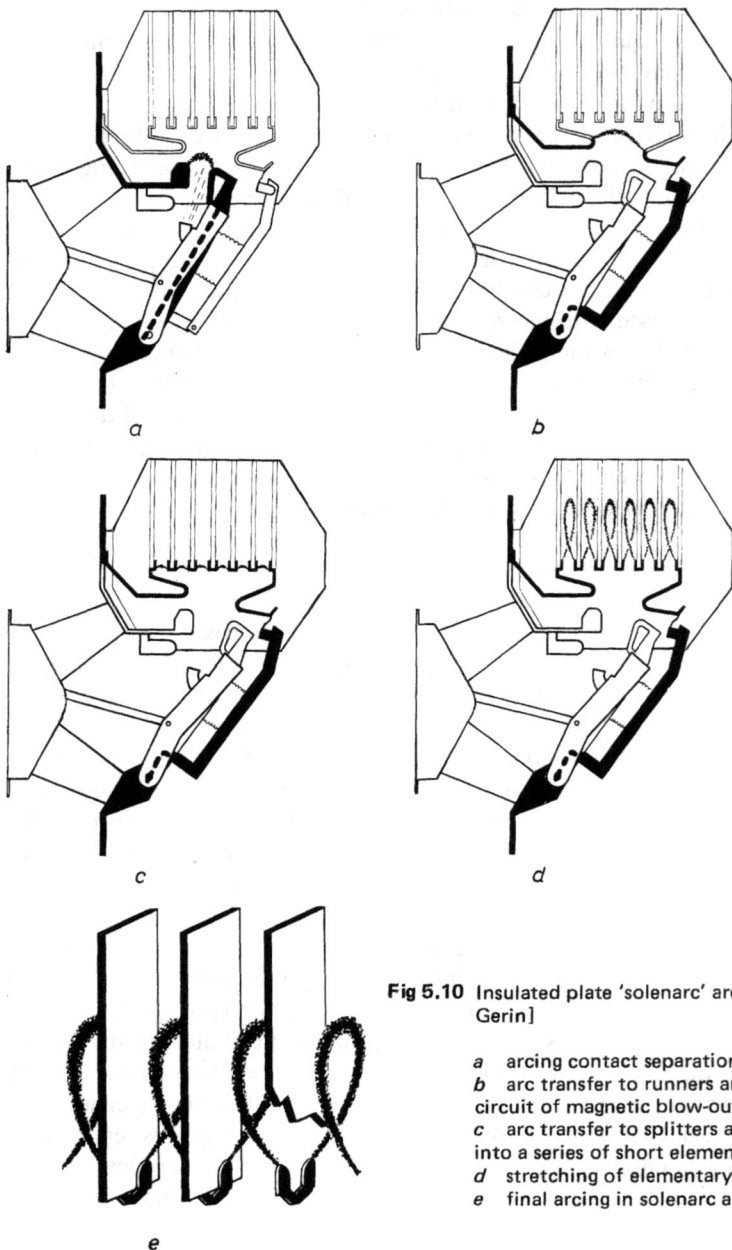

Fig 5.10 Insulated plate 'solenarc' arc chute [Merlin Gerin]

a   arcing contact separation; the arc strikes
b   arc transfer to runners and insertion into circuit of magnetic blow-out coils, if fitted
c   arc transfer to splitters and sectionalising into a series of short elementary arcs
d   stretching of elementary arcs into loops
e   final arcing in solenarc arc chute

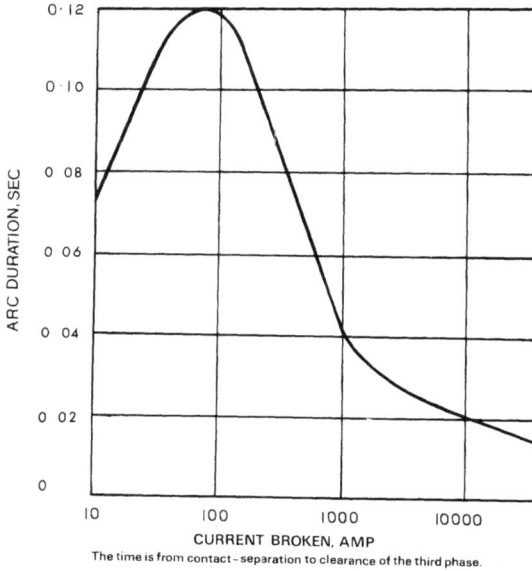

The time is from contact-separation to clearance of the third phase.

Fig. 5.11A   Arcing time/current characteristic of HV air-break circuit breaker [A. Reyrolle and Company]

5.2.3.2 Voltage/current characteristic

The characteristic that shows the voltage/current limits of performance of an insulated-steel-plate arc chute is illustrated in Fig. 5.11B. As previously described in Section 5.1.4, near the limit of performance of the arc chutes, the instantaneous recovery voltage for symmetrical currents is approximately proportional to the r.m.s. recovery voltage. The characteristic illustrated in Fig. 5.11B can be explained in terms of the instantaneous recovery voltage.

When a short-circuit-current arc is burning steadily in the chute, it has a mean zigzag path somewhere in the slots. As a current zero is approached, the arc core diminishes and the mean path moves during the post-zero current period, when the instantaneous recovery voltage is established across it mainly determines the clearance or failure of the chute, as follows:

(a) Below 20 kA, the residual arc path is near the top of the slots, and reignitions producing failure occur because the maximum zigzag arc path is too short to withstand more than a certain recovery voltage. The arc chute therefore has approximately a constant-voltage performance.

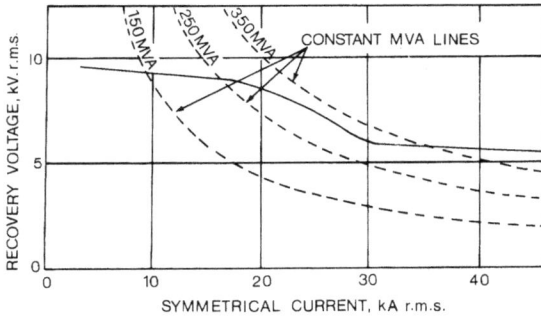

Fig. 5.11B Voltage/current characteristic of insulated-steel-plate arc chute [A. Reyrolle and Company]

(*b*) In the range 20-30 kA, the mean arc path is low down in the slots. The residual arc path will not move to the top of the slots because of the retarding effect of the greater amount of ionised gas that must escape from the chute. This path therefore becomes lower in the slots as the breaking current increases; thus failures occur at a recovery voltage that decreases as the current increases. In this range, the arc chute has approximately a constant MVA performance.

(*c*) Above 30 kA, the mean arc path will be at the crossover level or below, and the residual arc path may only just be in the slots. The movement of the residual arc path up the chute will be slow, and further increases in breaking current will further block the arc chute without affecting the mean arc path. In this range, therefore, the arc chute has approximately a constant-voltage performance.

All insulated-plate types of arc chute that work on the zigzag stretching-arc principle have very similar characteristics. However, metal-plate arc chutes tend to have a constant-voltage limit characteristic over the same range, as shown for the insulated plate, because their arc voltage is constant over this range.

5.2.3.3 Interrupting performance

Over most of the short-circuit range, arcing times are very short and a total break time of less than 3 cycles can be obtained from rated normal currents to rated short-circuit current. Arc energies of the order of a megawatt second may be expended when clearing the largest currents, giving a characteristic flash of reflected light and a high noise level.

Asymmetrical currents

The larger arcs that may occur during the clearance of asymmetrical short-circuit currents tend to overfill the arc chute, and impose a more severe duty on an air-break circuit breaker than for symmetrical short-circuit currents, not only because of the larger arc energies generated, but also because the instantaneous recovery voltage may be higher. Figs. 5.12*a*, *b* and *c* compare the performance of an air-break circuit breaker and a circuit breaker having a low-arc-voltage

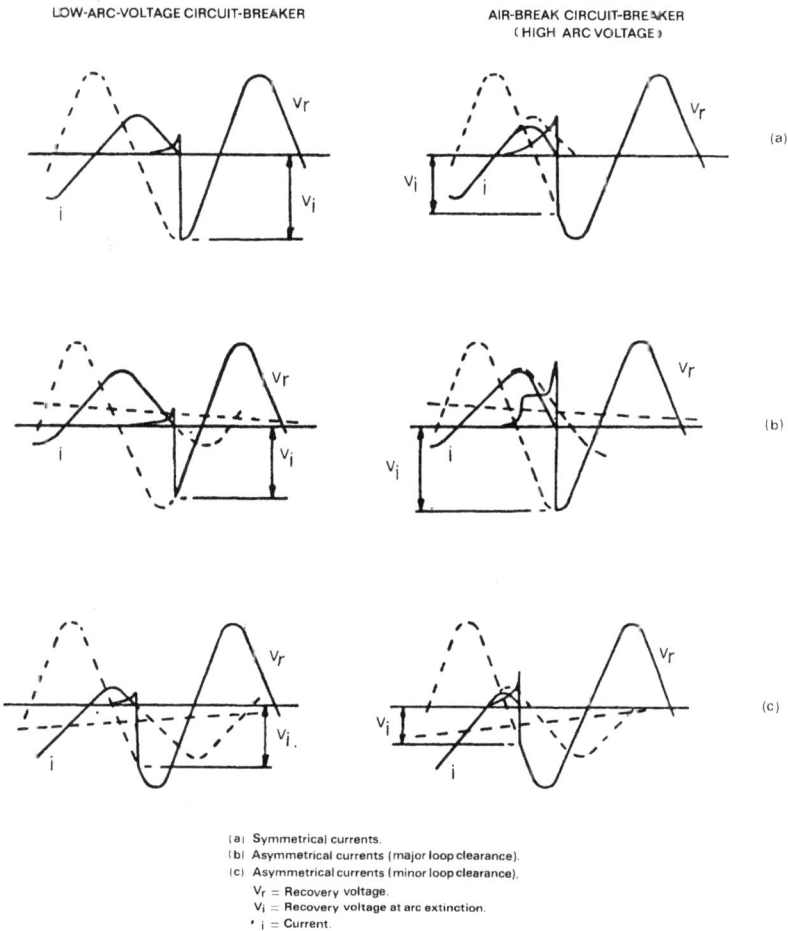

LOW-ARC-VOLTAGE CIRCUIT-BREAKER   AIR-BREAK CIRCUIT-BREAKER
                                             (HIGH ARC VOLTAGE)

(a)

(b)

(c)

(a) Symmetrical currents.
(b) Asymmetrical currents (major loop clearance).
(c) Asymmetrical currents (minor loop clearance).
V_r = Recovery voltage.
V_i = Recovery voltage at arc extinction.
i = Current.

Fig. 5.12 Comparison of instantaneous recovery voltages for a low arc voltage circuit breaker and an air-break circuit breaker [A. Reyrolle and Company]

characteristic when switching identical currents at the same power factor and point-of-contact separation. It will be noted that the instantaneous recovery voltage for a low-arc-voltage circuit breaker when switching symmetrical currents is approximately the peak phase voltage, although for the air-break circuit breaker, it is something less, owing to the current being forced to a premature zero. For asymmetrical currents on major-loop clearances, it is the air-break circuit breaker that is subjected to an instantaneous recovery of the peak phase voltage. Therefore when clearing the major loop of an asymmetrical current, the instantaneous recovery-voltage severity is a maximum and the ratio of asymmetrical to symmetrical severity is greater for air-break circuit breakers than for those that have a low-arc-voltage characteristic.

Live and dead-blade severity
In the short-circuit testing of air-break circuit breakers with contacts arranged unsymmetrically, it is required by specification that the supply be connected to the side of the circuit breaker that gives the more onerous conditions. Factors that govern the more onerous severity of test connections are dealt with in chapter 10, Section 10.10.8.

5.3 Construction

Constructions of air-break circuit breakers can be divided into two classes identified as open-type and cubicle-type, respectively. The former is usually a nonwithdrawable type, and the latter is both nonwithdrawable and withdrawable. The open-type nonwithdrawable circuit breaker was the type developed in the early days of air-break circuit breakers, and is characterised by its panel mounting. The contacts, arc chute and mechanism are all mounted on the front of the panel, with the incoming and outgoing connections projecting through the panel to the rear. Cubicle-type circuit breakers evolved because of the requirement for total enclosure of the flash and the arc from the arc chute, thereby giving more protection to the operator and, at the same time, presenting an earthed front screen. However, with nonwithdrawable circuit breakers in a cubicle, the room for maintenance is very restricted, and to provide reasonable access requires the cubicle to be made very large. For these reasons, the withdrawable circuit breaker was developed and nearly all switchgear is now designed on these lines, with the following advantages:

(a) totally enclosed dead front giving maximum security to the operator
(b) compact switchboards, because the circuit breaker cubicles are of minimum dimensions
(c) easy access for maintenance, because the circuit breaker can be withdrawn or removed
(d) interchangeability of circuit breakers, because, being removable, this gives maximum flexibility in maintaining security of supply.

5.3.1 D.C. circuit breakers

D.C. circuit breakers were the origin of the species from which all air-break circuit breakers were developed, and have changed very little. A large number of those in present-day manufacture are along traditional lines, i.e. nonwithdrawable open type. However, with the development of a.c. circuit breakers to the present being almost wholly withdrawable cubicle type, d.c. circuit breakers have had to follow suit and are now made of the same construction. Some are adaptations of the panel-mounted type on withdrawable carriages and some are new designs having close resemblance to their a.c. counterparts. Because d.c. circuit breakers are considered to have rather special applications, there have been no universally accepted standards for their performance characteristics. In practice, the selection of a d.c. circuit breaker for a particular application concerns the type of short circuit to be interrupted and the speed of interruption required. This has resulted in three general classifications being evolved over the years, which are recognised as follows:

(a) General-purpose circuit breakers: these are circuit breakers that may or may not interrupt quickly enough to limit the magnitude of the fault current.

(b) Semi-high-speed circuit breakers: these are circuit breakers that limit the magnitude of the fault current so that the cutoff current is reached within 30 ms from the start of the short circuit.

(c) High-speed circuit breakers: these are circuit breakers that limit the magnitude of the fault current so that the cutoff current is reached within 15 ms from the start of the short circuit.

Fig. 5.13A shows the current/time curves of performance for typical d.c. circuit breakers of the three classes, indicating the magnitude of the rate of rise of current likely to be seen by the circuit breaker, and its limiting effect on the current.

Systems employing solid-state rectifiers, when subjected to a short circuit at the d.c. terminals, do not produce the classic prospective exponential current wave shape. A typical current/time characteristic for this case is shown in Fig. 5.13B. It is made up of a peak followed by a steady static current having a peak/steady-state ratio of generally 1·67. A typical system with a prospective peak of 200kA could have a di/dt value of 30-36A x 10^6/s. At these levels it would be more usual to use high-speed circuit breakers which in turn would limit the let-through current as shown.

5.3.1.1 Insulated plate arc chutes

The majority of d.c. circuit breakers use arc chutes of the insulated splitter-plate type equipped with an external iron circuit energised by a coil or coils, either series or shunt connected or both. Series-connected coils are always in the main current-carrying circuit, and therefore must carry the rated normal current of the circuit breaker; these coils become rather bulky for the larger frame sizes. Shunt-connected coils are 'switched-in' during the interrupting operation, in a similar manner to

RATE OF RISE OF CURRENT $(R.R.C) = 15 \times 10^6$ A/SEC.

R.R.C $= 5 \times 10^6$ A/SEC.

INTERRUPTING CURRENT KA.

TIME m SEC.

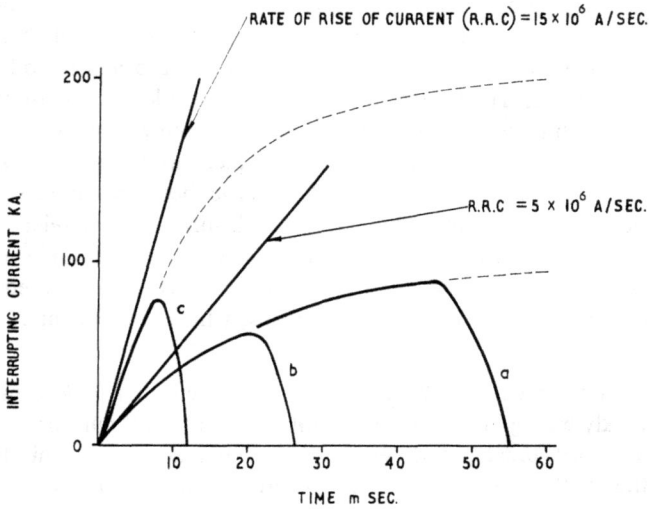

Fig. 5.13A   Interrupting performance curves for d.c. circuit breakers
a general purpose circuit breaker
b semi-high-speed circuit breaker
c high-speed circuit breaker

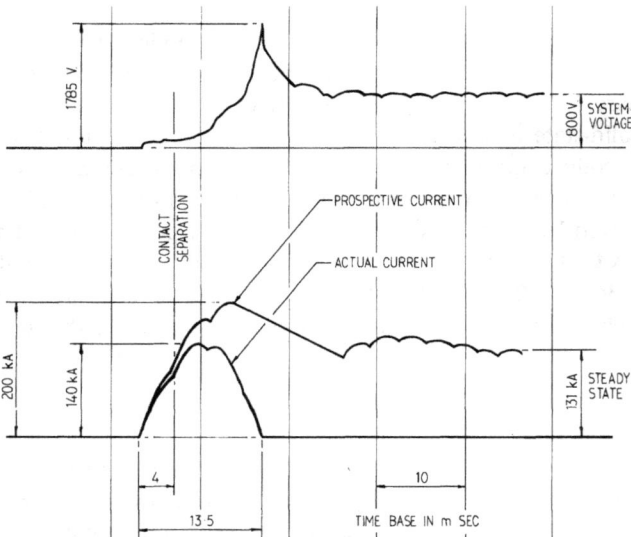

SYSTEM VOLTAGE

PROSPECTIVE CURRENT

ACTUAL CURRENT

CONTACT SEPARATION

STEADY STATE

TIME BASE IN m SEC

Fig. 5.13B   Solid-state rectified short-circuit current and high-speed interruption

those described in Section 5.2.2.3. for a.c. arc chutes. Because the principle of arc extinction of this type of arc chute depends on the stretching and cooling of the arc, to produce the necessary arc voltage, the arc chutes tend to be very large for the higher voltages, and, for this reason, the maximum rating used for d.c. circuit breakers is 3 kV.

The contacts of d.c. circuit breakers have one great advantage over those for a.c. circuit breakers in that the current is evenly distributed in the contacts for designs in which the contacts spread across the width of the pole, e.g. butt and laminated brush types. With a.c. contacts of the same type, the current is very unevenly distributed, and the majority of the current is carried in the outer contacts, with the inner ones used as heat sinks. Thus d.c. circuit breakers with normal current ratings of 10 000A and above are not uncommon.

Fig. 5.14A shows a traditional d.c. circuit breaker that can be classified as general-purpose or semi-high-speed open type, with all components mounted on a main panel. The arc chute is of the insulated longitudinal-splitter-plate type, equipped with an external iron circuit energised by a coil that is 'switched-in' during the interrupting operation when the current is transferred to the arcing contacts. It is of interest to note that the main contacts are of the laminated-brush type, a design that has given good service over the years, but has gone out of favour with new designs. With the laminated-brush contact, the contact pressure is obtained by means of the spring tension in the individual laminations, which is dependent on the deflection of the laminations. The contact material is usually bare copper, which has a maximum permitted temperature rise of 40°C; therefore these contacts operate at considerably lower current densities than are used with modern a.c. circuit-breaker contacts of the type described in Section 5.2.1, and tend to be rather bulky for their ratings.

Fig. 5.14B shows a traditional high-speed circuit breaker of the open type, with all components mounted on a pedestal. This same unit can be mounted on a withdrawable carriage and fitted into a cubicle. The arc chute is of the longitudinal-splitter-plate type, equipped with an external iron circuit energised by a series and a shunt coil. The series coil produces a strong field across the arc chute at the instant the arc is initiated, and the shunt coil is 'switched-in' as soon as the arc transfers to the fixed-contact runner and considerably increases the field strength. The mechanism is fitted with a high-speed release, so that the time from the initiation of the short circuit to contact separation is of the order 5 to 10 ms, depending on the rate of rise of the fault current.

5.3.1.2 Cold cathode arc chutes (Morton, 1980)
Arc chutes
Section 5.2.2.1 dealt with the theory of operation of the cold-cathode arc chute, with particular reference to the shape of the plates themselves and the final arcing. This Section is concerned with the design aspects of the arc chute and its associated contacts, explaining the principle of arc control from initiation to extinction, and

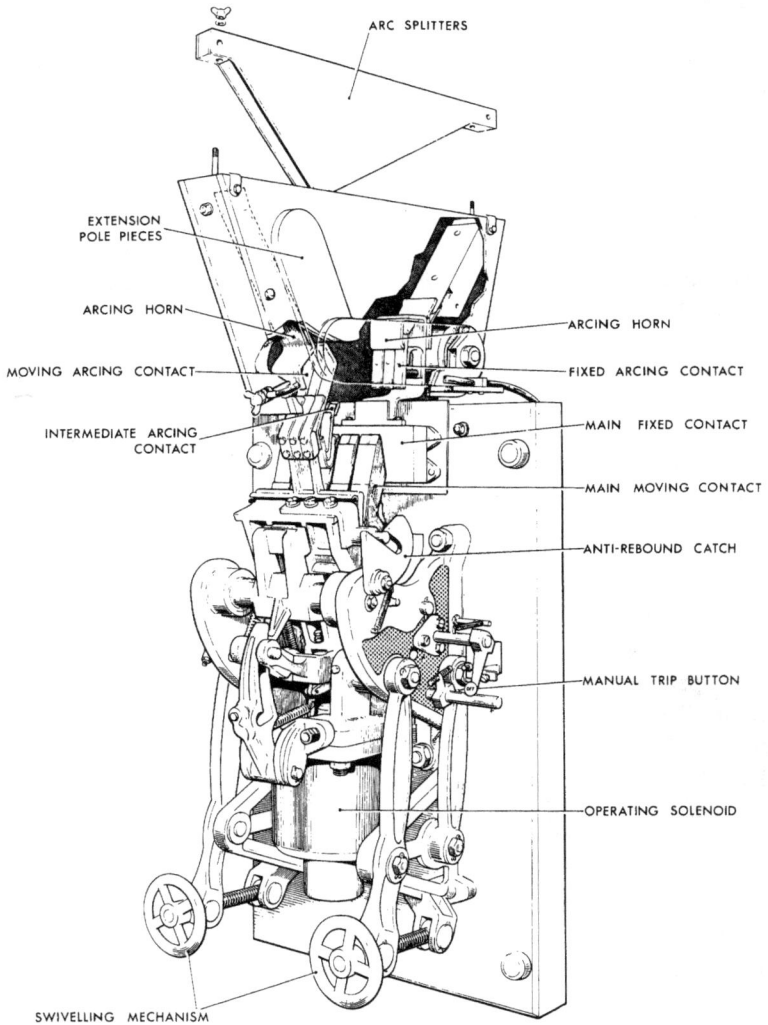

ARC SPLITTERS

EXTENSION
POLE PIECES

ARCING HORN

MOVING ARCING CONTACT

INTERMEDIATE ARCING
CONTACT

ARCING HORN

FIXED ARCING CONTACT

MAIN FIXED CONTACT

MAIN MOVING CONTACT

ANTI-REBOUND CATCH

MANUAL TRIP BUTTON

OPERATING SOLENOID

SWIVELLING MECHANISM

Fig. 5.14A Semi-high-speed open type d.c. circuit breaker [Whipp and Bourne Limited]

the effects that the design of the arc chute and contacts have on this process.

Fig. 5.14C*a* shows a schematic section view of a typical arc chute, indicating the design features associated with the plates, runners and contacts. Superimposed on this view is the arc-transfer sequence of a typical short-circuit interruption, which is shown on an oscillogram in Fig. 5.14C*b* to illustrate the arc-interruption process.

Fig. 5.14B High-speed open type d.c. circuit breaker [Whipp and Bourne Limited]

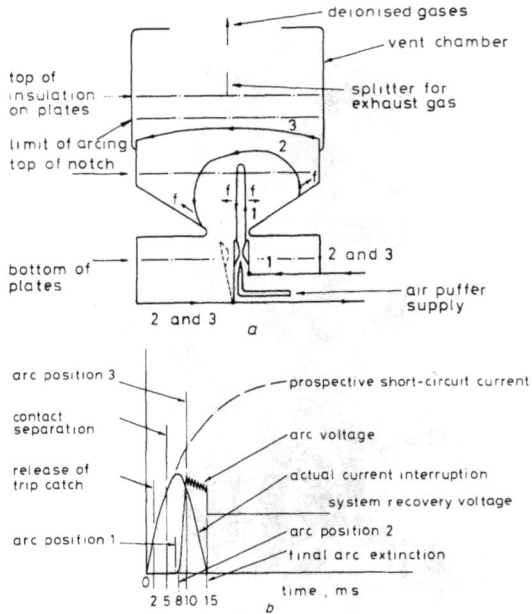

Fig. 5.14C   Typical cold-cathode arc chute
   a Schematic section view, showing arc transfer
   b Typical oscillogram, explaining arc control during short-circuit interruption

Contacts

For large short-circuit currents with high di/dt values, the contacts of a high-speed d.c. circuit breaker must separate approximately 0·005 s after the inception of the short circuit if the high-speed-interruption classification is to be achieved. The contacts illustrated are of the butt type; they have the shortest time to contact separation because of the minimum mechanical travel. Again, for high-speed operation this contact has the dual role of main and arcing contact, thus eliminating the need to transfer current from mains to arcing contacts, with the associated time delays, to achieve a clean transfer.

The combined main and arcing contacts must have sufficient conductivity, even with burnt contacts, to carry the rated normal current of the circuit breaker, and also to have a capability of withstanding the erosion associated with arc initiation for normal service duties, including several short-circuit interruptions, without detriment to their basic functions. This can be achieved by the use of a silver/tungsten insert into the butt contact face.

Arc initiation

The arcing contacts are situated between the legs of the metal plates, positioned so

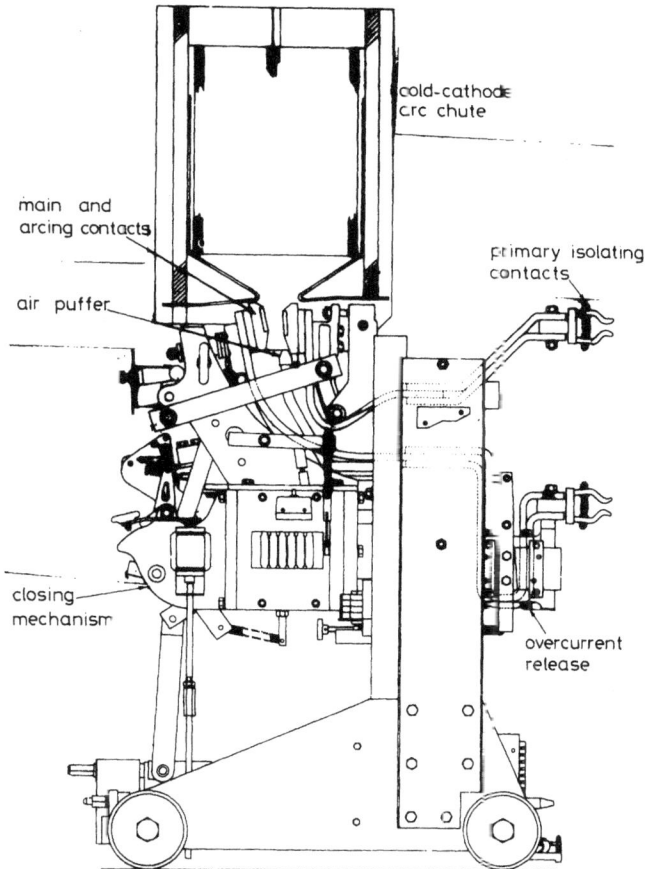

Fig. 5.14D   Withdrawable high-speed d.c. circuit breaker

that, at arc initiation, the arc is immediately within the magnetic field, so that the arc is drawn upwards into the arc chute (illustrated in Fig. 5.4B). Because the contacts are of the butt type, they are ideally suited to being designed to blow off from the tight loop of current which passes through these contacts at arc initiation. This automatically directs the arc up into the arc chute, independent of the sucker-loop effect from the magnetic circuit. The force on this arc and the moving contacts is in theory proportional to i^2, so that, as the short-circuit current increases, and at very high short-circuit current interruptions, the arc at its initiation is directed immediately upwards before its own arc roots have had time to move onto the runner system. Position 1 of Fig. 5.14Ca illustrates this; in practice, the arc voltage produced by this initiating arc is too great for the contact gap to withstand, so that multiple restriking takes place until a sufficient contact gap is established. It generally takes between 0·003 s and 0·005 s to clear this part of the operation in position 2 and then to position 3 with the arc roots established up in the arc chute.

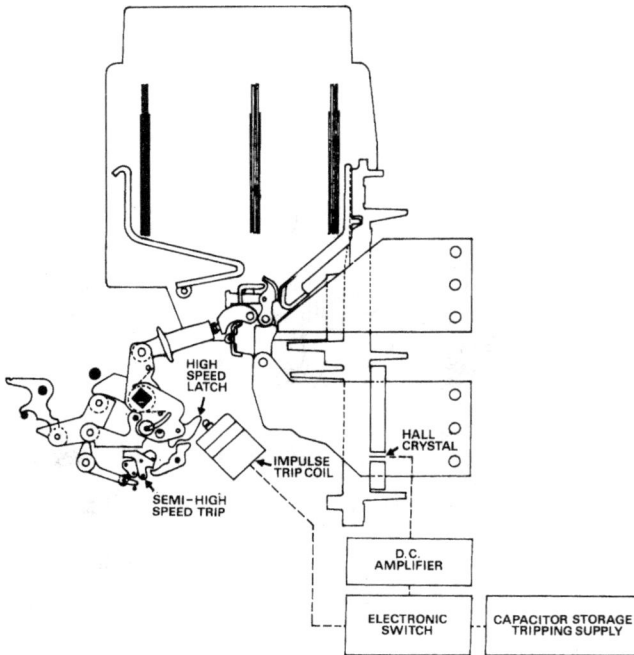

Fig. 5.14E   High-speed d.c. circuit breaker with electronic tripping and impulse coil release
[ITE Power Equipment]

Arc running
Once the arc roots have moved off the arcing contacts, which are themselves
designed to have divergent tail runners, they must transfer to the runner system and
then proceed up into the arc chute. To ensure that transfer is swift, the transfer
inductance must be kept to a minimum by keeping the paths in the transfer loop
also to a minimum, assisted by the electrical connection between the runners and
the moving contacts. The direction of current passing through the runners should
always be such that the force acting on the arc root is directed upwards along the
runner towards the final arcing position on the runner. Position 2 in Fig. 5.14C*a*
illustrates an intermediate stage in the process, and the direction of the force on the
arc near its root is always towards the end of the runner. The angle of the runners
and the arc-chute geometry are derived empirically: they are designed to give space
for the arc to travel up into the arc chute without causing down-strikes to the
runner system.

Final arcing
Position 3 in Fig. 5.14C*a* illustrates final arcing in the arc chute; it is at this instant
that the arc penetrates all the metal plates producing the maximum arc voltage.

As previously explained in Section 5.2.2.1, the arc voltage is a function of both the number of series arcs produced in the arc chute and the voltage drop across the arc roots themselves. The voltage drop across the arc root is also a variable and is dependent on its current density; thus the higher the current density, the higher is the voltage drop. The current density is kept high as long as the arc root is fast moving, and therefore a stationary arc on the metal plates must not be allowed.

Once the arc reaches the top of the notch in the metal plates and actually roots on to the plates themselves, there is no longer a magnetic field from the plates trying to drive the arc upwards further into the arc chute. Thus the arc penetrates the plates at a speed associated with the impetus received from the arc-running phenomenon, so that it is at the instant when full penetration is achieved, and while the arc roots are moving at their maximum speed on the plates that the maximum arc voltage is also produced. However, once the arcs are in their final arcing position, these short arcs are free to move independently of each other, with a separate existence, influenced mainly by the following effects:

(*a*) the loop effect of the current in the runner system providing a resultant driving force generally upwards into the arc chute
(*b*) the arc's own thermal forces providing an upwards driving force
(*c*) the interaction of adjacent arcs whose current is fed to their arc roots through the plates themselves, which act as individual runner systems. Thus, once the arcs are out of step with each other in the direction across the arc chute, the interaction forces tend to move the arcs further out of step. The phenomenon produced as a result of this is to make the arcs move up and down and from side to side as the arcing process continues, doing so in an apparently random manner but with the desired effect of keeping the arcs on the move.

The effect described above in (*c*) causes the arc voltage to have a drooping characteristic after the final arcing position is reached. This is considered to be due to the slowing down of the arcs themselves, associated with intermittent shorting out of arcs between the plates with newly established arcs at the top of the notch. The latter effect is caused by an arc between the plates moving downwards and out into the notch and then transferring its arc roots on to adjacent plates. However, the magnetic field is immediately re-established by the sucker-loop effect and the arc is drawn back into the metal plates. High-speed cine pictures show this phenomenon taking place; as time increases, the frequency of the shorting out of the plates by the arcs also increases, giving rise to a continuing drooping arc-voltage characteristic which races the falling current.

Arc extinction
Arc extinction is achieved by controlling the arc within the arc chute to its final arcing position, always keeping the arc voltage greater than the system voltage, so

that the resistance of the arc is always increasing until the current is forced to zero. Once the arc is in its final arcing position in the arc chute, the arc chute should then be virtually restrike-free if the cold-cathode design is being used, because the maximum stress across the arc chute occurred at the instant the maximum arc voltage was produced, i.e. the instant it reached its final arcing position. From here onwards, if a restrike does occur, it is associated either with excessive ionised-gas production in the plates themselves, or lack of arc control, and would indicate an approach to a limit of performance with the particular circuit parameters in use.

Because of the narrow spacings between the plates, which also assists with deionising of the gas, the amount of ionised gas produced is considerably less in the cold-cathode arc chute than in the insulated-plate arc chute. Thus a minimal amount of ionised gas is produced above the arc chute, and, by means of a suitable vent chamber, designed to minimise back pressure in the arc chute, the resulting exhaust gas can be made harmless. Because the arc chute is virtually restrike free, a minimal amount of ionised gas is produced at the bottom of the arc chute, mainly associated with good arc initiation, which, if it does not cause a restrike across the contact gap, is unlikely to strike elsewhere if the clearances are not less than those of the contacts.

During the final arcing period, the manner in which the arc is reduced to zero depends on two parameters:

(*a*)   the resistance of the arc being introduced into the circuit
(*b*)   the inductance of the circuit.

With the high short-circuit currents associated with the solid-state rectifiers the inductance in the circuit is provided mostly by the source itself, having a relatively low time constant, in the 0·01-0·02 s region, with a very high *di/dt;* for d.c. interruption purposes this is considered to be nearer the resistive end of the current-interruption spectrum. The maximum arc voltage in these conditions is rarely the maximum obtainable from the arc chute, because arc extinction is usually achieved before the arc has had time to reach its final arcing position. Sufficient arc voltage, and hence arc resistance, is introduced into the circuit to make the current decrease very rapidly to zero with the small amount of inductance in the circuit. With short circuits associated with time constants of the order of 0·1 s, the drooping voltage characteristic is clearly distinguishable on an oscillogram. The arc chute produces its maximum arc voltage and then must continually maintain the drooping arc voltage greater than the system voltage for extinction to be achieved. It is only when the current has actually fallen to approximately 100A that an increase in arc voltage is seen; this is in line with the characteristics of arcs in this region of current.

The interruption of short circuits which are highly inductive will eventually approach a limit of performance of the arc chute as the inductance is increased.

Air puffer
At very low values of current, when the magnetic field from the iron circuit is very

weak, it is desirable that the movement of the arc up into the arc chute should be assisted by external means, usually in the form of an air puffer. This is the accepted practice with a.c. air-break circuit breakers, and works equally well on d.c. Low currents which produce long arc durations are in the region of 10-200A.

Circuit breaker
Fig. 5.14D shows a side elevation of a single-pole 750V high speed d.c. circuit breaker fitted with a cold-cathode arc chute. The circuit breaker is on a with-drawable truck whose pedestal supports a panel-mounted circuit breaker. On the front of the panel is the mechanism, contacts and arc chute; at the rear are mounted the direct-acting short-circuit release and the main isolating contacts. The release is arranged with a tripping rod through to the front where the main trip catch is situated. Also illustrated is the means for manual closing of the mechanism for emergency and maintenance purposes; this feature is available for closing on to a live circuit, since no blow-off forces are taken on the closing handle for this particular design.

The circuit breaker employs a live-frame mechanism, so that all secondary connections to the mechanisms, i.e. coils, auxiliary switches etc. must be insulated to the full withstand voltage. An advantage of this type of design is the complete removal of any earthed metal from the region immediately below the arc chute.

For ease of maintenance, the arc chute is arranged to tilt backwards for contact examination, and the front of the circuit breaker is kept clear of all secondary controls, relays etc., giving easy access to all moving parts.

Circuit breakers of this design have been successfully used for rated voltages of 800V, 1600V and 3200V having short circuit interrupting ratings up to 250kA, 150kA and 40kA, respectively.

Another design of withdrawable circuit breaker (Brandt *et al.*, 1969) using a similar principle, which embodies a number of the design features developed from the a.c. circuit breaker, is shown schematically in Fig. 5.14E. It is of particular interest to note the high-speed release in the form of an electronic sensing and trip-ping system operating on to an impulse coil to the trip latch of the mechanism. Very much faster operating times over the conventional electromagnetic release are obtained by this method.

5.3.2 Low-voltage a.c. air-break circuit breakers
This class of circuit breaker is generally considered to consist of those having nominal voltages of less than 1000 V, and, for most cases, are identified for 400 or 600 V service. Circuit breakers of this class have used many varieties of interruptors, but most modern designs favour the bare-metal-plate or cold-cathode arc chute. The reason is its simplicity, its cheapness and its compactness, because so few plates are required at these voltages. Contacts have also varied considerably, and again most modern designs favour the butt contacts for both mains and arcing. The reason for this is that they are inherently self aligning, they are easily adaptable

to extensions for high current ratings and, the most important of all, they have the least length of stroke from arcing-contact touch to fully home of any contact arrangement, this being the most favourable arrangement for a circuit breaker to close and latch against the peak fault current. For this class of circuit breaker, these peaks are generally in the order of 100 to 150 kA.

Fig. 5.15 shows a section through a typical circuit breaker for use up to 630 V. The arc chute is the steel-splitter-plate or cold-cathode type, and the contacts, both main and arcing, are the butt type. The arcing contacts are the blow-on type, which maintain contact pressure until they separate, and the main contacts are the

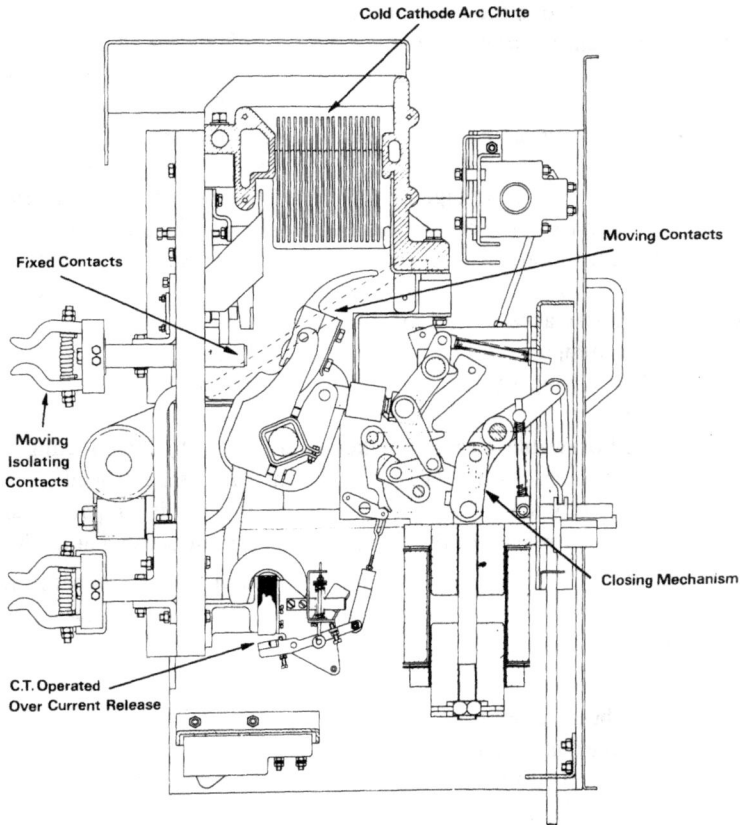

Fig. 5.15 Low-voltage air-break circuit breaker, 630 V a.c. cold cathode arc chute [Whipp and Bourne Limited]

'bridging' type. The mechanism is mounted in front of the moving contacts, and the overcurrent releases are mounted in the region near the moving-blade hinge.

Overcurrent releases are generally accommodated in circuit breakers of this class, for the following reasons:

(a) The armature of the overcurrent release can be mechanically coupled directly with the tripping mechanism of the circuit breaker, so that, for short circuits, the circuit breaker can be made to trip instantaneously. This is a requirement for feeder circuit breakers not required to discriminate with others down the line.

(b) The primary winding of the release can be the bottom conductor to the hinged blade of the circuit breaker, and may consist of one or more series turns of the main conductor.

(c) Because of the low voltages involved and the correspondingly smaller insulation clearances, positioning of the overcurrent releases below the moving contact arm is relatively easy.

(d) The fitting of overcurrent releases on the circuit breaker can eliminate the need for external relays and, in some cases, the need for current transformers.

Low-voltage circuit breakers with their integral protective equipment and control circuitry mounted on the moving withdrawable unit can be made so compact that they only require plugging onto the busbars and onto the circuit, enabling multi-tier arrangements to be easily obtained. For sizes up to 2000 A, this would generally be double or treble tier, the criteria being the cable accommodation and meeting the maximum permitted temperature rise.

It is interesting to note that many of the design features of high-voltage circuit breakers are not used or are ignored in the lower-voltage class. This is because these features become less effective or not so important at the lower voltage, and some of these and the reasons for their omission are listed, as follows:

(a) Air puffers are not fitted. The distance the arc has to travel from arc initiation to its final arc-extinction position is very short, and long arcing times do not occur. Further, the compactness of the contact configuration hardly allows room for fitting a puffer.

(b) Intermediate contacts are not used. If transfer arcing occurs at the main contacts during a break operation, restriking in the ionised zone beneath the arc chute does not follow, because the arc and recovery voltages produced in the arc chute are not as high as with higher-voltage circuit breakers.

(c) Ionised gases are not vented direct to atmosphere. The back pressure in the arc chute, associated with the exhaust gases impinging directly onto an insulated roof barrier, becomes less detrimental at lower voltages, and, in some cases, the back pressure prevents the arc restriking over the top of the arc chute. The ionised gases are deionised very quickly, within a few milliseconds of arc extinction, with the metal vapours condensing on the roof barrier, and very little other contamination occurring. Thus totally enclosed, but not sealed, metal enclosures are quite usual with this class of switchgear, because the arc energies do not produce high-pressure rises and back pressures that need vented cubicles, unless these are required for cooling purposes associated with the higher normal current ratings.

5.3.3 High-voltage circuit breakers

A typical cross-section through a high-voltage air-break circuit breaker is shown in Fig. 5.3, which indicates the relative position of the arc chute to the fixed and moving contacts and also the mechanism. Another typical cross-section is shown in Fig. 5.16, and it is interesting to compare the two layouts, both in the most popular form of construction, namely the horizontal-isolation arrangement. Vertical isolation arrangements have been made and have the same layout turned through 90°, with the isolating contacts pointing upwards and the arc chute exhausting towards the rear, but to lift a heavy circuit breaker of this type into its 'service' location requires additional handling equipment which can be expensive. The principal difference between the two arrangements of Figs. 5.3 and 5.16 lies in the positioning of the mechanism and its driving linkage relative to the moving blade, and the reason for the difference is due to the different types of main contacts used on each circuit breaker.

In Fig. 5.3, the circuit breaker is fitted with butt-type main contacts, which compress by 3 mm when the blade is fully home. Thus the forces tending to open the blade, such as short-circuit currents and contact pressure, must be opposed by a robust structure to ensure that the moving blade does not lose its contact pressure. The driving arm is therefore shown propping the moving blade directly behind it and nearer to the contacts than the hinge. The mechanism front frame, the chassis and the back frame are all tied together to form a robust structure, ensuring that contact pressure is always maintained.

In Fig. 5.16, the circuit breaker is fitted with finger-and-blade type main contacts, which have a constant contact pressure over the wiping travel just before the blade goes fully home. Thus movement of the blade would not lose contact pressure unless it were sufficient to reach the tapered faces of the blade. The contact-pressure forces are applied at 90° to the blade travel, and therefore the only opposing force to the blade is the short-circuit force on it. However, these forces have a beneficial effect on the contacts themselves, as, by being the finger-and-blade type, the fingers are pulled together with increasing currents, thereby increasing the contact pressure. The mechanism is shown mounted on the chassis, with the driving arm acting on the blade much nearer the hinge of the blade, giving a greater mechanical advantage than the arrangement of Fig. 5.3, but capable of containing any slight movement of the blade without harm to the contacts. It is of interest to note the different positions of the puffers, as both arrangements show them to be individually driven, each from their own moving

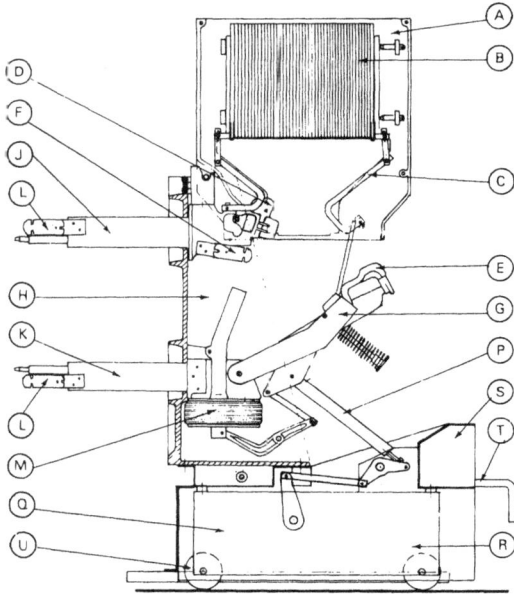

Fig. 5.16 Insulated plate arc chute and circuit breaker [Merlin Gerin]

A	arc chute	*K*	lower conductor
B	stack of refractory plates	*L*	disconnecting contacts
C	arc runners	*M*	puffer
D	fixed arcing contact	*P*	pole operating rod
E	moving arc contact	*Q*	roller truck
F	fixed main contact	*R*	GMh operating mechanism
G	moving main contact	*S*	local operation-mechanism enclosure
H	insulating support	*T*	locking lever
J	upper conductor	*U*	roller

blade. The puffer should be situated just below the main contacts directed up into the tunnel of the arc chute, and blowing just clear of the fixed contacts.

The arrangement shown in Fig. 5.3 has the additional feature of integral busbar and circuit earthing built onto the circuit breaker. This is done by swinging one set of isolating contact arms through 90°, so that a contact on the arm projects backwards midway between the two sets of isolating contacts. Thus, when the circuit breaker is put into the service position, the projecting contacts engage with earth contacts in the cubicle. Closing the circuit breaker will earth one side of its incoming connections. The other side can be earthed by resetting the isolating contacts and then swinging the other set of isolating contacts for them to engage with the earth contacts. A simple interlock ensures that only one set of isolating contacts can be moved at a time.

The arrangement shown in Fig. 5.16 can also have integral earthing, but on one side only, obtained by swinging the bottom set of isolating contact arms vertically downwards to engage with the earthing contacts on the back of the chassis.

The advantages of integral-earthing circuit breakers are in their safety features. Earths can be applied to a circuit quickly and correctly, and do not require any loose components, which may or may not be satisfactorily attached. By using the circuit breaker, the earth is applied by a switch capable of making onto the full short-circuit rating of the system, thus catering for the earthing of a live circuit in error.

5.4 Mechanisms

Mechanisms are divided into three main classifications identified by the source of their closing power: solenoid, spring and pneumatic. Solenoid mechanisms obtain their energy from an external battery or from a transformer rectifier connected to the live supply. Spring mechanisms obtain their energy either by manually charging the springs by means of a fixed or loose handle or by electrically charging the springs by means of a motor and gear train. The motor obtains its energy from the same sources as the solenoid, but at a very much reduced power. Pneumatic mechanisms are fitted with compressed-air storage reservoirs that operate the closing pistons when the closing control signal opens the closing valve. The valve is manually or electrically operated. For the latter, it requires the same power source as for motor charged springs. The first two types of mechanism are those in general use, and are discussed in Section 5.4.1.

5.4.1 Closing-mechanism design parameters
The following is a list of the major items that influence the design and type of mechanism to be fitted to a circuit breaker.:

(a) Voltage rating: This determines the clearance between the two sets of isolating contacts, the electrical clearances and, in particular, the clearance across the break. From these considerations, the stroke of the blade is determined with its angular movement.

(b) Current rating: This settles the size of the conductors, which in turn determines the mass of the moving blade. The temperature-rise limits have a direct influence on the contact size, British Standard BS3659, for high-voltage circuit breakers, sets a limit of 50°C rise and BS4752, for low-voltage circuit breakers, sets a limit determined solely by the class of the insulating material adjacent to the conductors. The equivalent IEC specifications state 65°C rise for high-voltage circuit breakers in IEC-56 and in IEC-157, for low-voltage circuit breakers, it is the same as BS4752.

(c) Short-circuit rating: This determines the strength of the mechanism, particularly when required to close and latch against the short-circuit current, which requires the blowoff forces to be calculated. The closing mechanism can make onto short circuits either by providing sufficient power in the mechanism so that it will close against the blowoff forces independently of its closing speed or by relying on the momentum of the moving parts achieved by a high speed during the closing stroke. Most circuit breakers, especially those of high current and short-circuit ratings work on the latter principle, since enormous powers would be required to close from standstill at contact touch.

(*d*) Shunt or series releases: These influence the design of the mechanism considerably, because circuit breakers fitted with shunt releases must be capable of closing and latching against the full rated short circuit. Those with series releases are only required to close against currents up to the instantaneous setting of the release. At currents above the setting, the series release will operate and immediately trip the circuit breaker, so the closing mechanism at these currents has only to hold the contacts made without latching until the trip operation takes over.

(*e*) Type of contacts: The type of contacts determines the extent of travel in the closing stroke between contact touch to fully home. This is very important to the closing mechanism because this travel, together with the travel associated with prearcing on 'make', determines the distance the blade must move from the instant current starts to flow to reaching fully home. Butt-type contacts give the shortest travel, with finger and blade giving the longest, as part of the travel takes place along the sliding faces of the blades.

(*f*) Mechanical duty: Some circuit breakers when in service are very rarely operated, and others are operated many times an hour, and the mechanical-endurance test required from the circuit breaker certainly influences the type of mechanism fitted. Applications vary considerably, and, for this reason, most designs of circuit breaker cater for two types of mechanism, namely the spring-charged mechanism and the solenoid mechanism.

5.4.1.1 Design problems

A solenoid mechanism relies on an external source for its closing power, which, for large installations, is obtained from a battery. Unfortunately, a battery has to be considered to be in various states of discharge, although in practice, with most reliable installations, the batteries are always fully charged. Most standards accept that the battery in its worst condition could be at 80% or 85% of its nominal voltage, and therefore the circuit breaker must be capable of its making capacity rating at this reduced voltage. The circuit breaker sometimes has to operate at 120% of its nominal voltage when the battery is fully charged. Since the closing power in kilowatts is proportional to V^2, the closing power at 120% of V is twice that at 80% V. This can have a damaging effect on the mechanism, and needs to be catered for in the initial design. The consolation to the user with such a mechanism is that, with the capability of latching at 80% of the battery voltage, it has a considerable safety factor in service, where the voltages are always on the top side.

With a solenoid mechanism, the static pull of the solenoid is inversely proportional to the airgap in the solenoid, and therefore the static pull increases as the airgap reduces. In practice, the dynamic pull does not increase in proportion to the static pull, because the movement of the solenoid itself produces a back e.m.f. that reduces the current in the solenoid. The effect of this is dependent on the speed at which the solenoid moves, which in turn is dependent on the mechanical advantage of the mechanism. The effect is to reduce the pull to below the static figures, but, since the mechanical advantage of the mechanism is at its greatest at the end of the stroke, the combined effects are designed to give the maximum closing force at the part of the stroke subjected to the maximum blowoff forces, i.e. from contact touch to fully home. Any hesitation occurring in the closing mechanism as a result of the blowoff forces is counteracted by an increase in the pull from the solenoid, because the current will immediately increase again, and, in most designs, the solenoid is able to pull in and latch.

With a spring mechanism, the closing force is known precisely for all positions in the closing stroke, there being no dynamic correction effects. The maximum force from the springs occur when they are fully compressed and the minimum is at the end of the stroke; however, with the mechanical advantage of the mechanism at its greatest, the maximum closing force on the contacts also occurs at the end of the stroke.

Two forms of spring mechanism are in general use (*a*) the independent manual-spring mechanism and (*b*) the spring-charged mechanism with the spring either hand or motor charged, or both. With the larger circuit breakers and the higher short-circuit ratings, the closing forces are assisted, and mostly rely on, the assistance given by the momentum of the moving contacts to close and latch against its rated short-circuit current. This means that there must be no hesitation in the closing stroke due to the blowoff forces, because the momentum will be immediately lost and latching may not take place. For this reason, spring mechanisms usually have higher closing speeds at contact touch than solenoid mechanisms. With some motor-charged spring mechanisms, the circuit breaker is held closed by means of the closing springs themselves and not by a latch, and, when the springs are charged, the mechanism sits back onto the latch. The mechanism can be fitted with a spring-recharging feature that can be done with the circuit breaker closed, thus storing the next closing power in the circuit breaker. With this feature, dependent on the type of mechanism, it is usual to fit a reverse-trip catch to the mechanism designed to operate in that part of the stroke between contact touch and fully home. The reverse-trip catch ensures that, if the latch does not engage, the charging of the spring, which would open the contacts slowly, operates the reverse-trip catch before or at contact separation, to give a normal-speed trip instead of a very slow break, with its dangerous consequences. This feature is not fitted to solenoid mechanisms because of their high safety factors and because a slow break does not occur if latching does not take place. The circuit breaker will open as soon as the solenoid has de-energised at a slightly slower speed than normal, because the tripping toggles have not collapsed.

With a spring-charged mechanism, interlocks are required to ensure that the closing springs can only be discharged when they are fully charged.

There is no question as to the simplicity of a solenoid mechanism over a spring mechanism, the former having fewer moving parts and no auxiliary linkage associated with interlocks. On this account, solenoid mechanisms are likely to prove more reliable than spring mechanisms where the application requires frequent operations.

5.4.2 Tripping mechanisms and design parameters

Most circuit breaker mechanisms work on the trip-free principle, i.e. the trip always overrides the close at any part of closing stroke. The opening springs that determine the speed of break are always charged up during the closing stroke, and give speeds usually in the range 2 to 4 m/s. The requirements for arc-free current transfer, best obtained with slow speeds, and for freedom from restrikes, best obtained with higher speeds, tend to be incompatible, and require some compromise. By using a trip-free mechanism, other interlocks normally associated with a circuit breaker can be easily obtained. A circuit breaker should not be capable of being closed in any intermediate position between the service and isolated locations. By holding the mechanism trip-free between these two locations, the desired interlocking is provided.

A trip-free mechanism having a number of interlocks that work on the trip-free principle is liable to some trip-free close operations. For such an operation, the closing mechanism operates without moving the blades, and therefore sees no opposing force and, unless the mechanism is designed to withstand this operation, severe damage can occur. No standards require such checks to be made, although most manufacturers cater for some trip-free operations during the mechanism-development testing. Solenoid mechanisms stand up to trip-free operations much better than springs, because of their slower closing speeds and because, on a trip-free operation, their starting ampere turns are usually less with the lighter load. On spring mechanisms, with their higher speeds, these are even higher on a trip-free operation, because the starting force is the same, independent of the starting load. For this reason, some designs of spring-charged mechanism are made of the fixed-trip type so that a trip-free close operation cannot occur, and tripping is possible only in the fully closed position.

5.4.3 Mechanism linkages

Typical mechanism linkages used with air-break circuit breakers are described and illustrated accordingly. They are only intended to point out the basic principles, without giving the design details. The illustrations are purely schematic, for purposes of clarity.

(*a*) Double-toggle closing mechanism with latch and trip catch: The two toggles are clearly shown in Fig. 5.17A, with the circuit breaker in the closed position. The closing force is applied in the direction shown, with a latch taking over and propping the circuit breaker closed when this force is removed. The trip catch

CONTACTS CLOSED

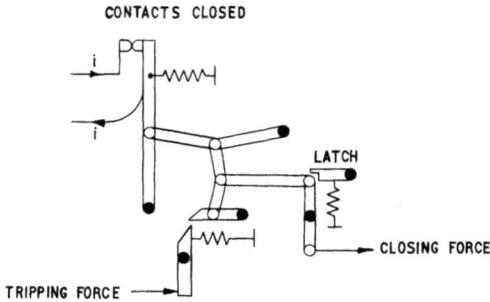

Fig. 5.17A Double toggle closing mechanism linkage with latch and trip catch

must be reset before a closing operation takes place, thus propping one end of the first toggle. If the trip catch is not reset, the end of the first toggle will move out of the way when the closing force is applied, and the moving blade will not close, giving a trip-free close operation. No details of the closing linkages from the power source to the first toggle are shown, as there are many variations, which include bell cranks, levers etc. that convert the power source from its sited position into a force applied in the correct direction and with the correct mechanical advantage. No details are shown of the close latch and the trip catch as, once again, there are many variations. It is important with a tripping mechanism to ensure that, during the tripping stroke, the closing latch is released and the trip catch reset before the circuit breaker completes its opening stroke. Meeting all these tripping requirements, together with those associated with the closing requirements, forms the complete design and usually results in a mechanical compromise within the design parameters.

(*b*) Cam-operated closing mechanism with latch and trip catch: The cam and roller are shown in Fig. 5.17B, with the circuit breaker in the closed position. A latch props the circuit breaker closed when the closing force is removed. The trip catch illustrated is a roller-hook trip toggle, which, when reset, converts the toggle into a solid link. The action of knocking the hook off the roller allows the toggle to collapse and the cam to rotate, letting the roller move down the cam face and opening the circuit breaker.

(*c*) A high-speed-tripping-mechanism latch type: This is illustrated in Fig. 5.17C. This circuit breaker has two moving contacts, identified as front and rear, and has a combined closing and tripping spring. The action of closing the circuit breaker is first to energise the solenoid, which charges the closing spring, resets the front contact together with the latch and the trip catch and, at the same time, moves back the rear contact in preparation for closure. When the

CONTACTS CLOSED

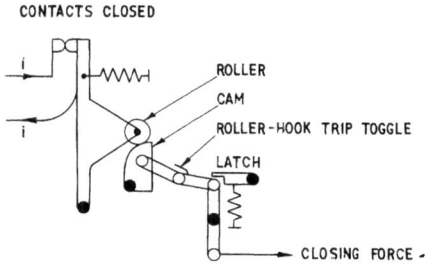

Fig. 5.17B Cam closing mechanism linkage with latch and trip catch

solenoid de-energises and resets, the rear contact follows the return of the solenoid by the action of the closing spring and the contacts close. The tripping action releases the trip catch first, which in turn releases the latch. The closing springs now become tripping springs and pull the front contacts away from the rear contacts at a high speed. The rear contacts are prevented from following the front contacts by a stop in the mechanism. High-speed opening is performed by the high spring tension moving only the light mass of one set of combined arcing and main contact. In (a) and (b), where toggles were involved, longer opening times occur because of the time taken for a toggle to collapse.

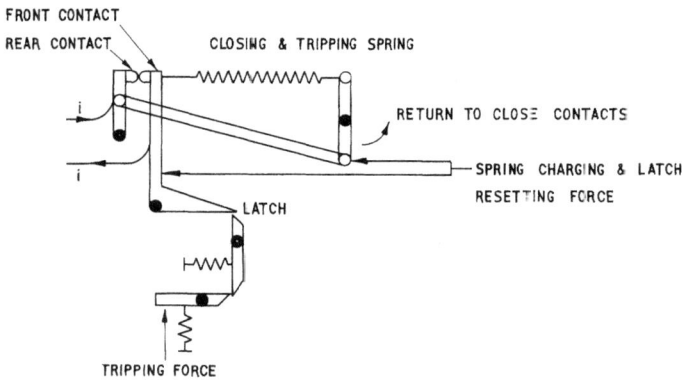

Fig. 5.17C High-speed tripping mechanism linkage with latch and trip catch [Whipp and Bourne Ltd.]

BUCKING BAR

EXTERNAL CONTROL SOURCE

CLOSING FORCE

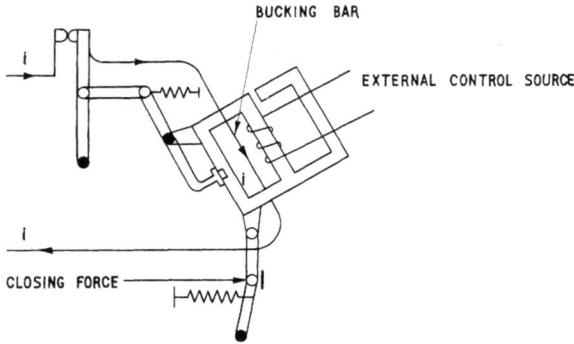

Fig. 5.17D High-speed tripping mechanism — electrically held [ITE Power Equipment]

High-speed mechanisms of type (c) can have a mechanical opening time of 0·003 s, to which must be added the operating time of the short-circuit release. Both forward- and reverse-current releases can be fitted on this type of mechanism as it is inherently nondirectional in its methods of operation. The electronic tripping and impulse-coil release, shown in Figs. 5.14C and D is ideally suited to this type of mechanism. They have release operating times of less than 0·001 s.

(d) A high-speed-tripping-mechanism electrically held type: This is illustrated in Fig. 5.17D. The armature is sealed to the magnet as long as the external control source is energised and the contacts are closed through the closing toggle, as shown, The conductor called the bucking bar carries the same current as the moving contact and produces a flux in the armature, to assist that provided by the external source. When fault current flows in the bucking bar, it must be in the opposite direction to the normal current flow, this produces a cancelling flux in the armature limb of the magnet, which releases the armature. The light-weight contacts are then free to open by the high-pressure pulloff springs. After tripping, the magnet resets, and resets the armature. Another variation on this type of mechanism is the use of a permanent magnet in place of the magnet produced by the external control source. The principle of operation and flux diversion is the same. Opening times of these mechanisms are of the order of 0·003 s inclusive of the release operating time. This type of mechanism is inherently directional in its high-speed operation, i.e. the direction of the tripping required determines the connection of the current to the bucking bar. Forward and reverse high-speed tripping can be obtained by using two separate armatures.

5.5 Fused circuit breaker

The fused circuit breaker (Allison *et al.*, 1972) has come into existence because of the need for high-speed protection on motor circuits, to limit the possible ensuing damage caused by a faulty motor terminal box. The position of the fuse to the circuit breaker is very important when considering the two interrupting devices connected in series, i.e. the fuse and the circuit breaker. Each device has a proved and accepted short-circuit rating independent of the other. The former being a high-speed device, which, because of its rapid current-forcing characteristic, can produce high arc voltages. The latter, being a high-arc-energy type does not normally produce overvoltages, because of its inherent resistance-switching properties. When considering both devices arcing at the same instant, the circuit breaker arc path could be subjected to the fuses' high arc voltage as a stress between the arc column and earth, while the circuit breaker is arcing This voltage stress may produce a flashover to earth in the circuit breaker, since its own proving series was not subjected to this stress. Therefore the safest arrangement is when the fuse is mounted on the incoming side of the circuit breaker.

Because the circuit breaker is always in the zone protected by the fuse, there is no longer a need for the circuit breaker to have a short-circuit rating as high as the fuse. The circuit-breaker interrupting rating can be reduced to a value not less than the takeover value of the maximum normal-current-rating fuse to be used with the circuit breaker. This is illustrated in Fig. 5.18A, which depicts the current/prearcing-time characteristic of such a fuse with the minimum-opening-time characteristic of the circuit breaker crossing at a takeover current of 11 kA. The interrupting rating of the circuit breaker need be no higher than this value, because it can never be called on to interrupt higher currents, since the fuse will always operate first with its shorter prearcing time. Relay times are ignored in this concept, thus catering for instantaneous earth-fault protection and the possibility of remaking onto a fault with the relay not reset.

The combination of the fuse and lower-rated circuit breaker has been termed a motor switching device, to distinguish it from a fused contactor, because of the large difference between their respective interrupting ratings. The curve shown in Fig. 5.18A illustrates this, with the contactor having an interrupting rating of 2·1 kA at a power factor of 0·3 maximum and the motor-switching device having an interrupting rating of 11 kA at a power factor of 0·15 maximum.

The specification and demand for motor switching devices has led to the development of specific economic designs meeting the above requirements, all having the features normally expected of a circuit breaker with the added features of a fuse. Fig. 5.18B shows a double-tier 3·3 kV motor switching device having the same panel dimensions as an equivalent single-panel fully rated circuit breaker. The particular features of the unit illustrated, in addition to those described above, are as follows:

(*a*) Top- and bottom-tier units are interchangeable.

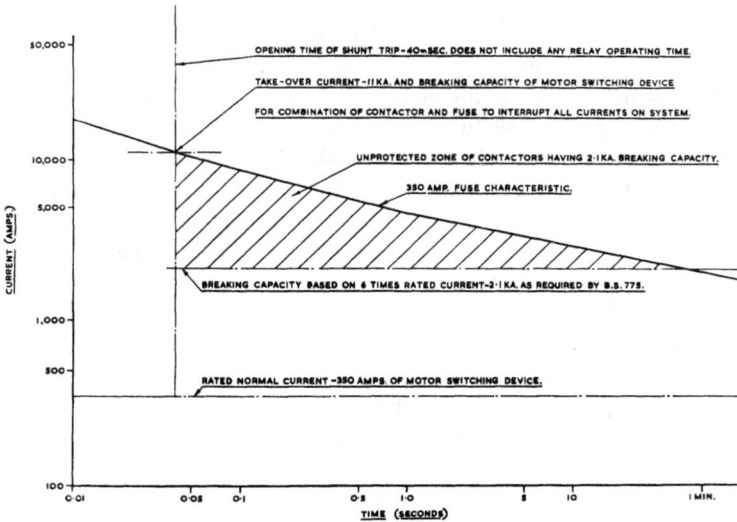

Fig. 5.18A Motor switching device characteristic curves [Whipp and Bourne Limited]

(*b*) Striker-pin fuses are accommodated, which operate auxiliary switches to trip, to lock out close, and to give fuse-blown indication and alarm services.

(*c*) A fully rated integral circuit earth switch operated through the front of the unit gives safe earthing, thus allowing its use on possible feedback circuits, i.e. reversing motors, step-down transformers etc.

(*d*) The moving carriage can be withdrawn on extensible slide rails to its full extent outside the cubicle, giving access to the fuses, circuit breaker and earth switch. It can also be tilted forwards to expose the interior of the cubicle and moving isolating contacts.

(*e*) Automatic individual shutters are provided for both busbar and circuit connections, the latter catering again for the use on possible feedback circuits.

(*f*) The circuit breaker and its arc chute, contacts, air puffer etc. are all similar, but of smaller size, to their equivalents used on a fully rated circuit breaker, and comply with the same design principles.

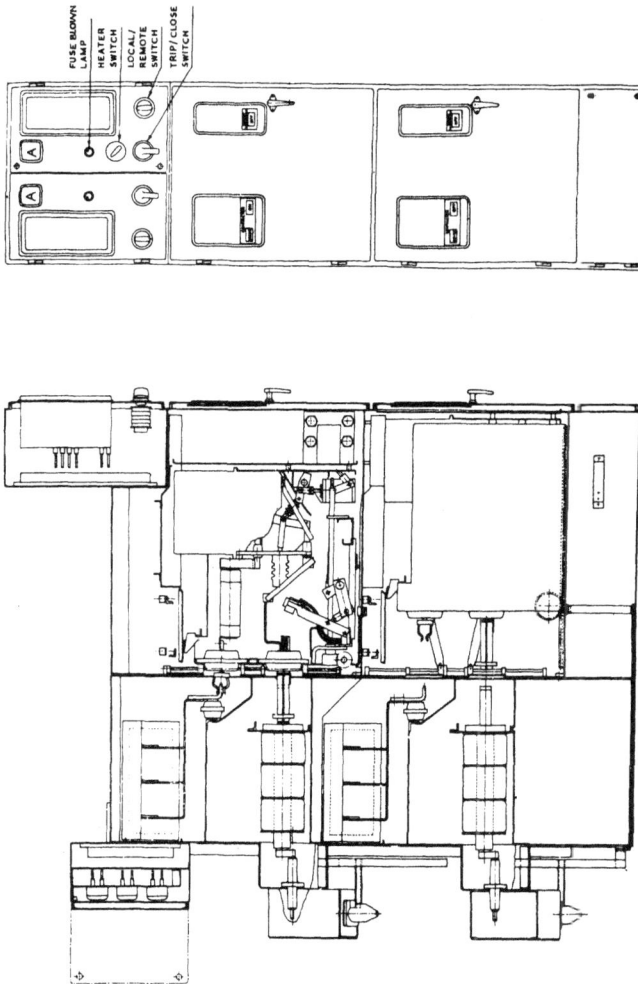

FUSE BLOWN LAMP

HEATER SWITCH

LOCAL/ REMOTE SWITCH

TRIP/CLOSE SWITCH

Fig. 5.18B Double tier 3·3 kV motor switching device [Whipp and Bourne Limited]

5.6 Acknowledgments

The preparation of this chapter has required information and illustrations freely given by A. Reyrolle & Co. Ltd., ITE Power Equipment, Merlin and Gerin, S.p.A. Costruzioni Electtromeccaniche, Westinghouse Electric Corporation and Whipp & Bourne (1975) Ltd.

5.7 References

ALLISON, E. H., and MORTON, J. S. (1972): 'Air break motor switching device' in 'Metalclad switchgear – Pt. 1' 81, *IEE Conf. Publ.* **83**, pp. 76-81

BRANDT, T. F., and NETZEL, P. C. (1969): 'New circuit breakers for modern d.c. traction power systems', ITE Power Equipment Publication SPC-MON-3.

FAY, F. S., THOMAS, J. A., LEGG, D., and MORTON, J. S. (1959): 'Development of high voltage air-break circuit breakers with insulated-steel-plate arc-chutes', *Proc. IEE,* **106A**, pp. 381-391

FRINK, R., and KOZLOVIC, J. M. (1957): 'A magnetic de-ion air breaker for 750 MVA 13·8 kV'. AIEE Conference Paper CP. 57-225

MORTON, J.S. (1980): 'Coil-less cold cathode arc chutes for high-speed d.c. circuit breakers for use on traction systems,' *IEE Proc.,* **127**, Pt. B, pp.34-45

YOUNG, A. F. B. (1953): 'Some researches on current-chopping in high voltage circuit breakers', *Proc. IEE,* **100**, Pt. II, pp. 336-348

Chapter 6

Air-blast circuit breakers

S. M. Gonek, C. Eng., F.I.E.E.

6.1 Introduction

This chapter deals with the technical aspects of the design of air-blast circuit breakers as seen in the early 1980s and does not establish cost or economic parameters for them (see chapter 13). It may therefore be worthwhile to bear in mind a number of factors which have been influencing their fortunes throughout their lifetime of some 40 years and will continue to do so in the future.

The most significant of these factors in a purely technical sense has been the well proven ability of the air-blast circuit breaker to meet each and every rating and performance requirement that has arisen so far in the power circuit-breaker field. Perhaps second to this has been the relative ease with which their design could be adapted to suit the various forms of switchgear from indoor cubicle through metal clad to indoor or outdoor live tank constructions.

However, as switchgear literature of the various periods would readily show, these impressive technical capabilities have, on the whole, been instrumental mainly in paving the way for the air-blast circuit breaker to be the first in the field in meeting most of the major upward changes in the technical requirements such as doubling or even trebling of the ratings, more stringent overvoltage limits or higher operating speeds.

Important as this undoubtedly has been at various times, the same literature would just as readily show that this technical ability of the air-blast circuit breaker to reach the forefront has not, by any means, been accompanied by an equal ability for it to survive commercially at every level it had first occupied. On the contrary, it would show that the steadily increasing knowledge of actual power system requirements and at times very rapid advances made in the circuit-breaking technologies of the other media have been creating a number of circumstances in which an increasing number of the technical requirements (see chapter 3) could be met by one or more of the other circuit-breaker types. Not surprisingly therefore, when and where such technical parities occurred and the other circuit-breaker types were shown to be economically more advantageous than the air-blast type, the latter had to regress from and, in some cases, even abandon the leading positions it had at first occupied.

Fortunately for the air-blast circuit breaker, this economic erosion process in some fields has been accompanied by further expansion of the power systems in other fields thus creating new demands for still higher ratings, higher operating speeds, more stringent overvoltage limits and so forth. These new demands and further exploration of the capabilities of compressed air have continued to create conditions in which air-blast circuit breakers could continue to readjust by moving up the technical scales to occupy the new fields and, in repeating the past patterns, re-establish their economic viability through meeting the newer and more arduous requirements.

This, in fact, had been the situation up to the early 1970s when two major factors intervened. The first of these was the emergence of the single-pressure type SF_6 circuit-breaker (Chapter 7) which overcame the complexities and the relatively poor r.r.r.v. performance of the earlier two-pressure types and was thus able to match the air-blast type up to and including the topmost fields. The second was a pronounced increase in the demand for the e.h.v. metalclad substations brought about by the environmental pressures and the trend towards maintenance-free installation which required the design of the circuit breaker to be integrated into and become fully compatible with the overall form of the substation employing SF_6 insulated busbars and connections.

Thus, whilst the first of these factors introduced a worthy challenger to the air-blast circuit breaker in its principal open-terminal form, the second factor placed several size and shape constraints on its metalclad form which could be better met by the SF_6 circuit breaker type. Consequently, the combination of the two factors forced the air-blast circuit breaker to regress once more from the leading position it had first occupied in the field, but, for the first time in its life of some 40 years, it now appears that this regression is unlikely to be followed by a recovery in accordance with its previous life patterns.

So, as this brief outline of the air-blast circuit breaker fortunes is intended to show, the essence of their design lay not so much in their relatively well known technical capabilities but in an evolution of the technically compliant constructions whose total costs to the user must be equal to or lower than those achievable through any other means or media. Therefore, the technical design parameters dealt with in this chapter need not be looked on in terms of the specific electrical or mechanical properties or capabilities of the compressed air medium, but in terms of their overall significance in relation to the complete development, manufacturing, installation and running costs of a circuit breaker in association with the other substation components.

Accordingly, and with the above qualification in mind, the principal advantages of the compressed air medium are:

(*a*) the availability of the medium in its natural form which disposes of the health hazard problems and limits the costs which could otherwise be incurred in its replenishment, renewal or reprocessing. This also permits the design principles to be based on its expendability in production and in service.

(*b*) the elasticity which, in contrast to the liquid media, permits supporting structures to be designed free of arc generated pressure transients and reactions

(*c*) the mobility (high wave and particle propagation velocities) which permits the medium to be easily ducted, allows it to be stored remotely from its actual working zones and enables it to follow closely the expansions and contractions of the arc column thereby contributing to short arcing times, low erosion of the contacts, reduced maintenance requirements and makes it eminently suitable for 'high speed' design applications

(*d*) the relative constancy of its characteristic gas properties over the full range of ambient temperatures and design pressures which eliminates the costs which could otherwise be incurred in the provision of ancillaries needed to maintain the medium in its useable working state

(*e*) the direct useability of the medium as a source of the easily amplifiable mechanical power needed to drive and control the circuit breaker mechanically.

(*f*) the nonflammability which, by widening the markets, can also contribute to reductions in the quantity dependent production costs

(*g*) the relative chemical inertness and the consequent compatibility with the readily available lower cost construction materials and an easement in the maintenance procedures

(*h*) the 'adjustability' of the electrical interrupting and insulating properties by means of changes in the design pressure

(*i*) the high voltage rating which, with proper measures, can be achieved on relatively few interrupters

Against this, and in the same economic context, the disadvantages are:

(*a*) the relatively high cost of compressor and (if any) dryer systems particularly in small installations or where each circuit-breaker has to be provided with its own, functionally self-contained, ancillary units

(*b*) the cost of silencers where noise level limitations make their use unavoidable

(*c*) the relative sensitivity of the medium's interrupting abilities to circuit severities (rates of rise of restriking voltage) which make the use of r.r.r.v. control resistors necessary at the higher levels of 'per-unit' ratings

(*d*) the constant pressure, constant interrupting effort, feature of the circuit breaker which at light current levels leads to 'chopping' of the current and makes the use of overvoltage control resistors mandatory in most cases

(*e*) the relatively high cost of pressure vessels and enclosures

The balance of these advantages and disadvantages indicates that air-blast circuit breakers ought to find their place principally in the higher voltage ranges where high speeds and stringent overvoltage limits are specified and also at the lower voltage ratings associated with very heavy short circuit or normal current requirements.

6.2 Properties of compressed air medium

6.2.1 General

Starting with atmospheric air, the electrical properties of which are described in chapter 12, its compression to an appropriate pressure, accompanied by cooling and condensation does, of course, leave the compressed air fully saturated with water vapour, and, as such, electrically unuseable. This mixture of atmospheric gases and vapour must therefore be further processed to reduce its moisture content by either chemical, adsorption, absorption or expansion drying to a degree which will ensure that precipitation of water on the internal insulation surfaces can not occur under service conditions.

For all practical design purposes this medium can, except for the presence of oxygen, be treated as a chemically inert gas compatible with a very wide range of constructional materials and imbued with a quantity of easily useable mechanical energy. The presence of the oxygen even in its compressed state does not, however, embarrass the design unduly as the constructional materials must, in any case, be suitable for coming into contact with it for extended periods during manufacture or in storage. Also, a low moisture content and humidity of compressed air can almost eliminate the usual corrosion processes associated with the electrolytic reactions between dissimilar metals.

Taking these advantages of the medium as read, its remaining properties of direct consequence to design can best be examined in relation to the two functions which it must perform in the air-blast circuit breaker, the electrical and the mechanical. The electrical properties can be conveniently examined in relation to the two working conditions of the medium; the 'static' properties when the medium is essentially at standstill and the 'dynamic' properties when the medium has to traverse the electrically stressed zones at relatively high velocities.

6.2.2 Static electrical properties

The static properties apply to any internal parts of the circuit breaker where the medium acts as an insulant and is therefore subject to the phenomena dealt with in chapter 12. Accordingly, with the modern air-blast circuit breaker pressures being in excess of 10 bars and with design voltage gradients reaching into the regions above 10–20 MV/m levels, the breakdown strengths of at least the most highly stressed zones can no longer be deduced from the Paschen law even for relatively infrequent and simple uniform field conditions. Instead, the design must take into account further reductions in the electrical strength due to nonuniform fields and a still further reduction brought about by the not easily accountable electrode surface macro and microirregularities, variations in the material's electron emission properties and the impurities which the compressed air is bound to contain in practice.

The most highly stressed compressed air zone in an air-blast circuit breaker of the live tank type occurs usually in the 'break' or one of the 'breaks' performing the duty of an isolator between the breaker terminals. In the dead tank or metalclad constructions where the internal terminal-to-terminal voltage strength should be greater than that to earth, this zone may well **occur** between a conductor and the tank outside the 'break' or 'breaks'. Taking the live tank conditions as being more pertinent and excluding any external 'breaks' of an open air isolator type, this isolating duty can be performed by either the interrupters or the isolators when pressurised in the open position. Within these, greater constructional restraints apply to the former where the isolating requirements must be combined with the interrupting and the circuit making (closing) requirements with the result that the electrostatic field utilisation factors (average to maximum voltage gradient ratios) tend to centre on values of the order of 0.5, seldom exceed more than 0.7, but may even be as low as 0.3. Fig. 6.1 shows typical 'conditioned' peak flashover levels obtained on 2·5 mm gaps in a uniform and a 0·5 utilisation factor fields at pressures up to 120 bars.

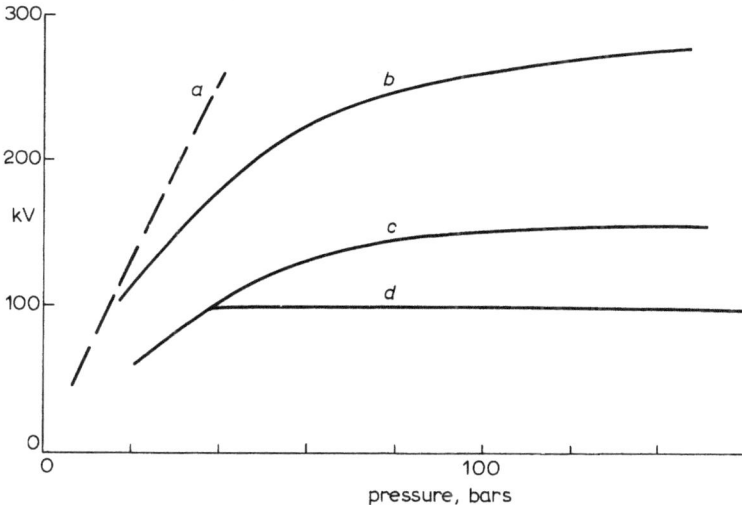

Fig. 6.1 Typical power frequency flashover levels for 2·5 mm gaps

   (a) Paschen's law
   (b) uniform field (conditioned level)
   (c) field factor 0·5 (conditioned level)
   (d) field factor 0·5 (1st flashover level)

It will be noticed that, as explained in chapter 12, Paschen's Law in this case ceases to apply to the uniform field configurations at about 100 kV pk achieved at approximately 15 bars. Likewise, that uniform field values can provide a useful guide to the 0·5 field factor values over the full range of pressures in that the latter tend to approximate to about 50% of the former. However, it will also be seen that the 'unconditioned' lowest flashover levels plotted for the 0·5 field factor gap intercept the 'conditioned' level increases at pressures of the order of 30—40 bars and that any further increases in the pressure do not raise the useable voltage withstand levels significantly above the 100 kV peak value.

In commercial interrupter contact systems where (for reasons dealt with later) the actual contact gaps are invariably greater than the 2·5 mm used in this example, the 'unconditioned' flashover level is quite considerably less than the pro rata value derivable from small gaps due, of course, to the changing geometry of the field as well as the increased applied voltages. Bearing in mind the very scanty information now available on the relationships between the breakdown levels of small and larger gaps in their 'unconditioned' state and realising that any empirical formulas derived for that purpose are bound to vary between the individual arrangements, the formula which appears to give a reasonable approximation for the typical electrode configurations with the field utilisation factor of about 0·5 can be written down as

$$V_{bd} = ad + bd^n \tag{6.1}$$

where

V_{bd} = breakdown voltage
d = gap between the electrodes
n = index
a, b = empirical constants

Over the range of the gaps most frequently used on the interrupters, say up to 50 mm, the formula may be further simplified to read

$$V_{bd} = cd^n$$
where $\tag{6.2}$

c = constant

For the electrode configuration used in the example given in Fig. 6.1 and with 'd' measured in millimetres, the formula was found to read

$$V_{bd} = 70 \, d^{0·4} \text{ kV (peak)} \tag{6.3}$$

giving approximately 100,200 and 250 kV first breakdown levels for gaps of 2·5, 15 and 25 mm, respectively.

At lower voltage stresses and air pressures where the first low breakdown phenomena do not arise, the breakdown level-gap relationships given in chapter 12 apply.

6.2.3 Dynamic electrical properties
The second group of the electrical properties, the dynamic, applies to circumstances in which the medium has to perform its electrical functions while flowing at

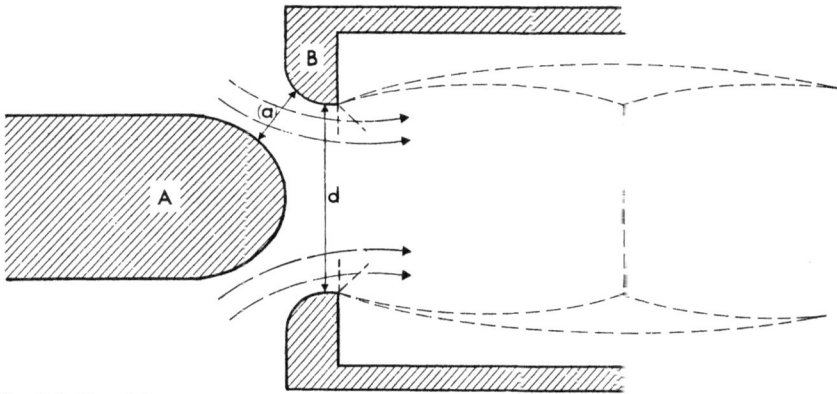

Fig. 6.2 Simplified interrupter contact system for monoblast

A plug electrode
B nozzle electrode
a air-flow approach area
d nozzle diameter

relatively high velocities, i.e. when the changes in its density and the flow phenomena approach or exceed the critical (Mach 1) conditions. In so far as the electrical performance of the circuit breaker is concerned, these conditions arise in the interrupter contact system zones which, for illustrative purposes and in absence of an arc, may be simplified as shown in Fig. 6.2.

In the single nozzle (monoblast) system as shown, the axisymmetrical electrode A in the form of a plug, and the other axisymmetrical electrode B in the form of a hollow cylinder provided with a 'nozzle' opening d, are placed in a chamber supplied with compressed air. The 'exhaust' passage through the hollow electrode B, connected to a zone of low pressure, causes the high pressure air to flow past the electrode A, through the conical approach areas a into the nozzle d and hence via the exhaust passage to the atmosphere. In alternative arrangements, where the plug electrode A is replaced by another nozzle electrode B (duo-blast systems), the approach areas a are formed by cylindrical areas defined by the axial distances which separate the two electrodes.

Bearing in mind that the pressure ratios between the high and the low pressure zones in an air-blast circuit breaker by far exceed the critical ratio for air (1:0·528) and that the approach areas a usually exceed the nozzle areas d by a factor of 1·5 or more, the cold air mass flow rate through the nozzle is directly proportional to the product of the working pressure and the area of the vena contracta formed just downstream of the nozzle. The flow through the vena contracta is sonic and a further expansion and acceleration of the air in the downstream exhaust duct increases the velocity of the flow to supersonic levels at which it is intercepted by the appropriate shock-wave patterns (as indicated by the dotted lines). The flow patterns and the air densities upstream of the vena contracta are governed by the flow approach areas a formed by the upstream shapes and dispositions of the electrode surfaces and, beyond this, also by the boundaries associated with the air supply ducts and vessels.

In actual designs of air-blast interrupters these electrode positions and the air-flow conditions are reached by prior opening of the appropriate inlet or exhaust valves and by movements of the electrodes or by-pass contact systems which bridge the two electrodes when the interrupters are closed. The full operating sequence of an interruption consists therefore of two phases; the 'opening' period during which the electrical clearances and the air-flow conditions alter in accordance with the movements of the electrodes or the by-pass contact systems, and the 'fully open' period when the electrode or contact movements have ceased and the air flow has built up to its full working rate. Under these conditions the voltage strengths of the interrupter are, of course, no longer determined by the 'static' (constant density, fixed geometry) laws dealt with previously but by a combination of the changing electrical clearances, electrical fields and air densities during the 'opening' period and, thereafter, by the changing air densities in the 'fully open' period.

Thus, if the air-flow conditions obtainable in the interrupters were completely uniform, the expected voltage strengths in the absence of any prior arcing (the 'no-load' strengths) could be determined theoretically from the densities derived from the isentropic gas equations and the electrostatic field gradients calculated from the geometries which gave the weakest (but not necessarily the shortest) dielectric breakdown paths. Such 'ideal' no-load voltage strengths, reaching up to about 50% of the equivalent 'static' strengths during the early opening and up to about 80% in the fully open positions of the electrodes are illustrated in Fig. 6.3 in which the later general downwards trend of the strengths may be assumed to be due to a falling air supply pressure.

However, design constraints, imposed by the achievable electrode and upstream passage configurations, do not permit such completely uniform air-flow conditions to be achieved in practice where the radial components of the air flow approaching the nozzle, the disturbances originating in the upstream air supply passages and the high pressure gradients just upstream of the vena contracta all combine to produce undesirable air-flow distortions of a transient or even persistent nature. These distortions may take the form of relatively minor air density disturbances which lower the voltage strengths by small amounts, or very severe distortions in the form of air vortices which, when fully developed, can form small diameter cores of very low pressure, low dielectric strength 'threads' between the two electrodes. Owing to this some electrode systems when first tested may exhibit rather startling no-load voltage weaknesses of the type indicated by shading in Fig. 6.3. The usual patterns obtained on such occasions, verifiable by high-speed Schlieren filming techniques, tend to display a relative absence of any very pronounced weaknesses during the early interrupter operating sequences and an increasing incidence of stronger and

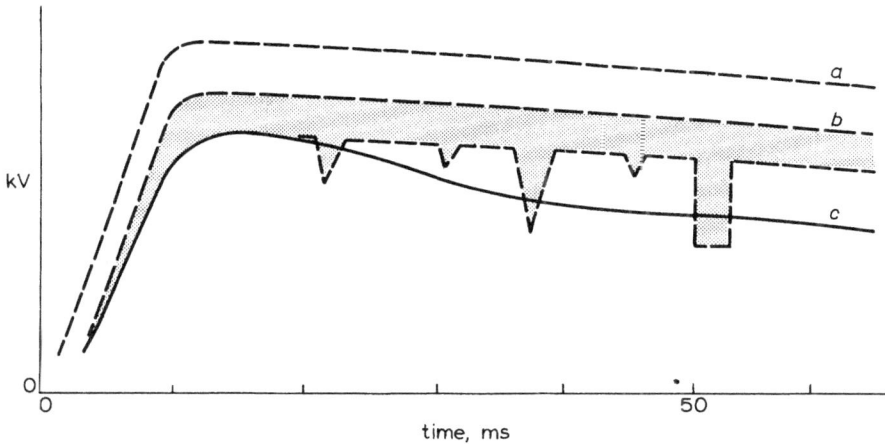

Fig. 6.3 'No-load' voltage strengths of interrupters

a static air
b flowing air — ideal
c flowing air — final

more persistent weaknesses as the duration of the flow increases. This time delay effect, explainable by a finite time needed for the vortices to start and develop fully, can be used as a guide to the efficacy of the counter-measures which, in the form of modifications to the flow boundaries, have to be evolved to suppress the vortices and so raise the voltage strengths to a consistent minimum and acceptable level.

Bearing in mind that it is not possible to eliminate all of the air-flow disturbances in practice due to the very nature of the constructions and that further minor voltage strength reductions must be allowed for to cater for the prior arc erosion of the electrode surfaces and other not yet fully identified minor phenomena, the resultant final no-load voltage strengths of the interrupters must invariably fall short of the ideal values by amounts that depend as much on the intrinsic properties of air as on the peculiarities of the individual designs. Typically, these final no-load strengths or, more accurately, the minimum breakdown values, may be expected to reach about 90% of the ideal strengths at times of the order of 10–25 ms from the start of the operation and to fall to about 80% at 40–60 ms as indicated in Fig. 6.3. It should, however, be noted that the stated relationships between the static and the ideal strengths can be altered at will by changing the ratios between the air-flow approach and the nozzle areas which materially affect air densities and hence also the breakdown levels and paths. Likewise, that the relationships between the ideal and the final strengths cannot be dissociated from the actual nozzle sizes due to increased air-flow control difficulties encountered at higher air-flow rates. It should also be noted that both the ideal and the final voltage strengths obtainable under flowing air conditions can be made almost equal to the static strength if the nozzle areas are made sufficiently small in relation to the air-flow approach areas.

Proceeding to the interrupting characteristics, it is, of course, the purpose of the movements of the electrode or the by-pass contact systems (mentioned in connection with Fig. 6.2) to draw the arc in the vicinity of the nozzle so that it is forced to transfer into and burn along the centre line of the nozzle. Depending on the chosen arrangement, one of its roots attaches itself to the centre of the plug electrode 'A' and the other to either the internal surfaces or a specially provided arc probe within the nozzle electrode 'B' on 'monoblast' or, alternatively, both roots will take up the positions described for the electrode 'B' on 'duo-blast'. In either case a portion of the arc will burn upstream and the remainder downstream of the nozzle.

It is therefore clear that, by comparison with the no-load flow conditions, the air will now be heated and expanded all along the length of the arc and, consequently, that both the nozzle and the exhaust passages must be of sufficient size to cope with the increased volumes of flow at any given design pressure, current and arc length. Failure to meet this condition must therefore result in a partial or complete stopping or even a reversal of the flow through the nozzle and a failure of the interrupter to establish the necessary interrupting conditions. This failure mode is usually referred to as nozzle 'chocking' or 'blocking', but inadequate exhaust passages can produce the same effect.

If the scavenging of the arc is adequate, i.e. if the size of the nozzle, the exhaust passages and the pressure are sufficient to establish an adequate flow of air at a given current and arc length prior to zero current, the interrupter can only clear the circuit in the immediate post zero period if its recovery rate along the post-arc column exceeds the rate of rise of restriking voltage (r.r.r.v.). Its inability to do so is usually referred to as 'r.r.r.v.' failure which can take the form of either an immediate re-establishment of the arc or a sudden collapse of the restriking voltage some way up along its front ('normal' r.r.r.v. failures) or, alternatively, a relatively slow transition from the restriking to the arc voltage values accompanied by heavy damping and flow of heavy post-arc currents ('thermal' failures, chapter 2). If the r.r.r.v. is within the capabilities of the interrupter, it can still fail to interrupt if the restriking voltage values exceed its voltage strength during the intermediate post-arc period when the post-arc hot spots, metal particles and vapours reduce its 'no-load' strength by amounts related to both the durations and the values of arc currents (voltage failures).

These three failure modes of a typical air-blast interrupter, operating at a short-circuit level close to its maximum rating, are illustrated in Fig. 6.4. In this the

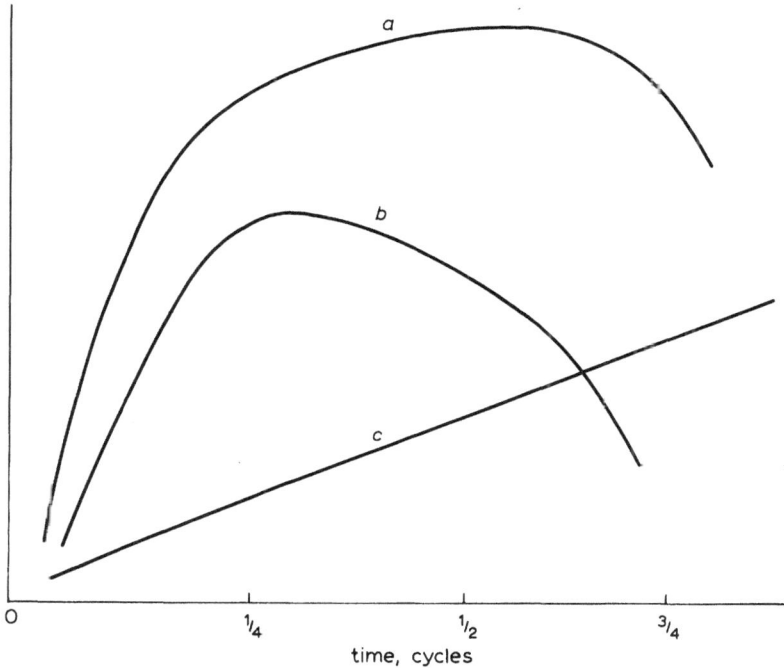

Fig. 6.4 Typical voltage and r.r.r.v. limits of interrupters at heavy s.c. currents

    (a) voltage
    (b) r.r.r.v.
    (c) electrode travel (position)

time scale represents arcing times from the first striking of the arc to a subsequent zero-current instant and the voltage and r.r.r.v. plots indicate the variations in the corresponding interrupting abilities in relation to the opening travel of the electrodes. On designs employing fixed electrodes this travel may be assumed to represent increasing electrode separation distances.

    Starting with a condition under which the electrodes part at or just prior to a zero current, their close proximity, the associated small air-flow area and the insufficient time available to transfer the arc from its starting position into the nozzle, limit both the voltage and the r.r.r.v. interrupting capabilities to a negligible level at first. Thereafter a progressively earlier parting of the electrodes in relation to the zero-current instant allows the several hundred microseconds needed for the arc to transfer into the nozzle and increases both the distance between the electrodes and the air-flow area, but, it should be noted, also increases the magnitude of the arc current and its duration to the first zero. The increasing inherent interrupting capability resulting from this in spite of the heavier duty imposed on the interrupter, raises the voltage and the r.r.r.v. limits quite rapidly up to a stage at which the air-flow approach area becomes equal to or slightly exceeds the nozzle area (as implied at the ¼ cycle time in the plot). Further separation of the electrodes beyond this point does not, of course, increase the air-flow areas but increases the length of the arc and its duration. Consequently, the increased energy input into the nozzle system and, in most cases, a deterioration in the air-flow patterns reduce the more sensitive r.r.r.v. ability quite rapidly while the increased distances between the electrodes permit the voltage strengths to rise, but at a progressively reducing rate. Thereafter, further travel of the electrodes and the consequential increased energy input upstream of the nozzle reduce, stop, and ultimately reverse the flow and lead to a partial and then complete deterioration of at first the r.r.r.v. and afterwards the voltage performance capabilities of the interrupter through the already mentioned 'nozzle chocking' mode of failure. So, as proscribed by these rising and falling r.r.r.v. and voltage characteristics, the travel of the electrodes has to be arrested (or their relative positions set) at a specific distance in order to obtain the best r.r.r.v. or voltage rating. Also, as implied by the different distances at which these maxima occur, a degree of choice may be exercised between the highest possible r.r.r.v. and voltage capabilities.

The low voltage and r.r.r.v. limits at small separations of the electrodes usually cause the interrupter to reignite if the first current zero occurs within a few milliseconds of the 'electrodes part' instant. For this reason the normal maximum arc durations expected of an air-blast interrupter on single-phase heavy current duties amount to these few milliseconds plus the duration of the following current loop.

Taking an electrode system in which the travel is arrested (or the electrodes are set) to give specific voltage and r.r.r.v. performance at a heavy short-circuit current, the constant air pressure, constant interrupting effort feature of the interrupter allow it to cope with progressively higher voltages and r.r.r.v. levels as the short-circuit current is decreased. This produces the well known air-blast characteristic illustrated in Fig. 6.5 in which voltage and r.r.r.v. limits are plotted to a short-circuit current base. It will be seen that the voltage limit of the interrupter rises steadily but relatively little as the short-circuit current is decreased to a small fraction of the full rating and then increases rapidly to terminate ultimately at the 'no-load' level. A typical order of values expected at, e.g. 100 and 25% of the full rating would be 0·6 and 0·75 of the 'no-load' level, respectively.

In contrast to these voltage limits, the r.r.r.v. performance changes quite markedly over most of the short-circuit current range where the well known hyperbolic approximation of

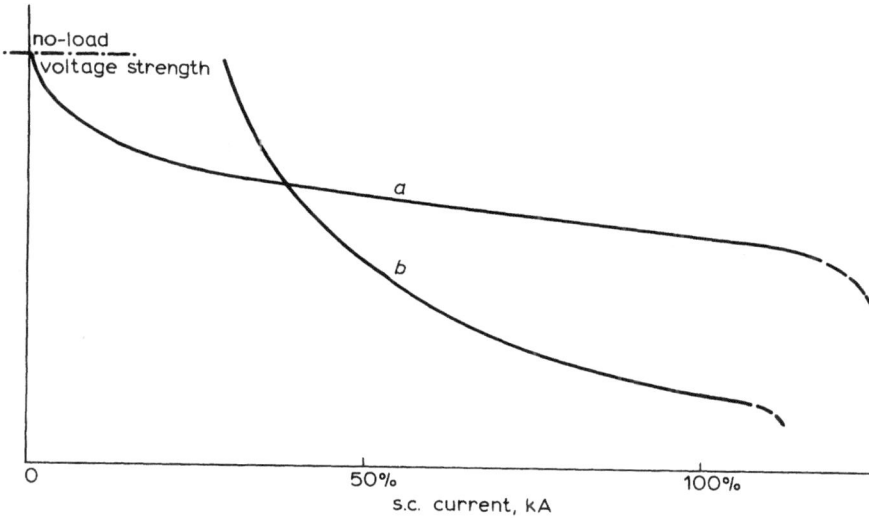

Fig. 6.5 Typical voltage and r.r.r.v. characteristics

    (a) voltage

    (b) r.r.r.v.

$$\text{r.r.r.v.} = \frac{K_1}{I^n} \quad \text{or} \quad \frac{K_2}{\left(\dfrac{di}{dt}\right)^n} \tag{6.4}$$

gives a useful guide to the typical air-blast characteristics (chapter 2). In this approximation the expressions K_1, K_2 and the index 'n' tend to remain constant for a given pressure and construction and, as the rate of change of current just prior to the interruption is not significantly affected at medium and heavy currents, the terms I^n and $(di/dt)^n$ are equivalent at the same power frequencies. Although most of the investigations and theories on interruption by air-blast have concentrated on the I-r.r.r.v. characteristics, the differences in test levels, test models and commercial designs do not permit definitive values to be assigned to K_1, K_2 and n. So, in spite of an enormous amount of good work carried out to-date, the quoted or derived values of K vary appreciably even if the nominal nozzle sizes and pressures appear to be the same. Doubts must therefore exist on whether these variations could be due to different arc lengths (Fig. 6.4) or to the not easily quantifiable differences in the shapes of the electrodes or the air-flow conditions.

On the whole, smaller variations appear to apply to the index n whose values tend to approximate to 1 but even here, perhaps for similar reasons, significant differences can exist.

The scanty information available on the effects of varying pressures appears to confirm the theoretical postulation that the r.r.r.v. capability is proportional to pressure if all other parameters remain unaltered. An even scantier information on different nozzle sizes and geometries suggests that the maximum short-circuit currents are approximately proportional to the nozzle areas at constant pressure if the upstream arc lengths are adjusted to hold the r.r.r.v. capability constant, but here precautions must be taken to ensure that the exhaust passages do not impede the flow at higher currents and flow rates.

It is, of course, possible to combine these pressure and nozzle area effects with eqn. 6.4 to relate them to currents and r.r.r.v.s in a more general empirical expression showing that the constants K are approximately proportional to the no-load mass flow through the nozzle. However, although this new expression has been found to provide a useful guide to the interrupter design parameters, its applicability is limited to similar constructional geometries of the nozzle systems and, within these, to the quantity ratios of not more than about 4:1.

When the current is reduced to a relatively light level, of the order of 0·5 to 2 or so kiloamperes, the constant pressure, constant interrupting effort of the air blast gives rise to a suppression of the natural current waveforms prior to the zero current instant. This, at still lower currents develops into the well known 'low current chopping' phenomena, which for all practical design purposes can be regarded as an instantaneous rupture of the circuit followed by either its clearance or by a series of successive restrikes and clearances ('multiple' chopping). Depending on the interrupter and the circuit, the instantaneous values of the chopped current may be as high as 100 A, i.e. well above the values at which high overvoltages, in excess of the circuit or the circuit breaker withstand levels, can be produced.

6.2.4 Mechanical properties

In contrast to the deficiencies in the fundamental knowledge of the arc interruption phenomena, the mechanical properties of compressed air are so well known that it is usually a matter of choice which of the numerous equations or formulas should be used to obtain a result with an adequate degree of accuracy.

Given the fact that the flow through the nozzle is sonic and therefore that the conditions downstream of it cannot influence the upstream flow, practically all of the compressed air supply systems for the air-blast nozzles can be represented by one of the three simplifications shown in Fig. 6.6. All of these contain a compressed air receiver (a), a nozzle (b) open to an unrestricted low pressure region and a fast acting valve (c), which for the purpose of these representations is assumed to open instantaneously and without any restrictions to the flow. In the first of these arrangements an integral combination of the nozzle and the valve is mounted directly on the air receiver wall. The second and the third arrangements

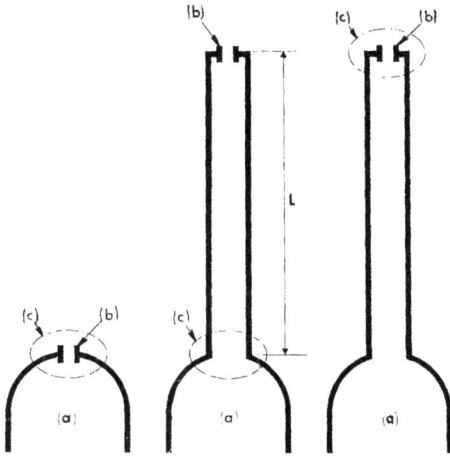

Fig. 6.6 Simplified compressed air systems

    (a) receiver
    (b) nozzle
    (c) fast acting valve
    (L) length of duct

are provided with ducts of a uniform area and a finite length L; the former has the nozzle mounted at the remote end and the valve at the receiver end, the latter has both the nozzle and the valve mounted at the remote end. In all three cases the air flow approach areas within the receivers are assumed to be sufficiently large to be neglected and the nozzles are treated as zero length apertures in one plane.

The assumptions that a uniform state of gas exists at each cross-section of the flow and that the friction forces due to the velocity gradients near the enclosure walls or due to changes of density may be neglected, permit the basic equations to be written as:

$$\frac{\delta V}{\delta t} + V \frac{\delta V}{\delta x} + \frac{1}{\rho} \frac{\delta p}{\delta x} = 0 \text{ (equation of motion)} \tag{6.5}$$

and

$$\frac{\delta \rho}{\delta t} + V \frac{\delta \rho}{\delta x} + \rho \frac{\delta V}{\delta x} = 0 \text{ (equation of continuity)} \tag{6.6}$$

in which the quantities refer to the values and incremental changes in velocity V, density ρ, pressure p, time t, and distance x.

Under conditions where there are no exchanges of heat or dissipative forces, the flow is isentropic and a further equation, gives the pressure–density relationship as:

$$\frac{P}{\rho^\gamma} = \text{constant c} \tag{6.7}$$

where γ is the ratio of specific heats.

This last equation does not hold in cases involving shock waves and must be replaced by:

$$\tfrac{1}{2}(p_1 + p_2)\left(\frac{1}{\rho_1} - \frac{1}{\rho_2}\right) = \frac{1}{\gamma - 1}\left(\frac{P_2}{\rho_2} - \frac{P_1}{\rho_1}\right) \tag{6.8}$$

in which the subscripts 1 and 2 refer to the downstream and the upstream sides of the shock plane.

The solution for the first of these simplified arrangements is, of course, completely straightforward in that apart from any almost macroscopic effects within the immediate nozzle zone, an instantaneous or a progressive opening of the nozzle flow area immediately establishes the flow at a rate determined solely by the solution of the eqns. 6.5 and 6.7 for the sonic flow rate:

$$\frac{dm}{dt} = 0.00405 \frac{C_d A P_r}{\sqrt{(T_r)}} \text{ kg/s} \tag{6.9}$$

where the subscripts r refer to the condition of the air in the receiver, C_d is the coefficient of discharge for the flow area A in mm^2, P is the pressure in bars and T the temperature in degrees K.

When used in conjunction with an air receiver volume this formula yields useful air receiver draining curves reproduced for an initial air temperature of 288 K in Fig. 6.7. The curves give both the isentropic (instantaneous) and the isothermal (final) ratios between the finishing and the starting pressures at time T in seconds for a receiver volume of 1 litre discharging through an effective area of 1 mm^2.

A further application of this formula and one derived for subsonic rates of flow from the same equations gives a useful air receiver filling curve reproduced in Fig. 6.8 for a constant pressure supply at 288 K feeding the same 1 l volume through the effective 1 mm^2 flow area. In this plot the finishing-to-starting pressure ratios are isentropic and the time T in seconds is referred to a completely evacuated receiver at $T = 0$.

The usefulness of these plots lies in the ease with which the draining or filling times or pressure ratios for any volume V litres, effective area $C_d A$ mm^2 and a starting temperature θ degrees K may be calculated using the conversion

$$\text{new time } t = \frac{VT}{C_d A}\sqrt{\left(\frac{288}{\theta}\right)} \text{ seconds} \tag{6.10}$$

In contrast to these straightforward solutions and formulas for the first arrangement, the solutions for the second and the third arrangements in Fig. 6.6 must be

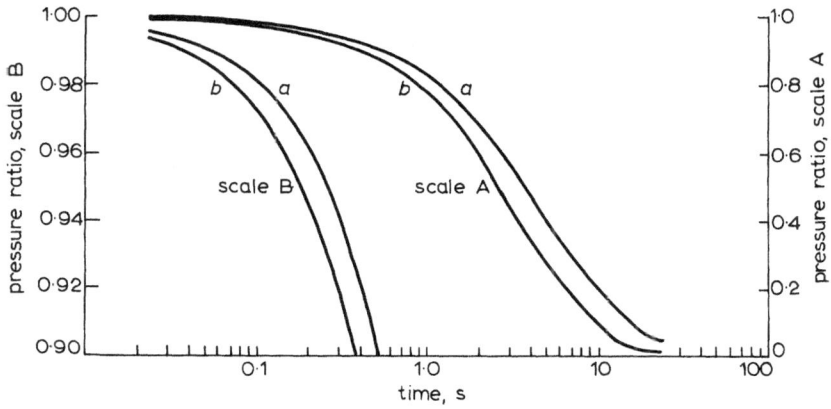

Fig. 6.7 Air receiver draining curve at critical flow rates
volume = 1 l, flow area = 1 mm^2, initial temperature 288°K (15°C)

    (a) final
    (b) instantaneous

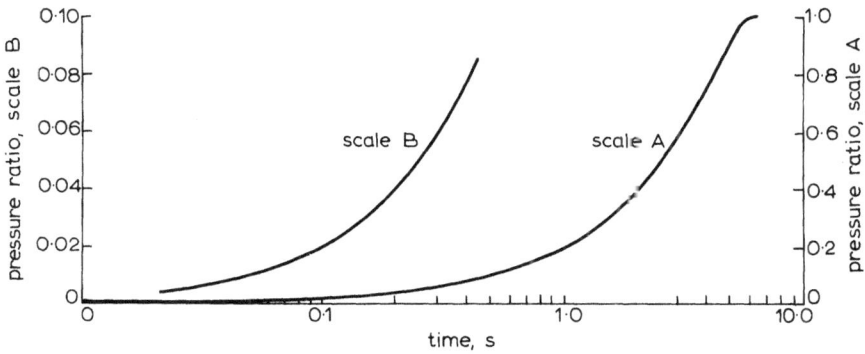

Fig. 6.8 Air receiver filling curve
volume = 1 l, flow area = 1 mm^2, Initial temperature 288°K (15°C)

based on laborious step-by-step integration methods employing eqns. 6.5—6.7 for simple waves and eqns. 6.5, 6.6 and 6.8 for shock waves. The processes must take into account both the wave and the particle motions, the opposite and the like polarity reflections of the waves at the open and closed ends of the duct, the duct lengths and all the volume, length and area ratios. Consequently, each specific set of conditions produces its own specific answer and, because of the assumptions and simplifications made in order to obtain these answers, their main value lies not in any readily useable design formulas but in an enhanced understanding of the pneumatic phenomena that do or might occur in air-blast circuit breakers.

In the case of the second arrangement, these calculations would show that the instantaneous opening of the valve would cause an intermediate pressure 'plug', led by a shock wave front, to race into the duct at the local velocity of sound until it reached the end of the duct. If the nozzle at the end of the duct were closed, a reflection of this step fronted pressure wave would raise the pressure at the end of the duct to a new reflected pressure level and this would then travel back towards the air receiver leaving behind it a column of static air. Depending on the ratio between the reflected and the receiver pressures, a new expansion or shock wave would be reflected from the open end of the duct at the receiver end and this process of reflections would, in the absence of any losses, continue ad infinitum.

However, the presence of the open nozzle at the end of the duct would both reduce the strength of these reflections and, in making up the loss of air through the nozzle, cause a separate rarefaction wave to travel towards the receiver where on meeting an open end it would be reflected back as a compression wave. So, instead of the perpetual reflections in a closed end duct, all pressures and waves in this arrangement would decrease until all compressed air has been discharged from the receiver.

A part of this process, associated with the effects produced by the nozzle, applies to the third arrangement where the instantaneous opening of the nozzle would cause a rarefaction wave to travel to the receiver whence it would return as a compression wave. Here again the magnitude of the subsequent reflections would decrease until the rate of flow through the nozzle became subsonic and all compressed air has been discharged from the receiver.

By comparison with the idealised conditions which permit all these phenomena to be calculated, the presence of bends, obstructions, losses and, above all, relatively slow opening of the valve or nozzle flow areas alter the conditions to such an extent that usually only mere shadows of any shock or steep fronted waves can be discerned in practice. This applies particularly to the second arrangement shown in Fig. 6.6 where, instead of a sudden increase of the first reflected pressure at the nozzle to a value several times greater than the receiver pressure, the actual pressure changes are gradual and the values seldom exceed the receiver pressure. Owing to the same reasons, any rarefaction waves caused by the nozzle in the second and the third arrangements tend to be heavily damped but the associated pressure drops at the nozzle are significantly greater than predicted by the idealised calculations.

It should of course be remembered that the presence of the arc in the nozzle during an interruption can materially affect the degree to which the nozzle can be considered to be open in terms of any pneumatic calculations. Also that the interruption which takes place within the first 15–20 ms of the flow period through the nozzle is comparable with the time needed for a pressure wave from the nozzle to travel to the receiver and back through a duct of 2·5–3 m in length, so that any pressure drop caused by opening a nozzle cannot be made up from the air receiver on higher voltage circuit breakers unless the receiver is placed in a close proximity to the interrupters.

6.3 Interrupter types

There are only two fundamental ways in which a blast of compressed air can be directed onto the arc; transversely to it as a 'crossblast', or longitudinally along its length as an 'axial blast'. Although the first of these is sometimes mentioned as a possibility for air-blast interrupters, its inherent inability to produce the required pressure and velocity conditions in the arc zones makes it unsuitable for high-power high-voltage applications other than as a low pressure auxiliary in the air-break designs (chapter 5). Accordingly, all modern air-blast interrupters employ only the 'axial blast' principle by forcing the arc to burn coincidentally with the axis of a nozzle.

As mentioned in Section 6.2.3, two basic axial blast nozzle systems are being used: a monoblast (single flow per interrupter) system in which the air subjects the arc to one unidirectional blast, or a duo-blast (double flow) system in which the air divides on approach to two nozzles of equal size and, while flowing in opposite directions, subjects the arc to two blasts per interrupter. A variant of the latter system, in which one of the nozzle areas is made smaller, is also being used and this is usually referred to as 'partial duo-blast' to distinguish it from the normal duo-blast having equal flow areas.

Referring to Fig. 6.9, let a monoblast system (a) and a duo or partial duo-blast system (b) be built into an insulating enclosure supplied with compressed air from a general direction indicated by the arrows R. Let the air supply to the nozzles be controlled by a 'blast valve' B placed on the upstream side somewhere between the nozzles and the supply source. Also, let the exhaust passages, downstream from the nozzles, be controlled by 'exhaust valves'. E. In order to meet the minimum requirement each of these arrangements must admit compressed air to the nozzles while the exhaust passages are open and then shut off the air supply to prevent high pressure reservoirs being exhausted.

This minimum requirement can be met in all cases by blast valves B placed somewhere in the duct leading to the nozzles as shown or across the electrodes just upstream from the nozzles. In the former case one blast valve may be arranged to serve a number of interrupters at the same time and the interrupter chambers are

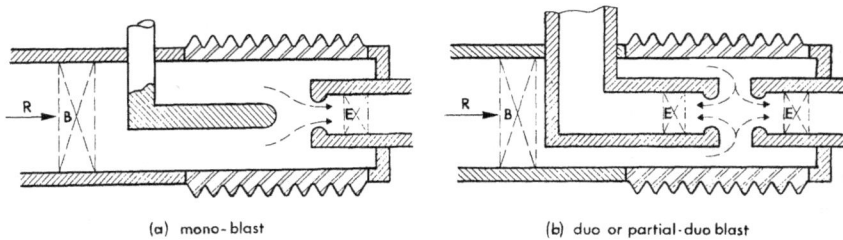

(a) mono-blast   (b) duo or partial-duo blast

Fig. 6.9 Compressed air controls for air blast interrupters

R = compressed air supply
E = exhaust valves (downstream)
B = blast valves (upstream)

not left pressurised outside the interrupting periods. In the latter case one blast valve can serve only one interrupter and the interrupter chambers are permanently pressurised up to the valve.

If it is required that the interrupters should be pressurised in open position only, both the blast valves B and the exhaust valves E must be used; the former to admit the air to the nozzles, the latter to stop the flow and hold the interrupter pressurised. One blast valve may again supply more than one interrupter at a time and one exhaust valve may be arranged to control two adjacent exhaust passages in a twin interrupter unit.

If it is required that the interrupter chambers should be permanently pressurised (i.e. in closed and open positions) either exhaust valves on their own or in combination with blast valves placed across the electrodes must be used. One exhaust valve may again serve two interrupters, but separate blast valves must in this case be provided for each interrupter.

So, as can be seen from the foregoing, all present-day air-blast interrupters employ axial-blast principles and, in so far as nozzle systems and pressurisation are concerned, can be divided into no fewer than nine types, i.e. mono or partial-duo or duo-blast, each pressurised in one of the three ways: (*a*) during interruption only, (*b*) during interruption and in open position and (*c*) permanently.

It should be mentioned that two other terms, axial flow and radial flow are sometimes used in descriptions or classification of air-blast interrupters. The former of these usually refers to interrupter constructions where, due to an arrangement of the upstream passages, the predominant direction of the air flow approaching the nozzles is parallel to the axis of the interrupter as in Fig. 6.9*a*. The latter of these terms usually refers to constructions where most of the air tends to approach the nozzles centripetally as, for instance, between the duo-blast electrodes in Fig. 6.9*b* or as would be the case if the monoblast electrodes in Fig. 6.9*a* were placed in the centre of an air receiver. Neither of these terms invalidates the axial-blast principle.

6.4 Circuit-breaker types and basic arrangements

Depending on the shapes and the materials used for any externally exposed insulators and the degree of weather protection afforded to the working parts, air-blast circuit breakers divide into 'indoor only' and 'indoor/outdoor' types. Each of these types subdivides into two further categories; first, an 'open terminals' category in which the terminals and possibly also other 'live' parts of the circuit breaker are exposed externally, and secondly, the 'metalclad' category in which the terminals and all other 'live' parts are enclosed in an integrally designed earthed metal cladding.

6.4.1 'Open terminals' type

Owing to the ease with which compressed air may be ducted to the interrupters, a great number of arrangements can be used to obtain the same functional solution for an 'open terminals' circuit breaker. Because of this it is difficult to describe or assign a specific name to any one of these arrangements other than in relation to Fig. 6.10, which, for the sake of simplicity, has been drawn for one pair of interrupters without any valves, isolators or other working parts which may have to be added in practice.

Bearing in mind that one interrupter may be used instead of two and that each of the arrangements shown can accept monoblast or duo-blast interrupters, the following general observations apply to Fig. 6.10:

Arrangement (a)

This is by far the most popular arrangement throughout all voltage and s.c. current ratings from a single interrupter 3·3 kV indoor to the highest voltage outdoor applications. The arrangement places the air receiver at earth potential where it can serve as a base for the interrupter support column or columns and provide a suitable platform or attachment points for the mechanisms, mounting feet, control cabinets etc. This solution offers full freedom in the choice of blast valves, exhaust valves and their combinations (Section 6.3) so that the interrupters and the support columns can either be permanently or intermittently pressurised. At lower voltages, one air receiver may serve all three phases of the circuit breaker through one blast valve. At higher voltages it may serve a number of support columns carrying two, four or even more interrupters at a time.

The physical distance between the interrupters and the air receiver tends to subtract from the attractiveness of this arrangement at high voltages due to time delays incurred in either sending the air from the receiver to the interrupters when the blast valves are mounted at the receiver end, or in making up pressure drops produced by the interrupter when the support columns and the interrupters are pressurised.

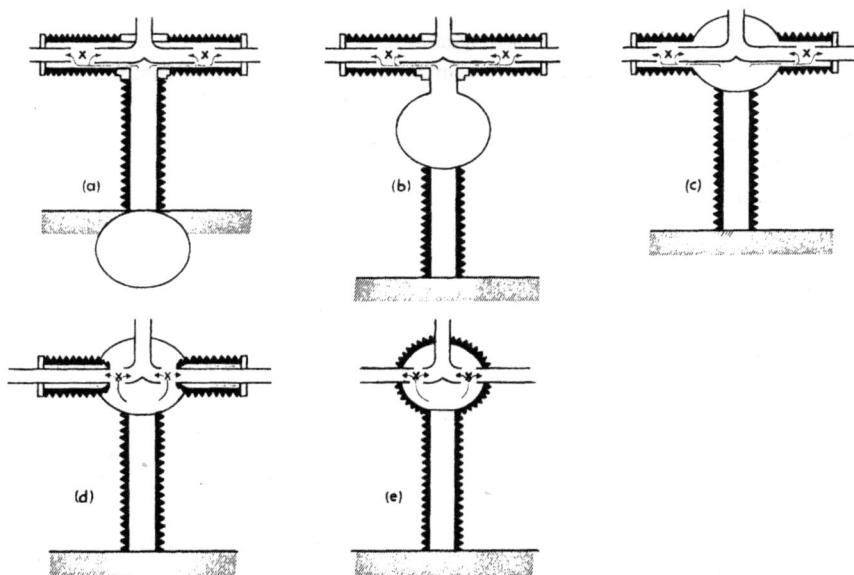

Fig. 6.10 'Open terminals' circuit breakers
Dispositions of receivers and interrupters

(a) metal receiver at earth potential
(b) metal receiver in proximity to interrupters
(c) metal receiver with partly built-in interrupters
(d) metal receiver with fully built-in interrupters
(e) insulating receiver with fully built-in interrupters

In its overall concept this arrangement permits the 'live' sizes and weights of the circuit breaker to be held to a minimum but with a penalty of having to provide relatively large bore compressed air ducts between the receiver and the interrupters.

Arrangement (b)
In this arrangement the receiver is mounted at high potential and in proximity to the interrupters. As in the case of arrangement (*a*), full freedom in the choice of blast valves or exhaust valves exists and the interrupters can be permanently or intermittently pressurised. The arrangement is generally intended for higher voltage ratings and more than one pair of interrupters may be served by one receiver if suitable insulators are interposed between them and the receiver. The overall dispositions and functions in such cases become reminiscent of arrangement (*a*).

Owing to greater proximity of the receiver, better pressure and flow conditions can be achieved in the interrupters, but with a penalty of increased 'live' sizes and weights. The main support insulators need no longer be designed to provide large bore compressed air ducts to the receiver.

Arrangement (c)

In this arrangement the air receiver is moved even further towards the interrupters and now encloses some of their working parts. Unless suitable by-pass connections are provided, the current carried by the interrupters has to pass through the receiver and nonmagnetic or only slightly magnetic shells or inserts may have to be used in its construction to prevent overheating.

The arrangement does not readily accept intermittently pressurised interrupters and is therefore used almost exclusively with exhaust valves, but blast valves of the type which bridge the electrodes can also be employed. As in the previous case, it is intended for higher voltage ratings, but only two interrupters can be conveniently supplied from one receiver. Nearly ideal pressure and flow conditions are available but again with a penalty in the 'live' sizes and weights.

Arrangement (d)

This is generally similar to arrangement (c) but the two hollow interrupter insulators are now replaced by bushings and the receiver encloses most of the working parts on the interrupters. The pressure and flow conditions are about the best that can be obtained on practical constructions. By-pass current connections are not possible with this arrangement.

Arrangement (e)

This shows perhaps the ultimate arrangement which can be achieved if the receiver is constructed of suitable insulating materials such as epoxide resin impregnated glass cloths or windings. Unfortunately, technical complications, associated with the necessary large through-wall openings and the sheds or convolutions needed to increase insulation creepage distances, make such arrangements uncompetitive, at least for the time being.

Dead tank arrangements

It is, of course, possible to adapt arrangement (d) Fig. 6.10 (by either dispensing with the central exhaust altogether with monoblast interrupters or providing an additional bushing in the case of duo-blast) so that the air receiver can be placed at earth potential while the terminals at the outer end of the bushings remain exposed and 'live'.This arrangement is very similar to metalclad designs except that it is permissible for the exhaust gases to be discharged while they are still ionised.

6.4.2 Metalclad type

In contrast to the 'open terminals' alternatives, the relative positions of the receiver and the interrupters on metalclad arrangements are limited to one or at most two

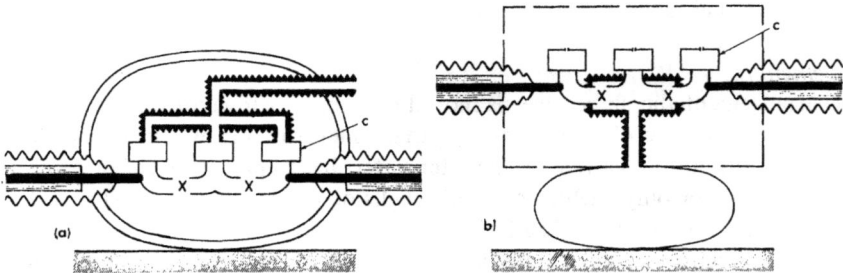

Fig. 6.11 Dispositions of receivers and interrupters on metalclad circuit breakers

(a) interrupters built into air receiver
(b) interrupters external to air receiver
(c) deionising chambers

basic dispositions dictated by the need to enclose all 'live' parts in an earthed metal cladding and deionise the exhaust gases fully before they leave the circuit breaker.

Using the same schematic pair of interrupters without any valves or isolators as in the previous examples, the following general observations apply to Fig. 6.11

Arrangement (a)
This arrangement can be best looked upon as a redevelopment of arrangement (d) in Fig. 6.10 in that the interrupters are fully built into the receiver and the main connections are taken out via the two bushings. Accordingly, it is well suited to accept permanently pressurised interrupters but the receiver may have to be constructed from nonmagnetic metals or be fitted with nonmagnetic inserts. It will be seen, however, that in order to deionise the exhaust gases the downstream passages no longer lead through any direct ducts to atmosphere but to three chambers C, one for each exhaust gas flow path. In each of these chambers the hot ionised gas fed during the arcing period is mixed with relatively clean and cold air flowing through the nozzles after interruption. The resultant mixture is passed through baffles and filters to further lower the temperature and remove all solid and most of the gaseous contaminants. The mixing, cooling and filtering of the gases and a definite period of time during which they are held in the chamber ensure that they can be safely passed through an arrangement of insulating ducts to atmosphere.

Arrangement (b)
It is possible, particularly at lower voltages, to enclose completely a 'live terminals' circuit breaker in an earthed metal cladding fitted with bushings providing again

that the exhaust gases are sufficiently deionised before they enter the cladding. In the arrangement shown, the chambers C can be assumed to be significantly simpler and smaller than those shown in arrangement (*a*), and the venting of the chambers to atmosphere need no longer be done through insulating ducts.

A variant of this arrangement, using through-wall bushings for the exhaust passages can also be employed to place the deionising chambers outside the first earthed metal barriers. The deionising chambers must, in such cases, be still enclosed in their own earthed metal claddings.

6.4.3 Isolators

Depending on the required rating, duty cycles, environment, and the type of interrupter adopted for the circuit breaker, one of three isolator types or isolating means must be used to permit the flow of compressed air through the nozzles to be terminated and enable the circuit breaker to maintain the required voltage strength between its terminals (chapter 9) after interruption. Noting that practically all of those isolators or isolating means are designed to cater also for the circuit closing ('making') functions and that more than one isolating 'break' may be used in series with the interrupters particularly at higher voltage ratings, the three types or means and their principal characteristics are:

External 'free-air' isolators Fig. 6.12a–c

This method employs an isolating break in free air externally to the circuit-breaker enclosures and can therefore be used only on circuit breakers of the 'open terminals' type. Fig. 6.12*a* depicts an arrangement in which the break is achieved by means of a blade or link rotating in the vertical plane. Fig. 6.12*b* indicates how the same function can be performed by a centrally held blade or link rotating about its centre in the horizontal plane. An interesting application of the free-air isolation, developed by AEG, is shown in Fig. 6.12*c* where a linear movement of the electrode (1), passing through a nozzle at n, is used to combine the interrupting and isolating functions.

As implied by this 'free-air' isolating method, all of these arrangements must employ relatively long 'breaks' to attain the necessary voltage strengths at atmospheric pressure and their contacts and conductors are exposed to the environment which, in the case of outdoor circuit-breakers, may include heavy pollution and icing conditions. Although shown to be mounted directly off an earthed base, arrangements *a* and *b* can be mounted on top of the circuit breaker to provide a number of 'unit breaks' in series analogous to arrangement *c.*.

All of the free-air isolators used to close the circuit cause the arc to be struck externally to the circuit breaker over distances determined by the dielectric strength of atmospheric air and voltages impressed across the break. Accordingly, the amount of ionised gas generated during closing operations and the degree to which the conductors or contacts are eroded by the arc are related to closing speeds, short-circuit currents and voltage ratings of the 'break'. This, together with

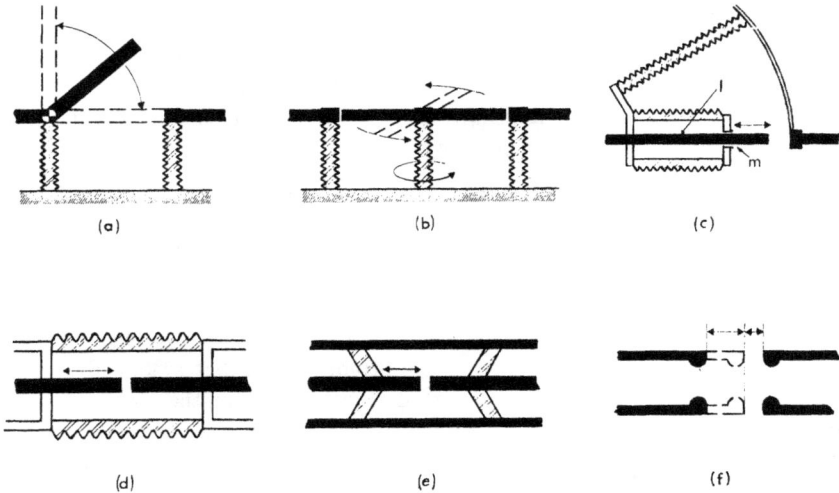

Fig. 6.12 Isolators and isolating means used on air-blast circuit breakers

(a), (b) and (c)   external 'free air' isolators
(d) and (e)   pressurised isolators
(f)   combined interrupter/isolator arrangement with single or two beat movements

increased sizes and inertias of the moving parts tends to limit the application of free-air isolating and circuit closing methods to lower voltage, lower MVA ratings, particularly in the case of arrangements *a* and *b* when mounted off an earthed base. The same applies, but to a lesser degree, to these arrangements when mounted on top of the circuit breaker and to arrangement *c*.

Pressurised isolators Fig. 6.12d–e
In order to reduce prearcing on closure of the circuit breaker, increase the operating speeds, protect the moving parts and contacts from adverse environmental influences and increase voltage or MVA ratings, linear movement isolators, permanently immersed in compressed air, can be used. The usual form these take on 'open terminals' circuit breakers is presented graphically under Fig. 6.12*d* where the 'break' is made within an insulated enclosure, and, except for special cases, is positioned at or near the geometrical centre of the enclosure to take advantage of the natural electrostatic field configurations. In metalclad or dead tank applications, the insulating enclosure is replaced by metal and the supports for the

isolator take the form of conical, disc or post type insulators. A diagrammatic representation of such an isolator, supported by conical insulators, is shown in Fig. 6.12e.

In all these arrangements the flashover distances are determined by air pressure and electrostatic field stresses. Prearcing on closure can be reduced by high speed movements and use of multiple breaks in series, so that the arc current values do not rise to more than, say, 50% of the asymmetrical current peak before a physical contact between the conductors is made. Owing to this, the amount of ionised gas and the erosion of the electrodes can be held within acceptable limits at the highest voltage and current ratings demanded so far or envisaged in the foreseeable future.

Interrupters with built-in isolating and closing means (Fig. 6.12f)
Owing to increased voltage strengths inherently available between the interrupter electrodes under static (no flow) conditions dealt with in Sections 6.2.2 and 6.2.3, it is usually possible to achieve the necessary voltage withstand levels for isolating purposes by holding the electrodes pressurised in their normal arc extinguishing position. This facility enables the interrupting and isolating functions to be incorporated in the design of the interrupter providing, of course, that the interelectrode gap is held pressurised by means of suitable exhaust valves. An addition of arc metal inserts and separate normal current carrying contact surfaces permits such interrupters to close the circuit without undue prearcing or erosion if the closing speed is sufficiently high.

In relatively rare cases where the gap between the electrodes in the arc extinguishing position is inadequate or when an additional further movement of the electrodes suits the design solution mechanically, two-beat movements can be used. The first movement places the electrodes in the arc extinguishing position; the other, after a slight pause, increases the electrical gap to raise its strength or shut off the flow, or both.

The isolating and circuit-closing performance of these interrupters is equivalent to pressurised isolators, but additional precautions must be taken to ensure that any ionised gases generated on closure against heavy short-circuit currents, do not affect the interrupting capability particularly on 'CO' duty cycles.

6.4.4 Voltage control impedances
There are five main cases that may require voltages impressed on or induced by air blast or other high performance circuit breakers to be controlled by means of added shunt connected impedances:

(a) during interruption of small inductive currents to reduce overvoltages produced by current chopping

(b) during interruption of circuits by multi-interrupter circuit breakers to ensure that each interrupter takes its due share of restriking (high rate of change) voltage transients (*c*) during the intermediate postinterruption period and the following 'open-circuit' period on multi-interrupter or multi-isolator circuit breakers to ensure that each 'break' takes its due share of recovery and open-circuit (power frequency) voltages

(*d*) during closing onto open-circuited transmission lines to reduce closing overvoltages

(*e*) where it is more economical to employ r.r.r.v. damping resistors than provide larger numbers of interrupters or raise the working pressures.

In all of these cases all damping of overvoltages and r.r.r.vs can only be performed by means of low-inductance resistors connected either across the interrupters or (surge-diverter-wise) to earth. All sharing of voltages amongst the breaks can be controlled by either similar resistors or capacitors connected across the breaks to act as potentiometers. Because of their thermal rating limitations and, in many cases, unacceptable power losses that would otherwise ensue, all resistor circuits are only intermittently energised and rated. Some means to switch them in or out of the circuit must therefore be provided in the form of either resistor current interrupters or suitably adapted isolators.

Many formulas, equations and computer programs are available to calculate the effect of these resistors or capacitors on the circuit or the circuit breaker under 'idealised' conditions in which all circuit inductances, capacitances and resistances are assumed to be constant, the circuit breaker behaves as a perfect switch and the circuits themselves are quite significantly oversimplified. As a result, very few of such calculations can be applied directly in design without some allowances being made for, as appropriate, nonlinearities of some of the circuit 'constants', the mutual interactions between the circuit and the circuit breaker and, where known, the probable effects of any secondary circuits left out of the equations. It is therefore an accepted practice in design to use the results of such idealised calculations as a guide rather than an answer to the practical behaviour of the circuit breaker. The following observations are therefore intended to cover only the trends rather than the detailed substance of the various phenomena or issues which must be taken into account when dealing with design of voltage control impedances:

(a) Small inductive currents

In the oversimplified circuit shown in Fig. 6.13*a*, the constants C, L and R are assumed to represent a highly inductive load connected through a circuit breaker shunted by resistor R_1 to an infinite source of supply having a peak voltage value of E.

Assuming that the supply voltage is constant during the period under consideration, the oscillatory solution for a single current chop occurring on falling current (when the voltage due to the chop adds to the voltage on inductance L) is given by

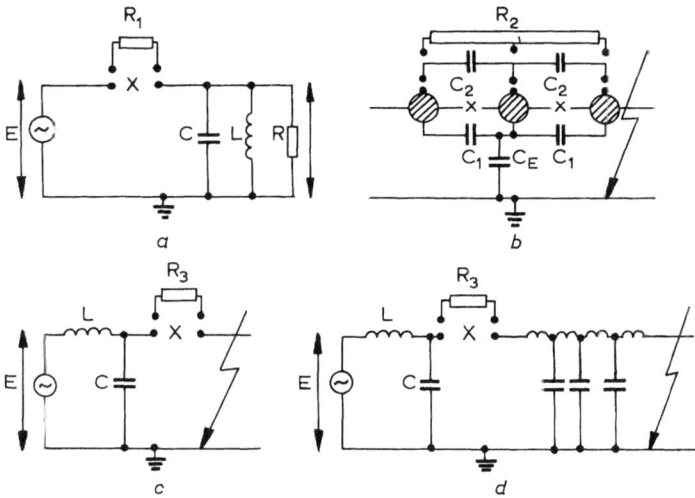

Fig. 6.13 Simplified single-phase circuits

    (a) small inductive currents
    (b) two-interrupter circuit breaker
    (c) short circuit at circuit breaker terminals
    (d) short-distance fault (s.d.f.)

$$v = (A \sin nt + B \cos nt) \exp(-at) \tag{6.11}$$

where

$$a = \frac{1}{2R'C}$$

$$R' = \frac{R_1 R}{R_1 + R}$$

$$A = \left(\frac{E}{R_1} - \frac{v_0}{2R'} + i_1 \right) \frac{1}{nC}$$

v_0 = voltage across inductance at $t = 0$
i_1 = current through inductance at $t = 0$

$$n^2 = \frac{1}{LC} - a^2$$

$B = v_0$
E = peak supply voltage

and the maximum voltage across the inductance is given when

$$\tan nt = \frac{An - Ba}{Aa + Bn}$$

If all damping of this circuit is omitted, i.e. when R and R_1 are infinitely high, eqn. 6.11 becomes

$$v = E \sqrt{(m^2 \sin^2\theta + \cos^2\theta)} \qquad (6.12)$$

where

$$m = \frac{f_n}{f_e}$$

f_n = resonant frequency
f_e = supply frequency
θ = chopping angle measured from $i = 0$

The maximum voltage across the inductance in this case is given by:

$$v = i \sqrt{\frac{L}{C}} \qquad (6.13)$$

when $\theta = \pi/2$, that is when the chop occurs at an instant when the current through the inductance is at its maximum. It follows therefore that the highest voltage that could be produced under this condition would occur when the product of the peak chopped current and the surge impedance of the highly inductive load $\sqrt{L/C}$ is at its maximum within the range of currents which the circuit breaker is able to chop.

It is equally clear that if the circuit breaker is fitted with a resistor R_1, the ultimate voltage which it could produce on a single chop at peak current would be:

$$v = i_{max} R_1 \qquad (6.14)$$

where i_{max} = the maximum chopped peak current

By comparison with the simplified conditions used to obtain these solutions, the actual circuits are mostly 3-phase with mutual inductances and capacitances. Their 'constants' L and R are not independent of currents or frequencies and the capacitances and inductances exist in both lumped and distributed forms. The connections to and from the circuit breaker itself also include capacitances and inductances in one form or the other and the supply is by no means infinite, but should be represented by a further set of L, R and C networks. Quite apart from these complications, the circuit breaker produces multiple chopping which may resonate with one or more sections of the circuit. Although the circuit breaker may clear on one or other of the high-frequency oscillations, its maximum current chopping capability is, to a significant degree, dependent on the circuit and the point on wave at which its contacts open.

So, in their overall effects these differences introduce so many complexities that, instead of attempting to account for them case for case in design, somewhat inaccurate but seemingly quite satisfactory short cuts are taken to arrive at a decision. Although these methods are bound to vary from case to case, they do generally recognise ,that the maximum current chopping capability of the circuit breaker is virtually independent of nozzle sizes, and does not increase very much with the number of interrupters in series and only moderately with pressure. By inference of eqn. 6.14 and the fact that the number of interrupters in series tends to be proportional to service voltages, much the same overvoltage ratios can generally be expected across a range of circuit breakers as long as the same values of resistors are used per interrupter and the circuits are generally similar. Accordingly, the usual (linear) resistance values per phase tend to be proportional to service voltages, and their typical values, usually sufficient to limit the overvoltages to between 2·5—3·0 p.u. in average circuits, tend to be of the order of 25—50 ohms per 3 phase kV (r.m.s.) of system voltage.

(b) Sharing of restriking voltages
When an ideal multi-interrupter circuit breaker interrupts a 3-phase to earth short circuit, its 'first-to-clear' phase can be represented by a network of inherent capacitances between each of the interrupters on that phase with respect to each other, to the earthed interrupters on the other phases, and to earth and any other earthed structures. When viewed from the side opposite to the short circuit, this network is equivalent to the first inter interrupter capacitance being in series with a combination of series/parallel inter-interrupter and interrupter-to-earth capacitances of the second, third, etc. interrupters. The effective series capacitance of this combination can, even in simple cases, be several times greater than the first inter-interrupter capacitance with the result that the voltage and r.r.v. impressed on that interrupter would, unless corrected, amount to anything from about 140 to 300 or more percent of the pro rata value based on equal voltage division.

Taking for example one of the simpler cases of an ideal two-interrupter circuit breaker shown in Fig. 6.13b, the typical equivalent inter-interrupter capacitances C_1 and the like capacitance C_E to earth could be of the order of 20 and 40 pF, respectively. The effective series capacitance across the second interrupter in this case would therefore be 60 pF and the voltage and r.r.v. impressed on the first interrupter would amount to 150% of its pro rata value.

The addition of a voltage control capacitor C_2 across each interrupter would, in this case, alter this percentage in accordance with

$$\frac{v}{V_{pr}}\% = \left(\frac{C_1 + C_E + C_2}{C_1 + \frac{1}{2}C_E + C_2}\right) \times 100 \qquad (6.15)$$

where

v = voltage or r.r.v. impressed on the first interrupter

V_{pr} = pro rata voltage or r.r.v. per interrupter

An addition of C_2 of 200 pF would therefore reduce the inherent percentage from 150 to about 108·5 or only 8·5% above the pro rata value. It should be noted that in an idealised case this control would be instantaneous and quite independent of voltages or r.r.r.v.s applied to the circuit breaker. Also, it should be noted that the equivalent capacitance network for the circuit breaker is by no means constant but alters with switching conditions in accordance with relative potentials of the earth plane with respect to the interrupters. A good example of this occurs on out-of-synchronism switching where the earth plane is effectively at the midpoint of the interrupters. The inherent capacitance network of the circuit breaker in such cases is represented by the series inter-interrupter capacitances shunted to the midpoint by the capacitances to earth with, of course, the midpoint to earth capacitance no longer present. A two-interrupter circuit breaker in these cases is represented by two inherent and (usually) equal inter-interrupter capacitances which would distribute voltages and r.r.r.vs uniformly without any added capacitances.

The addition of a voltage control resistor across each interrupter of a multi-interrupter circuit breaker changes its circuit from purely capacitive to one consisting of a network of series and parallel connected resistors and capacitors. This network no longer responds to voltage changes instantaneously but exponentially in accordance with its time constant and its instantaneous or terminal values which can be predominantly capacitive or resistive (at high and low rates of change of voltage). Owing to the effect of the time constant, the very initial sharing of voltages and r.r.r.v.s is determined solely by capacitances but all later sharing of the actual voltage and r.r.r.v. values impressed on the interrupter can be significantly modified by the resistors. Although quite straightforward solutions can be obtained for most of these cases, the number of possible conditions is far too great to be dealt with here. Recourse is therefore taken to the example of an ideal two-interrupter circuit-breaker used in the preceding Section (Fig. 6.13b). Assuming, for illustrative purposes, that a step function voltage is applied to the circuit-breaker circuit composed of inherent capacitances and resistors R_2, the percentage share impressed on the first interrupter, again expressed in its pro rata form, is given by:

$$\frac{v}{V_{pr}}\% = A + B \exp(-\alpha t) \tag{6.16}$$

where

A = final percentage share determined in this case by resistors and equal to 100
B = additional percentage share due to capacitance, in this case equal to 50
α = exponential index, in this case equal to

$$\frac{2}{R_2(2C_1 + C_E)} \tag{6.17}$$

It will be seen that the initial share impressed on the first interrupter at $t = 0$ is independent of resistors and equal to 150% as in the case of a purely capacitive circuit. Also that the speed at which this share is brought under control by the resistors is determined by the exponential index α which, in this simple example, is inversely proportional to the resistance R_2. For a time constant T in microseconds the value of this resistance is given by:

$$R_2 = \frac{T}{40} \times 10^6 \text{ ohms} \tag{6.18}$$

and the resistance values for time constants of 1 and 3 μs are therefore 25 000 and 75 000 Ω, respectively. Assuming that the voltage rating of a two-interrupter circuit breaker could be 145 kV, these resistance values are about 10 times greater than needed for overvoltage control purposes and the latter would therefore be able to fulfil both functions in this case. It is not necessarily so on bigger circuit breakers with a larger number of interrupters so that a crossover of the two resistance values can occur well within the present range of system voltages. A stricter control, of particularly r.r.r.v.s on the most highly stressed interrupter, may have to be imposed in such cases either by lower resistance-values or by balancing of the capacitance network with added capacitors or both.

(c) Sharing of recovery and open-circuit voltages
Essentially all factors discussed in the preceding Section apply in this case except that the low rate-of-charge of voltages, associated with power frequencies, allow the resistors to predominate during the circuit recovery period. The resistors must, however, be switched out of the circuit after this period and sustained control of voltage sharing is therefore exercised by means of added capacitors.

It is worthwhile to note that in the example of a two-interrupter circuit breaker used in the previous Sections, resistances of about 20 MΩ per interrupter would suffice to reduce the inherent 150% pro rata share to about 105% at 50 Hz.

(d) Closing onto open-circuited transmission line
In order to limit overvoltages which can be produced by random closing onto an open-circuited transmission line (chapter 3) preinsertion resistors have to be used (either on the circuit breaker itself or separately). Depending on the specified overvoltage limits, line lengths, other equipments connected to its circuit and the interphase timing and autoreclosing duties of the circuit breaker, single-stage or even multistage resistors may have to be used. Their optimum values can only be calculated for each specific combination of those parameters but even then the result cannot be presented as a single value but statistically. However, the typical resistance values tend to approximate to surge impedance of the line for single stage and to about 65 and 300% of that impedance for the first and second stage of a two-stage preinsertion.

The important design aspect to be noted in connection with their use on circuit breakers (apart from their thermal rating requirements) is that although these resistors need not be connected to the interrupters potentiometer wise, such a connection can be beneficial in that it may produce a gradual shorting out of the resistor. On the debit side, such a connection may lead to very onerous thermal and voltage stressing of the last resistor section or sections to be shorted out by the interrupters under short-circuit or out-of-synchronism switching conditions.

(e) R.R.R.V. and restriking voltage peak damping resistors

Two main cases must be considered in relation to the r.r.r.v., and hence also the restriking voltage peak damping resistors; one for a short circuit at or in close proximity to the circuit-breaker terminals, the other for a short circuit at a short but significant distance from those terminals.

The first of these is often simplified as shown in Fig. 6.13c for which four solutions exist. Assuming that the interruption takes place at a natural zero of a symmetrical fault current loop and that the significant changes occur in a time sufficiently short for the supply voltage E to be regarded as substantially constant, the first solution, for an ideal circuit breaker, without any damping, is given by

$$v_t = E(1 - \cos nt) \tag{6.19}$$

where

$$n = \sqrt{\frac{1}{LC}}$$

The remaining three solutions, an oscillatory one for subcritical damping, one for critical damping and one for overcritical damping by resistors R_3 are given by:

$$v_t = E \left\{ 1 - \exp(-\alpha t) \left(\cos nt + \frac{1}{2R_3 Cn} \sin nt \right) \right\} \tag{6.20}$$

where

$$\alpha = \frac{1}{2R_3 C}$$

$$n = \sqrt{\left(\frac{1}{LC} - \frac{1}{4R_3^2 C^2} \right)}$$

$$R_3 > \frac{1}{2} \sqrt{\frac{L}{C}}$$

$$v_t = E \left\{ 1 - \exp(-\alpha t)(1 + \alpha t) \right\} \tag{6.21}$$

where

$$R_3 = \frac{1}{2}\sqrt{\frac{L}{C}}$$

$$v_t = E \left[1 - \frac{1}{2}\left(1 + \frac{1}{n}\right)\exp\{\alpha(1-n)t\} - \frac{1}{2}\left(1 - \frac{1}{n}\right)\exp\{-\alpha(1+n)t\} \right]$$

$$(6.22)$$

where

$$n = 1 - \sqrt{\left(4R_3{}^2 \frac{C}{L}\right)}$$

and

$$R_3 < \frac{1}{2}\sqrt{\frac{L}{C}}$$

In the second of these cases, short distance faults or S.D.F. (chapter 3) are often simplified as shown in Fig. 6.13d where the same supply-side circuit is connected through the circuit breaker to a length of short-circuited line.

In the absence of damping the r.r.r.v. and the peak voltage v_p of the saw-tooth oscillations on the line side of an ideal circuit breaker interrupting a symmetrical fault current are given by:

$$\text{r.r.r.v.} = \frac{di}{dt} Z_0 = I\omega Z_0 \qquad (6.23)$$

$$v_p = \eta(E - I\omega L)$$

$$= \eta E \left(\frac{I_0 - I}{I_0}\right) \qquad (6.24)$$

$$= I\omega Z_0 t_p \qquad (6.25)$$

where

Z_0 = surge impedance of the line = $\sqrt{\dfrac{l}{c}}$

l, c = distributed line inductance and capacitance
I = peak s.d.f. current
$\omega = 2\pi f$
f = power frequency
η = excursion factor (rated peak factor)

E = peak supply voltage
L = inductance on supply side
I_0 = peak short-circuit current at circuit-breaker terminals
t_p = time to peak voltage v_p

$$= \frac{2x}{u} \tag{6.26}$$

x = distance to fault

u = wave propagation velocity $= \dfrac{1}{\sqrt{(lc)}}$

The supply side transient associated with this undamped condition is given by eqn. 6.19 modified to

$$v_{t'} = E'(1 - \cos \omega t) \tag{6.27}$$

where

$$E' = I\omega L \tag{6.28}$$

The total voltage and the r.r.v. impressed on the ideal circuit breaker under undamped conditions are obtained by an addition of the supply-side and line-side transients with the polarity of the latter inverted at each half cycle of the saw-tooth oscillation. This voltage and the r.r.v. at any time 't' up to the first peak of the s.d.f. transient are therefore given by:

$$v_t = I\omega \underset{t=0}{\overset{t=t_p}{\Big[}} Z_0 t + L(1 - \cos nt)] \tag{6.29}$$

$$\text{r.r.v.} = I\omega \underset{t=0}{\overset{t=t_p}{\Big[}} Z_0 + L \sin \omega t] \tag{6.30}$$

The addition of a resistor R_3 to the circuit breaker modifies the line-side r.r.v. and v_p by a damping factor d given by

$$d = \frac{R_3}{R_3 + Z_0} \tag{6.31}$$

which holds for subcritical, critical and overcritical damping given by $R_3 > Z_0$, $R_3 = Z_0$ and $R_3 < Z_0$, respectively, so that eqns. 6.23 and 6.24 for the first part of the oscillation become:

$$\text{r.r.v.} = \frac{di}{dt} dZ_0 = I\omega dZ_0 \tag{6.32}$$

$$v_p = \eta d(E - I\omega L)$$

$$= \eta dE \left(\frac{I_0 - I}{I_0} \right) \tag{6.33}$$

The time to peak voltage, given by eqn. 6.25 does not alter as both sides of this equation are multiplied by η.

The total damped voltage and r.r.r.v. values appearing across the ideal circuit breaker under these simplified conditions may be obtained to first approximation by addition of the damped supply-side values given by eqns. 6.20, 6.21 or 6.22 but with E replaced by $I\omega L$ and, subject to wave propagation and damping laws being applied, the damped line-side values given by eqns. 6.32 and 6.33. Alternatively, if the supply-side transient contributes relatively little during the period up to the first peak of the line-side oscillation, an equally acceptable approximation may be obtained for that period from s.d.f. equations only if Z_0 is increased to a value corresponding to the sum of the inherent transients at that peak and this 'apparent' new surge impedance is used directly in these equations.

For instance, if this latter method is used for calculation of the damping resistance value required to double the s.d.f. capability of the previously mentioned ideal two-interrupter circuit breaker fitted with 200 pF condensers, eqn. 6.31 when modified to include any changes in sharing of r.r.r.v.s on the two interrupters, gives the solution as

$$d = \frac{1}{2} \frac{\phi_2}{\phi_1} = \frac{R_3}{R_3 + Z_0'} \tag{6.34}$$

where

ϕ_1 = r.r.r.v. (or voltage) sharing factor for the circuit breaker when fitted with 200 pF capacitors, equal to the ratio of the nominal (pro rata) share per interrupter to the actual share taken by the most highly stressed interrupter.

ϕ_2 = the same factor for the circuit breaker when fitted also with resistance R_3

Z_0' = the 'apparent' surge impedance combining the inherent supply and line side transients

Since by inference of the previous r.r.r.v. and voltage sharing example ϕ_2 may be assumed to approximate to unity and $\phi_1 = 0.5/0.5425$, the required value of R_3 for a typical $Z_0' = 1:03 Z_0$ and $Z_0 = 450 \, \Omega$ would be approximately 550 Ω or one order of values less than required for the average overvoltage control purposes.

This latter method also allows the typical air-blast characteristics given in eqn. 6.4 to be combined with eqn. 6.32 which, in conjunction with the r.r.r.v. sharing factor ϕ and a number of interrupters N, proscribe the s.d.f. current capabilities of the ideal multi-interrupter circuit breaker as

$$I^{n+1} = \frac{\phi N K_1}{\omega Z_0{'}} \frac{R_3 + Z_0{'}}{R_3} \tag{6.35}$$

where K_1 = characteristic constant (eqn. 6.4) at 'per-interrupter' value.

This shows that for the index n approximating to unity, the maximum s.d.f. current ratings of the ideal multi-interrupter circuit breaker, fitted with resistors giving a constant ϕ, are approximately proportional to the square root of K_1 and N when $Z_0{'}$ is constant and R_3 either remains unaltered or is made high in relation to $Z_0{'}$. Likewise, that relatively constant s.d.f. current ratings across a range of N can be achieved if values of R_3 are made intentionally low in relation to Z_0, and more than proportional to N.

Although the equations given so far for resistors R_1, R_2 and R_3 can be used to portray the right order of values or changes encountered in practice, at least two modifications must be made in order to calculate their effects or values with an acceptable degree of accuracy in design of air-blast circuit breakers.

The first of these modifications must be applied to circuits which even in their simplest practical forms must be assumed to be lightly or moderately damped with voltages reduced to suit subtransient characteristics of the supply source and with inductances and damping related to natural frequencies or rates-of-change. The same factors must also be taken into account in more complex circuits consisting of a number of individual supply sources and load circuits, which, depending on their distances from each other and the circuit-breaker, may introduce not only significant distributed 'constants' but also definite time-lag effects in their influence on and response to the operation of the circuit breaker.

The second of these modifications must cater for the nonideal behaviour of air-blast interrupters which produce significant damping effects particularly at heavier short-circuit currents and high r.r.r.v.s. These effects vary not only with switching conditions but also with time during the immediate postarc period and thus introduce an additional current − r.r.r.v. − time dependent resistance which, depending on the circumstances, might be adequately represented by an averaged 'effective' constant for a range of current-r.r.r.v. levels or, more accurately, by a function of all three quantities at the same time. In so far as the circuit breaker resistors are concerned, these two modifications require their values to be calculated from an already partly damped condition of the circuit in which some of this damping is due to the circuit itself and the remainder is produced by the nonideal behaviour of the interrupters. The resistance or resistance functions equivalent to damping produced by the interrupters affect also the division of voltages or r.r.r.v.s on multi-interrupter circuit breakers particularly during the immediate postarc period following interruption of high-current, high-severity short circuits, and since these effects may swamp the influence of added resistors or capacitors, uniform instantaneous behaviour of all interrupters is essential.

Having established the required resistance value, its element, shape and enclosures must be designed to withstand the highest voltage that can appear across its terminals and the worst power and energy input and dissipation conditions. Except for cases where a single interrupter and resistor can withstand the highest voltages produced in the circuit, the lowest design voltage withstand value for resistors connected directly across individual interrupters must not be less than the maximum interrupter no-load voltage withstand capabilities under flowing air conditions. The same design voltage value also determines the necessary power intake capabilities of the resistor, but this rate of energy intake need only be reached and maintained in accordance with the longest single voltage waveform which can appear across the resistor between cessation of arcing on the interrupter and the first resistor current zero. The total energy input capacity of the resistor must be matched to duty cycles specified for the circuit breaker multiplied by the duration of the resistance current flow during each operation, advisably with at least one, preferably the last, of these operations including the maximum power input condition. The energy dissipation requirements are determined by the frequency at which the required duty cycles must be repeated or, alternatively, by air consumption of the circuit breaker and the compressed air storage and pumping capacities. Depending on values of current passed through the resistors, resistor current interrupters may take the form of simple spark gaps pressurised to interrupt light resistor currents through to miniature air-blast type interrupters for low ohmic value resistors. In the latter case overvoltage control problems are transferred from the main to the resistor interrupters and throttling of the latter's exhaust passages and relatively high upstream arc length-to-nozzle diameter ratios can be used to reduce their current chopping abilities to an acceptable level. It should, however, be noted that these resistor current interrupters must retain a definite I-r.r.r.v. interrupting capability to deal with the worst current-r.r.r.v. conditions when the inductance of the main circuit becomes comparable with the ohmic value of the resistor and interruption must take place at power factors significantly below unity and at relatively high natural frequencies.

The element, shape and enclosures of voltage control capacitors must be designed to withstand not only the highest voltages that can appear across the individual resistors during interruption or the specified system withstand levels but also the highest voltages which can appear across the last open or partly open 'break' or 'breaks' during closing of multigap circuit breakers. This design voltage level should preferably be equal to the internal-voltage strength of the fully open 'break' and this strength, in its turn, should be higher than the flashover level of the adjacent external insulation. The element must, of course, be continuously rated and capable of withstanding repeated short circuiting.

6.4.5 Contacts
Contact materials used on air-blast circuit breakers are chosen on the basis of chemical, physical and electrical properties which offer the necessary combinations

of mechanical durability, high electrical and thermal conductivities, good resistance to erosion by arcs, low electron emissivity at high temperatures, good resistance to spot welding, etc. in dry, compressed air, environment with high oxygen concentration.

Because not one single material can combine all of these properties at the same time, a degree of specialisation and selection must be employed with the result that the overall range of available materials is divided at its extremes into 'arcing' contact materials, such as sintered copper or silver tungstens, whose conductivities and contact resistances are high and 'current-carrying' contact materials, such as hard drawn copper or copper alloys, which have poor resistance to arc erosion or spot welding but good conducting properties particularly when silver plated or fitted with silver inserts. The middle of this range is occupied by insert materials, such as silver–nickel or silver–cadmium oxide, which possess moderate resistance to erosion by arcs and spot welding in combination with relatively high conductivities.

Although the physical and mechanical properties of these materials are generally quite satisfactory for butting contact applications involving little or no rubbing between the surfaces, they are all prone to scuffing on sliding applications. This disadvantage can be overcome quite readily on arcing contacts which can be designed for butting only or with relatively small sliding movements but not so on current-carrying contacts which transfer the current from stationary conductors to the main moving components of the interrupter where electrical requirements for high interface pressures are directly opposed to low pressure mechanical and large area heat transfer requirements. Because of this and aside from any flexible braids, laminations or rolling contacts which either occupy too much space or are prone to produce untoward thermal, mechanical or electromechanical effects, multiple but relatively lightly loaded contacts are used for such applications. Preference is often given to line rather than point type contacts on account of their better mechanical wear properties and relatively shallow curvatures are employed for the same reason and to increase conduction of heat. Area type contacts do not appear to offer any advantages in this application and are more prone to pitting on passage of heavy short-circuit currents as well as to trapping of metallic swarf.

The overall size and arrangement of contact systems are dictated by thermal, electromagnetic, air-flow and electrostatic requirements. Low contact and conductor power losses and good heat transfer and conduction paths are needed to meet temperature rise limits economically while low losses and sufficient thermal capacities must be provided to enable the arrangements to carry short-circuit currents for specified periods of time without overheating. The electromagnetic forces tending to lift the contacts due to current pinching or local changes in the current flow paths have to be either overcome by mechanical loads or preferably compensated by opposing electromagnetic forces which increase contact pressures at heavy currents but do not significantly affect these pressures under normal conditions. Where by-pass contacts are used, the main and the auxiliary contact

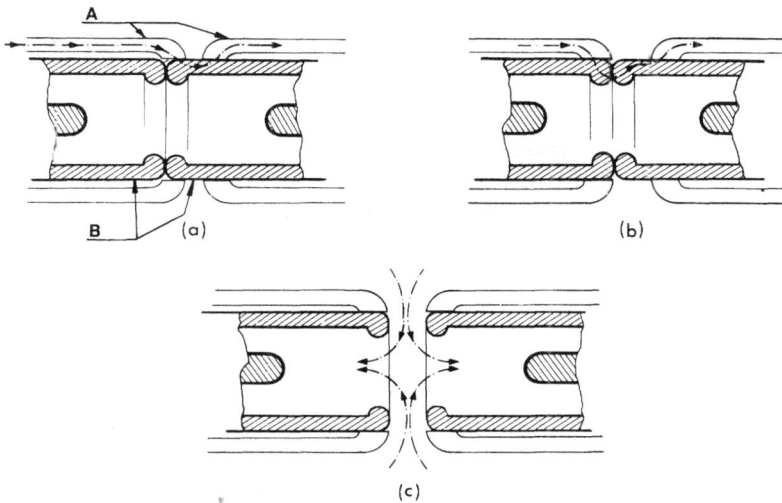

Fig. 6.14 Typical interrupter contact system

    (A) stationary contacts
    (B) moving nozzle contacts
    (a) closed
    (b) opening
    (c) open

current paths should be arranged so that least possible inductive energy has to be dissipated on commutation. The positions and overall contours generated or affected by the contacts must not produce untoward effects on air flow or on electrostatic fields.

A typical interrupter contact system, illustrating some of these points is shown in Fig. 6.14 where two stationary contact sets A formed by concentric clusters of elastic fingers are in contact with and enclose movable nozzle contacts B. As indicated by arrowed lines, the current in the closed position (*a*) flows along the fingers of one of the stationary contact clusters into one of the movable nozzle contacts and hence onto the other cluster. A small difference in diameters keeps the first set of fingers clear of the first nozzle contact. A small relative movement of both nozzle contacts to the right would cause the first set of fingers to momentarily bridge both nozzle contacts and, with these still touching, allow the current to be commutated into the path shown in (*b*) with relatively small involvement of inductive energy. This current path produces the desired electromagnetic effect at the point of contact between the two nozzle contacts in that it would drive the arc towards its correct extinguishing position. It should be noted that the flow of current in each set of finger contact clusters produces centripetal forces tending to increase contact pressures. With the first nozzle contact stopping at the position shown in (*b*) and the other continuing to the position shown in (*c*) high contact separation speed can be achieved, and, as can be inferred from the sequences, the main current carrying contacts are well protected from the arc.

If this arrangement is also used for closing, the reverse sequence applies except that the first closing movement is executed by the r.h.s. nozzle contact only and the arc is struck when it has travelled a distance towards the other nozzle contact. The arc is short circuited by physical contact as indicated at (*b*) but a virtually bounce-free impact must be achieved under this condition.

It is, of course, possible to simplify this arrangement for lighter short-circuit current applications by dispensing with the l.h.s. cluster of fingers and thus produce a single butt-type contact system. Similarly the r.h.s. cluster of fingers may be replaced by a solid tube and the sliding contact connection may be established by spring loaded segments, piston rings or obliquely formed garter springs housed in a groove cut in the wall of the nozzle contact. Alternative arrangements, employing stationary nozzle electrodes can also be used with the interelectrode gap bridged by an axially movable set of contacts whose cage may serve as an upstream blast valve or a radially moving set of contacts which can be moved entirely out of the air stream. All of these systems can be applied to mono, partial-duo and duo-blast interrupters.

6.4.6 Valves and air engines

Even when the great variety of various stop, check, nonreturn, safety etc. valves used to control or protect compressed air supplies is ignored, the remaining varieties of valves used specifically on air-blast circuit breakers appear to be still too numerous to divide them into a few meaningful types or categories identifying not only the principal functions (e.g. blast valves, exhaust valves), but the manner in which these functions are performed. It is, for instance, quite possible to design and use a mechanically, pneumatically or hydraulically operated valve which in its reset position closes off one port, keeps a second port open and at the same time interconnects a number of other pneumatic lines at either atmospheric or air-receiver pressures. A movement of this valve member may changeover the position of the two ports and reconnect the pneumatic lines to quite a different pattern of connections so that two, three or more specific functions are performed at a time. The valve member itself may also be self-resetting so that a removal of the controlling force causes it to return to its original position, or non-self-resetting whereby the removal of that force and a specific application of a further force are needed for that purpose.

Bearing in mind the multiplicity of ports and pneumatic lines which one valve member may control, the simplest single function that a valve may perform is to admit or drain compressed air to or from a volume in response to a mechanical, pneumatic or hydraulic control and to reset or reclose when this control is cancelled. Valves performing this function are usually referred to as 'one-way' valves and are quite often employed as blast valves and exhaust valves.

The next simplest valves, performing a similar function but draining compressed air admitted to a volume or readmitting it to a drained volume on resetting, are usually referred to as 'two-way' valves to indicate that they control the output both 'in' and 'out', and are therefore provided with two output ports. Valves of this type are frequently used to operate 'one-way' valves, such as blast valves or exhaust valves, and, when connected in series with each other, allow a relatively small signal effort to be multiplied pneumatically to any desired power level for operation of large valves or air engines. The smallest of these valves are used as the first stage 'pilot' valves operated by electromagnetic devices translating electrical closing or opening signals to the circuit breaker.

A further valve of either one-way or two-way type may be used to interpose an accurately controlled time delay into a pneumatic line by causing the first pressure change in the line to bias the valve towards its closed position and then to reverse this bias at a predetermined time by metering a quantity of compressed air into a chamber fitted with the valve operating piston. Valves of this type may be used to sequence the operations of, for instance, main interrupters and resistor interrupters or isolators where the main pressure changes in the passages occur too soon or too gradually.

As inferred from before, it is quite possible to control all movements and operating sequences of the air-blast circuit breaker pneumatically as long as the time delays inherent in pneumatic circuits allow the overall operating times of the circuit breaker to be held within the required limits. Although this condition can usually be met on smaller circuit breakers, any high operating speed requirements on these and even more so on physically bigger high-voltage circuit breakers may not allow compressed air to travel along a pneumatic circuit and reach a control point in sufficient time to operate, e.g. the interrupters or the exhaust valves. It is therefore necessary in such cases to localise or concentrate the pneumatic circuits in the immediate vicinity of the principal operating components and to transmit the signals or the movements over longer distances by means of mechanical linkages or hydraulic lines. The only differences which this change need make in design of the valves may be confined to a substitution of their pneumatic piston drives by direct mechanical or hydraulic drives.

Although valves used on air-blast circuit breakers vary in sizes from as little as 3 mm diameter to an equivalent diameter of 200 mm or more, most of them employ one of two port types shown in open and closed half sections in Fig. 6.15. The first of these, the disc type, is formed by a valve head closing off the full diameter of the passage so that the initial force needed to pull the head off its seat is equal to the product of the effective seat diameter area and the pressure difference. The second of these is formed by a thin wall tube or sleeve in contact with a valve seat at its upper end and with its bore closed off by a stationary piston

(a)   (b)

(c)

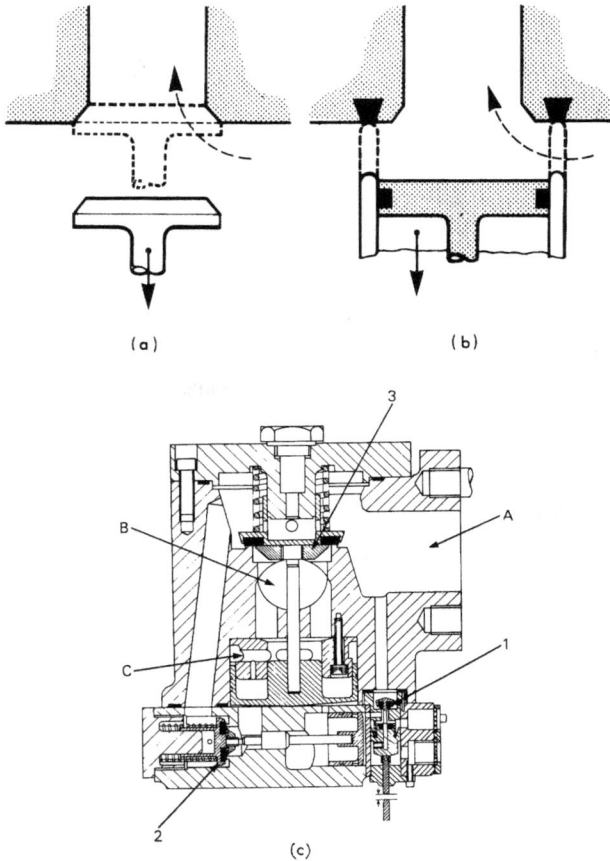

Fig. 6.15 Valves
    (a) Typical disc-type port
    (b) Typical sleeve-type port
    (c) Example of 3-stage valve assembly

equipped with a sliding air-tight seal. The initial force needed to open this valve is equal to the product of the pressure difference and the annular area contained between the effective diameter of the seat and the bore of the sleeve so that, by comparison with the previous type of ports, considerably reduced forces and energies to open the valve are needed.

Valve seats employed on these and other air-blast valves can be of metal-to-metal, metal-to-plastic or plastic-to-plastic type as shown in (a) or (b). The materials used for these purposes must have high mechanical strength and impact properties throughout the range of design temperatures, be chemically inert and dimensionally stable. Stainless steels, bronzes and some specially treated ferrous or aluminium components are being used for metal parts and high-quality elastomers, polyamides (nylons) or polycarbonates can be used as seat materials. The last of these two can also be used for valve heads in lighter duty applications.

Sliding seals of the type shown in (b) are usually formed by p.t.f.e. rings backed by internal elastomer toroids of varying cross-sectional shapes designed to exert the required radial and axial pressures. The surfaces of the sleeves in contact with such seals must be both very smooth and fully protected against scoring.

An example, showing how a number of sequentially operated valves may be used to multiply a low energy mechanical input signal in order to open a large bore valve, and then to hold this valve open for a required period of time after the signal has been removed, is shown in Fig. 6.15c. In this arrangement, a 3-stage valve assembly is connected to a compressed-air supply through a large-bore inlet port (A). Its output is lead out of a large-bore port connected to chamber (B). The inlet port (A) communicates directly with the up-stream sides of the first-stage valve (1), the second-stage valve (2) and the final, third-stage, valve (3).

With all of these valves closed and reset as shown, the downstream side of the first-stage valve communicates directly with atmosphere and with the driving end of the second-stage cylinder. The downstream side of the second-stage valve communicates directly with the driving end of the third-stage cylinder and through this indirectly with atmosphere via a hole leading to the idling end of the second-stage cylinder, and hence to atmosphere. The idling end of the third-stage cylinder, the chamber (B) and hence also the output port communicate with the atmosphere through passages (C) provided in the idling end of the third-stage cylinder. Thus, when the rod operating the first-stage valve is depressed, its movement closes off the passage between the downstream side of this valve and atmosphere and, at the same time, admits compressed air to the driving end of the second-stage cylinder. The pressure now acting on the second-stage piston drives the second-stage valve into the open position while the movement of this piston blanks off the hole leading from the driving end of the third-stage cylinder to the idling end of the second-stage cylinder. The movement of the second-stage valve admits compressed air to the driving end of the third stage cylinder, and this causes the third-stage piston to open the third-stage valve and blark off the passages (C). This completes the opening movements of the 3-stage valve assembly and admits compressed air into chamber (B) and hence through the output port to the next piece of the apparatus to be operated.

A straightforward reversal of these movements occurs on reclosure of the 3-stage valve assembly except for the resetting of the third-stage valve whose movement is, in this case, intentionally delayed by a controlled leakage rate through the hole leading from the driving end of the third-stage cylinder to the idling end of the second-stage cylinder and hence to atmosphere.

As it can be readily deduced from the details given in this Figure, each stage-valve assembly is spring loaded towards its closed position, and its resetting movement is assisted by the like orientated pressure forces. Other arrangements and uses of the disc-type valves shown in this example and of the sleeve-type valves depicted in Fig. 6.15*b* may be found in the illustrations associated with Section 6.6.

Design of air engines for driving valves, interrupters, isolators or any link mechanisms generally follow the usual diametral or annular piston-cylinder practices, except that the effective piston areas are frequently calculated to operate at about 50% of the working pressure to ensure that high rates of change of pressure are available to minimise the effects of any differences or variations in the opposing forces. Depending on the requirements, the air engines may be of single-acting type to drive only in one direction or double-acting to drive in either direction. Differential pistons in which only a proportion of the driving area is initially exposed to pressure changes in starting position are used when, by virtue of very rapid pressure changes, a high degree of consistency in timing of the movements can be obtained. Quick-return piston designs employ pressure forces in opposition to the working stroke preferably applied towards the end of the stroke to reduce the size of the driving piston.

When the driving forces generated by air-engines and the kinetic energies gathered by moving masses are very high, special measures must be taken to arrest the movement towards the end of the stroke without damage and without rebounds which could detrimentally affect the operation. Dissipation of this energy can be achieved pneumatically by feeding or otherwise trapping a metered quantity of compressed air in the air engine so that it becomes highly compressed by the piston and exhausted through a calibrated outlet towards the end of the stroke.

6.4.7 Silencers

Although the time during which compressed air is discharged from modern circuit-breaker interrupters can be as short as 50 or so milliseconds, their almost simultaneous operation, particularly on high rupturing capacity multi-interrupter designs, produces an impulsive noise comparable with heavy gun reports. The overall sound pressure levels, spread over all audible frequencies, can reach pain and even damage thresholds for human ear in proximity to the circuit-breaker and acute to annoying nuisance strengths as the distances increase.

Depending on whether only the health hazard must be removed or whether this must be accompanied by a substantial reduction in the nuisance levels, rudimentary or high efficiency silencers are used. Both of these are based on the same acoustic principles of reducing the efflux velocity and of weakening the sound by absorption or filtering or both.

The rudimentary silencers can consist of little more than multichannel divergent area passages with some changes in direction and lined with sound absorbing materials able to withstand (or protected from) the scouring and high temperature effects of the exhaust gases. The total impedance to flow of gases at maximum short-circuit current levels must not impede the flow through the nozzles so that the air or gas mass flow rates at outlets remain essentially unaltered.

At the other end of the scale high efficiency silencers are provided with an expansion volume which traps most of the exhaust gases without impeding the flow through the nozzles and then, over a considerably longer period of time, allows the gases to escape to the atmosphere either through beds of sound deadening material or sound absorbing passages or both. The gas or air mass flow rates at the outlets from this silencer can be reduced to 30% of the unimpeded rates quite readily and to less than this if necessary.

Although most of the noise is generated by efflux of air from the interrupters, its overall level can be significantly increased by operation of pneumatic mechanisms at the base of the circuit breaker, by transmission of internally generated noises through the walls of the enclosures including the silencers and by 'ringing' or resonance of the enclosures and structures. All of these additional noise sources must be taken into account in reducing noise levels of circuit breakers and sound deadening materials or bracing struts or pads may have to be provided in strategic places.

There are no internationally or nationally agreed standards on acceptable impulsive noise levels either in relation to health hazards or nuisance values, but it seems that, subject to noise frequency spectra and sound pressure levels being controlled over most of the audible range, overall sound pressure levels of the order of 105 dB (linear) may be acceptable from the health hazard point of view. The acceptable nuisance levels, involving the quality as well as the strength of sound waves and subjective judgments are much more difficult to assess and field trials are generally unavoidable at the present time.

6.5 Operating sequences and requirements

6.5.1 General

Operating sequences of circuit breakers consist of a series of alternate closing (C) and opening (O) operations in which the former may be immediately followed by the latter (CO) but the latter does not have to be followed by the former in less than a definite minimum time interval t determined, in most cases, by high-speed automatic reclosing 'dead-time' requirements $(O - t - C)$. Complete cycles of these operations consist of a combination of such sequences repeated at specified intervals but with a proviso that the circuit breaker must always be able to terminate them with an opening operation and with unimpaired interrupting capabilities. Pressure switches are therefore used in closing and opening control circuits and the former are set to operate at higher pressures.

The number of times that the circuit breaker is required to perform opening operations in rapid successions determines its air storage volumes and the frequency at which any subsequent operations can be carried out is determined by storage capacities, pumping rates and feed rates of the compressor installation. The CO sequences require closing controls to either reset rapidly or to be of the 'free-to-trip' type in order not to impede opening operations and 'antihunting' or 'antipumping' interlocks must be provided to ensure that the circuit breaker will stay open even if the closing circuit continues to be energised. The overall 'circuit-closed' time on this sequence should be as short as possible and any stringent time limits for it may constrain the principles and the overall form of the control scheme designs. The same stringent time limits and implications tend to coincide with short 'short-circuit break' times, which, in their turn, are more frequently specified for physically bigger, high-voltage circuit breakers. Hardly any design constraints arise from the $O - t - C$ sequences in which the shortest dead times, of the order $0.2-0.3$ s, are quite long by comparison with the inherent capabilities of modern designs.

Structural considerations play an increasingly important role as sizes of circuit breakers increase with short-circuit and voltage ratings particularly since the choice of structural insulating materials and shapes becomes all but limited to cylindrical porcelain insulators. Internally generated reactions on the structure, produced principally by moving masses, changes in air-flow directions in exhaust passages and at the outlets, require the circuit breaker to be treated as mechanically 'live'. A considerable concentration of masses and external areas at its upper end make the structure inherently sensitive to wind loads and earthquakes.

6.5.2 Closing

Main closing sequences for a typical air-blast circuit breaker are shown in Fig. 6.16a where line XX denotes a start of the operation at an instant when the closing coil solenoid has become energised as indicated by trace A. Depending on whether the mechanism is of a mechanical linkage type or pneumatic, a movement of the solenoid armature operates a latch or a pilot valve in order to allow the linkages or a further second-stage valve to multiply the effort to the required output level while all further control of the latch or valves is taken over by the mechanism. This output is then transmitted across the main insulation system to high potential as a mechanical movement through insulated rods or pneumatically via an insulated tube as indicated at B. Compressed air fed through this tube or through a valve opened by the rod system operates preinsertion resistor switch air engines and, with a suitable time delay, main contact air engines as indicated by traces C and D or, if preinsertion resistors are not fitted, only the latter as indicated by trace D. When a satisfactory completion of the closing strokes has been assured, all valves driving the air engines are reset automatically by mechanical or pneumatic signals to leave the circuit breaker in a reset and ready-to-open position. The closing coil current may be cutoff by auxiliary switches driven by the mechanism or, if it cannot interfere with subsequent operations, by a relay or release of hand operated switch at a later time. The preinsertion resistor switch may be left in the closed position if the same resistor is used also as an opening resistor, otherwise it is returned to its open

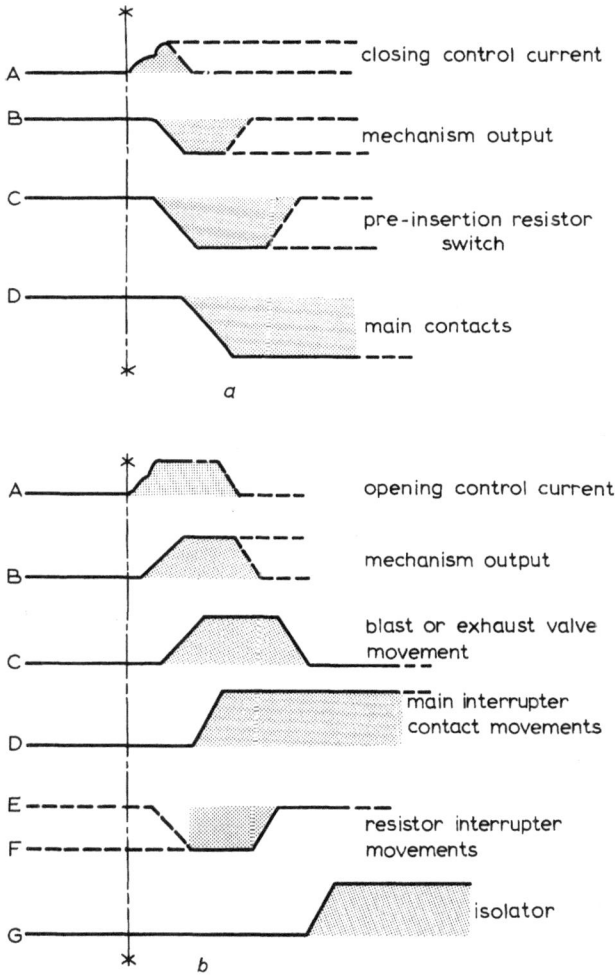

Fig. 6.16 Main closing and opening sequences
  (a) Closing
  (b) Opening

position soon after the main contacts have closed and before they can open on a CO sequence.

Closing speeds of contacts should be high to limit erosion and generation of ionised gas on prearcing, but very high contact speeds on multibreak circuit breakers require higher degrees of synchronisation to be achieved on each phase to ensure that excessive voltages, capable of striking long arcs between the contacts or of causing flashovers outside the intended path, are not impressed on any one break. The interphase synchronisation is not critical in so far as satisfactory operation of the circuit breaker is concerned, but a small spread of closing instants is helpful in control of closing overvoltages.

6.5.3 Opening

Main opening sequences for a typical air-blast circuit breaker are shown in Fig. 6.16*b* with line XX denoting the instant at which the opening coil electromagnet has become energised as indicated by trace A. Again depending on whether the mechanism is of a mechanical linkage type or pneumatic, a movement of an armature causes the signal effort to be multiplied mechanically or pneumatically to a required output level while all further control of the circuit breaker is taken over by the mechanism. This output (trace B) is then transmitted to blast valves at earth or high potential or exhaust valves at high potential (mechanically or pneumatically) to operate their air engines directly or through a further stepup valve (trace C). A pressure rise or drop induced by the operation of the blast or exhaust valves, or a direct coupling of the interrupter contact systems to these valves, causes the contacts to open as indicated by trace D while the voltage or r.r.r.v. control resistors shunt the interrupters (trace E). After a period of time determined by maximum arc durations, usually of the order of 20—25 ms from commencement of arcing on main interrupters, resistor circuit interrupters open as indicated by trace E followed by any isolators as in trace F. Sequential closures of all air-flow passages take place at the earliest possible instants as determined by circuit-breaker and interrupter types (Sections 6.3 and 6.4) to limit air consumption to a minimum and leave the circuit breaker in open position with all controls reset to receive and next closing command.

Opening speeds must generally enable the contacts to reach their arc extinguishing position in about 8 or so milliseconds to achieve restrike-free performance and minimum arc durations. A sufficient degree of synchronisation between their movements on each phase of multi-interrupter circuit breakers must be achieved to ensure that similar interrupting conditions are achieved on these interrupters at the same time. The interphase synchronisation is not critical except for protection requirements.

6.6 Examples of typical circuit breakers

As already mentioned in the introduction to this chapter, circuit breakers of the air-blast type have proved themselves capable of meeting all rating and performance requirements that have arisen so far, and their designs have been adapted to suit various forms of switchgear, from indoor cubicle and metalclad to outdoor live-tank applications. Likewise, owing to the relative ease with which compressed air may be ducted and used as a source of mechanical energy, numerous basic design forms and arrangements have been, and still are being, used to achieve the same desired ends.

There is, however, one major area in which the air-blast circuit breaker has been outstanding and another area in which it has met the requirements but with some difficulties. The latter case arose in the e.h.v. metalclad applications where, as indicated in Fig. 6.11, the need to deionise the exhaust gas, prior to its discharge to atmosphere, necessitated special cooler/mixer chambers to be fitted within the earthed enclosure with significant size and cost penalties. The former case applies to the open-terminal, high s.c. rating, high-performance transmission system applications up to the highest rated system voltages used nowadays. Its particular forte in those applications is, undoubtedly, the economically desirable 'unit construction' feature which permits one basic interrupter 'unit' to be used across a wide range of service voltages by holding their 'per-unit' voltage ratings approximately constant. Allied to this, and equally desirable, is the facility provided in its 'unit construction' solutions to add and vary the values of the r.r.r.v. or overvoltage control resistors to, as appropriate, increase the short-circuit current ratings up to a maximum, or decrease the closing or opening overvoltages.

So, although there are other circuit-breaker designs for more specialised applications at lower voltages, the examples of the typical air-blast circuit breakers chosen for this chapter refer to these 'unit-constructed' ranges. By the same token, and to avoid unnecessary repetitions, each example given here should be taken to represent but a sample from such ranges which, depending on the number of the interrupters, the types and values of the resistors etc., can be made to cover much the same voltage and short-circuit rating requirements.

Confining the remarks to the main principles and features, examples are listed in alphabetical order of the manufacturers on pp. 286-302.

'Free-jet' circuit breakers
AEG Co., W. Germany

This circuit breaker, shown in its four interrupters per phase form in Fig. 6.17a, is of the outdoor 'open-terminals', live-tank type. Compressed air for the operation of its twin interrupter units (Fig. 6.17b) is stored within each unit up to the sleeve-type blast valve (1). As already mentioned in Section 6.4.3, the linear travel of the moving contact (2) is used to combine the interrupting and isolating functions, so that the former is attained at atmospheric pressure by the clearance between the fixed contact (3) and the insulating nozzle disc (4), when the moving contact has moved into its fully open position (shown dotted). The same fully open position of the moving contact is used to interrupt the arc when the blast valve is open, and the flow of compressed air through the insulating nozzle and the hollow stem of the moving contact produces the 'partial duo-blast' interrupting conditions. To close the circuit breaker, the moving contact (2) is driven from its fully open position into engagement with the fixed contact (3).

a

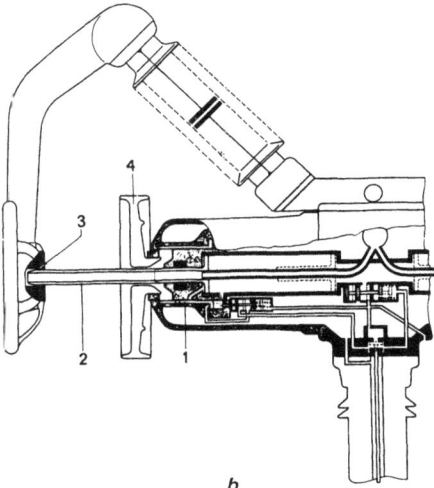

b

Fig. 6.17 'Free jet' circuit breaker, AEG Co.
(a) Circuit breaker
(b) Interrupter

DLF type circuit breaker
Brown-Boveri Co., Switzerland
This outdoor, 'open terminals' circuit breaker is shown in its six interrupters per phase form with preinsertion closing resistors in Fig. 6.18a. Compressed air for the operation of its twin interrupter units is stored in the vertical air receivers at earth potential, and throughout all of its support columns up to the sleeve-type exhaust valves placed at the outboard ends of the three exhaust passages of each unit [Fig. 6.18b, (1) and (2)]. The interrupters are of the permanently pressurised, duo-blast type. The moving contact (3) employs a two-beat movement, the first to its normal interrupting position when the exhaust valves are open and then, after a suitable pause, to the isolating position. The circuit breaker is closed by driving the moving contact (3) into engagement with the fixed contact (4) while the central exhaust valve (2) is held open to vent the intercontact zone to atmosphere.

Fig. 6.18 DLF type circuit breaker, Brown-Boveri Co.
    (*a*) Circuit breaker
    (*b*) Interrupter

Metalclad circuit breakers
Coq Co., The Netherlands
An arrangement of this circuit breaker in a duplicate busbar indoor installation is shown in Fig. 6.19*a*. The circuit breaker consists of a stack of interrupters in series with separate isolators all insulated within and immersed in the vertical compressed-air storage receivers. The interrupters (Fig. 6.19*b*) are of the duo-blast type, and are fitted with the sleeve-type blast-valve (1) placed just upstream of the fixed nozzles (2) and the tubular bridging contact member (3). An opening movement of the blast valve, effected by the control rods (4), admits compressed air to the bridging-contact member piston, which drives the contact into its open position and thereby draws the arc and admits compressed air to the nozzles. After a predetermined pause, the series isolator (not shown) opens and the control rods (4) drive the bridging contact and the blast valve into their normal closed and reset position. The arc products generated during interruption are, of course, passed through the de-ionising and silencing chambers (5) and then exhausted to atmosphere. The closure of the circuit breaker is effected by the pressurised series isolators.

a

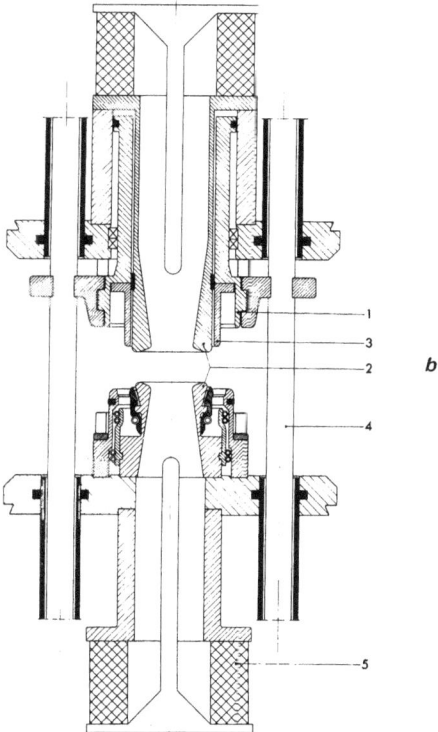

b

Fig. 6.19 Metalclad circuit breaker, Coq Co.
    (a) Circuit breaker
    (b) Interrupter

PK type circuit breakers
Delle-Alsthom Co., France
This outdoor, 'open terminals' circuit breaker is shown in its 12 interrupters per phase form in Fig. 6.20a. Compressed air for the operation of its twin interrupter units (Fig. 6.20b) is stored in the air receivers at earth potential and throughout the support columns and the interrupters up to the exhaust valves, (1) and (2), placed downstream from the duo-blast type nozzle system. The opening movement of the moving contact (3) is controlled by the exhaust valve (2), and the two exhaust valves reclose after interruption to hold the circuit breaker in its open and isolated position. The closure of the circuit breaker is effected by driving the moving contact (3) into engagement with the fixed contact (4).

a

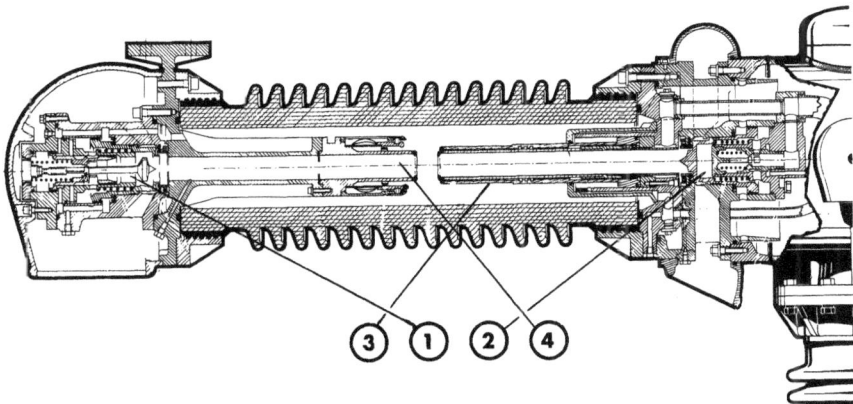

b

Fig. 6.20 PK type circuit breaker, Delle-Alsthom Co.
    (*a*) Circuit breaker
    (*b*) Interrupter

AS type circuit breaker
General Electric Co., Great Britain
A six interrupters per phase arrangement of this circuit breaker is shown in Fig. 6.21*a*. The air receivers at earth potential, the support columns and the twin interrupter units are permanently pressurised. The opening movements of the main contacts (2) (Fig. 6.21*b*) are controlled by one sleeve type exhaust valve (1). The mono-blast contact system shown in the illustration can be changed to duo-blast by substitution of an alternative, second nozzle and exhaust valve, assembly in place of the mono-blast arcing contact assembly (3). The opening movement of the main contacts (2) charges the spring (4) which drives the main contacts on closing.

a

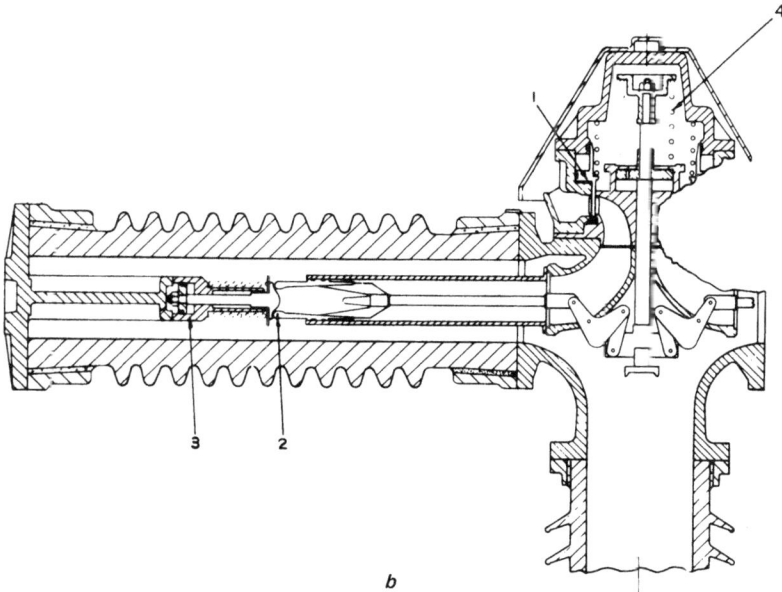

b

Fig. 6.21 AS type circuit breaker, General Electric Co., Great Britain
(*a*) Circuit breaker
(*b*) Interrupter

ABT type circuit breaker
General Electric Co., USA

An interesting feature of this outdoor 'open-terminals' circuit breaker (Fig. 6.22a) is the inclusion of an SF_6 insulated current transformer in one of its support columns. Compressed air for the operation of this six interrupters per phase circuit breaker is stored in the unit interrupter air receivers (Fig. 6.22b) up to the single-sleeve-type exhaust valve (1), which serves the two unit breaks at the same time. The interrupters are of the mono-blast type with fixed arcing contacts (2) and separate blade-and-finger type main current carrying contacts (3), which wipe past the former to draw the arc on interruption and 'make' the circuit on closure of the circuit breaker. The exhaust gases are passed into an integrally designed cooling and mixing chamber, which also acts as a silencer. The unit receiver houses the resistors (4), and the electrical connections are brought into the receiver through two gas insulated bushings (5).

a

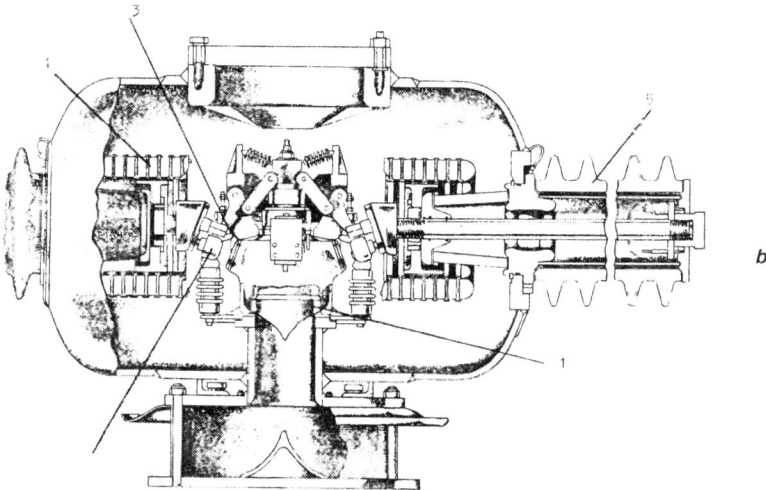

b

Fig. 6.22 ABT circuit breaker, General Electric Co., USA

(a) Circuit breaker
(b) Interrupter

PP type circuit breaker
Merlin Gerin Co., France

A six interrupters per phase arrangement of this circuit breaker is shown in Fig. 6.23a. Compressed air for its operation is stored in unit air receivers mounted in close proximity to the interrupters (Fig. 6.23b) and through to the exhaust valves (1) and (2). This permanently pressurised interrupter is of the duo-blast type, and the opening movement of its main moving contact (3) is controlled by the outboard exhaust valve (2). The other contact (4) can be either fixed or, with the addition of suitable control features, made to execute a movement after interruption to increase the intercontact gap on arrangements which require higher per-unit insulation levels. The closure of the circuit breaker is effected by driving the moving contacts into engagement with each other.

a

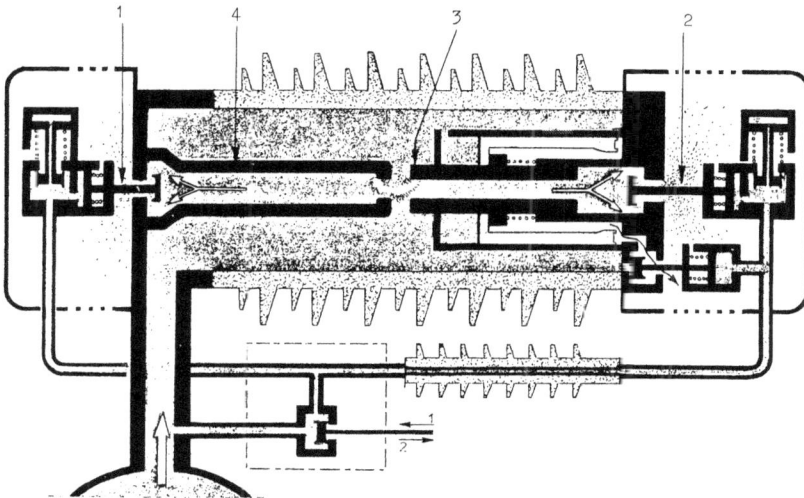

b

Fig. 6.23 PP type circuit breaker, Merlin Gerin Co.
(a) Circuit breaker
(b) Interrupter

OHBR type circuit breaker
A. Reyrolle & Co., Great Britain

A six interrupters per phase arrangement of this outdoor type circuit breaker is shown in Fig. 6.24*a*. Compressed air for the operation of its twin interrupter units is stored in the unit receivers (Fig. 6.24*b*), which are fitted with solid insulation bushings (1). Each of the twin interrupters of the mono-blast type is pressurised up to its sleeve-type exhaust valve (2) mounted just downstream of the nozzle. The exhaust gases from the interrupters are passed into an integrally designed silencer (3), and thence to atmosphere. The finger type interrupter current carrying and arcing contacts (4) form an inverted U crossbar bridging the two fixed contacts (5), and their movement controls the operation of the exhaust valves on interruption. The resistors (6) and the resistor interrupters (7) are mounted within the receiver, and a downwards movement of crossbar contact assembly is used to close the circuit breakers.

a

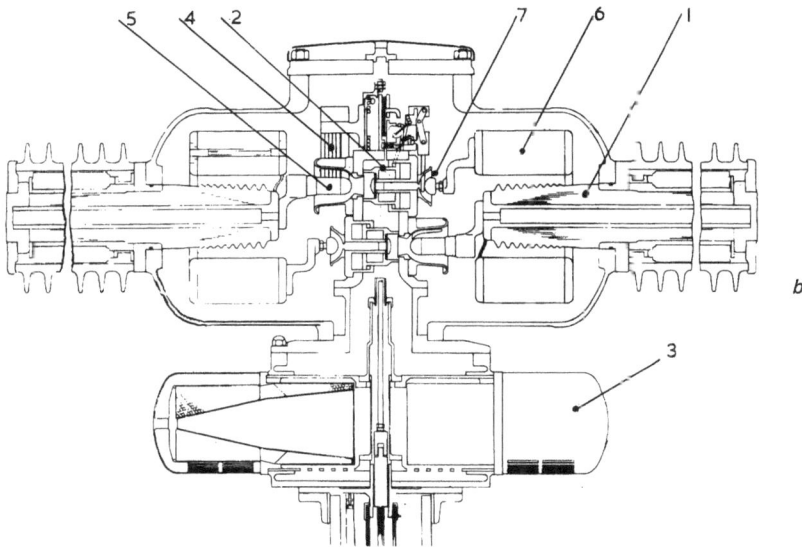

b

Fig. 6.24 OHBR type circuit breaker, A. Reyrolle & Co.
    (*a*) Circuit breaker
    (*b*) Interrupter

6.7 Acknowledgments

The author acknowledges the help and advice received from his colleagues, the permission of A. Reyrolle & Co. to use their facilities to prepare the scripts and sketches, and the help and permission received from this and the other companies to reproduce their photographs, sketches and to publish the extracts from their literature on the circuit breakers.

Chapter 7

SF$_6$ circuit breakers

Part I *E. Maggi, Dr. Ing.*   Part II *C. A. Fawdrey, B.Sc., C.Eng., M.I.E.E.*

Part I

7.1 Historical

The first application of SF$_6$ as an insulating medium dates from 1940 and was covered by a US Patent (Cooper, 1940), while its utilisation as an arc-extinguishing medium was initiated in about 1952 (Lingal, 1953b). The special properties of this gas as an arc quenching medium were also studied in Europe prior to 1940 in several countries, as is shown by a number of patents issued in Germany, France, Switzerland and Sweden. However, there does not appear to have been any published development of actual circuit breaker designs, so that SF$_6$ has become of importance in circuit breaking only during the last 20 years.

The first disclosure based on actual experimental work of the outstanding interrupting ability of SF$_6$ in comparison with air dates from 1953 (Lingal *et al.*, 1953a). This marks the beginning of intensive research into the special properties of the gas as an arc extinguishing medium, as a dielectric and as a heat conductor, which properties have facilitated considerable increases in voltage and current ratings in SF$_6$ circuit breakers relative to air circuit breakers, without resorting to extreme gas pressures or large numbers of break in series.

7.2 Properties of SF$_6$

7.2.1 Physical properties

Pure SF$_6$ is colourless, odourless, nontoxic and nonflammable: its specific density at 20°C and at a pressure of 760 torr is 6·135, i.e. five times the specific density of air. Its boiling point is −60°C at a pressure of 760 torr (Fig. 7.1) (Delevoy, 1960), and its global thermal transfer coefficient, which includes convection effects, is 0·034, or 1·6 times the coefficient for air. Its vapour pressure at 20°C is 24 atm, making it of great interest for insulation in the million-volt range (Phillips, 1963).

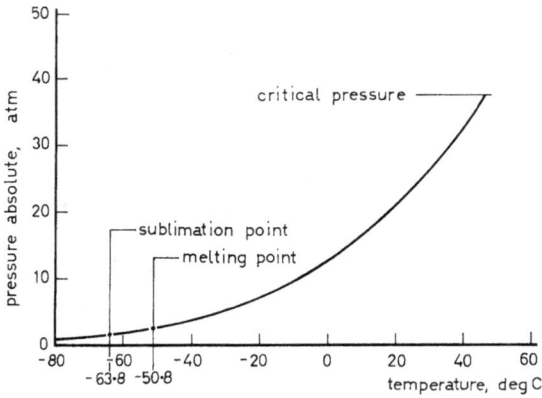

Fig. 7.1 Vapour pressure curve for sulphur hexafluoride (SF$_6$)

The sonic conductivity of SF$_6$ is remarkably low; in fact, the speed of sound in SF$_6$ at 30°C and 760 torr is 138·5 m/s; i.e. 41% of that in air.

7.2.2 Chemical properties

SF$_6$ is chemically inert up to 150°C and will not attack metals, plastics and other substances commonly used in the construction of high-voltage circuit breaker components. However, at the high temperature caused by power arcs, it decomposes into various components which, according to Gutbier (1966), are principally SF$_4$ and SF$_2$, together with small amounts of S$_2$, F$_2$, S, F, etc. which are in part corrosive to both glass and metals in the presence of moisture. The substances formed by the combination of such elements with vaporised metals appear as a whitish powder which has good insulating properties. The breaker contacts must therefore by designed with a wiping action to ensure self cleaning of the co-operative current-carrying surfaces.

The complete absence of carbon in the SF$_6$ molecule is a major advantage for an arc-interrupting medium as it permits more freedom in structural design. For instance, oil circuit breakers must not use horizontal insulating parts, to prevent accumulation on them of conductive carbon deposits, and the oil must be changed when the dark colour it gradually acquires in service indicates the presence of increasing amounts of carbon particles, which may endanger also the vertical insulating parts.

All the chemically active impurities formed by the arc at various temperatures, such as the ones indicated above, recombine in the extremely short time of $10^{-6} - 10^{-7}$ s after extinction of the arc, thus eliminating the risks to personnel that would otherwise arise during maintenance. The remaining traces of impurities can be eliminated by means of absorbing materials, such as activated alumina, located inside the breaker pressure vessel. Alumina also absorbs moisture and contributes to the SF_6 stability.

The absence of air eliminates contact oxidation, and contact abrasion when breaking high currents is extremely small compared with contacts in air, with the result that contact service life is greatly increased and replacement is rarely necessary.

7.2.3 Electric properties

The dielectric strength of SF_6 under static conditions is given in Fig. 7.2a; it may be observed that at a pressure of 3 atm absolute the dielectric strengh of SF_6 is comparable to that of the insulating oils in common use.

The dielectric strength of SF_6 under impulse conditions is given in Fig. 7.2b (Browne Jun., 1960). For realistic designs of contacts used in high-voltage circuit breakers, when the configuration is normally a cylindrical moving contact and a tulip-shaped fixed contact arranged to take full advantage of the properties of SF_6, Figs. 7.2c and 7.2d show the comparison of the behaviour obtained for SF_6 gas and air at various pressures (Einsele, 1964).

The remarkable performance of SF_6 as an interrupting medium, which exceeds the expectations one might deduce from its relatively modest superiority as an insulator, is due largely to its ability to recover dielectric strength quickly after the arc passes through current zero, and this may be explained as follows. It is well known that one of the most important characteristics of an arc-quenching medium is represented by its thermal time constant at the instant when the current flowing in the arc passes through its natural zero; this value should be very small (Browne Jun., 1960, Yoon, 1959, and Gozna 1964).

The absolute value of the thermal time constant of SF_6 flowing at high speed has been deduced from various experments. One of these methods (Fukuda, 1966) is represented in Fig. 7.3a, where the values measured during tests with various mixtures of air and SF_6 have been plotted in curve II as a function of the proportion of air in SF_6, down to the limit of measurement of the equipment available. From such curves, the value of the time constant for pure SF_6 flowing at sonic speed can be estimated to be of the order of one or two millimicroseconds, on the assumption that the same exponential relationship shown by the experiments made with high proportions of air are also valid for pure SF_6. Such an assumption is confirmed by the similarity of the relationship existing for still gas as shown by curve I, which was extended experimentally down to pure SF_6.

The temperature distribution of arcs in SF_6 is also very important if the characteristics of this gas, represented in Fig. 7.3b, are considered. The unusual correlation between the specific heat C_p, which may also represent the thermal conductivity K of the gas, and its electrical conductivity σ is a special property of SF_6 at high temperatures.

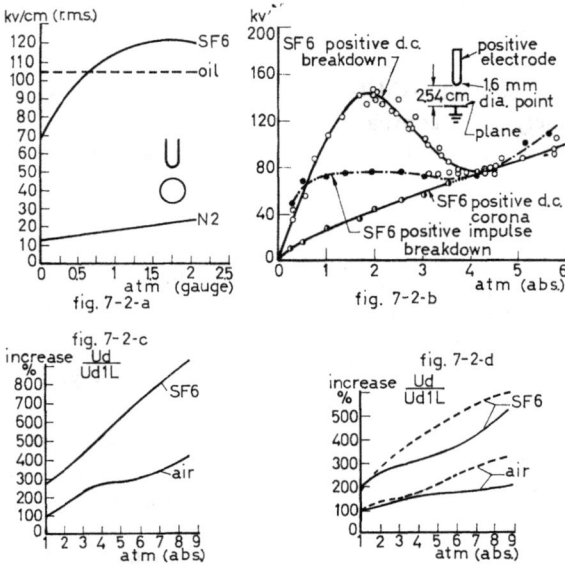

Fig. 7.2 Dielectric strength curves for SF_6 and others insulating media

(a) comparative curves of the dielectric strengths of SF_6, nitrogen and insulating oil at 50 Hz and at various gas pressures

(b) breakdown and initial corona voltages in SF_6 at various gas pressures for a point-to-plane gap having a length of 2·54 cm

(c) breakdown voltages for SF_6 and air at 50 Hz and at various gas pressures between specially designed contacts having a gap length of 5 cm

Ud breakdown voltage of SF_6 at 50 Hz

Ud1L breakdown voltage of air at 50 Hz

(d) breakdown voltages for SF_6 and air with 1/50 μsec impulse waves at various gas pressures between specially designed contacts having a gap length of 5 cm

———— positive impulse polarity of stationary contact

- - - - - negative impulse polarity of stationary contact

Ud breakdown voltage of SF_6 with 1/50 μsec impulse wave

Ud1L breakdown voltage of air with 1/50 μsec impulse wave

It is interesting that in Fig. 7.3b the increase in C_p appears at the rather low temperatures of 2000–2500 K, where σ becomes almost negligible. The peak of C_p in this temperature range is caused by the presence of the gas thermal dissociation reactions which take place within the same range of temperatures and which are almost complete at a temperature beyond 4000 K.

Frind (1960a) has shown that, for arcs having a cylindrical shape, the so-called time constant (θ) is expressed as a function of the radius of the arc core as follows:

$$\theta = C\pi r_0^2$$

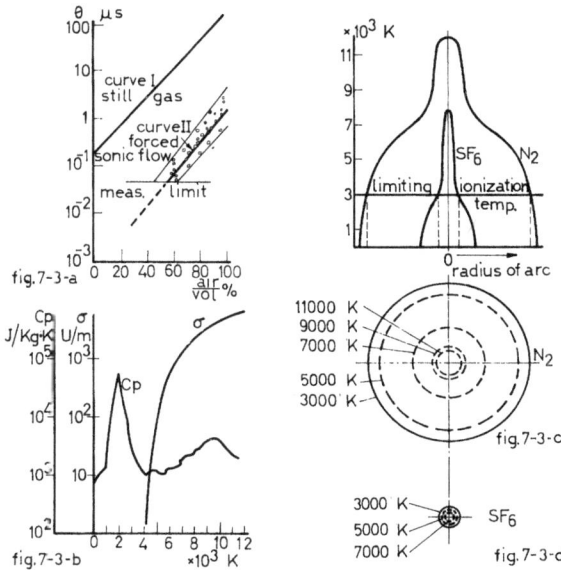

Fig. 7.3 Arc characteristics in SF$_6$

(a) variation of the thermal time constant with the percentage of air in SF$_6$
θ = thermal time constant measure in microseconds
meas. limit = measurable limit
(b) calculated results of specific heat and electrical conductivity of high temperature SF$_6$
C_p = specific heat
σ = electrical conductivity
(c) radial temperature distribution in cylindrical arcs not subjected to blowing (schematic distribution) in nitrogen and SF$_6$, according to Frind
(d) radial temperature distribution in SF$_6$

where C is a constant and r_0 is the radius of the arc core. The radius, or the cross-section, of the arc core can be estimated from the calculated profile of the temperature distribution, taking into account the experimental results for the axial potential gradient of the arc.

The temperature distribution of arcs in SF$_6$ can be calculated, when a cylindrical symmetrical arc in a steady state is assumed, by solving Elenbaas–Heller's equation

$$\sigma E^2 + \frac{1}{r_0}\frac{\partial}{\partial r_0}\left(r_0 K \frac{\partial T}{\partial r_0}\right) = S(T)$$

where E represents the axial potential gradient, $S(T)$ is the radiation loss from the arc column, and T, r_0, σ and K denote the temperature, radius, electrical and thermal conductivities of the arc, respectively. The results of the calculations are plotted in Fig. 7.3c. It is evident that in SF_6 heat produced by the arc is generated in the region along the arc axis, where very high temperatures in the range 10 000-30 000 K are produced. Owing to the very low thermal conductivity of the gas at such temperatures (Fig. 7.3b), the arc heat can be dissipated radially only by the effect of a very steep temperature gradient. This means that the temperature in layers even close to the arc axis will be relatively low and will fall below the ionisation temperature of the order of 3000 K; this space will have little or practically no electrical conductivity owing to the great affinity of SF_6 for free electrons, while its thermal conductivity will be very high.

It therefore appears that an arc current approaching zero is carried in SF_6 by a very thin arc core having a high temperature and surrounded by a nonconductive layer. Thus, after a critical current zero, the dielectric strength of the circuit breaker will recover very quickly, and very high rates of rise of the recovery voltage can be withstood (Swarbrick, 1967).

With an arc burning in nitrogen (N_2) the conditions are much less favourable for quenching the arc because, owing to the good thermal conductivity of the gas in the temperature range 7000–3000 K, the arc does not take the shape of a thin core and the temperature gradient is rather low; the arc radius remains large and the corresponding time constant is also large in proportion to the square of the radius. The external layers having temperatures below the ionisation level remain electrically conductive owing to the absence of the electronegative property in nitrogen.

Figs. 7.3c and d show schematically the radial distribution of arc temperatures in nitrogen and in SF_6, respectively. According to Frind, the ratio of the conductive cross-sections of nitrogen and SF_6 for a current of 1 A is approximately 50:1. The thermal time constant measured for arcs of 1 A is extremely small for SF_6 (1 μs) while for nitrogen it is 100 μs.

The unique combination of thermal and electrical characteristics of SF_6 provides, as already mentioned, the basis of its remarkable properties as an effective arc quencher, since they make it capable of handling high r.r.r.v. without the use of resistors and also capable of interrupting transformer magnetising currents without overvoltages.

7.3 SF₆ circuit breakers

7.3.1 Methods of current interruption

The principal methods in use at present for heavy-current interruption in SF_6 are the so-called 'double pressure' method and the 'impulse' (or 'puffer') method. Other methods based either on using the thermal effect of one section of the arc to generate sufficient gas pressure to produce a jet of gas across a second section of the arc existing within a separate chamber or on the free interruption of load currents in a still SF_6 atmosphere under moderate pressure have been limited in practice in the first case to experimental low capacity load interrupters or to sectionalising reclosers for medium distribution systems and for the second case to high voltage disconnecting switches.

7.3.2 Metal-enclosed double-pressure types

The first power circuit breakers using SF_6 gas were developed in the USA maintaining the general structure conceptions of the bulk oil circuit breaker having the extinguishing chambers and the contacts of each phase housed in an earthed steel tank as shown in Fig. 7.4a (Fredrich, 1959).

Because of its cost, SF_6 is not discharged into the atmosphere as in airblast circuit breakers, and, consequently, completely sealed and self contained unit construction has been adopted for all SF_6 circuit breakers. The sealing between the different sections in such breakers is effected by means of Neoprene gaskets developed specifically for this purpose, while the sealing for each rotating shaft,

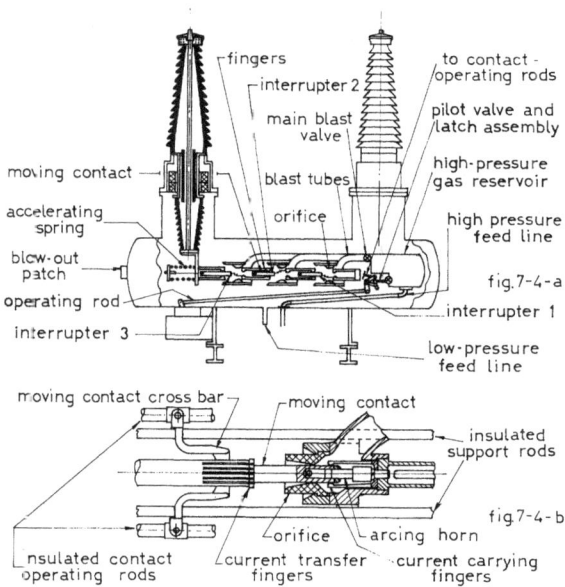

Fig. 7.4 SF_6 double-pressure pole unit for a 15 GVA, 230 kV circuit breaker of the dead tank type [Westinghouse]

(a) interrupters tank and terminal bushings

(b) longitudinal cross-section of interrupting unit

transmitting driving forces from the external operating mechanism, is made with several Teflon rings acting in series and having a 'V' shaped cross-section. Such Neoprene gaskets have a round or rectangular cross-section and are mounted in circular slots machined in the metal flanges. They have been used since 1960 and have caused no deterioration (Van Sickle, 1965), as they are fully capable of resisting arced SF_6, water penetration, ozone deterioration, high and low temperatures, etc. (Ushio, 1971). Because dielectric and arc quenching properties of SF_6 are not significantly reduced by arcing, and contact designs have been developed that can be subjected to repeated arc interruptions equivalent to many years of service, it now seems reasonable to consider that hermetically sealed SF_6 circuit breakers need to be opened for inspection and maintenance only at long intervals, perhaps of the order of five to ten years.

The contacts are shown schematically in Fig. 7.4*b* and are immersed for insulation purpose in an atmosphere of SF_6 at a pressure of 3 atm absolute. The bushing internal conductors are insulated from the steel tank enclosure by the same insulating atmosphere. Contact construction is to minimise erosion due to arcing on the portions that conduct current in the closed position of the breaker. To achieve this the current path is arranged through the side walls of the fixed contact and a set of fingers co-operating with the moving contact. An arcing horn, located within the finger cluster, projects a short distance beyond the end of the fingers and into a cavity in the end of the moving contact. On opening, the arc quickly transfers from the end of the finger cluster to the centrally located arcing contact and to the end of the moving contact; both surfaces are faced with arc resisting material.

The interrupting function is performed by a high-velocity flow of SF_6 through the orifice of a nozzle made of Teflon, (a special arc resisting insulating material) and located inside the arc extinguishing chamber. The gas is maintained at approximately 16 atm absolute within a high-pressure reservoir during normal operation. At the start of contact movement, on an opening operation, the blast valve opens under control of a pilot valve and allows high-pressure gas to flow through an insulating tube to the interrupting orifice, thereby extinguishing the arc as the moving contact moves to the open position. As the contact linkage reaches the open position, the pilot valve closes the main blast valve and conserves gas for the next operation.

After each interruption a compressor system pumps the low pressure gas from the circuit breaker tanks to the high-pressure reservoir via a filter containing activated alumina. Since at 16 atm absolute the gas liquifies at approximately $10°C$, a heating arrangement is provided around the high-pressure reservoir to keep its temperature above this point.

Uniform distribution of voltage across each of the three breaks is obtained by capacitor assemblies. Electrostatic shields around the metal portions of the assembly control the electric field between the interrupter and the tank.

The interrupting capacity of the typical extinguishing chamber described above was tested during 1959 up to 52 kA at 22 kV and has permitted the construction of a circuit breaker capable of interrupting 15 000 MVA at 230 kV within three cycles, by using three breaks for each phase, as illustrated schematically in Fig. 7.4*a*. This class of metal enclosed circuit breaker is readily applicable to full e.h.v. metalclad substations, and has been widely adopted for this purpose.

7.3.3 Insulated-construction double-pressure types
The same principle of interruption has been applied also to circuit breakers having the interrupting units mounted on porcelain columns. Figs. 7.5*a* and *b* illustrate one phase of a European 15 000 MVA circuit breaker for 220 kV, having two double break units supported by two porcelain columns, an example of the so-called modular construction (Einsele, 1964). In this breaker, the contacts are actuated mechanically be a pneumatic operating mechanism fitted on each pole driving a mechanical linkage and bell crank drives mounted at the top of the columns. These bell crank drives are the mechanical link with the contact mechanism, and they also operate the gas blast valves. Compression type accelerating springs are mounted close to the contacts driving the breaker to the open position, latched by a roller trip system in the operating mechanism.

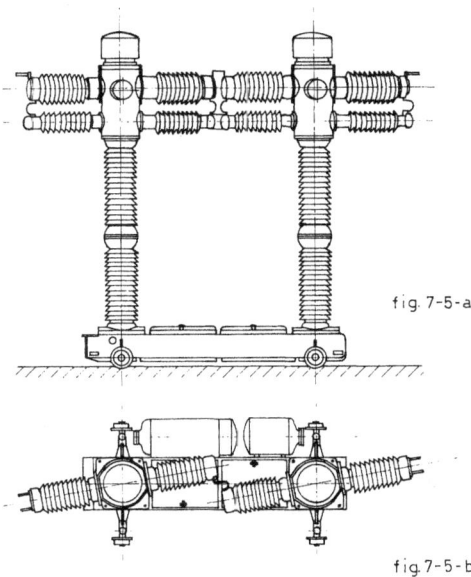

fig. 7-5-a

fig. 7-5-b

Fig. 7.5 *SF*$_6$ double pressure pole unit for a 12 GVA, 230 kV circuit breaker of the modular construction type [Siemens]

   (*a*) vertical view
   (*b*) plan view

The interrupting units and the hollow porcelain columns are filled with SF_6 at a pressure of 4 atm absolute, constituting the low-pressure system. A high-pressure reservoir operating at 17 atm is accommodated in the breaker chassis at earth potential and is connected via a high-pressure pipe, which runs through the hollow column with a receiver tank located in the distribution head.

The blast valves which are mounted in the upper receiver tanks are opened when the breaker is tripped, and the high-pressure SF_6 flows at a high velocity through short pipes to the interrupter nozzles and into the low-pressure chambers. A compressor located in the breaker chassis pumps the gas back to the high-pressure reservoir through filters containing activated alumina. The reservoir is heated to keep the gas temperature at 10°C with an ambient temperature down to −35°C. The arcing time ranges from 5–10 ms at full breaking current, associated with arc lengths of 15–50 mm.

The modular construction facilitates using the same interrupting unit with higher voltage breakers: a good example of this principle is represented by the circuit breaker built in the USA for three and two cycles interruption for 35 GVA at 500 kV having three double-break units supported by three porcelain columns. Figs. 7.6a and b illustrate one such unit. The increased capacity and voltage rating of the unit are obtained by an alternative design of extinguishing chamber (Van Sickle, 1965). The tank forming the central sections of the modular unit is part of the low-pressure system at 4 atm absolute: a door on its side provides access for assembly and maintenance. Inside the tank and opposite the access door is a high-pressure reservoir at 17 atm absolute which provides gas storage close to the interrupters. A blast valve is mounted on the reservoir and supports the bearings for a rotating contact arm which surrounds it. The crossarm is hollow, forming the gas passage to the two interrupting gaps and surrounds a section of the blast valve containing high-pressure gas. When the blast valve is opened, it discharges SF_6 radially outward, directly through the hollow crossarm to the contacts and interrupting chambers, so that the pressure drop is small.

The moving contacts rotate horizontally through arcs of approximately 40° and engage stationary contacts mounted on the inner ends of two insulating bushings; the inner ends of the bushings also support resistor assemblies. These resistor elements are in series with the inner ends of the bushings, and the resistor contacts are mounted near the main stationary contacts. The resistors are in circuit only during closing operations, when they damp switching surges.

This is achieved by means of a resistor switching system comprising a projecting contact carried on the main contact crossarm which engages with a pivoted auxiliary resistor contact biased to the extended position by springs and mounted on the fixed resistor assembly. During closing, the projecting crossarm contact makes contact with the extended pivoted resistor contact prior to closing the main contacts, and the resistor is therefore inserted during the final part of the closing stroke. During the opening movement, the main crossarm accelerates faster than the

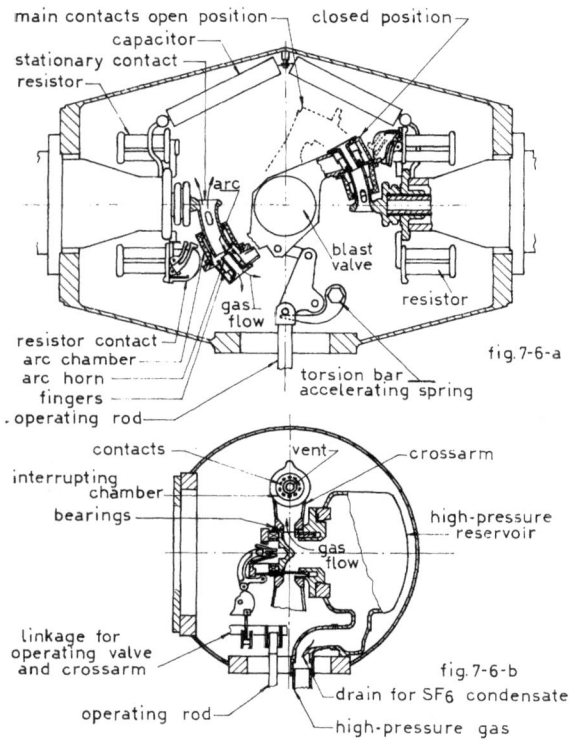

Fig. 7.6 SF₆ double pressure interrupting unit for a 35 GVA, 5̄0 kV circuit breaker of the modular construction type [Westinghouse]

(a) longitudinal cross-section
(b) transversal cross-section

auxiliary resistor contact, and this results in the resistor circuit being open circuited throughout the opening cycle.

A torque bar, located in the modular unit linkage system, provides a high-force low-inertia energy source for the high initial acceleratior of the crossarm and blast valve toward the open position. The blast valve opens rapidly and remains open until the contacts approach the fully open position. The blast valve does not operate during a closing operation.

Gas released by the blast valve flows through the crossarm and into an insulating chamber in which the arc is drawn. The gas, after passing through the arc interrupting area, is discharged through vents in the centre of the stationary and moving contacts. The gas blasts sweep the arc roots away from the fingers where they are initiated and into the interior of the vent passages. The gas is then discharged into the surrounding chamber, which is the principal container for the low-pressure gas; and the fine dust particles formed from the materials vaporised by the arc are also deposited in this chamber. A fine metallic filter at the top of the vertical porcelain column confines the dust to the modular unit.

Filled with low-pressure gas, the support column acts as a part of the low-pressure reservoir in addition to the tube through which the SF_6 gas, at low pressure, returns to earth potential. The column also contains and protects the vertical operating rod and an insulating tube which conducts high-pressure gas from the large reservoir at earth potential to the smaller reservoir in the modular unit.

The high-pressure reservoir is equipped with heaters and with thermal insulation, and the heaters of the three reservoirs are controlled by thermostats on the bottom of each tank. As the thermostats respond to temperatures of both the tank and of any condensate existing in it, they are set to turn on the heaters at a temperature somewhat higher than that at which condensate is formed.

An interrupting time of three cycles is obtained with a normal pneumatic operating mechanism. An alternative mechanism with a high-speed latching device, arranged to reduce the inertia of the parts which must be moved during the tripping operation and equipped with flux diverting trip gear, is used to provide a two cycle interrupting time (Bratkowski, 1966).

Analysis of the performance of 2-module 345 kV and 3-module 500 kV circuit breakers of the type just described in laboratory tests and in service over five years have led to the conclusion that, with certain modifications in the module and associated breaker components, a 500 kV breaker can be built with two modules and a 765 kV breaker with three modules per phase (Leeds, 1970).

Systems requirements indicate that three levels of switching surge control may be required; 2·0 per-unit level can be obtained by providing a single step resistor insertion; 1·7 per-unit level can be obtained by providing a two step resistor insertion; 1·5 per-unit level can be obtained by providing either a three-step resistor insertion or by a two-step resistor insertion in combination with point-on-wave single-phase timing.

Fig. 7.7*a* shows the cross-section of an interrupting unit similar to the one just described but equipped with resistor contacts capable of inserting successively two groups of resistors in the circuit of the breaker for definite lengths of time, as shown in Fig. 7.7*b*.

7.3.4 Gas drying for double-pressure types

Experience with all double pressure types of SF_6 circuit breaker has shown the need for maintaining a very high level of dryness. In the case when unheated high-pressure SF_6 gas is stored in the live assembly, as referred to previously when the ambient temperature falls below 5°C some of the SF_6 in the live receiver will condense and cause a continuous flow of liquid SF_6 down the vertical supply pipe

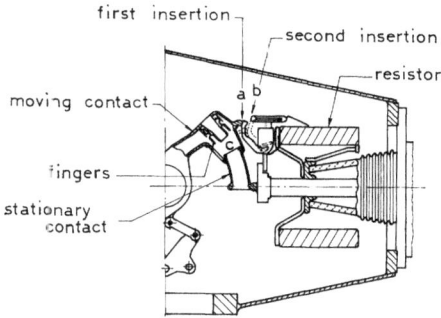

fig. 7-7-a

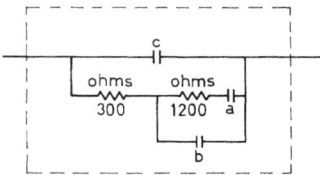

fig. 7-7-b

Fig. 7.7 SF$_6$ double pressure interrupting unit for an EHV circuit breaker of the modular construction type [Westinghouse]

(a) longitudinal cross-section
(b) diagram illustrating closing sequence (a—b—c) for a two step resistor equipped breaker unit

connecting to the heated high-pressure receiver mounted at earth potential. There will then be a balancing counter flow of SF$_6$ gas upwards to the live receiver. This gaseous SF$_6$ is capable of carrying with it more water vapour than the downward flow of liquid SF$_6$ can return, so that a concentration of water vapour could build up, resulting in condensation of free water in the upper receiver. The use of SF$_6$ gas with water content restricted to 30 parts per million in the high-pressure system, in conjunction with filters mounted in the SF$_6$ flow and accurate control of temperature in the earth potential high-pressure receiver have proved adequate precautions to eliminate the risk of insulation failure from this cause.

The moisture level in the low pressure system must be maintained below 150 p.p.m. The progress made by the chemical industry, which is now able to supply almost pure SF$_6$ gas dried to a level which corresponds to a moisture content of only 7 p.p.m. has contributed greatly to the solution of this problem.

7.3.5 High-voltage impulse (or puffer) types

This class of circuit breaker has been called, perhaps improperly, 'single pressure' because the SF$_6$ remains in the circuit breaker for most of the time at a constant pressure in the range of 3—6 atm absolute with the function of insulating the parts of the circuit breaker having different potentials. During the opening operation, only the gas contained inside a part of the circuit breaker is compressed by moving a·cylinder supporting the contact system or by a piston which forces the SF_6 through the interrupting nozzle. The 'impluse' of the sudden gas flow across the arc space, caused by the 'puffer' arrangement as described justifies the other names, impulse or puffer, mentioned above which were used initially in the USA for forced blast oil circuit breakers, and for SF_6 circuit breakers of moderate interrupting capacities at relatively low voltages (Easly, 1964).

Fig. 7.8 shows the longitudinal and transverse cross-sections of a single pole of a 3-phase circuit breaker installed during 1957 for service at 46 kV with an interrupting capacity of 500 MVA. The three individual poles are mounted, in the horizontal position, on a steel structure housing a pneumatic operating mechanism together with its compressor unit and accessories. The simplicity of the internal arrangement of the active parts of the breaker is clearly apparent from the figure which shows the moving contact and its supporting cylinder in the position corresponding to the arcing period of the interrupting operation.

The concept of modular construction has constituted the basis for developing efficient modular units allowing SF_6 impulse circuit breakers to be built for voltages of 765 kV, and possible 1050 kV, the interrupting capacity of which can be demonstrated by both unit test and synthetic test methods. The possibility of

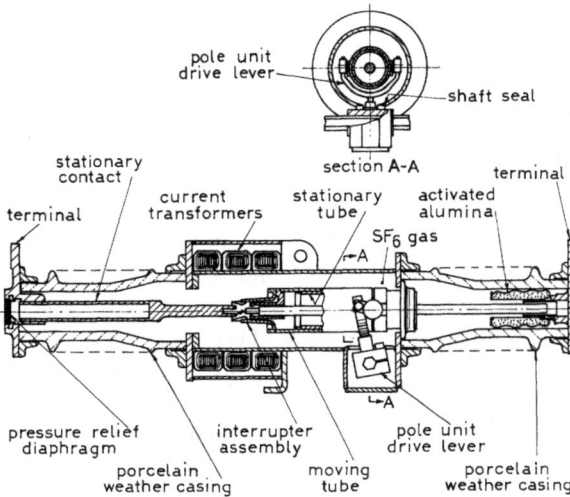

Fig. 7.8 SF_6 single pressure pole unit for a 500 MVA 46 kV circuit breaker [Westinghouse]

eliminating the complicated high-pressure system of the double pressure SF$_6$ circuit breaker together with their requirements for both temperature and dryness of the gas and of obtaining independence from the ambient conditions for SF$_6$, which at the pressure of 5 atm absolute is able to withstand temperatures as low as $-30°$C, has also been an important contribution to simplification. The performance, combined with these advantages, has encouraged a number of manufacturers to study new ranges of e.h.v. circuit breakers based on these principles.

The mechanical force required by such breakers for compressing the gas in all the units to the proper pressure during the very short time available during partial opening of the contacts has to be supplied by powerful mechanisms which represent an expensive part of the circuit breakers: the advantages of exploiting the self regulating property of the impulse arrangement has resulted in most designers adopting the elastic action provided by the pneumatic mechanism, as discussed later. The pneumatic mechanism has acquired a good record of reliability in addition to the evident economy of its operation: it has practically replaced the solenoid mechanism in the field of high-voltage circuit breakers on account of its higher speed, its ability to provide high energy levels, and of its independence from the need for large storage batteries. It requires only dry and clean air under moderate pressures in the range 12-20 atm, requirements which are met by modern compressor systems.

An example of a self contained unit (Calvino, 1968) capable of interrupting 31 kA at 105 kV (synthetic tests have demonstrated later on a similar unit interruption of 70 kA at 145 kV) is represented in Fig. 7.9.

The Figure shows two cross-sections of a modular unit where the moving contact appears in the open and closed positions, respectively. It consists of a sealed insulated cylinder filled with SF$_6$ at 5 atm absolute; a distinctive feature of the interrupting unit is the total absence of any blast valve. The seals are of the static type using Neoprene O ring gaskets. The fixed contact has the shape of a hollow tube as more clearly shown in Fig. 7.10, when the successive positions of the moving contact from closed (Fig. 7.10a) to open (Fig. 7.10d) are illustrated.

When the contacts separate, an arc is established across the arcing contacts, as shown in Fig. 7.10b. If the current intensity is not very high it is extinguished at the first zero passage due to the high deionising effect of SF$_6$ and by the gas blast caused by the piston. The impulse effect is initially slight as the contact distance at the instant of interruption is small and the gas pressure is low. This feature is important for the interruption of small inductive currents which takes place without chopping and therefore without high overvoltages. In the case of small capacitive currents, which are interrupted at the first zero after contact separation, the recovery voltage maximum value appears after one half cycle, that is after sufficient time for the contact separation to reach a length that will be capable of withstanding this voltage and prevent arc reignitions.

If the short-circuit current is high the interruption may not occur at the first zero passage and the arc may produce a blocking effect on the nozzle preventing

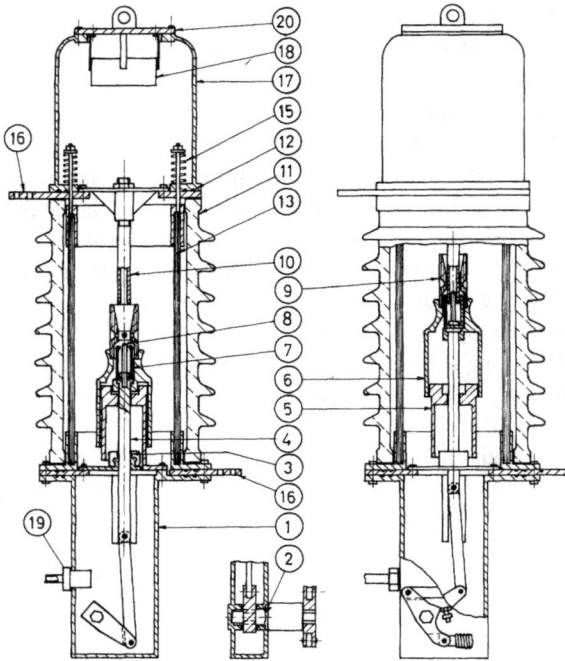

Fig. 7.9 SF$_6$ single-pressure modular unit capable of interrupting 31 kA at 107 kV [Magrini]

1 level system case
2 sealed-shaft transmitting motion from external level to internal level system
3 finger-cluster sliding contact
4 moving-contact stem of silver-plated copper
5 stationary piston of aluminium alloy
6 moving cylinder of aluminium alloy
7 finger-cluster type moving contact
8 arcing contact with tip of tungsten alloy
9 Teflon nozzle
10 stationary contact stem
11 porcelain weather casing
12 type 'O ring' gaskets ensuring permanent sealing
13 protective insulating tube
15 springs suitably adjusted to maintain the unit tight by stressing the procelain weather casing only by compression through the 'O ring' gaskets
16 terminals
17 upper cap
18 basket containing absorbing substances
19 plug-in joint for filling the unit and controlling the internal pressure
20 removable cover

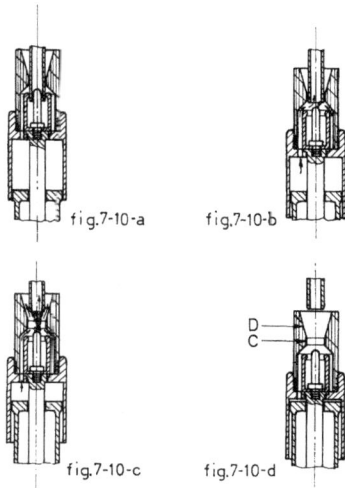

fig.7-10-a fig.7-10-b

fig.7-10-c fig.7-10-d

Fig. 7.10 Sequence of an opening operation in the modular unit of Fig. 7.9 [Magrini]

    (a) contacts closed
    (b) contacts initiate separation, arc established
    (c) contacts continue separation, gas blast through arc
    (d) contacts fully opened, arc extinguished

SF_6 from flowing freely out of the cylinder. The pressure inside the cylinder will increase to a high value (Fig. 7.10c). At the second or third natural passage through zero, depending on the current intensity, on the rate of rise of the recovery voltage dV/dt and on the instant of contact separation, the gas pressure inside the cylinder will reach a value sufficient to produce a gas blast across the arc space which will ensure extinction of the arc.

It is evident from the above description that the gas pressure adjusts itself according to the task to be carried out; it is also clear that if the opening force is applied by an operating mechanism using compressed air for the opening movement, an adaptable means will be available for interruption by the circuit breaker, especially since in such a system the instantaneous force applied by compressed air to the mechanism piston is automatically adjusted to match the opposing factors to be overcome.

Another important feature of this type of circuit breaker is the considerable residual interrupting capacity it retains with reduced pressure of the SF_6 gas. The unit just described has been tested with an internal pressure of 1 atm absolute, which represents the lowest limit of pressure for the case of a leaky unit, and, in this condition, has interrupted 20 kA at a power factor of less than 0·1. This permits the circuit breaker to be opened in service even if inspection of the individual gas manometer indicates abnormal low pressure in one modular unit.

During an opening operation, the mechanism charges the closing springs of the circuit breaker which are located in the chassis at earth potential. Owing to the light weight of the parts to be operated the closing spring effort is relatively small and is not influenced by prearcing across the closing contacts; in fact the arrangement of the gas passages in the interrupting chamber can cause any pressure rise that might be produced by high current prearcing to result in an additional thrust in the direction of contact closing. The levers operating the different modular units constituting a single-phase pole of the circuit breaker are mechanically connected to the operating mechanism through insulating rods maintained constantly under tension and all moving contacts are therefore operated simultaneously.

Repeated short-circuit tests at currents covering the entire range of values from small inductive to full capacity and adding up to 400 kA have demonstrated no appreciable alteration of the dielectric and interrupting properties of SF_6 contained in the tested unit.

The insertion of single-step or double-step resistors during the closing operation is arranged by the addition of light external units containing both the contacts and the resistors, the final closing of the main circuit breaker contacts shunts the resistors and causes the resistor contact latches to be tripped after approximately 10 ms from resistor insertion.

Similar solutions have also been adopted for the construction of SF_6 circuit breakers incorporated in metalclad substations for 225 kV and having four breaking units per phase identical with the one illustrated in Fig. 7.11 and an interrupting capacity of 12 000 MVA (Maury, 1971a) the operating mechanism used in this case is of the oil pneumatic type providing direct control of the circuit breaker for both the closing and the opening operations, and employing oil and compressed gas at high pressure (300–330 atm). Originally conceived for actuating suspended e.h.v. circuit breakers (Maury, 1971b) this class of mechanism is now extensively used, partly because of the convenience with which high energy levels can be stored in the gas cylinder.

High-pressure oil servomechanisms are, however, precision devices requiring specialised manufacturing and maintenance procedures to ensure the accuracy in machining and cleanliness during assembly that are essential for reliable service.

7.3.6 Distribution impulse types

Because of its simplicity, the SF_6 impulse circuit breaker offers an attractive solution for circuit interruption, but its application to switchgear at distribution voltages has been much slower mainly because traditional solutions represented by magnetic, oil and low oil circuit breakers have shown themselves to be satisfactory for their duty and also because of the relative cost. In other words the simplifications introduced in high-voltage circuit breakers by the use of SF_6 may well justify the increased cost of the more powerful operating mechanism the

Fig. 7.11 SF₆ single pressure interrupting unit for a 12 GVA 225 kV circuit breaker incorporated in metalclad switchgear [Delle-Alsthom]

impulse circuit breaker requires, but such an advantage becomes less important with breakers of the distribution voltage class, so that their application then appears to be justified only for installations presenting very special conditions as for instance, mines, chemical plants or hazardous atmosphere locations, where the danger of fire or of exposure to a corrosive atmosphere may suggest the elimination of both oil and open arc air breakers.

An example of a recent design of an SF₆ impulse circuit breaker for distribution voltages of 7·2-24 kV and capable of interrupting currents up to 31·5 kA is represented in Figs. 7.12 and 7.13. Each pole is enclosed in a sealed cylindrical container at earth potential (24) made by cold forming a nonmagnetic alloy tube; this choice of material eliminates the danger of gas leakage arising from the use of materials which may be porous, as castings or welded steel. The top cover (1) is bolted to the container and supports the filling valve (3) with its cap and the basket (4) containing chemical substances capable of absorbing the products resulting from the SF₆ decomposition due to the action of the arc. The bottom steel cover (17) is similarly fastened to the container and supports, together with a stationary piston (41), a sliding stuffing box (20) for the moving contact operating rod. The working parts of the pole are also supported by this bottom cover, through a robust insulator (16), comprising: a tubular moving contact (40) constituting the puffer cylinder actuated by the operating rod; a lower set of finger contacts (23) fastened

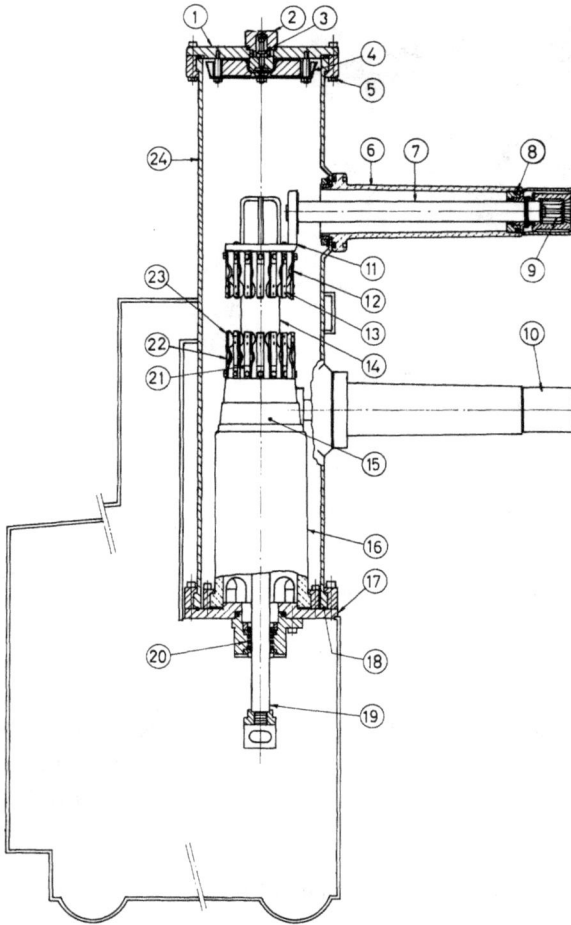

Fig. 7.12 Vertical cross-section of an SF₆ single pressure pole unit for a 1200 MVA 24 kV, horizontally withdrawable circuit breaker [Adda, S.p.A.]

1 unit top cover
2 filling valve cover
3 filling valve
4 basket containing absorbing substances
5 cover tightening bolts
6 primary bushings insulation
8 gasket
9 tulip type finger contact
10 tulip contact assembly
11 upper contact support
12 finger contact spring
13 upper contact fingers

14 blast tube
15 lower contact support
16 support insulator
17 unit bottom cover
18 gasket
19 operating rod
20 gaskets
21 moving contact cylinder
22 finger contact spring
23 lower contact fingers
24 metallic container

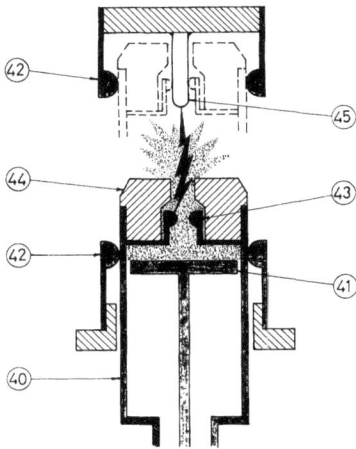

Fig. 7.13 Schematic interrupter unit for circuit breaker represented on Fig. 7.12 [Adda S.p.A.]

40 moving contact cylinder
41 stationary piston
42 main contacts fingers
43 arcing contact fingers (moving)
44 nozzle (moving)
45 arcing horn (stationary)

to a cast support (15); a blast conduit which contains internally both fixed (45) and moving (43) arcing contacts.

An upper set of finger contacts (13) is also fastened to a cast support (11). Each cast support is fastened to one of the horizontal terminal conductors (7) protected by external insulating bushings (6).

The arcing contacts effectively protect the main contacts from the arc as they separate later and confine the arc within an inner region completely surrounded by the gas blast expelled by the stationary piston from the moving cylinder during the opening operation; the moving cylinder carries on its upper part a nozzle (44) made of insulating material capable of resisting the action of the arc and housing, in a central cavity, the moving arcing contact (43).

The opening operation is affected by means of springs loaded during the closing operation accomplished by the operating mechanism closing springs.

The operating rod causes the successive opening of the main contacts and of the arcing contacts as the cylinder is moved towards the stationary piston; the SF$_6$ gas thus compressed flows out of the cylinder through the nozzle in the form of an axial blast which surrounds the arc and ensures rapid extinction. The operating mechanism is of the spring stored energy type and is charged by a motor in approximately 30 s to provide sufficient energy for a complete autoreclosing cycle.

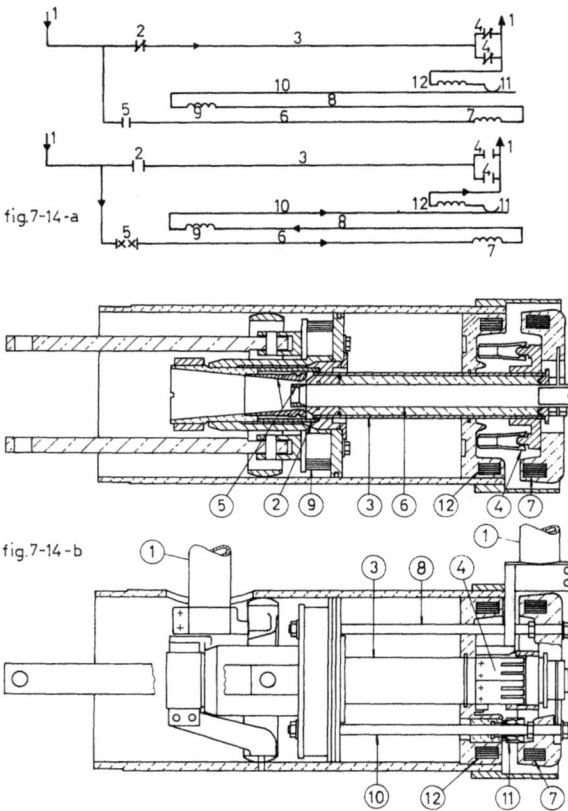

Fig. 7.14 SF$_6$ single pressure interrupting unit for a 1500 MVA 34·5 kV vertically withdraw-able circuit breaker using magnetically driven pistons [Westinghouse]

(a) schematic diagrams showing electric operation of interrupter in the closed position (upper diagram) and open position (lower diagram)
(b) cross-section (upper drawing) and internal view (lower drawing) of interrupting unit with contacts in the closed position

  1 primary bushings
  2 front main contact
  3 main moving contact tube
  4 rear main contact
  5 arcing contacts
  6 moving arcing contact tube
  7 driving coil (moving)
  8 upper guide rod
  9 piston coil (moving)
10 lower guide rod
11 sliding ball contact
12 cylinder coil (stationary)

An alternative solution to the problem of powerful operating mechanisms for impulse SF_6 circuit breakers is shown in Fig. 7.14 and uses the fault current energy to drive the moving contact and the puffer element magnetically, thus forcing the SF_6 through the arc space (Frink, 1967 and 1968).

This breaker is capable of interrupting 1500 MVA at 23 and 34·5 kV facilitating the use of such higher distribution voltages in suburban substations, which are often located in densely populated areas where space is restricted and low operating noise level is important. It consists of three single-pole units mounted on a structural steel frame, housing also a spring stored energy mechanism.

During closing the mechanism charges the opening springs and the closing springs are immediately recharged by a small motor, which requires about 10 s. The additional magnetic driving force is proportional to the square of the fault current, so that the piston driving force is proportional to interrupting requirements. At low fault current the magnetic assistance is negligible, but, at the maximum fault rating, 10% of the piston driving force is supplied by the spring loaded operating mechanism and 90% by the magnetic coils. Since the magnetic coils only come into the circuit after arc transfer, they are not energised during breaker closing.

Fig. 7.14 shows two longitudinal cross-sections of a pole in the closed position made along two planes at 90°, together with schematics of the breakers internal components and magnetic coils.

The interrupter mechanism in each pole unit is operated by insulating links attached to a bell crank. The crank shaft passes through a seal in the tank to an external lever. It must be noted that in this breaker the puffer cylinder is stationary while the main moving contact (3) and the moving arcing contact (6), together with the driver coil (7) and the puffer piston carrying its own coil (9), are actuated initially by the operating mechanism through the insulating links and the two guide rods (8) and (10).

The electrical operation of the breaker may be understood by the following description referred to in Fig. 7.14: with the breaker closed, the normal current path is from the front bushing (1) to the front main contact (2), to the moving tubular contact (3), to the rear main contact (4), to the rear bushing (1). As the breaker begins to open under the action of the opening springs arcs are drawn between the main contacts (2) and (4), because both the moving contacts (3) and (6) together with the piston, the guide rods (8) and (10) and the driver coil (7) with its support are moved backward. Movement of the piston causes gas in the cylinder to be driven through the arc at the main contact (2) and across the arcing contacts (5) (contacts (2) and (5) are in close physical proximity). The arc quickly transfers to arcing contacts (5) and inserts the interrupting circuit putting the magnetic coils (7), (9) and (12) in series with the arc through the following circuit, from the front bushing (1) to the arcing contacts (5), to the moving arcing contact (6), to the driver coil (7), to the lower guide rode (8), to the piston coil (9) to the upper guide rod (10), to the sliding ball contact (11), to the stationary cylinder coil (12), to the rear bushing (1).

The piston coil (9) is attracted by the cylinder coil (12), and the driver coil (7) is repelled by the cylinder coil (12). As the piston moves it forces SF_6 from the cylinder, through an annular space in the piston, and through the arc space (5) extinguishing the arc.

When the fault current is interrupted, the magnetic driving force ceases and the compressed gas behind the piston acts as a cushion to bring the piston to rest.

7.4 Nozzles for SF_6 circuit breakers

The nozzles used on SF_6 circuit breakers at the present time are usually made of Teflon (Polytetrafluorethylene). The characteristics of this insulating resin are rather unique as it is mechanically strong, easily machinable and capable of resisting relatively high temperatures; it has the property of vapourising under the direct action of electric arcs of short duration without carbonising, although its molecule contains carbon.

The interruption process of the electric arc appears to be favourably influenced by the presence of such material while other refractory materials, as ceramics, besides being mechanically fragile, tend to lose their insulating properties under the action of the hot electric arc and are attacked by the SF_6 dissociated components.

The reduction in dimensions of the interrupting units, several of which are often mounted in one container, has also favoured the use of axial nozzles made of insulating material in preference to metal ones, but the main reason for such preference seems to be the following: the metal nozzles, as used in most air-blast circuit breakers, need to be protected against the direct action of the arc by a high speed and continuous flow of the interrupting medium obtained by a high pressure jet of air discharging to free atmosphere and not impaired by the back pressure existing in a closed environment. SF_6 gas has a much higher density than air, and on account of its cost is used in circuit breakers that do not discharge the gas externally; its speed of discharge is much lower and might be insufficient to protect metal nozzles from the direct action of the arc. However, Teflon nozzles have proved capable of resisting satisfactorily such action by withstanding without serious damage a very large number of repetitive interrupting operations with heavy short-circuit currents reaching 70 kA.

The shapes of the nozzles used in double pressure SF_6 circuit breakers are broadly similar to those used in air-blast circuit breakers, that is the well known Ruppel type which ensures a continuous flow of the gas stream avoiding the formation of vortexes capable of disturbing the gas flow. This type of nozzle has been generally adopted in air-blast circuit breakers because in addition the ionised air needs to be swept away very rapidly from the arc space in order to obtain good insulation between the separating contacts: such a condition is not however as essential in the double pressure SF_6 circuit breaker because of the inherent properties of the gas which recovers its insulating characteristics much more rapidly than air after current zero and does not need to be entirely replaced by fresh gas.

The shapes of the nozzles used in the impulse type or single pressure circuit breaker are then based on principles rather different to those applying for air-blast and double pressure SF₆ circuit breakers. Owing to their position with respect to the moving contact, which surround the arc during its development, the nozzles have the shape of a double funnel having its smaller transverse section limited by the diameter of the moving contact passing through it.

The length of the cylindrical intermediate section of the double funnel C (Fig. 7.10*d*), where the deionising action of the gas blast takes place, and that of the discharging section D in the same figure, are closely related to the voltage rating of the interrupting unit.

The decompression vents provided in the cylindrical section and the annular grooves cut in the inner surface of the discharging section have dimensions and shapes conveniently selected in order to ensure both proper flow of the gas and sweeping away completely of the decomposition products generated by the nozzle material at the high temperatures imposed by the electric arc (Calvino, 1967).

7.5 Future developments in SF₆ circuit breakers

The use of SF₆ as an insulating and interrupting medium has certainly proved very successful, and the results obtained during the relatively short period of time elapsed since its first applications to circuit breakers has permitted it to be used for capacities exceeding the limits reached by the traditional means of interruption.

The problems which had to be solved were of importance to transmission system engineers who wished to increase both system voltage and short-circuit rating. The methods adopted for solving the initial problems have resulted from concentrating the minds of the first generation of SF₆ circuit-breaker designers on the shortest way to attain the most urgent goals, and these were evidently suggested by the existing techniques already adopted successfully with traditional circuit breakers. It is the author's conviction that the methods now in use for exploiting the special properties of SF₆ gas will make available further experimental data and knowledge of its real possibilities, and other new methods basically different from the present ones may prove better adapted to meet conditions that have been neglected during this initial period of anxious effort dedicated to the quick production of high capacity circuit breakers for high and extra high voltages.

The new generation of designers may resume in the near future the study of the data made available from the use of a large number of SF₆ circuit breakers in all types of service and discover methods for making a better use of SF₆. The heavy mechanical stresses imposed on the present circuit-breaker structure in order to obtain rapid flow of this high density gas may be, for instance, drastically reduced by displacing the electric arc rapidly in a still atmosphere of gas under moderate pressure, and examples of some possible solutions are already being studied.

One such new method utilising static means for driving the arc (Poltev, 1967) makes use of a radial electromagnetic field generated by the short-circuit current which is made to flow in two coils at the time of interruption; double main contacts are necessary in order to insert these coils temporarily into the circuits. These contacts are located at both ends of the arc gap, and the arc length is controlled by the arrangement of the circular arcing contacts on which the arc rotates rapidly under the action of a concentrated radial field, before its extinction.

A second method (Maggi, 1964 and 1971) permits substantial simplification of the structure and the operation of the circuit breaker by using transverse fields generated by permanent magnets located conveniently along the arc path which is not then restricted in its length or shape; the high resistivity of modern sintered ferrite magnets virtually eliminates the major part of the difficult insulation problems previously existing in the use of metallic alloy magnets for similar purposes.

The displacement of the arc by magnetic means can be made at speeds higher than that of sound in SF_6 and reaches a maximum at the peak of the current, whereas the flow of the gas in the present breakers is greatly reduced by the blocking effect of the volume of the arc to be extinguished, leaving a very short time for the breaker to develop its interrupting ability as the current approaches zero.

The brutal action of the forced gas blast on the arc, permanently adjusted for the maximum current to be interrupted could be replaced by a self regulated action driving the arc towards its extinction at the most convenient speed.

Acknowledgements see 7.10

Part II
Recent developments in SF_6 circuit breakers (1981)
C.A.Fawdrey, B.Sc., C.Eng., M.I.E.E.

7.6   Methods of current interruption

Although the double-pressure system is, to some extent, still used in the USA for the higher ratings, in Europe the puffer reigns supreme for all except the lower voltages. At low voltages, magnetic rotation of the arc is established as an interrupting means, and the extension of this method to higher voltages is under active development. Pressure generation by direct arc heating, to produce an extinguishing gas flow, has also been employed. Related development using electro-negative fluids, akin to o.c.b. principles, is being pursued.

Remarkable improvements in performance of puffer interrupters have been attained, such as 50kA, 420kV on two interrupters, and 40kA, 245kV on a single interrupter. Nevertheless, it is interesting to note that at least one manufacturer of such breakers also lists preferential ranges of low-oil-content circuit breakers for applications which include high rated short-circuit breaking currents and extremely high i.t.r.v. values; also for missing current zero, where the higher arc voltages of the low-oil circuit breaker can more readily force an asymmetric current through zero.

The dv/dt performance of SF_6 plotted against di/dt shows a much steeper characteristic than air or oil; so in spite of the fact that SF_6 provides the better performance at lower values of *di/dt,* there is a problem in using SF_6 at values of

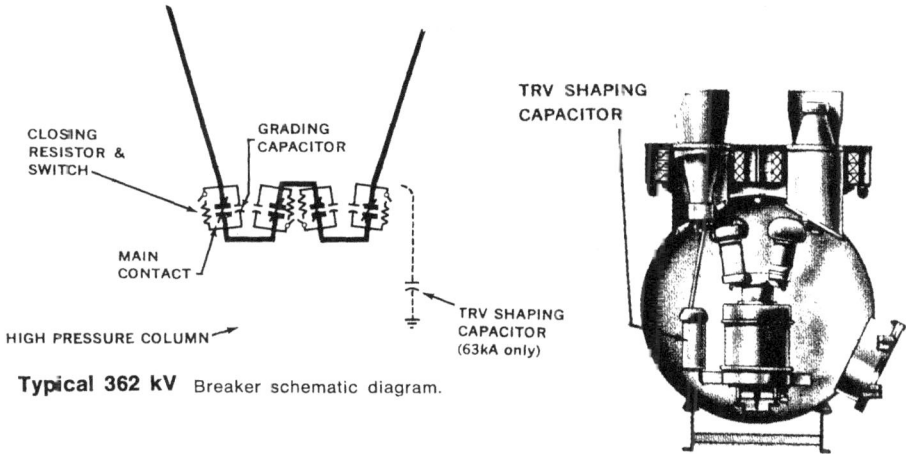

Typical 362 kV Breaker schematic diagram.

121, 145 kV One module with two breaks in series.

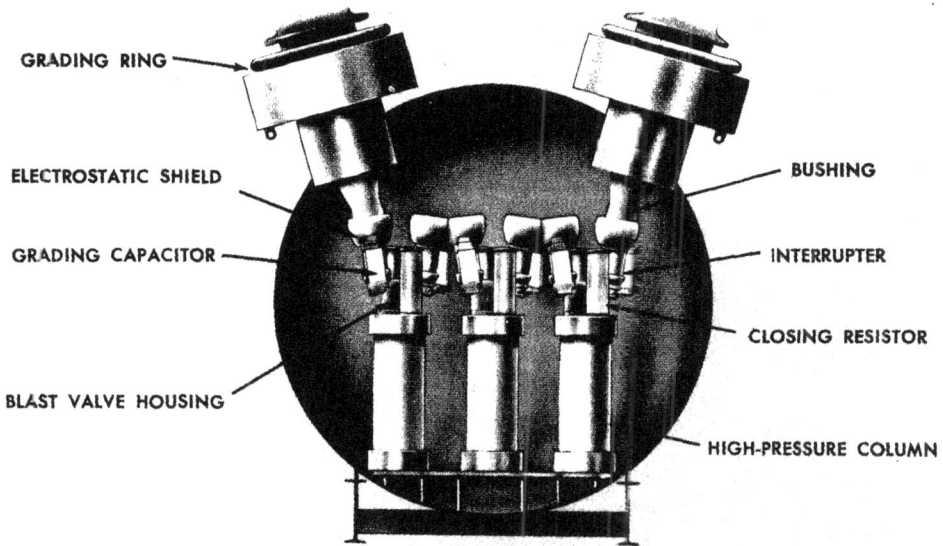

550 kV Three modules with six breaks in series.

Fig. 7.15 SF_6 double-pressure circuit breaker (ITE)

di/dt above the order of 30 A/μs. This is one of the major points of attack of interrupter development.

7.7 SF₆ circuit-breaker types

7.7.1 Metal-enclosed double-pressure types

The basic elements of 145 and 550kV versions of a range of dead-tank circuit breakers are shown in Fig. 7.15. Gas is stored at 18 bar for interruption, with insulation at 3 bar. Each of the double-break interrupter modules is mounted on a blast valve, in turn supported on an insulated column containing high-pressure gas. A high-pressure storage tank is mounted externally.

Closing is by pneumatic mechanism, and tripping by springs, with mechanical coupling to the blast valves and contacts. Although the basic mechanism is conventional, an exceptional variety of couplings are available. In addition to 3-pole gang operation, by using separate close and trip linkage, both independent pole operation, and independent pole trip with 3-pole gang closing, can be provided. The independent pole version can further be equipped with an electronic timing device which, using monitored busbar voltage, closes each pole so that current flow through each pre-inserted resistor is zero when the main contacts close.

A further feature of this design is that the blast valves are opened when the breaker is closed, which is claimed to reduce contact erosion; the gas flow also purges the gap.

Since short line-fault interruption is a function of current, however, lower-voltage breakers, with their reduced number of interrupters, can present problems. By fitting line-to-earth capacitors, symmetrical ratings up to 63kA are achieved over the voltage range 121 to 362 kV. Low ohmic opening resistors further increase the rating to 80kA at 169 and 242kV. Current commutation is employed, with two of the four breaks shunted, the remaining two breaks being used to interrupt the resistor current.

The closing resistors fitted for line charging necessitate additional pre-insertion switches.

The breaker may be equipped with gas-filled porcelain bushings, or connected to a metal-enclosed gas-insulated system. In both cases, c.t.s are incorporated in the breaker bushings.

7.7.2 E.H.V. puffer types

The most widely adopted design, (Fig. 7.16), is the type already described where a p.t.f.e. nozzle is attached to the end of a movable puffer cylinder, together with the moving contact. During interruption, the contacts are enveloped by the p.t.f.e nozzle. Gas is exhausted between this insulating nozzle and the fixed conducting nozzle, and usually through the fixed nozzle. The moving contact may either be a solid poker or hollow with provision for gas discharge (Figs. 7.17 and 7.18). For the higher ratings, the puffer cylinder engages with a fixed array of load-current contact fingers.

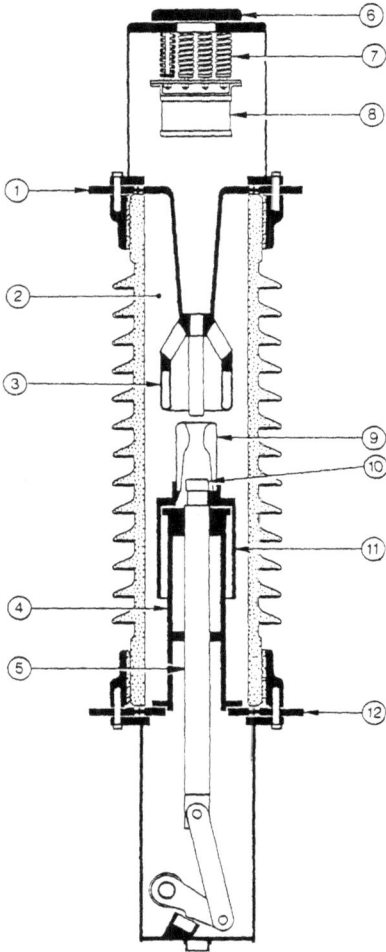

Fig. 7.16 SF₆ single-pressure puffer-type
interrupter (Magrini Galileo)

1 Top terminal pad
2 SF6 gas
3 Stationary contact
4 Stationary piston
5 Movable contact
6 Safety plate
7 Calibrated springs
8 Container with molecular sieve filters
for life-long moisture absorption
9 Interrupter
10 Arcing contact
11 Puffer cylinder, integral with the
movable contact
12 Lower terminal pad

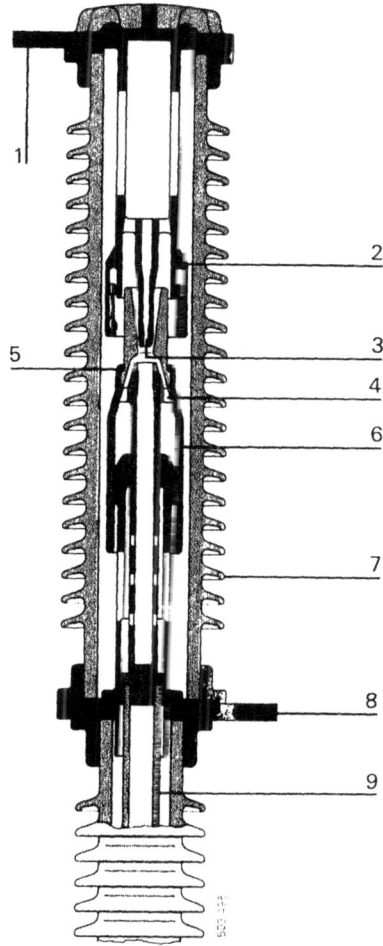

Fig. 7.17 SF₆ single-pressure puffer-type
interrupter (Brown Boveri)

1 Upper connection flange
2 Fixed continuous-current contact
3 Fixed arcing contact
4 Moving arcing contact
5 Moving continuous-current contact
6 Puffer cylinder
7 Puffer piston
8 Porcelain insulator
9 Lower connection flange
10 Operating rod

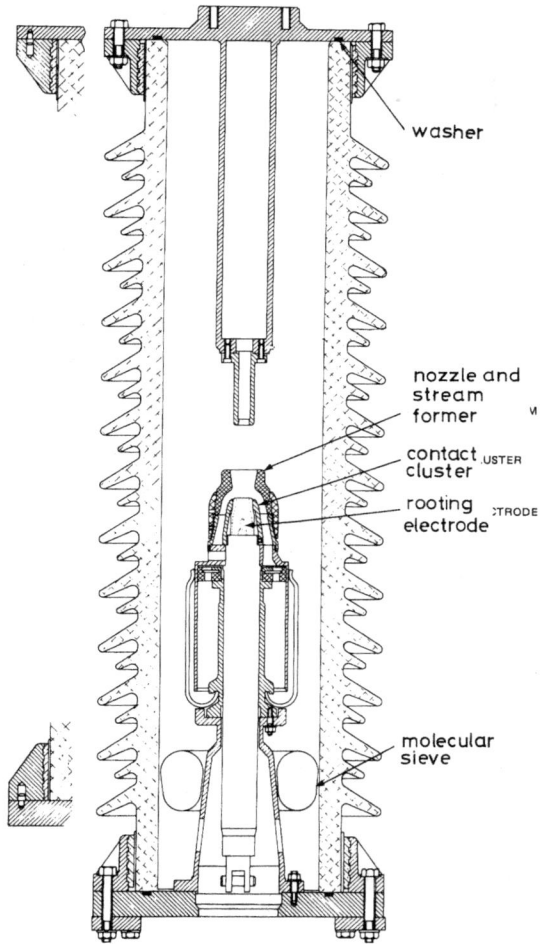

Fig. 7.18   SF₆ single-pressure puffer-type interrupter (Reyrolle)

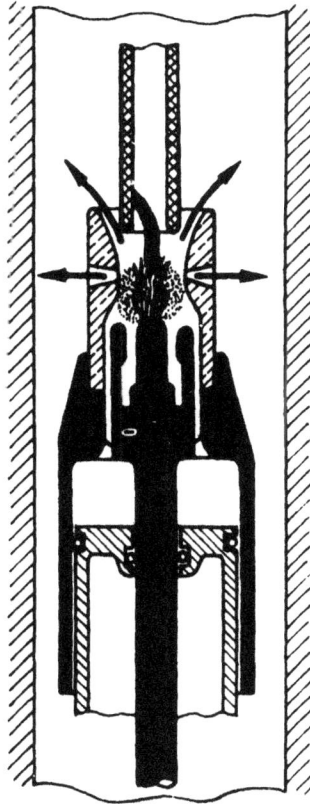

Fig. 7.19   SF₆ single-pressure puffer nozzles (Westinghouse)

A variant on this basic theme is shown in Fig. 7.19 where the main p.t.f.e. nozzle is equipped with radial ports, which, although covered by the fixed contact in the initial opening movement, begin to discharge before axial flow develops in the main nozzle.

An alternative arrangement is shown in Fig. 7.20 in which two conducting nozzles are held at a fixed distance, and are bridged by a tubular contact. This moving contact is carried within the puffer cylinder. During the first stage of the opening movement, the moving contact prevents gas flow from the puffer cylinder into the nozzle. The arc, drawn as the moving contact leaves the outboard nozzle, is driven into the nozzles. In the fully open position, the puffer cylinder does not bridge the nozzle gap. The fixed nozzle arcing contacts are of graphite, which not only has a particularly high resistance to arc erosion, but also, of course, cannot evolve metallic contaminants.

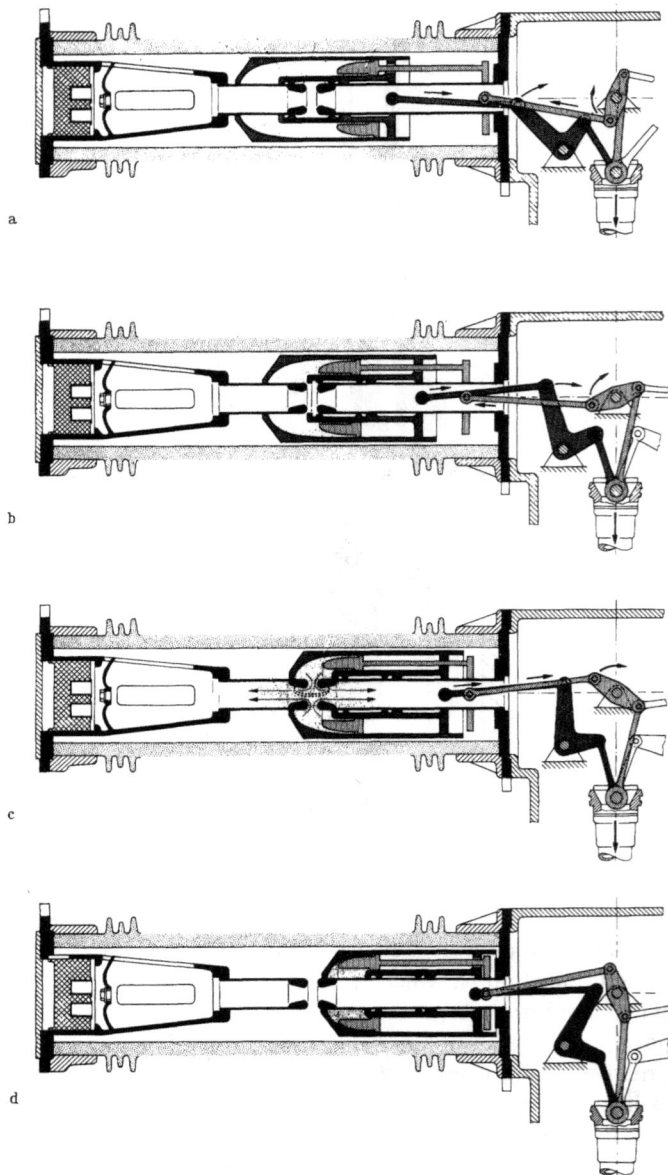

a

b

c

d

Fig. 7.20 SF_6 single-pressure puffer-type interrupter (Siemens)

The illustration shows the high-speed version, in which the piston which closes the puffer cylinder is driven by a crank, which first advances and then retracts it. The more rapid compression of the gas gives 2 Hz interruption, instead of the more usual 3 Hz. This design takes full advantage of gas-pressure enhancement by current blocking and arc energy, since in the latter part of the stroke mechanical compression is reduced.

In the lower-speed version, the blast cylinder only is moved. An alternative method of enhancing puffer performance is to apply suction to the exhaust side of the nozzle, thus increasing the pressure drop across the nozzle.

Open terminal versions of all the foregoing types use porcelain enclosures, of either parallel or barrelled shape. The latter configuration reduces porcelain internal surface voltage stress adjacent to the contact gap, whilst minimising flange diameter, and hence pressure loading on the flange (Fig. 7.21).

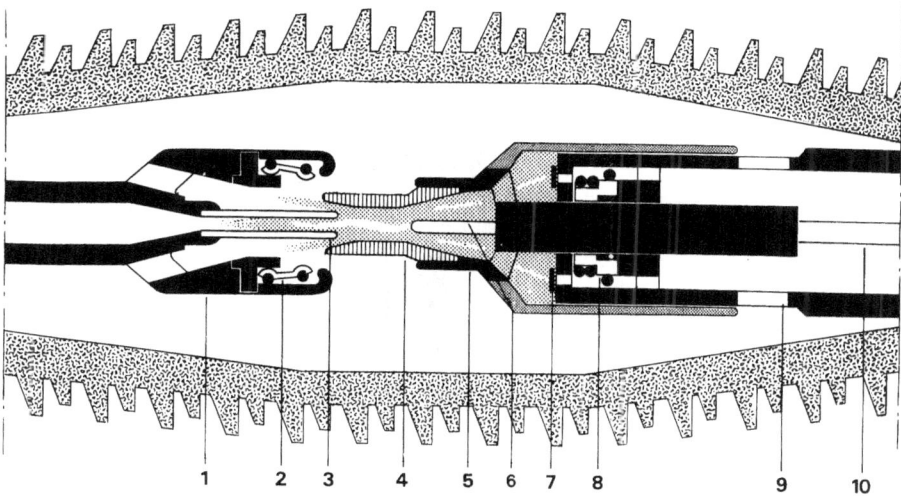

Fig. 7.21   SF₆ single-pressure puffer-type interrupter (Merlin Gerin)

    1   Fixed contact
    2   Fixed contact clusters
    3   Fixed arcing contact
    4   Insulating nozzle
    5   Moving contact and moving cylinder
    6   Moving arcing contact
    7   Check valve
    8   Moving-contact clusters
    9   Fixed piston
   10   Interrupter operating rod

7.7.3   H.V. puffer type for distribution voltages

Fig. 7.22 shows an interrupter design used for 12 to 36kV, up to 31·5kA, which combines the basic features of the two preceding arrangements. A cast resin housing provides interrupter and earth insulation, and also forms the puffer cylinder. This means that the puffer piston is the moving component. It is solidly connected both to a p.t.f.e. nozzle and to the moving contact nozzle; also to an outer cylindrical component which engages with the load-carrying fingers surrounding the fixed nozzle contact.

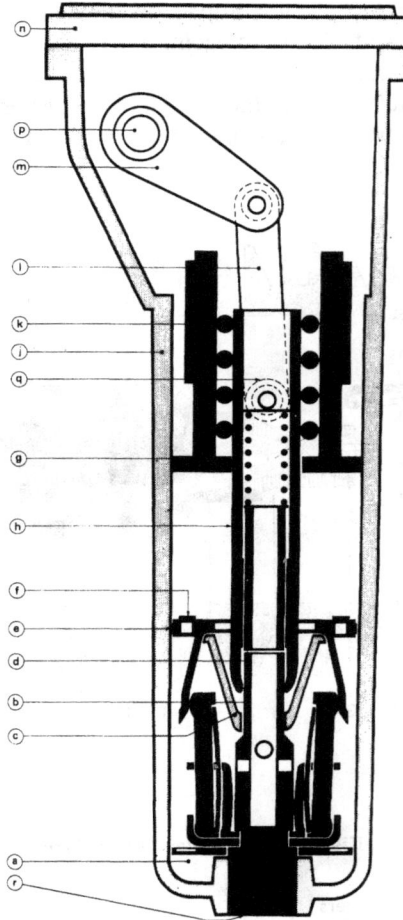

Fig. 7.22   SF₆ 12-36kV single-pressure puffer-type interrupter (Merlin Gerin)

a	SF₆ regenerating material	h	Contact Stem	r	Lower curreet terminal
b	Fixed contact	j	Mégélit sealing envelope		
c	Insulating nozzle	k	Conical rollers		
d	Moving contact	m	Crank		
e	Piston	n	Cover		
f	Valve	p	Filler Plug		
g	Bottom	q	Upper current terminal		

A configuration is shown in Fig. 7.23 which in addition incorporates current transformers in a similar fashion to a bulk-oil circuit breaker, except that the c.t.s are on one side of the interrupter. However, these c.t.s can be removed without disturbing the bushings. The interrupter design is basically similar to that already described, with the puffer piston running directly in the insulator. This is comprised of an epoxy-filament wound tube, encapsulated in epoxy weather sheds. Epoxy is not only highly chip and shatter resistant, but also has the characteristic of naturally 'chalking' on the surface, which eliminates the need for application of grease, and provides for self-cleaning by normal rainfall.

A self-contained pneumatic system, with storage for five operations, provides closing energy, tripping being by spring. The interrupting rating is 28kA, 34·5kV, using SF₆ at 5 bar.

Like their e.h.v. counterparts, these distribution-type circuit breakers require considerably less inspection and maintenance than o.c.b.s. One manufacturer, for example, quotes 20 full rated faults before maintenance, as compared to 5 for an o.c.b.

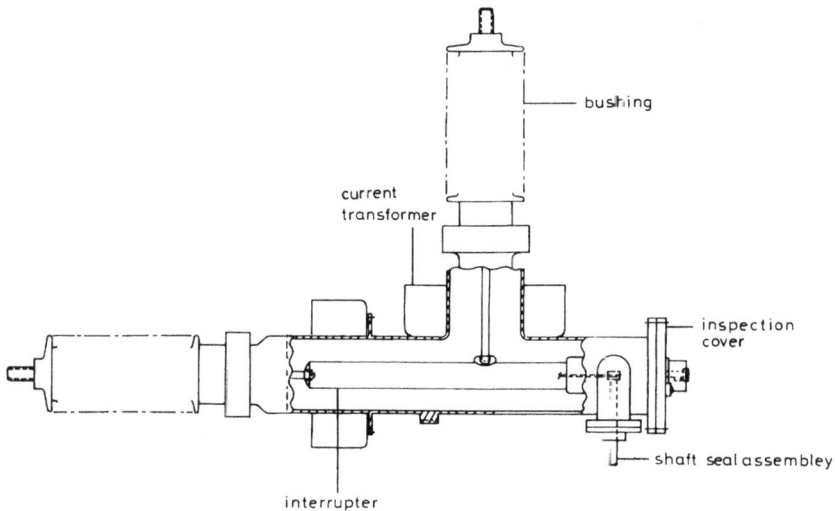

Fig. 7.23 SF₆ 34·5kV single-pressure puffer circuit breaker incorporating current transformer (Westinghouse)

7.7.4 Metalclad gas-insulated types

Undoubtedly the growth of SF₆ switchgear has been most dramatic in the metalclad field, because it has provided a solution to the problem of providing for the conflicting demands of more power together with reduced environmental impact. A gas-insulated metalclad installation occupies only some 10% of the area of its outdoor equivalent, so that substation location not only within cities, but also in buildings, is readily achievable.

Generally, the interrupters used for open terminal designs are adopted, because of the obvious advantages in using common components. Since the interrupters in e.h.v. c.b.s are usually double-break modules, with a common drive, this arrangement is frequently translated to the metalclad application, (Fig. 7.24). However, to suit the system layout, the interrupters may be coupled in tandem, with the mechanism at one end of the tank (Fig. 7.25). In this case, the considerable force required to drive three or more interrupters may necessitate a hydraulic mechanism, although similar open terminal double-break modules may also be driven pneumatically, or with a spring mechanism.

At the lower voltages the three phases can be in a single enclosure. Fig. 7.26 shows an example of a 145kV design. Again, standard interrupter modules are used, driven by a single hydraulic mechanism.

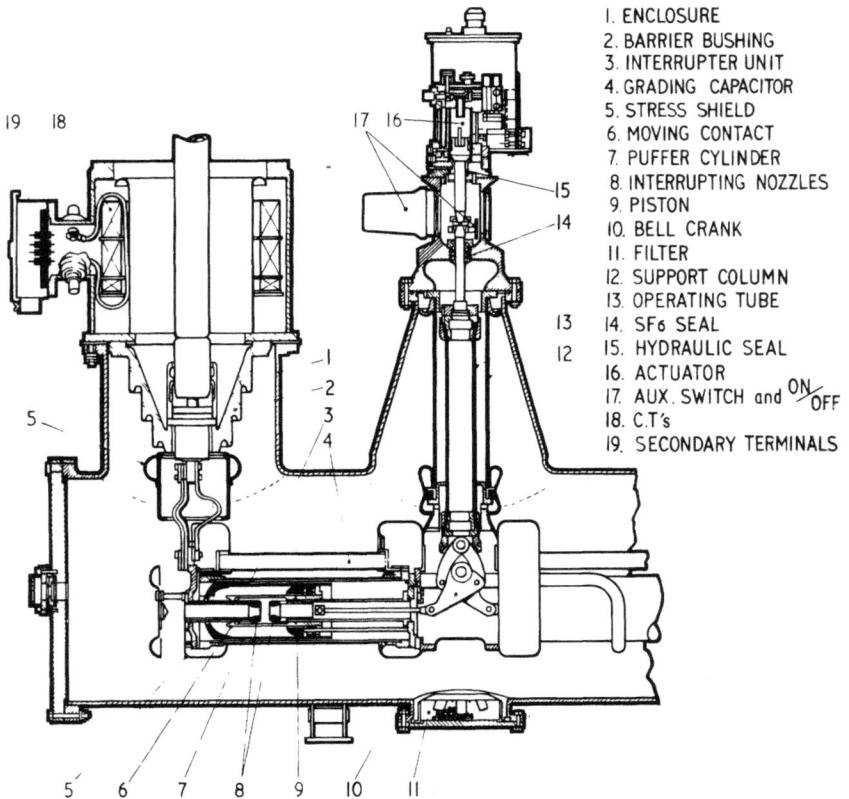

1. ENCLOSURE
2. BARRIER BUSHING
3. INTERRUPTER UNIT
4. GRADING CAPACITOR
5. STRESS SHIELD
6. MOVING CONTACT
7. PUFFER CYLINDER
8. INTERRUPTING NOZZLES
9. PISTON
10. BELL CRANK
11. FILTER
12. SUPPORT COLUMN
13. OPERATING TUBE
14. SF₆ SEAL
15. HYDRAULIC SEAL
16. ACTUATOR
17. AUX. SWITCH and ON/OFF
18. C.T's
19. SECONDARY TERMINALS

Fig. 7.24   SF₆ metalclad circuit breaker (GEC)

1   End-shield
2   Grading capacitor
3   Extinction chamber
4   Truck guide
5   Operating rod
6   Supporting insulator
7   Moulded resin insulator
8   Current connector
9   Enclosure
10   Breaker stand containing hydraulic operating mechanism

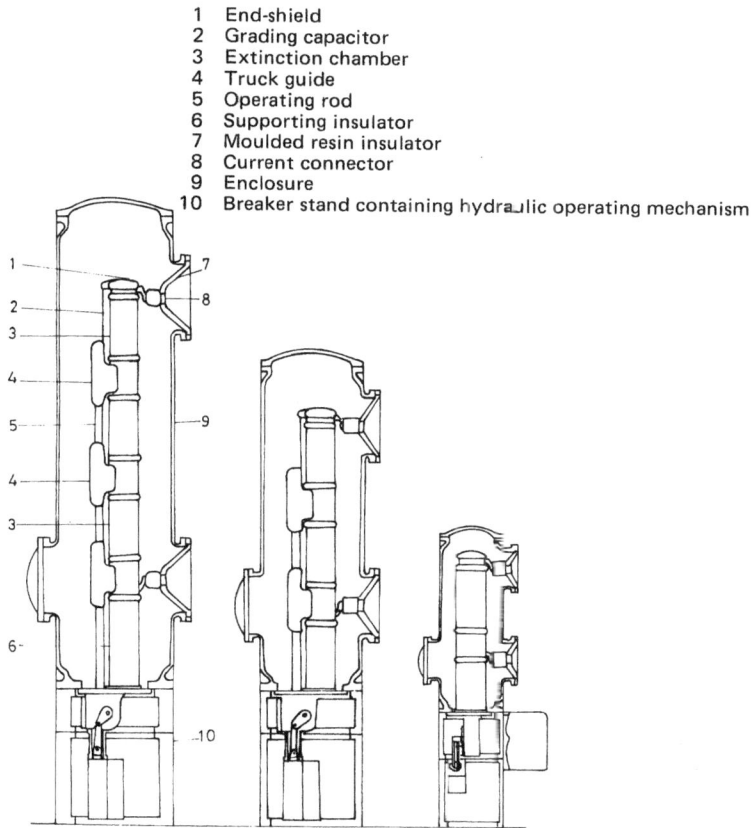

Fig. 7.25   SF$_6$ metalclad circuit breaker; tandem arrangement of interrupters (Brown Boveri)

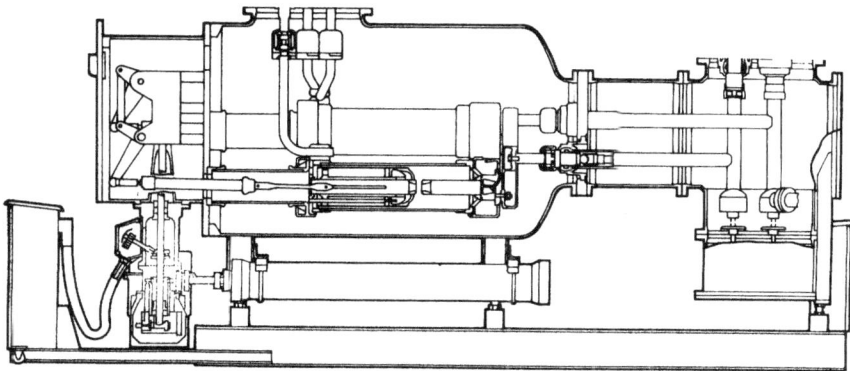

Fig. 7.26   SF$_6$ 145 kV metalclad circuit breaker (Siemens)

7.7.5   Self-extinction types

Utilising the current itself to effect extinction is an old principle; it has been exploited very successfully in o.c.b.s where the generation of hydrogen provides an extinguishing blast, and in air-break gear, where the arc is lengthened by being driven into an arc chute by a self-induced magnetic field. The attraction is that a much less powerful mechanism is required; hence lower cost, improved reliability and less maintenance. Not surprisingly, the adaptation of these principles to SF_6 has been widely investigated. There have been some quite ambitious programmes, but so far the designs marketed have been largely limited to 36kV.

Many configurations have been examined, including some akin to air-break, but ideas appear to be coalescing around that shown in Fig. 7.27. In this arrangement an arc is drawn between two rings, and the circuit is completed through a coil. The coil current sets up a magnetic field driving the arc around the gap between the rings; it also induces in the ring adjacent to the coil a large current which is out of phase with the coil current. In consequence, as the arc approaches zero, the magnetic field is sustained by the ring current. The phase shift depends upon the thickness and material of the ring, with optimum value between 30° and 60°.

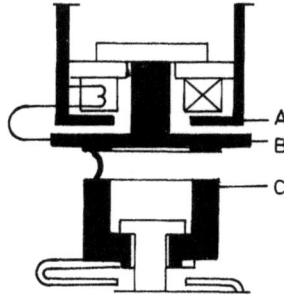

Fig. 7.27   SF₆ self-extinguishing interrupter (schematic) (Merlin Gerin)

A further advantageous factor derives from the high insulation values of SF_6, which enables the arc to loop itself without shorting out to adjacent turns or to the arcing rings. This, in combination with arc-root movement, which in turn is modified by the material of the arc rings, results in considerable lengthening of the arc.

The sequence of events of the arrangement illustrated is that a time $t=0$ contacts A B part, and progressive initiation of an arc shorts the load current through the coil. An electromagnetic field appears and the arc rotates. At $t=1$ all the current flows through the coil and arc A B is extinguished. Contacts B G part shortly after A B, and the arc B C is rotated by the field to extinguish at current zero, $t=2$.

Fig. 7.28 shows another application of the magnetically driven arc, but in this device the rotation of the arc is used to heat a volume of gas, the thermal expansion creating an extinguishing axial blast through the nozzle. A small puffer assists in interruption of low currents.

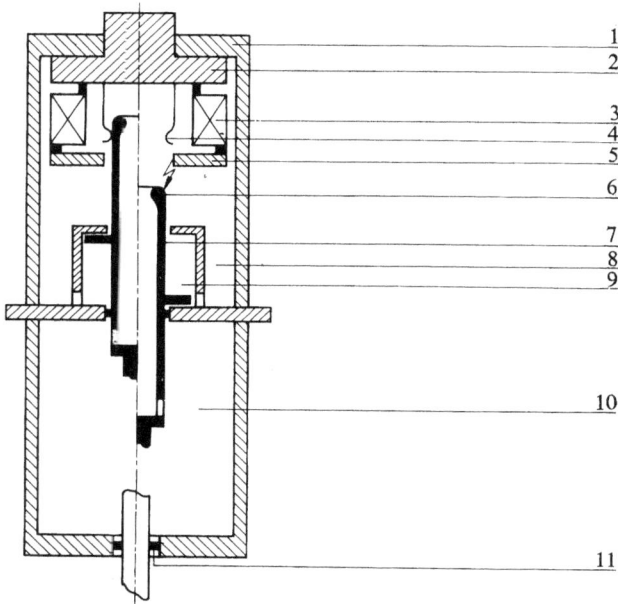

Fig. 7.28 SF$_6$ self-extinguishing interrupter (schematic) (Brown Boveri)

1	Tubular extinction chamber	7	Moving contact
2	Fixed contact	8	Pressurised space
3	Cylindrical coil	9	Auxiliary blast facility
4	Contact fingers	10	Exhaust space
5	Annular electrode	11	Seal against atmosphere
6	Nozzle		Left of centre-line: breaker contact closed; right, open

7.8 Operating mechanisms

Three basic types of mechanism are in general use, namely spring, pneumatic and hydraulic. Spring mechanisms dominate the low-power application because of their low cost, the absence of monitoring devices, and the reliability which results from their basic simplicity. High-voltage puffer circuit breakers require very powerful mechanisms, and here pneumatic and hydraulic mechanisms pre-dominate. Nevertheless, spring mechanisms are also used, the manufacturer's choice not unnaturally being influenced by experience on other types of circuit breaker.

7.8.1 Spring mechanisms

The basic characteristic of the spring does not appear to offer a good match with the compression load of the puffer. However, correction can be achieved by suitable manipulation of linkage velocity ratios, combined with transfer of kinetic energy, even to the point of stalling.

There are two distinct boundaries to an arc; the core boundary inside which the current flows in ionised gas, surrounded by hot gas with its thermal boundary. Even when the thermal boundary reaches the nozzle wall, gas heated by the arc continues to flow through the nozzle, but the dense cold flow is blocked, and upstream pressure therefore increases more rapidly. When the core boundary reaches the nozzle wall, gas discharge through the nozzle practically ceases, and upstream pressure rises even more rapidly.

A further complicating factor is that as the arc length increases, its core diameter increases, so that stalling can delay or even prevent the onset of core blocking. The resultant reduction of hot arc gases in the upstream volume can provide improved dielectric performance, even though the interrupting gap is smaller and the gap pressure is lower. It is not possible to achieve precise balance between spring force and arc-generated pressure, particularly since both single and three phase faults must be catered for, reversal of contact movement can occur, but this immediately lowers arc energy, and the system is thus self balancing to some extent.

By exploiting these factors, one 31·5kA, 36kV puffer circuit breaker uses tripping springs delivering only 250 joules.

At 50kA the puffer forces and energies are considerable, ranging typically for

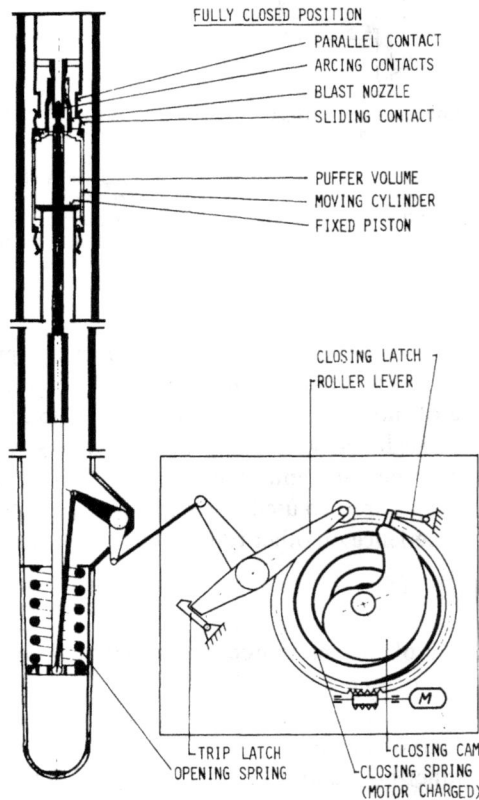

Fig. 7.29   Spring mechanism for single-pressure SF₆ puffer circuit breaker (Sprecher & Schuh)

one interrupter from the order of 10kN 1kJ on no load to 25kN, 3kJ at 50kA. A very powerful mechanism is therefore required.

Such a mechanism is shown in Fig. 7.29, which is used on a range of e.h.v. puffer circuit breakers. An interesting feature of this mechanism is the use of a clock-type closing spring. This lends itself to a simple coupling system, in which the spring output is clockwise, and re-winding is also clockwise, so that the motor wind is permanently engaged. With the breaker open, the first operation is to charge the closing spring when release of the closing latch closes the breaker and simultaneously charges the tripping spring. Latching of the operating mechanism in the closed position causes simultaneous recharging of the closing spring through the winding motor, enabling the circuit breaker to perform an open-close-open cycle. Should the breaker be closed onto a fault, it can, of course, open immediately.

Although the coefficient of utilisation of materials of the clock spring, like the compression spring, is approximately one-third that of a uniformly stressed tension rod it occupies a smaller volume than an equivalent compression spring.

From 123 to 245kV, one interrupter per phase is used, coupled to a common closing mechanism, but each interrupter having its own tripping spring. Single-phase auto-reclosure, of course, necessitates a closing mechanism on each phase. For 420kV, each phase consists of two interrupters mounted upon a single column equipped with closing mechanism and tripping spring.

7.8.2 Pneumatic mechanisms

Two of the major advantages of pneumatic mechanisms are that inertia is low, even for large outputs, and energy storage for a large number of duty cycles can be readily provided. Also the mechanism can have a degree of self-regulation; as the load increases and the velocity of the driving piston falls, cylinder pressure increases to achieve a new state of balance. As with the spring mechanism, in some instances stalling or near-stalling occurs.

A pneumatic power unit used on a range of e.h.v. puffer circuit breakers is illustrated in Fig. 7.30. The space between the two pistons is permanently pres-

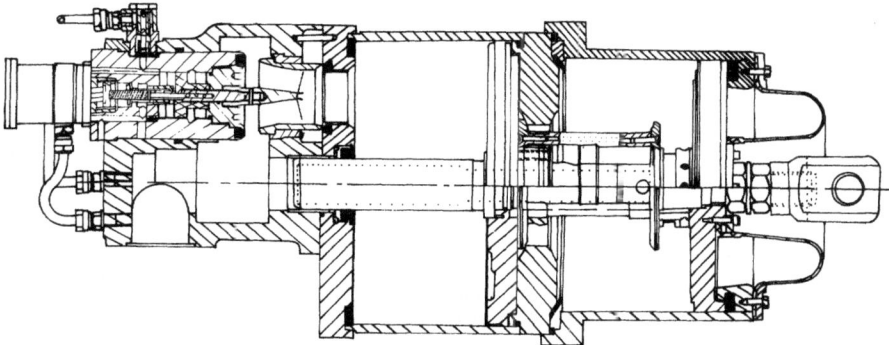

Fig. 7.30 Compressed-air mechanism for single-pressure SF$_6$ puffer Circuit breaker (Reyrolle)

surised, and to trip the circuit breaker, compressed air is released into the larger-diameter cylinder by a double-acting valve. The reversal of the valve exhausts air from this space to close the breaker. Air damping is achieved by closing off the port in the barrier between the large and small cylinders at either end of stroke.

The operating pressure when driving a 145kV circuit breaker is 11 bar (lock-out) and this is increased to 20 bar to drive a complete three-phase 420kV circuit breaker. Careful selection of materials and protective finishes enables the power unit to be used with wet air, so that a simple single-stage air system can be employed.

Fig. 7.31 shows another example, in which air is admitted to either side of the main piston to drive the interrupter(s). To achieve high-speed opening, there are two stages of valve amplification.

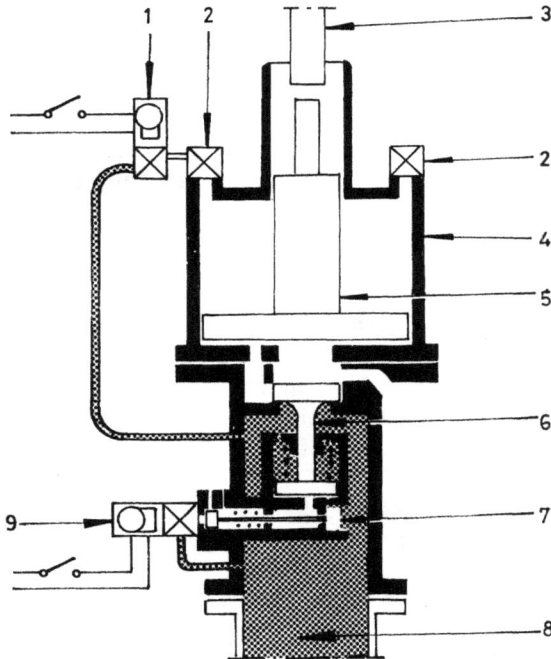

Fig. 7.31   Compressed-air mechanism for single-pressure SF₆ puffer circuit breaker (Magrini Galileo)
   1   Closing control valve
   2   Pressure relief valves
   3   Connecting rod
   4   Main cylinder
   5   Main piston
   6   Main, pneumatically-operated valve
   7   Auxiliary, pneumatically-operated valve
   8   Compressed-air supply
   9   Opening control valve

7.8.3   Hydraulic mechanisms

With standard components available for working pressures up to 320 bar, the hydraulic mechanism is particularly compact. Since the energy storage is compressed nitrogen, like the pneumatic system it can be readily extended.

A schematic of a hydraulic mechanism is shown in Fig. 7.32. Energy is stored in an accumulator, where a free piston separates the nitrogen and oil. The basic mechanism consists of an actuator and control valves in one combined unit, on which the low-pressure tank is mounted. Oilways are drilled within the valve blocks, minimising potential leakage points.

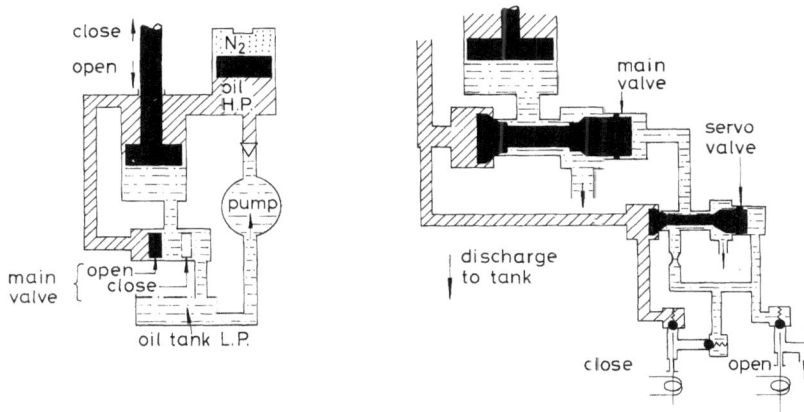

Fig. 7.32   Hydraulic mechanism for single-pressure SF₆ puffer circuit breaker (GEC)

High-pressure oil is fed to the cylinder between the actuator piston and piston rod, and also to the valves. To close the circuit breaker, high-pressure oil is admitted to the piston by the main valve, producing a resultant force on the piston rod. In the other position the main valve shuts off the high-pressure oil and connects the actuator cylinder to the l.p. tank. The resultant force on the piston opens the circuit breaker. The main valve is controlled by a double-acting servo valve which functions in the same manner, in turn controlled by single-acting close and trip valves. A bypass across the servo valve piston maintains it in either position.

7.9   Gas mixtures

Over the typical filling range for SF₆ puffer circuit breakers of 5 to 7 bar at 20°C, the onset of liquefaction is from −38 to −20°C, respectively, limiting the working temperature range of these designs. The temperature rise due to adiabatic compression in the puffer means that liquefaction does not occur when operating at minimum temperature. Heating has of necessity been used in double-pressure breakers, but heaters and associated monitoring are additional elements which can have an adverse effect upon reliability, apart from the not inconsiderable energy requirements of several kilowatts. This has given a major impetus to research into gas mixtures, which has shown that, for example, replacing SF_6 by nitrogen has little effect upon the breakdown strength of small gaps, down to 50% of nitrogen by volume. With helium this holds good down to 60%.

Dielectric strength is, of course, only one of the considerations, other factors being much more complex in their effect upon interrupting capability. For example, the lighter the gas, the higher its velocity, and hence its attack upon the arc, but higher velocity means lower pressure, so lower dielectric strength, a typical conflict. Full-scale tests, in fact, show some diminution of interrupting capability.

The large SF_6 molecule breaks down under arcing into a considerable number of lighter particles, and an essential feature of any admixture is that it should not provide for recombination into harmful products. In addition to its other attributes, nitrogen fulfils this requirement.

7.10   Acknowledgments

The preparation of this chapter, and the 1981 section, has required in addition to the references listed, information freely given by the following organisations.

Brown Boveri
Delle Alsthon
GEC
ITE
Magrini Galileo
Merlin Gerin
Mitsubishi Electric Corporation
Reyrolle
Siemens & Halske
Sprecker & Schuh
Westinghouse Electric Corporation

7.11   References

AIREY, D.R., KINSINGER, R.E., RICHARDS, P.H., and SWIFT, J.D. (1975): 'Electrode vapour effects in high current gas blast switchgear arcs' *IEEE* PES Summer Meeting, 20-25, July F75 468-9

BOUDENE, C., CLUET, J.L., KEIB, G., and WIND, G. (1974): 'Identification and study of some properties of compounds resulting from the decomposition of SF6 under the effect of electrical arcing in circuit breakers,' 48/RGE – Special issue

BRATKOWSKI, W. V., FISHER, W. H., and RADUS, R. J. (1966): 'High speed trip mechanism for extra high voltage circuit-breakers', *Westinghouse Eng.*, pp. 51–55

BROS, E. (1981): 'H.V. circuit breakers with SF6/N2 gas mixture for extremely low ambient temperatures,' CEA Paper 10

BROWNE, T. E. Jun., and LEEDS, W. M. (1960): 'A new medium for circuit interruption'. CIGRÉ Report 111

CALVINO, B. J. (1967): Italian Patent 791 651

CALVINO, B. J., and PEZZI, P. (1968): 'Interruption characteristics of an SF₆ self extinguishing circuit-breaker: influence of the initial rate of rise of the transient recovery voltage and of the zero current phenomena' CIGRÉ Report 13-04

COOPER, F. S. (1940): USA Patent 2 221 671

DELEVOY, V., and CASSELETTE, F. (1960): 'Les disjoncteurs haute tention a hexafluorure de soufre' *Révue ACEC* 2, pp. 2−17

EASLY, C. J. and TELFORD, J. M. (1964): 'A new design 34·5 kV to 69 kV intermediate capacity SF₆ circuit-breaker', *Trans. Am. Inst. Electr. Eng.*, PAS-83, pp. 1172−1177

EINSELE, A. (1964): 'The new Siemens F circuit-breaker for 220 kV, 15 GVA', *Siemens Rev.*, 31, pp. 219−223

ELECT. REV. 1971 'SF6 solves transmission amenity problems' (Compares open terminal and metalclad substations)

EPRI (1978): 'Development of distribution and subtransmission SF6 circuit-breaker and hybrid transmission interrupter.' *Project 661-1*, Report EL 810

EPRI (1977): Report EL 284

FAWDREY, C.A. (1979): 'SF6 circuit breaker mechanisms' IEE International Conf. on Developments in Design and performance of EHV Switching Equipment.

FERSCHL, L., KOPPLIN, H., SCHRAMM, H.H., SLAMECKA, E., and WELLY, J.D. (1974): 'Theoretical and experimental investigations of compressed gas circuit-breakers under short-line fault-conditions.' CIGRE 13-07, 21-29

FREDRICH, R. E., and YECKLEY, R. N. (1959): 'SF₆ circuit-breaker − a new design concept', *Westinghouse Eng.*, pp. 51−55

FRIND, G. (1960a): 'Uber der Abklinken von lichtbogen I', *Zeitschrift fur angewandte physic*, 12, pp. 231−237

FRIND, G. (1960): 'Uber der Abklinken von lichtbogen II', *ibid.*, 12, pp. 515−521

FRIND, G., and RICH, J.A. (1974): 'Recovery speed of axial flow gas blast interrupter: dependence on pressure and di/dt for air and SF6.' IEEE PES Winter Meeting. N.Y. 27 Jan-1st Feb, Paper T74 183-0

FRIND, G. (1977): 'Experimental investigation of limiting curves for current interruption of gas blast breakers.' BBC Symposium. Extensive bibliography

FRINK, R. (1967): 'SF₆ magnetic-puffer circuit-breaker for 34·5 kV, 1500 MVA', *Trans. Am. Inst. Electr. Eng.*, Paper 31, TF 67-414

FRINK, R. and KOZLOVIC, J. M. (1968): 'An SF₆ circuit-breaker for metalclad switchgear', *Westinghouse Eng.*, pp. 71−75

FUJIWARA, K., and ONO, S. (1973): 'Rotating arc driven by magnetic flux in SF6 gas.' 2nd Int. Symposium, Switching Arc Materials, Pt II, Lodz

FUJIWARA, K. (1974): 'Arc quenching behaviour of rotary arc in SF6.' 7th Int. Conf. Electrical Contact Phenomena

FUKUDA, S., USHIO, T., ITO, T., and MIYAMOTO, T. (1966): 'Recent studies on arc behaviour around current zero in SF₆'. CIGRE Report 13-03

GARZON, R.D. (1976): 'Rate of change of voltage and current as functions of pressure and nozzle area in breakers using SF6 in the gas and liquid phases.' F76 186-7 IEEE

GOZNA, C. F., PALMER, R. T., and VAREY, R. H. (1964): 'Arc interruption in electronegative gases', CEGB Report 40032

GUTBIER, H. G. (1966): 'Massenspektrometriske untersuchungen der Zerstezung von SF₆ in einer Glimmentladung', *Vortrag auf der tagung der Arbeitsgemeinschaft massenspektrometric am*, Mainz

HAGINOMOVI, E., YANABU, S., MURAKAMI, Y., and OISHI, M. (1978): 'Recent progress in SF6 gas circuit breaker technique,' *Toshiba Review* No. 117

HENRY, J.C., PERRISIN, G., and ROLLIER, C. (1978): 'The behaviour of SF6 puffer circuit-breakers under exceptionally severe conditions.' CIGRE 13-08 30th Aut-7th Sept

HERTZ, W., MOTSCHMANN, H., and WITTEL, H. (1971): 'Investigations of the properties of SF6 as an arc quenching medium,' *Proc. IEEE*, **59**, (4)

JACOB, T., SCHADE, E., and SCHAUMANN, R. (1977): 'Self-blasting, a new switching principle for economical SF6 circuit breakers.' CIRED, London, IEE Conf. Pub. 152

KOPAINSKY. J., and SCHADE, E. (1976): 'Investigation of a rotating, magnetically driven SF6 arc.' 4th Int. Conf. on Gas Discharges, IEE Conf. Pub. 143

KRENICKI. A., and SCHADE, E. (1977): 'Recent investigations on arcs in SF6 gas-blast breakers and application of the results to the development of h.v. switchgear.' World Electrotechnical Congress, Moscow, 21-25 June

LEEDS, W. M., FRIEDRICH, R. E., WAGNER, C. L., and BROWNE, Jun. T. E. (1970): 'Application of switching surges, arc and gas flow studies to the design of SF₆ circuit-breakers'. CIGRÉ Report 13-11

LINGAL, H. J., STROM, A. P., and BROWNE, Jun. T. E. (1953*a*) 'An investigation of the arc quenching behaviour of sulphur hexafluoride', *Trans. AIEE*, Part III, 72, pp. 243–246

LINGAL, H. J. and OWENS, J. B. (1953*b*): 'A new high voltage outdoor load interrupter switch', *ibid.*, Pt. III, 72, pp. 293–297

LUHRMANN, H., LUXA, S., and SCHRAMM, H. (1979): 'Development of a 765kV SF6 puffer-type circuit-breaker.' IEE Symposium, pp.27-31

MAGGI, E. (1964): Italian Patents 736 176, 809 296 and corresponding UK Patent 1 082 453

MAGGI, E. (1971): Italian Patent 811 744 and the corresponding UK Patent 1 235 300

MAURY, E. (1971*a*): 'Postes blindés 225 kV au SF₆, Delle Alsthom', *Schweizerischen Technischen Zeitschrift*, 67, pp. 554–559

MAURY, E. (1971*b*): 'Evolution des disjoncteurs des réseaux de transport', *Révue Général d'Electricité*, Numero special, pp. 102–117

MURANO, M., NISHIKAWA, H., KOBAYASHI, A., and IWAMOTO, T. (1974): 'Heavy current clogging phenomena in SF6 gas arc.' IEEE C74 185-5 Jan 24-Feb 1

MURANO, M., NISHIKAWA, H., KOBAYASHI, A., OKAZAKI, T., and YAMASHITA, S. (1975): 'Current zero measurement for circuit breaking phenomena,' *IEEE Trans.*, PES-T75 070-8, 26-31 Jan

NISHIKAWA, H., KOBAYASHI, A., OKAZAKI, T., and YAMASHITA, Y. (1976): 'Arc extinction performance of SF6 gas blast interrupter,' *IEEE Trans.*, PES F76 057-0 25-30 Jan

PHILIP, S. F. (1963): 'Compressed gas insulation in the million volt range: a comparison of SF₆ with N₂ and CO₂,' *AIEE* Paper 63-27

POLTEV, A. I., PETINOV, O. V., and MARKUSH, C. D. (1967): 'Electromagnetic extinction of the arc in SF₆', *Electrichestvo*, 3, 59–63, pp. 159–169

RIEDER, W. (1976): 'Actual circuit-breaker arc research.' Nicola Tesla Symposium, Yougoslav Academy of Science (extensive bibliography)

SCHMIDT, W., HOGG, P., and EIDINGER, A. (1977): 'On-site high-voltage testing of SF6 metalclad switchgear installations and compressed gas insulated cable,' *Brown Boveri Rev.* 11-77

SWANSON, B.W., and ROIDT, R.M. (1970): 'Arc cooling and short line fault interruption,' *IEEE Trans.*, 70TP 585-PWR 12-17 July

SWARBRICK, P. (1967): 'Characteristics of an arc discharge in sulphur hexafluoride', *Proc. IEE*, 114, pp. 657–660

TOMINAGA. S., KUWAHARA, H., YOSHINAGA, K., and SAKUMA, S. (1978): 'Estimation and performance investigation on SLF interrupting ability of puffer-type gas circuit breaker,' *IEEE Trans.* PES F78 149-7 29 Jan-3 Feb

USHIO, T., SHIMURA, I., and TOMINAGA, S. (1971): 'Practical problems on SF₆ gas circuit-breakers', *IEEE Trans.*, **PAS-90**, pp. 2166–2174

VAN SICKLE, R. C., and YECKLEY, R. N. (1965): Discussion on Paper 31, T.P. 65–87, *Trans. Am. Inst. Electr. Eng.*, **PAS–84**, pp. 892–901

YOSHIOKA, Y., and NAKAGAWA, Y. (1978): 'Investigation of the interrupting performance of a puffer type gas circuit breaker,' *IEEE Trans.*, **PES** A78 597-7 16-21 July

YOON, K. G., and SPINDLE, H. E. (1959): 'A study of the dynamic response of arcs in various gases', *ibid.*, Pt. III, 78, pp. 1634–1642.

Chapter 8

Vacuum circuit breakers

M. P. Reece, B.Sc., Ph.D., C.Eng., F.I.E.E.

8.1 Introduction

The vacuum circuit breaker has fascinated the switchgear designer for many years, primarily because of the great advantages of vacuum interrupters, which are:

(*a*) They are entirely self contained, need no supplies of gases or liquids and emit no flame or gas.

(*b*) They require no maintenance and in most applications their life will be as long as the circuit breaker in which they are applied.

(*c*) They may be used in any orientation.

(*d*) They are not flammable.

(*e*) They have a very high commutating ability and need no capacitors or resistors to interrupt short line faults.

(*f*) They require a relatively small mechanical energy to operate them.

(*g*) They are silent in operation.

These advantages were undoubtedly the driving force which spurred the development of vacuum interruption forward (Cobine, 1962; Selzer, 1971; Reece, 1970). The main disadvantage of the vacuum interrupter was that its cost was somewhat higher than that of conventional interrupters, but present evidence is that not only can this be reduced, but consequential savings in the rest of the switchgear will result from the use of vacuum interrupters for arc extinction.

The possibility of interruption in vacuum was first proposed in the nineteenth century, but the first serious study of vacuum interruption was undertaken by Sorensen and Mendenhall at the California Institute of Technology in 1923-26, as referred to in chapter 1, when a current of 900 A was interrupted at 40 kV (Sorensen and Mendenhall, 1926). Nothing came of this pioneering work, because of the technical difficulties of maintaining a good vacuum in sealed interrupters at that time.

During the 1930s very small vacuum switches were developed and these were capable of switching a few amperes at a kilovolt or so. The size of this type of interrupter was steadily increased until during the 1950s interrupters of this type were capable of interrupting about 4 kA r.m.s. at 20 kV r.m.s. These interrupters were only capable of switching loads, and at this stage there was no such thing as a vacuum circuit breaker. Typical uses of these small interrupters were the switching of high voltage supplies to radio transmitters, radar sets and similar applications.

During the 1950s the position changed substantially. Ross, using the largest type of load breaking interrupter then available, started applying vacuum switches to power systems (Jennings, *et al.*, 1956). Circuit breakers could not, of course, yet be made because fault currents were beyond the capacity of the interrupters available, but Ross, putting six interrupters in series, was able to switch the load and capacitance currents in 230 kV systems (Ross, 1961).

In 1953 the Electrical Research Association in the UK commenced a programme of research into the vacuum interrupter with a view to turning it from a load breaking device into a circuit breaking interrupter, when in the same year it so happened that the General Electric Company of the USA also started a similar study.

By about 1957 the ERA had elucidated enough of the basic physics of the vacuum arc to have shown beyond any doubt that a vacuum circuit breaker was possible. The cause of the limitation of the interrupting ability of vacuum interrupters with plain butt contacts to about 10 kA peak had been discovered. The effect of the thermal properties of the contact materials on the stability of vacuum arcs had been brought to light, so that current chopping, which in the early days had been a considerable problem with vacuum interrupters, could now be satisfactorily controlled. As a spin off of the ERA work on vacuum interrupters, in 1958 a bimetallic contact system was developed which has proved of considerable use in vacuum contactors (Reece, 1959 and 1963).

The General Electric Company of the USA was working on the problem of power vacuum interrupters at the same time, and in January 1962 they announced the development of the first power interrupter which could/interrupt system fault currents (Fig. 8.6). This interrupter was rated at 12·5 kA 650 A 15·5 kV, and it was initially applied to outdoor type reclosers and later to metalclad switchgear at 15·5 kV (Lee *et al.*, 1962).

In the UK the work initiated by ERA was continued in industry, and in 1967 AEI announced the development of a 132 kV 3500 MVA 1200 A vacuum circuit breaker using eight breaks per phase. Shortly after this, a joint development programme of English Electric and Reyrolle also produced power vacuum interrupters, and in 1969 a trial installation of a 25 kV single phase breaker was installed by English Electric for British Rail. At the present time in the UK the main emphasis is on the development of 3·3-11 kV distribution vacuum circuit breakers. A considerable programme of vacuum circuit breaker construction for 25 kV railway electrification is also in hand.

Vacuum circuit breakers for distribution at voltages up to 36 kV have developed very rapidly in Japan, where some five companies are manufacturing power interrupters for this type of circuit breaker, mostly under licence from the American General Electric Co. The very rapid growth of the vacuum circuit breaker for this type of duty in Japan has, it is thought, been connected with the high price levels of distribution circuit breakers in this voltage range in Japan, which has made the vacuum circuit breakers competitive in Japan from the beginning.

In the USA, at least five companies are producing power vacuum interrupters for vacuum circuit breakers and these have initially been applied to distribution duties at voltages of 15·5 and 38 kV. Recently the first American vacuum circuit breaker for 138 kV has been announced, and experimental vacuum circuit breakers up to 40 kA at 760 kV are being built by General Electric (Electrical Review, 1973).

The vacuum interrupter is at present somewhat more expensive than the interrupting devices in some other circuit breakers, and its cost is very sensitive to production volume. There is, therefore, a vicious circle to be broken before these interrupters can really be widely applied. It must, however, be remembered that the switchgear user pays for a complete circuit breaker, and the greater cost of the interrupters may be offset by simplifications in the circuit breaker itself arising from their advantageous properties. For the vacuum interrupter to become commercially successful, the first applications must meet two conditions, these are:

(*a*) The market for the prospective application should be large.
(*b*) The type of equipment must be such that the use of vauum interrupters will lead to cost savings in the rest of the circuit breaking unit.

At a later stage of course, the first condition can be relaxed since then all that is required is that the total market for a particular interrupter should be large. A careful study of the UK switchgear market has shown that metalclad distribution switchgear appears to satisfy both of these conditions in the UK. The market is large, some 10 000 circuit breakers per annum, and the installation, which at present has certain features needed to allow maintenance of oil circuit breakers, can be simplified and cheapened by the application of vacuum interrupters (Gibbs, 1972).

In the field of high-voltage circuit breakers the position is rather less clear. In general, the frequency of faults is low on e.h.v. transmission systems, and other types of circuit breakers, which would need much more maintenance than vacuum circuit breakers in low-voltage distribution services, may well require little maintenance in e.h.v. transmission duties. This means that one of the advantages of vacuum circuit breakers over other types may well be less important in e.h.v. applications. Nevertheless vacuum interrupters are being considered for e.h.v. metalclad switchgear, but it seems likely that further progress in research and development to increase the voltage and current ratings of interrupters and, in particular, to reduce specific volumes and specific weights of interrupters, will be required if vacuum interrupters can economically enter the e.h.v. end of the switchgear field (Flurscheim, 1969).

8.2 Arcing phenomena in vacuum related to contacts for vacuum interrupters

8.2.1 Basic processes

The basic process of arcing in vacuum and of the extinction of an a.c. vacuum arc has been discussed in chapter 2. In this Section the relevance of these processes to the design and operation of vacuum interrupters is considered.

It has been shown that the diffuse vacuum arc has a very high interrupting ability, due to the extremely short thermal lag (much less than 1 μs) in metal vapour emission from cathode spots, because of their very small physical size; it has also been seen that once magnetic constriction of the diffuse arc into a constricted discharge occurs, large heated regions with very long thermal and vapour emission time constants (which may be greater than 100 μs) may be formed. Such large heated areas will, when the emission of vapour from them persists beyond the current zero, cause the interrupting ability of the vacuum arc to fall virtually to zero.

The consequence of this onset of magnetic arc constriction to the vacuum interrupter designer is that plain butt contacts, which have good interrupting ability up to peak currents of 8–15 kA, depending on the contact size and material, cannot be used for circuit breaker interrupters which must interrupt a minimum of say 28 kA peak. The plain butt contact is thus limited to the interruption of currents not exceeding about 4–8 kA r.m.s., and such contacts are therefore used in interrupters for load breaking duties. Typical performance data for a plain butt contact are given in Fig. 8.1 where it can be seen that once serious anode spot formation with long time constants occurs interrupting ability rapidly falls effectively to zero.

To overcome this limitation of the plain butt contact, a number of special designs of contact have been evolved. All depend for their operation on the interaction of the arc and a magnetic field to keep the arc in rapid motion, and prevent heated regions with very long cooling times developing. There are many designs in the patent literature, some of which use contact systems with permanent magnets, some with separate coil systems external to the interrupter, but two contacts are widely used. These are the 'spiral petal' contact developed by GE (USA) (Lee *et al.*, 1962), which cause a constricted arc to rotate, and the 'contrate' contact developed by AEI (UK) that prevents constriction of the arc, which remains in a diffuse form throughout the arcing period (Reece, 1973).

8.2.2 Spiral petal contact

The basic design of the spiral petal contact is shown in Fig. 8.2. The contacts are essentially discs, slightly coned, or with an annular ridge near the centre by which

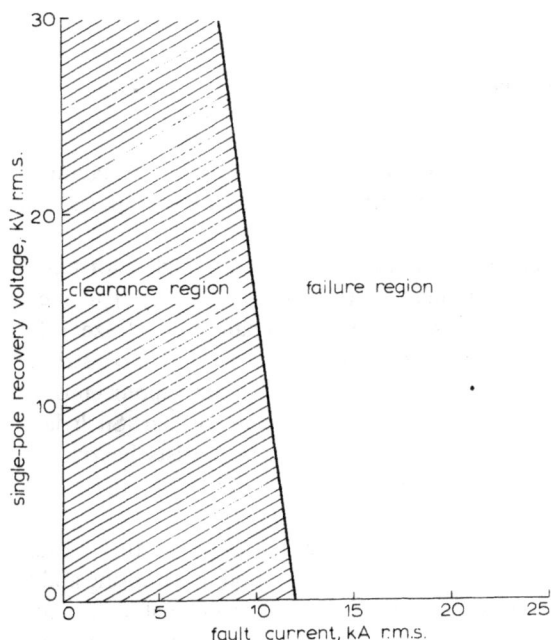

Fig. 8.1 Sharp cutoff to interrupting ability due to magnetic arc constriction on plain butt copper contacts

contact is made, and through which normal current flows when the contacts are closed.

The outer portion of the discs do not touch even when the contacts are closed, and they are cut into spiral shaped 'petals' by a series of spiral slots, the slots being opposite handed in each contact so that when the contacts are juxtaposed the rotation of the slots is the same in each contact.

This design of contact has been widely used in many interrupters, usually under licence from GE (USA).

The operation of this type of contact is as follows: when the magnitude of the current to be interrupted is small, such that a diffuse vacuum arc is formed, separation of the contacts leads to the usual process of current constriction in the metal at the last point or points to separate, the formation of liquid bridges and the rupturing of the liquid metal bridges in the usual way. The arc formed then develops rapidly into a diffuse arc having a number of cathode spots, which move about on the contacts. Some move over the edge of the contacts and are extinguished, and others on the contacts split so that the number of spots remains

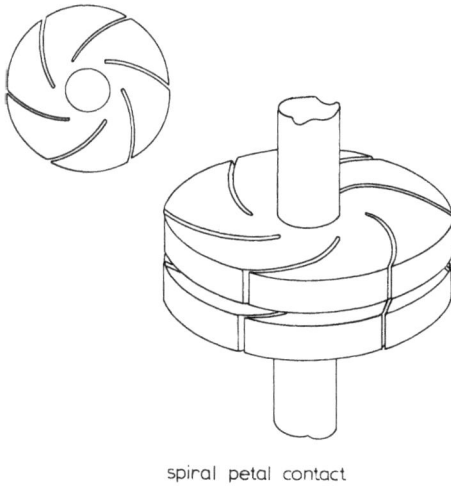

spiral petal contact

Fig. 8.2 Spiral petal contact originally developed by GE (USA)

roughly proportional to the current. The cathode spots tend to spread across the whole or at least a considerable part of the area of the contact disc.

Should the magnitude of the current to be interrupted be much higher, in the constricted arc range, the contact operates in a rather different way. At the rupture of the liquid metal bridge or bridges (there may be more than one) a single constricted arc is formed. Cathode spots continuously leave the region of the arc, because of retrograde motion, but are almost immediately extinguished because their plasma regions are pinched magnetically back to the main arc. The constricted arc now moves bodily away from the centre region of the contact, because of the magnetic effects of the blowout loop of the current in the contact. It should be noted that the constricted arc, the movement of which is plasma dominated, moves in the Amperian and not the retrograde direction. The arc moves in a radial direction out to the edge of the contacts, and remains bowed out between them — when the arc is bowed between the periphery of the contacts, however, the current feed to it has a circumferential component, because of the spiral slots, and thus sets up a radial magnetic field which drives the arc around the periphery of the contacts in the direction dictated by the handing of the spirals.

The effectiveness of the arc rotation on these contacts depends on the angle of the spiral slots; slots which are strongly spiral being more effective than those which are more nearly radial.

The general effect of the constricted arc is to heat the contacts locally, but because of the movement of the arc roots, the arc does not remain in any one position for very long, and the heat does not penetrate deeply into the metal. The cooling time constant of such shallowly heated regions, is commonly in the order of hundreds of microseconds, and thus as the current falls at the end of the loop of arcing the constricted arc reverts to the diffuse form when the current has an instantaneous value of a few kiloamperes, which typically may be 400–500 μs before current zero, depending on the magnitude and symmetry of the current. This gives sufficient time for the heated surface layer of contact metal to cool to a temperature at which its vapour pressure is sufficiently low, and the final interruption is achieved in the diffuse mode by cathode spots.

Various detailed designs have been proposed for the spiral petal contact. In some, metal-to-metal contact occurs at the centre, and in others an annular contact ring is provided, having a diameter of about a third of the overall diameter, so that the radial magnetic blow out force comes into play immediately the contacts are separated. There is little evidence that these details affect the performance of the interrupter significantly, probably because in most cases external magnetic fields always provide sufficient perturbation of the magnetic field between the contacts to induce the radial movement from the contact region in the centre to the arcing region at the edge.

8.2.3 Contrate contact

The other design of contact used for power vacuum interrupters is the contrate contact (Fig. 8.3). In this design the contact is shaped like a very thick walled cup,

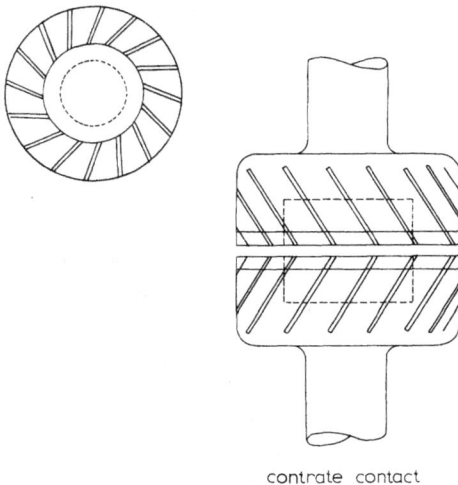

contrate contact

Fig. 8.3 Contrate contact originally developed by AEI (UK)

and the contact support stems are attached to the bottom of the cup. Contact is made through the rims of the cup. Angled slots which approximate to sections of helixes are cut in the walls of the cups, the rotation being of opposite hand in the two cups.

When the contacts separate, molten bridges are formed at a number of points around the periphery; the number of bridges depends on the magnitude of the current. When the bridges rupture, arcs are formed, and the fact that the transition can be made between bridges and arcs at a considerable number of sites in parallel is accounted for by the small but finite inductance of the individual fingers.

As soon as arcs form, the strong radial magnetic fields set up by the circumferential components of the currents in the angled fingers, cause the plasma of the discharge to move circumferentially around the contact, in the Amperian direction, while the cathode spots move slowly backwards in the opposite direction. In this way a hollow ring of arc is produced, and the transition from a diffuse to a constricted arc is prevented. The arc in this design of contact is thus constrained to burn throughout the period of current flow in the place where it is initially formed, i.e. between the rims of the two cups, and no radial movement is required of the arc. Design features of this contact include the provision of a contact tip, having nonwelding and low current chopping properties.

8.3 Design and construction of power vacuum interrupters

The true interrupting volume of a vacuum interrupter lies in the space between the two contacts, the length of the contacts themselves and the space between the contacts and the centre shield.

Because of the relatively high electric strength of vacuum, an interrupter could be quite small internally, but electric strength must also be provided outside the interrupter, and it is largely the external electric strength which determines the length of the insulator of an interrupter – some interrupters have been specially designed for immersion in an insulating fluid, and these have been considerably shorter than interrupters for use in air.

A large number of possible arrangements of vacuum interrupters have been considered and Fig. 8.4 shows three of the most common. In Fig. 8.4a, probably the most common arrangement, the contacts are surrounded by the main or centre shield which serves to condense metal vapour, and two end shields which prevent metal vapour from being reflected back from the end plates on to the envelope. The envelope, which may be of any suitable impermeable inorganic insulating material, surrounds the whole assembly and supports the main shield. The bellows is inside the envelope, and is thus mechanically protected, and a bearing to guide the moving contact stem may or may not be fitted, depending on the design philosophy of the manufacturer. This design of interrupter has a length somewhat greater than the diameter. It has relatively short contact stems which simplifies the mechanical and

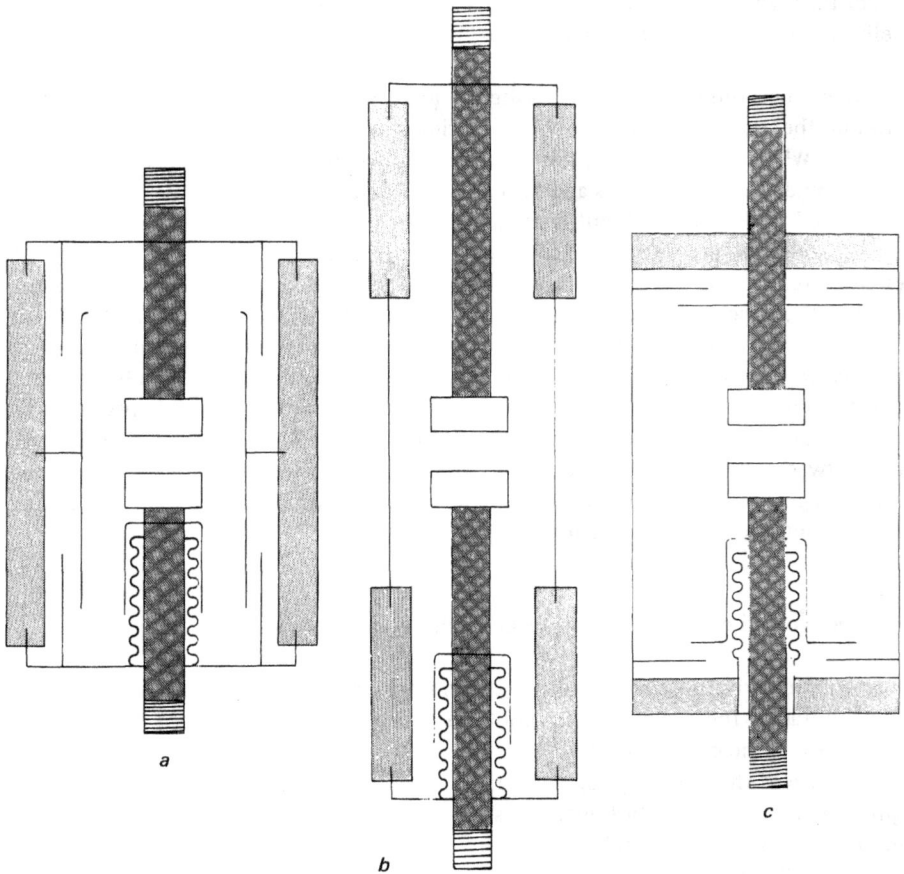

Fig. 8.4 Interrupter designs, each with the same contact and shield sizes (so far as possible)

(a) and (b) are commercial designs
(c) has been postulated but not used widely

thermal design of the stems. Fig. 8.4b shows an alternative design in which the diameter has been reduced at the expense of length. The main shield becomes part of the envelope and the insulation is now in two complete halves, one piece at each end of the interrupter. Fig. 8.4c shows a design which has frequently been postulated, but which is not used commercially for power interrupters. The insulation is now radial, and cheap compression seals can be employed. It is, however, difficult to obtain economically sufficient creepage and string distance for voltages above 3 kV with this type of design.

Many other designs have been postulated and made experimentally, but in general most interrupters commercially available are of the form shown in Fig. 8.4*a* or Fig. 8.4*b*.

The geometry and size of the electrodes are related both to the fault and to the normal full load current. The geometry of the contacts has been discussed in Section 8.2. The contact area which is subjected to arcing, must be large enough to absorb the arc energy without becoming excessively heated, and the contact must give enough contact points of sufficient area (determined by contact hardness and applied force) to give an acceptable power dissipation on full load current.

The design of the power interrupter must allow for the relative softness of those metal components which are affected by heating applied to the interrupter during construction, exhaust and processing, and must ensure that the design is such that scrupulously clean assembly is possible.

The metal—glass or metal—ceramic seals are exactly as used in the power tube industry and need no further comment. Usual assembly methods as used in the tube industry suffice for the components of the interrupter.

Various chemical cleaning processes are followed by vacuum furnace degassing and then furnace brazing. Final assembly of the interrupter is usually either a metal—ceramic braze or argon arc welding of metal to metal joints. Interrupters are then exhausted, baked and sealed off.

Contact materials for power vacuum interrupters must have the following properties:

(*a*) *The correct interrupting ability:* this must be neither too low, since the interrupter may fail to interrupt short line and similar faults, nor too high, since the interrupter may tend to give rise to excessive overvoltages during reignitions when contacts part shortly before current zero.

(*b*) *A low current chopping level:* this is necessary to prevent excessive overvoltage generation when small inductive currents are interrupted.

(*c*) *Low welding forces:* the clean surfaces of contacts in vacuum tend to promote welding, and materials which develop only very low weld strengths are thus needed.

(*d*) *Good electric strength:* the electric strength must be maintained even when contacts have been deconditioned by mechanical operations, weld breaking or fault interruption.

(*e*) *High electrical conductivity:* obviously switch contacts need to have reasonably high conductivities to carry large normal full load currents.

(*f*) *Ease and economy of manufacturing:* economics are crucial to the successful application of vacuum interrupters to switchgear, and so materials must not be costly metals or be difficult and thus expensive to manufacture.

Now it will be apparent that many of these properties are somewhat contradictory; for example, single metals with high electric strengths such as molybdenum, tend to have high current chopping levels, high weld strengths, etc.

Much effort has been put into the development of suitable contact materials, which tend to be binary, ternary or quaternary alloys.

There is a very considerable literature on the subject of contact materials for vacuum circuit breaker interrupters, but information on actual contact materials is naturally enough proprietory and hard to come by. The two most important contact materials in use today are probably as follows:

(*a*) Copper-bismuth alloys (GE, USA)
(*b*) Copper-chromium alloys (VIL, UK)

A very recent development in copper-beryllium alloys by GE (USA) has recently been announced, but is only just coming into service.

Bellows are almost without exception made of stainless steel. Two types are available, those with rounded profiles to their convolutions, which are either rolled or are hydraulically formed, or those with Vee shaped convolutions made by welding stacks of 'washers' together at the inner and outer circumferences alternatively. The choice of bellows depends on a number of factors, but where a large bore bellows with a high compliance is required the welded plate bellows has considerable space advantages.

Leak testing may be done on subassemblies during manufacture by normal mass spectrometer methods, or the complete interrupter may be leak checked during or after exhaust.

A fairly common procedure is for the manufacturer to measure the pressure after the interrupter is sealed, and on a number of occasions spaced over a period of about one month. The pressure can be measured using the interrupter as a magnetron, with an axial magnetic field and a radial electric field. The current that flows between contact and main shields under these conditions is a measure of pressure within the interrupter.

The interpretation of the pressure readings is highly complex, since the pressure in the interrupter is determined by a number of processes in dynamic equilibrium. Each different interrupter will have its own particular characteristics with regard to the pressure and time relationship, depending on the materials and the processing of the interrupter, but correct interpretation of the pressure readings taken over a period of a month can give complete quality assurance.

Most manufacturers give the shelf life of their vacuum interrupters as 20 years, but it must be appreciated that this is very much a lower limit and much longer shelf lives may be expected in most cases. When interrupters are operated the gettering effect of the metal vapour film deposited on the shield and contacts usually causes pressures to fall, and even longer lives than those quoted are likely to be obtained.

Pressure in interrupters usually in the range $10^{-4}-10^{-7}$ torr, although, as explained previously, considerable variation may occur during the life of an interrupter, and not necessarily in the direction of increasing pressure. It has been shown in chapter 1 that vacua of $10^{-3}-10^{-2}$ torr are sufficient for interruption and voltage withstand, but pressures as high as this are not used in interrupters largely because of the difficulty of measuring down to sufficiently low leakage rates for quality assurance at these high pressures.

Three typical interrupters are shown in Fig. 8.5. These interrupters, all with glass—ceramic envelopes, are designed on the general principle illustrated by Fig. 8.4a.

The V2 interrupter, on the left, has a plain butt contact, and is therefore limited on fault current to 6 kA r.m.s.; it is used primarily for switching duties at 11 kV.

The interrupter in the centre, type V8, is rated at 350 MVA 11 kV 1250 A and can also be used at 15 kV. Type V5, to the right, is rated at 600 MVA 11 kV 2000 A, and again can be used at 25 kA at 15 kV. Both of these interrupters have contrate contacts.

All three interrupters may be mounted from either the fixed or moving contact ends, and have built in guides for the moving contact stems.

Fig. 8.6 shows a 12.5 kVA 650A 15 kV interrupter with spiral petal contacts.

Fig. 8.5 Vacuum interrupters manufactured by Vacuum Interrupters Ltd.

Left to right:	V2	11 kV	6 kA	400 A
	V8	15 kV	13 kA	1250 A
	V5	15 kV	25 kA	2000 A

[Courtesy Vacuum Interrupters Ltd.]

Flexible   Insulating   Electrical   Vacuum
Metallic   Vacuum   Contacts   Chamber
Bellows   Envelope
Assembly

Stationary
Electrical
Terminal

Movable
Electrical
Terminal

Metal to Insulation   Metal Vapor   Electric Arcing   Metal to Insulation
Vacuum Seal   Condensing Shield   Region   Vacuum Seal

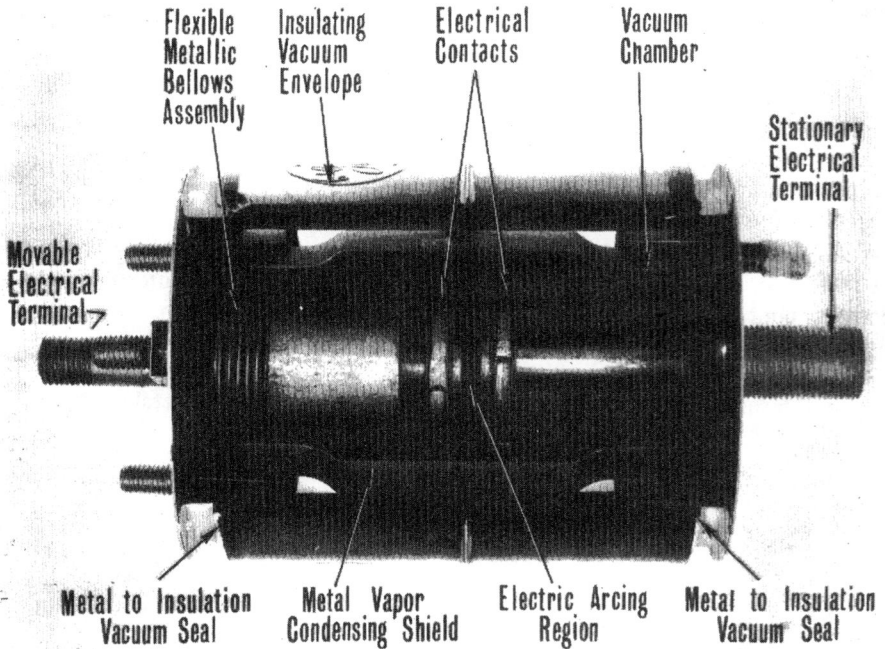

Fig. 8.6 Vacuum interrupter type PV01 manufactured by GE (USA) [Courtesy GE]

8.4 Interrupting ability of the vacuum interrupter

8.4.1 Interruption

The vacuum interrupter has two entirely different limits to interrupting ability. The first can be defined as the ultimate interrupting ability, or commutating ability, and is primarily related to the ability of a single (the last) cathode spot to extinguish in circumstances where the bulk of the contacts is cool. This, as we have seen in chapter 2, is related to the thermal properties of the cathode material. A cathode with a low vapour pressure and high thermal conductivity has a very high commutating ability, and one with a high vapour pressure and low thermal conductivity has a very low commutating ability. This means that contacts with a high commutating ability can interrupt when

$$\frac{-\mathrm{d}i}{\mathrm{d}t} \times \frac{\mathrm{d}v}{\mathrm{d}t}$$

is very large, and for many metals to the right hand side of Fig. 2.13, this commutating ability or inherent interrupting ability is very much larger than required for normal power systems.

The second limit to interrupting ability concerns not the rate of fall of current and rate of rise of restriking voltage, but the magnitude of power frequency current which can be interrupted. This depends on the contact geometry (the contact geometry must be one which prevents a constricted arc remaining in one place), the contact size and the contact material.

For example, even assuming a contrate contact which causes the arc to remain in the diffuse mode throughout the current loop, the contact must be large enough and/or the thermal properties of the contact material, such that even the uniform energy input from arcing does not heat the whole area of the contact face to a temperature, so that its vapour pressure rises above 10^{-3} torr at current zero. Thus a contact of either the petal or the contrate form has a maximum power frequency interrupting current which depends on its size and the thermal properties of its material.

A plain butt contact, of course, has an absolute upper limit of interrupting ability which varies from about 10 to 17 kA peak, depending on the material and the power frequency, and this current is substantially independent of the contact size once it is greater than about 3 cm in diameter.

Providing the vacuum interrupter contact is working within its capacity of commutating factor, power frequency current and power frequency recovery voltage, it can be expected to interrupt at the first current zero at which a reasonable contact gap is reached (say 3–4 mm at 11 kV and 5–6 mm at 15 kV), and this characteristic is maintained across the whole range of current from a fraction of the load current to the full short circuit current.

8.4.2 Reignition in vacuum interrupters

When a vacuum interrupter reignites at an early current zero, because the contact gap is too short to withstand the full recovery voltage, an arc may not be reformed at the instant of the restrike, as occurs for instance in arc chute circuit breakers. Because of the high commutating ability of the vacuum arc, the transient high-frequency oscillatory current which flows from the local LC circuits through the interrupter at the instant of reignition may be interrupted at a high-frequency current zero. The interrupter recovers, only to reignite again when the system recovery voltage reaches the breakdown level of the short gap once more. This can repeat itself a number of times, until the instantaneous level of the power frequency current building up is greater than the peak of the transient oscillatory current, when no further high-frequency current zeros occur, and the full arc is re-established. This process can produce undesirable transient pulses, and the degree to which it occurs in service depends on the high frequency properties of the circuit being interrupted, and, to a much greater extent, on the commutating ability of the vacuum interrupter (Itoh *et al.*, 1972; Greenwood *et al.*, 1971). The commutating ability of the contact material should not be too high, and metals based largely on tungsten or copper may tend to be undesirable in this respect. Providing that the correct contact material is chosen, this phenomenon does not prove a problem. It is worth noting that the process occurs in other types of circuit breakers, and has been reported by Reece and White (1957) in the oil circuit breaker.

8.4.3 Capacitance current switching
The very rapid build up of the electric strength of vacuum interrupters gives good performance on capacitative current switching, but contact speeds should not be too low since should a restrike occur, the classical voltage multiplying process is likely to occur.

8.4.4 Interruption of small inductive currents
All vacuum interrupters (as do all circuit breakers) interrupt current shortly before a current zero due to arc instability at low currents. In vacuum interrupters, the unstable current level or chopping current level is related to the commutating ability of the contact material, and again for this reason the commutating ability should not be too high.

The chopped current (the mean value of which varies from less than 1 A for a metal like bismuth to about 15 A for copper and 30 A or so for pure tungsten) depends, like the arcing voltage, on thermal properties of the contact materials. Sintered mixtures may have chopping levels lower than that of any one constituent alone.

The chopped current is usually statistical in nature, the distribution being approximately Gaussian, so that a mean and a standard deviation can be specified. Further, arc instability is usually increased by the presence of parallel capacitance.

The overvoltage generated in circuits by current chopping in vacuum interrupters depends on the chopped current, the surge impedance and damping of the circuit, and any possible limitation by breakdown of the interrupter contact gap. The performance varies widely from one type of interrupter to another, but with proper selection of contact materials, chopping gives rise to no problems in most normal distribution, utilisation and transmission circuits. Circuits with very high surge impedances can give rise to problems with most types of interrupter, but these circuits are usually of such a nature that precautions would also need to be taken with other types of circuit breakers.

8.5 Design of vacuum circuit breakers

8.5.1 General form and geometry of vacuum circuit breakers
The self-contained nature of the vacuum interrupter is such that it can be applied in a multiplicity of ways. It is therefore difficult to be at all specific as to the general

layout of vacuum circuit breakers, since they can take almost any form that the designer desires; another difficulty is that the subject is so new that designs have not yet frozen sufficiently for simple statements of the position to be made.

In this Section only those features of the switchgear special to vacuum interrupters are considered. Thus mechanisms are not discussed, since once the stroke, speeds and forces required to operate an interrupter are known, the design of the mechanism becomes entirely conventional switchgear engineering, and any type of mechanism may be used. It goes without saying that in choosing mechanisms the designer should endeavour to match the zero maintenance requirement of the interrupter with a similar low or zero maintenance mechanism. The mechanism should ideally also be capable of rapid operation to use the short and constant interrupting time of the interrupter to the best effect.

It must not be thought that the use of vacuum interrupters reduces the power required of a mechanism to negligible proportions. Although the stroke of a vacuum interrupter is short, the forces involved are somewhat similar to the forces involved in other types of interrupter, and the mechanism energy may be about one fifth or so of that needed for other types of circuit breaker – this is by no means negligible.

The designer of vacuum circuit breakers needs to consider carefully the following points in Section 8.5.

8.5.2 Interrupter mounting

An interrupter may be mounted either by the fixed contact stem, or by studs or other fixing means on the end plate at the moving contact end of the interrupter. Most interrupters may be mounted in any orientation.

If the interrupter is mechanically supported from the fixed contact stem, the impact when the contacts close is transmitted directly from the fixed contact to the circuit breaker structure, and the envelope is stressed only be second-order forces arising from its acceleration (vibration) due to the compliance of the mounting point in the circuit breaker.

If on the other hand the interrupter is mounted from fixing means at the moving contact end, the closing forces are transmitted to the circuit breaker structure, via the interrupter envelope, which must be sufficiently strong for this. It can be taken that if the interrupter manufacturer has provided fixing means at the moving contact end then the envelope will be strong enough for the duty imposed.

The interrupter should not be rigidly supported at both ends, since any inaccuracy in the fixing, or movement in the frame of the circuit breaker, may apply excessively large stresses to the interrupter envelope. In cases where the interrupter is mounted from the fixed contact stem and studs or other fixing means are provided by the manufacturer at the moving contact end of the interrupter, it may be acceptable to provide some lateral support at the moving contact end of the interrupter to assist in withstanding side thrust, mechanical or electromagnetic, applied to the moving contact stem.

8.5.3 Mechanical-drive arrangements

Where the system voltage is such that a single interrupter per phase is sufficient, the coupling between the interrupters and mechanism will usually take the form of a simple insulating rod, which will carry as its end the lost motion and overtravel device referred to in Section 8.5.11.

Where two interrupters per phase are used, they will often be mounted along the head of a tee, with the moving contact stems directed inwards and driven with toggles or bell cranks from an insulating rod forming the stem of the tee.

Where more breaks per phase are required, the interrupters will usually be mounted in line and operated by crossheads bridging two operating rods, which run the whole length of the interrupter stack. Two groups of interrupters, rather than merely two interrupters, can then be set up to form the head of a tee again, operated by toggles or bell cranks from a drive rod up the stem of the tee. Hydraulic operation has often been considered and is attractive because of the short stroke requirements of the interrupters.

8.5.4 Moving-contact guidance

In some interrupters the moving contact stem is guided in a sleeve supported from the end plate at the moving contact end of the interrupter. This guide is preferably fitted with an insulating low friction liner.

In other interrupters, no integral guidance for the moving contact stem is provided and the circuit breaker operating mechanism must provide the guidance.

8.5.5 Connections to moving-contact stem

The designer should ensure that connections to the moving contact stem made during manufacture of the circuit breaker, or made during the on site replacement of an interrupter, do not subject the moving contact stem to unacceptably large rotational force or angular movements which might damage the bellows.

8.5.6 Current feed to the interrupter

Care should be taken, in particular when interrupters are mounted from studs at the moving contact end, to see that the current always flows to the moving contact stem and that no substantial current flows to the fixing studs. Since the fixing studs are electrically connected to the moving contact stem, even when an insulated guide is fitted, via the bellows, if currents are allowed to flow through the bellows the thin material may be damaged by overheating. To this end the use of insulating bushes on the fixing studs may be necessary if there is risk of stray current flow to the interrupter.

8.5.7 Contact pressure

The contact pressure in a vacuum interrupter must be sufficient (*a*) to give an acceptably low contact resistance, (*b*) to close on to fault current, and (*c*) to remain closed during the passage of fault current. The contact pressure required to give satisfactory normal current carrying capacity, is normally of the same order, or somewhat less than, that required for closing on to fault and maintaining the contacts closed during the passage of fault current, and so it is usually the latter which dictates the contact pressures required. The electromagnetic repulsion forces between plain butt contacts in a vacuum are not significantly different from those between the same contacts in air, since they are due to the magnetic pressure in the pinched region of current flow at the contact interface. When arcing, the rather more complex structure of some vacuum interrupter contacts leads to slightly higher electromagnetic throwoff forces than for the equivalent plain butt contacts. However, the multiplicity of contact points in this type of interrupter results in a compensating reduction of the electromagnetic throwoff forces. Forces due to pressure in the plasma between the contacts are in general negligible compared with the electromagnetic forces.

During closing operations there are no electromagnetic forces until the start of prearcing, which is typically only for a fraction of a millimetre or so, and therefore the additional work to be done in closing is in general quite small. Even after prestrike occurs the force developed is only the normal electromagnetic throwoff force, and there is no development of anything like the hydraulic forces caused in oil circuit breakers by prestriking.

8.5.8 Contact stroke
Because of the high electric strength of vacuum gaps vacuum interrupter strokes are short. Typical strokes are between 8—12 millimetres for 11 kV and 15 kV vacuum interrupters down to 2 mm for 3·3 kV vacuum contactors. Many interrupters could be operated on fault interruption at shorter strokes than these, but the figures allow for impulse strength and capacitance switching, and it is these applications in general rather than short circuit interruptions which will determine the stroke required of the interrupter.

8.5.9 Contact speed (closing)
Contact speeds in closing must satisfy two conflicting requirements:

Speeds should be low to reduce mechanical stresses and shocks, bearing in mind that vacuum interrupter components will have been brazed and outgassed at fairly high temperatures and will frequently not be so hard and resilient as components in more conventional switchgear. A low speed of closing also reduces the mechanical surging of the bellows and increases bellows life. Finally a low impact velocity reduces the problems of controlling bounce on make.

On the other hand a higher speed of make reduces the duration of prearcing, and thus the amount of contact wear, the tendency to weld, and possible generation of repeated pulse voltages due to unstable sparking during the prearcing period. Speeds normally employed are of the order of 0·6—2 m/s at the instant of make (Boehne and Low, 1969).

8.5.10 Contact speed (opening)

Opening speeds required in vacuum interrupters depend primarily on the following two factors:

(a) The arcing time should be reasonably short. The maximum arcing time should not be greater than about 1·5 loops.
(b) Capacitance switching must be restrike free so that when current zero occurs near the instant of contact separation the electric strength of the interrupter in the open position is sufficiently great after half a cycle of the power frequency wave.

Typically some 50–80% of full stroke will normally be expected to be attained in one loop of power frequency.

8.5.11 Accelerating forces

Most vacuum interrupter contacts may weld to a very slight degree, and the design of a suitable contact material requires among other things that the strength of welds which form is low enough for them to be easily broken. This has been achieved in all commercially available interrupters to the extent that the energy required to break any welds which form is usually a small fraction of the energy necessary to accelerate the contacts to the recommended opening speed.

A typical arrangement for opening the contacts of vacuum interrupters is to arrange the mechanism to accelerate a mass (which should be about equal to the moving contact mass but may be considerably larger) to a velocity somewhat in excess of the opening velocity required. The mass then snatches off the contact, the severity of the snatch being controlled by the compliance of the coupling between 'hammer' and contact. This coupling sometimes incorporates an elastomer or may alternatively be of metal with a designed elasticity.

8.5.12 Contact wear

Vacuum interrupter contacts wear principally because of the erosion caused by arcing and only to a negligible degree by no-load operation. In most interrupters the erosion rate in terms of grams per coulomb is not a constant but increases with increasing current. The best commercial interrupters will interrupt their full short-circuit current some hundreds of times, and will give many thousands of operations on normal full load current.

Material is lost from the contacts in three ways; these are: (a) melting globules resulting from the destruction of liquid metal bridges drawn between the contacts at the instant of contact separation, (b) metal vapour evaporated from cathode spots, some of which recondenses upon the opposite electrodes, and (c) liquid droplets thrown out of a molten metal film on the surface of the electrodes at high currents — these droplets are usually ejected by the magnetohydrodynamic forces set up by the interaction of the arc current and its own magnetic field.

Metal lost from the contact surfaces is deposited partly on other parts of the contact and partly on the shield.

The limit to the permissible contact wear is set either by the shape of the contacts changing to such a degree that the magnetic control of the arc is lost, or because if all the contact material is eroded severe welding may then occur between the contact stems or backing which usually will not be of a nonwelding material.

The other possible limitation to the wear is the increased travel of contacts in the closed position overstraining the bellows.

For one or other of these reasons a limit is necessary to the permissible contact wear measured by the amount of additional movement to the moving contact stem when the contacts are closed.

8.5.13 Interrupter cooling

Because of the thermal insulation between contacts and envelope provided by the vacuum, substantially all the heat dissipated at the contacts and in the contact stems must be removed by conduction along the stem. About 50% of the specified contact dissipation must be removed from each terminal of the interrupter, while maintaining the terminal below the specified temperature.

The contacts will of course be a few degrees above the terminals, depending on the thermal resistivity of the contact stems. Since, however, the contacts are completely protected from oxidation, this presents no problems.

8.5.14 Mechanical biasing

All power vacuum interrupters tend to close under the action of atmospheric pressure on the bellows. In some circuit breakers this force has been used to resist the closing springs, and the mechanism then pulls the interrupter open when the breaker is tripped. Alternatively, biasing springs may be fitted as close as it is desired to the interrupter, so that it becomes a 'push-to-close' instead of 'pull-to-open' arrangement. The choice is not fundamentally significant, but the presence of a local bias spring reduces the numbers of components which may in principle, by mechanical failure, allow the interrupter to close when it should be open.

8.5.15 Insulation

It is probably true to say that the electric strength to any given wave form of a vaccum interrupter can vary more with time and with interrupter condition than that of other types of interrupter. In particular, it is usually possible by 'conditioning' an interrupter with a series of controlled applications of high voltages, to achieve breakdown levels very considerably above the rated electric strength of the interrupter. This has little or no significance in single break vacuum circuit breakers, but where interrupters are used in series, as in all multibreak circuit breakers, some care must be given to the question of insulation co-ordination between the contact-to-contact gap internally, (and any other internal path) as well as the external breakdown path, so that on closing, or in the event of an impulse being applied during opening or closing, no external breakdowns occur across the outside of any one single interrupter in the series stack.

8.6 Vacuum tests for vacuum interrupters

The integrity of the vacuum envelope may be established by a simple dielectric strength test carried out either immediately after the short circuit or other tests, or after some suitable interval, should it be thought that small leaks due to testing may be a possibility.

The pressure in vacuum interrupters can be measured using the interrupter itself as a magnetron type ion gauge. With this arrangement, the contacts may be closed or open and an axial magnetic field is applied parallel to the contact axis. A voltage, some 5 or 10 kV, is then applied between the contacts and the centre shield. In the presence of the correct value of electric strength and magnetic field, this low voltage can give rise to a breakdown, since electrons in the gap have a Larmor radius of a few centimetres, so that they can continue to revolve around the contact axis many thousands of times, and thus very long mean free paths may be achieved. The total current flowing under these conditions is a measure of the quantity of gas available to be ionized in the interrupter.

In general, however, this is a factory test and is not so convenient for the field.

The usual practice recommended is the use of a single high potential withstand test to demonstrate that the pressure in the interrupter is satisfactory. This, of course, does not enable a measure of pressure to be made but experience indicates that it is completely adequate for normal field testing. Typical power frequency test voltages will be in the range 15–60 kV r.m.s. for 11–15 kV interrupters and 45–100 kV r.m.s for 33/38 kV interrupters. It should be noted that although the X-ray output of presently available vacuum interrupters is virtually undetectable at their rated voltages, small but finite X-ray output occurs under test conditions, and if interrupters are to be tested at voltages significantly above 28 kV r.m.s., X-ray shielding should be provided.

If severe contact damage is suspected and it is not possible to open the interrupter for inspection, then X- or gamma-ray photography can give an indication of the state of the contents.

The pressure within the vacuum interrupter changes during the life of the interrupter in a complex way, and depends on the manufacturing process and on the history of the interrupter.

It can be said that in a correctly processed interrupter pressure tends to fall with storage over a few years. Furthermore, any arcing within the interrupter tends to reduce further the pressure, and the life of the vacuum interrupter, whether used or not, should be considerably longer than the 20 year period which it has become customary to quote in commercial literature.

8.7 Vacuum circuit breaker types

Vacuum circuit breakers have been in quantity production since 1962, when GE (USA) first manufactured a 15 kV 12·5 kA 650 A outdoor type circuit breaker/recloser.

They are now being manufactured in quantity in the UK, the USA, Germany and Japan, and experience to date with their performance has been excellent.

Vacuum circuit breakers are being applied in two main areas, as follows:

(a) Switchgear for distribution and utilisation
These equipments cover the range from 3 to 36 kV. This is the field of application where the vacuum circuit breaker has been most widely applied, and where the relatively large amount of maintenance needed by conventional types of equipment has favoured the vacuum circuit breaker (Streater *et al.*, 1962).

This type of application, which has developed extremely rapidly in Japan, where some 50–70% of new distribution switchgear now uses vacuum interrupters, was pioneered in the USA, and is now developing very rapidly in the UK (Cooke and Klint, 1969) (Reece and Ellis, 1971) (Armstrong, 1972) (Headley, 1972) (Auton *et al.*, 1972) (Passant and Ruddock, 1972).

(b) Transmission switchgear
These circuit breakers may be considered to cover the range from 100 kV upwards, although nowadays many circuits at 100–150 kV are used for distribution rather than transmission. Progress in these voltages ranges has been rather slower than at the lower voltages, possibly because e.h.v. switchgear usually operates much less frequently than distribution switchgear, and, therefore, the near zero maintenance requirement does not so significantly benefit vacuum circuit breakers vis-a-vis conventional circuit breakers at these voltages.

This type of application of vacuum interrupters was pioneered in the UK, but it now seems unlikely to develop further.

8.8 Examples of vacuum circuit breakers

8.8.1 Circuit breakers for distribution and utilisation
Development of 11–15 kV distribution switchgear has taken place in two phases. In the 'first generation' switchgear, vacuum interrupters merely replaced conventional interrupters, the general form and arrangement of the switchgear being otherwise unchanged. The major exception to this was the early GE (USA) vacuum breaker/recloser (Fig. 8.7).

The 'first generation' distribution circuit breakers retained the plugging features of oil breakers in the UK and air magnetic breakers in the USA. It was appreciated (particularly in the UK, because of the low cost of the oil circuit breaker), that

Fig. 8.7 Arrangement of interrupters in breaker/recloser for 12.5 kA 650A 15 kV [Courtesy GE]

'first generation' vacuum circuit breakers were not economic, but they had the advantage that they could be quickly modified to use vacuum interrupters, and thus service experience with vacuum interrupters could be quickly gained without the delay and expense of designing complete new equipments.

This philosophy, which has been fully justified by the service experience obtained, was behind the design of the distribution circuit breaker shown in Fig. 8.8. Because the vacuum interrupters were somewhat larger than the arc control pots of the original oil circuit breaker, they were arranged in trefoil, and the steel tank replaced by a moulded epoxy tank. Apart from these changes, and some slight modifications to the mechanism, the circuit breaker is identical to the oil circuit breaker, and plugs into the same switchgear.

A similar modification has given rise to the recloser shown in Fig. 8.9. In this, since clearances are small, oil has been retained as an insulating medium, but since it does not become carbonised, no maintenance is needed.

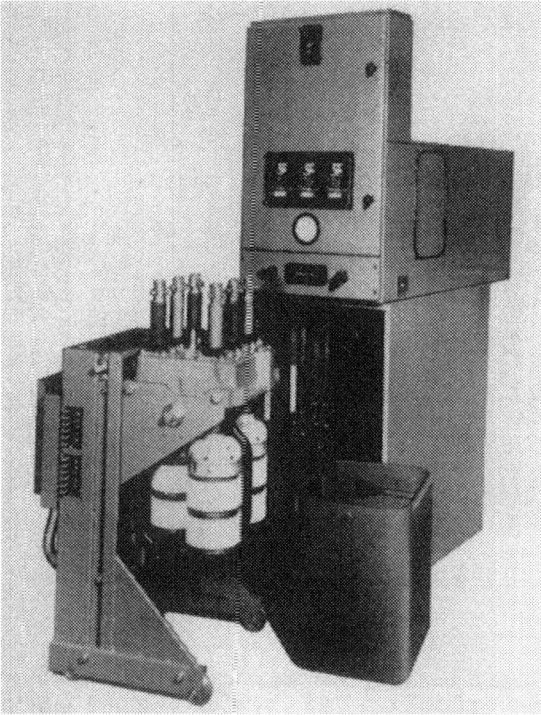

Fig. 8.8 Type LMV 11 kV 250 MVA 600 A vacuum circuit breaker panel [Courtesy A. Reyrolle & Co. Ltd.]

Both of these equipments are capable of some hundreds of full fault interruptions, and many thousands of full load interruptions, and have now been in service for five years.

Very similar developments have taken place in the USA, with plugging metal clad breakers available up to 34·5 kV, using two interrupters in service per phase, although single interrupters are now becoming available for the more common ratings at this voltage. Because of the relatively high cost of the air-magnetic circuit breaker in the USA, the 'first generation' vacuum circuit breaker appears to be persisting there longer than in the UK.

8.8.2 'Second generation' distribution circuit breakers

All the major switchgear manufacturers in the UK have designed or are designing their second generation version of metalclad distribution switchgear.

Fig. 8.9 Type OYV 11 kV outdoor vacuum recloser [Courtesy A. Reyrolle & Co. Ltd.]

'Second generation' vacuum circuit breakers take advantage of the long life and zero maintenance of the vacuum interrupter, which is bolted solidly into the equipment, isolation and circuit selection being achieved by blade type selector switches on the busbar side or in some cases on both sides of the interrupter. This design saves the cost of bushings and spouts and of building the circuit breaker on a removable truck.

The use of solidly mounted interrupters is central to all the designs of 'second generation' vacuum circuit breakers; in other respects designs differ. Busbars are situated either above or below the interrupters and mechanisms are spring mechanisms (in their various forms) and solenoids. Mechanism powers tend to be between 20–40% of those in equivalent oil circuit breakers.

Fig. 8.10 shows one version of the FV equipment made by Brush Switchgear Ltd. arranged with selector isolation on both sides of the interrupter, the circuit breaker having solenoid operation.

Fig. 8.11 shows the BVAC circuit breaker manufactured by GEC Switchgear Ltd., which has a horizontal mounting arrangement of the interrupters and the interrupter drive mechanism. The circuit breaker is spring operated. In this equipment the busbars are above the circuit breaker, and the equipment may be double tiered by mounting a second circuit breaker virtually upside down on top of the lower one, with the common busbar between the two.

Fig. 8.10 Type FV 11 kV 250 MVA 650 A vacuum circuit breaker with single selector (side plate removed) [Courtesy Brush Switchgear Ltd.]

Fig. 8.11 Type BVAC 250 MVA 11 kV vacuum circuit breaker showing horizontal mounting of interrupters [Courtesy GEC Switchgear Ltd.]

8.8.3 25 kV railway trackside circuit breaker

Trackside circuit breakers that feed the catenary are usually placed in rather inaccessible substations along the route of the railway and low maintenance as well as low first cost is an important criterion. Furthermore, fast operating circuit breakers reduce arcing damage to the caternary when faults occur, and for these reasons British Rail decided to use 25 kV single-phase vacuum circuit breakers for the extensions of the west coast electrification from Crewe to Glasgow. The general specifications to be met were as follows:

normal full load current = 800 A
system voltage = 25 kV (single phase)
fault current = 12 kA
impulse level = 250 kV
asynchronous interruption = 6 kA at 44 kV (1·7 amplitude factor)

Owing to possible contamination from moisture and brake dust, a long creepage path (71 cm) was specified on insulation.

The circuit breaker uses two VIL type V8 interrupters in series, and because of the high impulse level and amplitude factor, ceramic capacitors are used to control voltage distribution. Because of the long-creepage path specified the interrupters are sleeved with rippled porcelain tubes; the interspace being filled with a suitable elastomer.

Fig. 8.12 shows two bays of a trackside substation. The interrupters are mounted in a tee headed geometry with the heads of the tees vertical, the mechanical drive up the stem of the tee being taken via a glass reinforced plastic drive rod alongside the solid support insulator. The circuit breaker is operated by a conventional motor wound spring mechanism, which gives fast operation, the total break time being 46 ms.

Because the interrupters require negligible maintenance, isolators have been omitted from both sides of the interrupters. Air-to-air bushings in the roof of the steel substation enclosure connect the outgoing side of the circuit breaker to the overhead conductor system. These bushings allow ring type current transformers to be used, exactly as in an oil circuit breaker. The busbars run along the lower part of the switchgear housing, the circuit breakers being directly connected to the busbars.

8.8.4 Railway rooftop circuit breaker

A vacuum circuit breaker, also using two V8 interrupters in series, but in this case without capacitors for voltage distribution, has been developed by GEC Traction Ltd. Two VIL type V8 interrupters are operated by a compressed air mechanism, air being taken from the locomotive air main.

Circuit breakers, rated 25 kV 12 kA 800 A, with a 170 kV impulse level, have been in experimental service for some years, and a large number are now in use by British Rail.

8.8.5 E.H.V. circuit breakers

E.H.V. vacuum circuit breakers were first built in the UK and installed by the CEGB in 1968.

Fig. 8.13 shows two 132 kV vacuum circuit breakers commissioned at the West Ham substation of the CEGB in 1968. These were rated 15·3 kA at the first phase to clear voltage of 125·5 kV, both terminal and short line faults, an asynchronous condition of 4 kA at 168 kV, and earth faults of 18·5 kA at 87 kV. The normal full load current required was 800 A, but the circuit breakers were rated at 1200 A.

Eight AEI interrupters were used in series per phase, the interrupters having contrate contacts; these interrupters were especially designed for use in e.h.v. circuit breakers, and were 13 cm in diameter and 20 cm long.

The interrupters were mounted in groups of four in series in each half of the tee head of the circuit breaker (Fig. 8.14). They were supported in a linear array on a framework of insulating rods, and were operated by crossheads connected to the moving contacts, themselves driven by two rods, one down each side of the interrupter stack. To give a high degree of insulation coordination within the circuit

Fig. 8.12 25 kV railway trackside circuit breakers [Courtesy GEC Switchgear Ltd.]

Fig. 8.13 Type VGL8-AC 132 kV 3500 MVA 1200 A vacuum circuit breaker [Courtesy GEC Switchgear Ltd.]

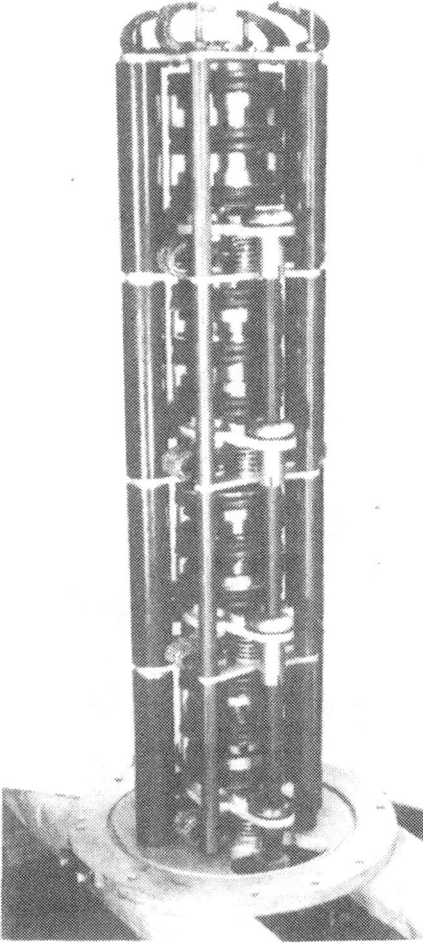

Fig. 8.14 Arrangement of four type SC1-11 vacuum interrupters in series in VGL8-AC1 132 kV vacuum circuit breaker [Courtesy GEC Switchgear Ltd.]

breakers, the interspace between the interrupters and the porcelain housing was filled with SF_6, at a pressure slightly above atmospheric pressure.

The interrupters were normally held closed by springs acting directly on the crossheads and by the pressure of the SF_6 on the bellows, and were pulled open by the mechanism when the circuit breaker was tripped. The mechanical drive was taken from an interphase coupling rod via bell cranks to fibre-glass drive rods in the vertical support insulators.

The mechanism was a standard AEI hydraulically charged spring mechanism, as used on 132 kV oil circuit breakers, with its normal output stroke geared down to suit the vacuum interrupters. Because of the short travel of the interrupters two cycle performance was achieved without any special design approach. Four of these circuit breakers have been in satisfactory service since 1968.

Five years were to elapse before the next development which was in the USA. A 110 kV vacuum circuit breaker for 12·5 kA fault levels were introduced by H. K. Porter, in which the circuit breaker rotates for isolation and can then be lowered to the ground for maintenance. This was made necessary because parallel blades operating under oil carry normal current and close, the interrupters merely serving to break current.

A further advance has been announced recently by GEC (USA). 138 kV 10 000 MVA vacuum circuit breakers have been ordered in commercial quantities by American Electric Power, who have also ordered single experimental vacuum circuit breakers from the same company rated at 345 kV 40 kA and 760 kV 40 kA. It is reported that this latter circuit breaker is metalclad, and may have up to 14 breaks per phase. The knowledge of vacuum interrupters and experience of up to eight breaks per phase in the UK suggests that there is no fundamental reason why any of these circuit breakers should not be completely successful technically. The economic justification may however depend on life costing, including maintenance, at the present state of the art.

8.9 Vacuum interrupters since 1975

8.9.1 Developments in vacuum interrupters
Ratings of interrupters commercially available have increased since 1975, and physical sizes and costs have been reduced.

Most makers of interrupters seem to have been concentrating on developing the economic manufacture of interrupters, and on their application in mass-produced circuit-breakers, rather than investing large sums in research and development in new contact structures and materials.

8.9.2 Developments in contact structures
So far as is known, most interrupters continue to be based on either the 'spiral petal' or 'contrate' contact geometries. Two manufacturers, one in Continental Europe and one in Japan, have continued to develop contacts with axial-magnetic-field arc stabilisation, but the use of the technique does not appear to be spreading.

Two methods of generating the axial magnetic field, which causes the arc to remain in the diffuse form (with many cathode spots burning in parallel) even at high values of current, have been used. In the simplest, the contacts are plain butt contacts, and single- or multi-turn coils are built into the circuit breaker around

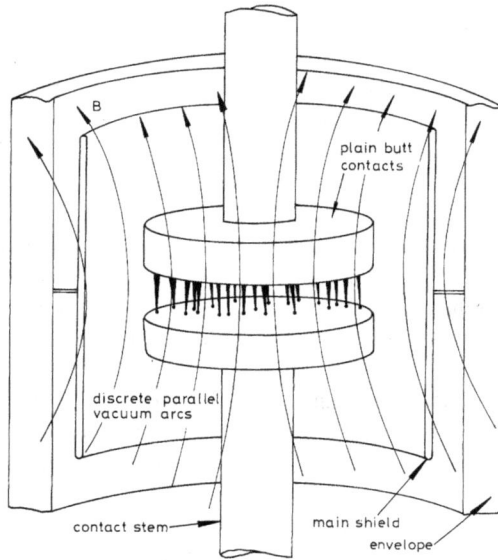

Fig. 8.15 Plain butt contact showing high-current vacuum arc maintained in the diffuse form by an axial magnetic field. The axial field may be generated by series-connected coils surrounding the interrupter or by coils built into the back of the contacts

each interrupter, and are connected in series with it. To obtain a satisfactory insulation level, the coils are usually encapsulated in cast resin.

In the second arrangement, the contacts are formed in such a way that the current-carrying paths feeding current from the stem to the flat contact face form a solenoid, the magneto-motive force of the coils in the two contacts being additive, so that an axial magnetic field is set up in the arc space between the contacts. This arrangement removes the necessity for an external coil, but leads to a rather complex construction for the contact.

A company in the USA is developing a 'multifinger' contact for very high-current interruption. There appear to be problems of contact erosion when contact separation takes place at high instantaneous currents, and such contacts may well be costly to manufacture. This type of contact was first described by Reece in 1959. In this arrangement a pair of main contacts open and close in the usual way, drawing an arc between them when they separate whilst carrying current. Surrounding these moving main contacts are a series of intercalating fixed 'finger' electrodes, which form the auxiliary contacts. The arc formed at the separation of the main contacts transfers to these fixed electrodes very soon after its initiation. This contact geometry maintains the arc in a diffuse condition up to high levels of short-circuit current. The problem with these original multifinger contacts was, however, that considerable damage could be done to the main contacts when they

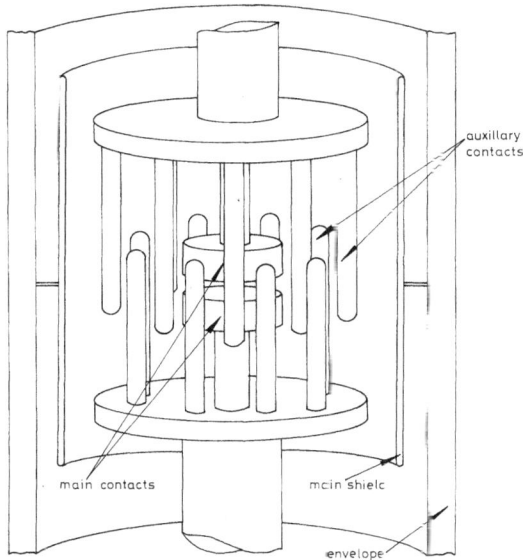

Fig. 8.16 Multifinger contacts for vacuum interrupter

separated with a high instantaneous current flowing, even in the short time which it took to transfer the arc on to the auxiliary fixed contacts.

Generally, it seems that interrupter contact geometries will remain much as they are at present for a considerable time to come.

8.9.3 Developments in contact materials
There has been little disclosure of new contact materials. Materials in common use are still based on either copper-bismuth alloys or copper-chromium alloys.

Mention has been made of the use of cobalt instead of chromium (a possibility covered incidentally by Robinson in the original UK copper-chromium alloy patent), but the high cost of cobalt and its relatively high ductility have militated against its use.

Probably the most significant development has been the use of smaller proportions of chromium in the copper-chromium alloys than was originally used: this has, however, resulted in a requirement for increased contact pressures to offset the increased tendency to weld, which has in turn called for more powerful mechanism when this type of contact material has been employed.

8.9.4 Interrupter ratings currently available
It is difficult to be precise about the interrupter ratings which are currently available, since lists, catalogues and published claims do not always seem to correspond

with the range of equipments which manufacturers are prepared to supply. Further, some published ratings have been withdrawn; the data below are believed to summarise the situation.

8.9.5 Interrupter unit voltage ratings

Interrupters have recently been supplied with unit voltages up to 84kV r.m.s. (3-phase line-to-line system voltage) with an associated short-circuit current interruption capacity of 40 kA r.m.s. It is believed, however, that these interrupters are no longer available. Interrupters with this voltage rating are very large, and require a long stroke (of the order of 60mm). They have been said to be prone to restrikes when switching capacitance current and sensitive to mechanical shock-induced breakdown.

The upper limit of rated unit voltage offered by most interrupter manufacturers appears to be settling down at between 24 and 36 kV. At these voltages interrupters can still retain a conveniently short stroke, of the order of 10-15mm, can give restrike-free capacitor switching performance, and remain relatively insensitive to shock-induced breakdown: a form of breakdown in which charged microparticles are loosened from an electrode by a mechanical disturbance, are accelerated across the gap by the electric field, and are evaporated and cause a breakdown when they strike the other electrode.

A further advantage of limiting unit voltages to the 24-36kV region is that there is no possibility of detectable X-ray emission in service.

Associated with these voltage ratings of 24-36 kV are maximum short-circuit currents of between 25 and 40 kA.

The majority of interrupters manufactured continue to be rated at 7·2, 12, 15 or 24 kV.

8.9.6 Continuous current ratings

The upper limit of rated continuous current available is 3150 A, but this is exceptional. Most manufacturers' upper limits are 2000 or 2500 A. However, it continues to be true that the continuous current rating which can be assigned to a given interrupter depends on the cooling means employed by the circuit-breaker designer. Provided that heat sinks and/or fans are used to remove the heat from the interrupter stems increases of current rating in the region of 30% are attainable.

8.9.7 Short-circuit current ratings

Interrupters are now available for virtually every rating needed for the manufacture of distribution switchgear; the most common upper limit is 40kA but 50kA is available at 15kV.

8.10 Vacuum circuit breakers since 1975

8.10.1 Developments in vacuum circuit-breakers

During the period under review the vacuum circuit-breaker has left the development phase, and, furthermore, its proper field of application has become clearly established.

In many parts of the world it has established a position as the leading type of oil-less circuit-breaker for use in distribution systems, and in a few places it has displaced the oil circuit-breaker virtually completely at distribution voltages.

By the same token it has been clearly shown that the v.c.b. is not economically competitive at e.h.v. for transmission systems; the changeover from vacuum to SF_6, air-blast or s.o.v. seems to be taking place somewhere between 36 kV and 100 kV. That this is so may be demonstrated by the fact that in the UK in 1968, in the USA in 1974 and latterly in Japan, e.h.v. v.c.b.s were introduced, but did not go into production. In the UK and USA they were quickly found to be uncompetitive, and it is being reported that a similar situation is now being found in Japan.

8.10.2 Vacuum circuit-breakers for distribution systems

The vacuum circuit-breaker can be said to have established itself as the circuit breaker par excellence for power distribution systems.

For example, in Japan, the v.c.b. commands about 70% of the distribution switchgear market, mainly at 7·2 and 12 kV, although v.c.b.s are widely used up to and including 36 kV.

In the UK, v.c.b.s have taken about 40% of the total distribution switchgear market at 11 kV. The market share is higher than 40% in industrial applications, where circuit breakers tend to be worked harder than in the utilities systems. All the leading UK circuit-breaker manufacturers are now offering vacuum circuit-breakers for distribution applications.

V.C.B.s have been particularly successful in 25 kV, 50Hz railway traction service, and virtually no other type of circuit breaker is now being supplied for trackside use.

In the USA, two major circuit-breaker manufacturers have now standardised on vacuum for 15 kV withdrawable indoor circuit-breakers, in place of their previous air-magnetic circuit-breakers, and other manufacturers are offering vacuum or air-magnetic as alternatives.

In Germany, all the major switchgear manufacturers have developed or are developing v.c.b.s, and the same appears to be happening widely in Eastern Europe. V.C.B.s are now well established in South Africa and Australia, and are being well received in India.

The original UK concept of the medium-voltage v.c.b. as non-withdrawable equipment has, in general, not been followed by the rest of the world, where the large majority of v.c.b.s are withdrawable: this has the advantage of making them interchangeable with the circuit breakers in existing switchboards. It appears possible that the UK may move back towards the withdrawable concept in future designs.

Philosophies with regard to switching overvoltages have not changed. Users of interrupters with high chromium-alloy contacts are continuing to find that overvoltage generation is not a problem, and users of interrupters with copper-bismuth contacts are tending to build zinc-oxide type nonlinear resistors into their circuit breakers as a matter of course, and are also not experiencing overvoltage problems.

8.10.3 Service experience

In the UK, with an estimated 15 000 circuit-breaker years of experience of vacuum circuit-breakers in distribution systems, service experience has been outstandingly good. This is also reported to be the case in Japan and USA. In the UK there have been reported five cases in which vacuum circuit-breakers have failed in some way in operation and suffered damage as a result. In only one case was the failure due to a vacuum interrupter, and it was considered that this was due to loss of vacuum. Four of the five failures led to 3-phase faults within the circuit breakers, and in no case was there any structural damage to the rest of the switchboard or to the sub-station.

It is significant that, in a number of interrupters examined after between six and ten years service, no deterioration in vacuum has been found, nor of course, would any deterioration be expected. In some arc-furnace applications, 150 000 switching operations have been performed with negligible contact wear and with no maintenance of the circuit-breakers. It can be concluded that service experience with the vacuum circuit-breaker is fully supporting the original claims of its protagonists.

8.11   Acknowledgment

The author acknowledges the assistance of friends and colleagues, and companies which have supplied drawings and photographs. Thanks are due to Vacuum Interrupters Ltd. for permission to publish this chapter.

8.12   References

ARMSTRONG. T. V. (1972): 'The future of vacuum interrupters in distribution switchgear'. *IEE Conf. Publ.,* 83, pp. 91–97

AUTON, G. MIDWORTH, N. C., FORD, F., and ROWLAND, F. G. (1972): 'Heavy duty metalclad switchgear'. *IEE Conf. Publ.,* 83, pp. 163–168

BARKAN, N. P., LAFFERTY, J. M., LEE, T. H., TALENTO, Z. (1970): 'Development of materials for vacuum interrupters'. AIEE conference paper 70 TP 179 PWR, Winter Power Meeting

BOEHNE, E. W., and LOW, S. S. (1969): 'Shunt capacitor energisation with vacuum interrupters – A possible source of overvoltage', *IEEE Trans.,* PAS-90, p. 1424

COBINE, J. D. (1962): 'Research and development leading to the high power vacuum interrupter – an historical review'. AIEE Winter General Meeting, paper CP 62–67

COOKE, S. C., and KLINT, R. V. (1971): 'A new generation of metalclad switchgear using vacuum interrupters', *IEEE Trans.*, **PAS-90**, pp. 1589–1597

ELECTRICAL REVIEW (1973): 'Worlds most powerful vacuum interrupter developed by GE', *Electr. Rev.*, pp. 857–858

FARRALL, G. A., LAFFERTY, J. M., and COBINE, J. D. (1963): 'Electrode materials and their stability characteristics in the vacuum arc', *Trans. Am. Inst. Electr. Eng.*, **CE-66**, pp. 253–258

FLURSCHEIM, C. H. (1971): 'High-voltage switchgear', *Electron. & Power*, **17**, pp. 477–483

GIBBS, F. S. (1973): 'Switchgear – engineering science as a profitable business', *ibid.*, **19**, pp. 382–385

GREENWOOD, A. N., KURTZ, D. R., and SOFIANEK, J. C. (1971): 'A guide to the application of vacuum circuit breakers'. IEEE Winter Power Meeting, New York, paper 71

HEADLEY, P. (1972): 'Distribution switchgear with vacuum circuit breakers', *IEE Conf. Publ.*, **83**, pp. 98–100

ITOH, MURAI, OHKURA, and TAKAMI (1972): 'Voltage escalation in the switching of the motor control circuit by the vacuum contactor', IEEE Winter Meeting, New York, paper T 72 053–2

JENNINGS, J. E., SCHWAGER, A. C., and ROSS, H. C. (1956): 'Vacuum switches for power systems', *Trans. Am. Inst. Electr. Eng.*, **75**, pp. 462–468

KUROSAWA, Y. (1980): 'Vacuum circuit-breaker electrode generating multipole axial magnetic field and its interruption ability.' IEEE PAS Winter Meeting 3-8 Feb.

LEE, T. H., GREENWOOD, A., CROUCH, D. W. and TITUS, C. H. (1962): 'Development of power vacuum interrupters'. AIEE Winter General Meeting, paper 62–151

PASSANT, W. L., and RUDDUCK, A. J. (1972): 'The future of vacuum circuit breakers in the electricity supply industry', *IEE Conf. Publ.* **83**, pp. 7–13

REECE, M. P. (1959): 'The vacuum switch and its application to power switching', *JIEE* **5**, pp. 275–279

REECE, M. P. (1963): 'The vacuum switch – Pts. 1 and 11, *Proc. IEE*, **110**, pp. 793–811

REECE, M. P. (1973): 'A review of the development of the vacuum interrupter', *Phil. Trans. R. Soc.*, **275**, pp. 121–129

REECE, M. P. and ELLIS, N. S. (1971): 'Good prospects for vacuum interrupters in circuit breakers', *Electr. Rev.*

REECE, M. P., and WHITE, E. L. (1957): 'Measurement of overvoltages caused by switching out a 75 MVA 132/33 kV transformer from the high voltage side'. British Electrical & Allied Industries Research Association, Report G/T 313

ROSS, H. C. (1961): 'Vacuum power switches: five years of application and testing'. AIEE Conference Paper CP 51–135

SELZER, A. (1971): 'Switching in vacuum: a review', *IEEE Spectrum*, pp. 26–37

SORENSEN, R. W., and MENDENHALL, H. E. (1926): 'Vacuum switching experiments at California Institute of Technology', *Trans. AIEEE*, **45**, pp. 1102–1105

STREATER, A. L., MILLER, R. H., and SOFIANEK, J. C. (1962): 'Heavy duty vacuum recloser'. AIEE Winter General Meeting, paper 62–120

Chapter 9

Special switching systems

M. N. John, B.Sc., F.Eng., Sen. Mem. I.E.E.E., F.I.E.E.

9.1 Introduction

In addition to their development for the solution of new problems, special systems in any technological area arise because required equipment ratings exceed the present levels or because new techniques and/or materials become available. The switching systems described in this chapter exhibit some or all of these origins.

The systems are of two main types; first, there are those which assist conventional switchgear by reducing or eliminating the potential fault current, and, in this category, come fuses and the more sophisticated current-variation devices (so-called to distinguish them from current-limiting fuses) which return automatically, or can be made to return to, the prefault condition. Also in this first category of switchgear 'assistors' are techniques of synchronised switching and methods for the creation of artificial current zeros for d.c. switching. The second category is devoted to new methods of switching, principally the application of thyristors in power 'static' circuit breakers.

Some of these systems will survive only where special expenditure is justified, others will find general application as further development reduces costs, and some will disappear as ratings of existing equipment are yet again extended or as new special systems are developed. It is also possible that some will give rise to new branches of power-system technology as has already happened with h.v. d.c. transmission. In selecting the examples quoted here, consideration has been given to those systems which have already found application or which are, at this point in time, generally regarded as practicable. Nevertheless the likely degree of continuing success of any particular technique is, perhaps, an interesting speculation for the reader.

9.2 Power fuses

9.2.1 High-rupturing capacity fuses

Power fuses are the lowest cost means of protecting against fault current and there are two basic types in common use, namely the h.r.c. fuse and the expulsion fuse (Cameron and Smith, 1971; Feenan, 1971 and 1980).

The well known h.r.c. fuse comprises an element, normally of silver, wound on a supporting member and surrounded by a quenching medium such as silica sand, the whole encapsulated in a ceramic body with copper end caps. Operation is achieved, often within several milliseconds, by the melting of the element to limit the current to a value substantially less than the prospective peak value. Thus the magnetic and thermal effects proportional to $I^2 t$ or let-through energy are drastically reduced.

The important criterion in considering the load current-carrying capability of the fuse is temperature, and temperature is also a main factor on which current rating is determined. It is simple to specify a limiting temperature that corresponds to the steady-state and stabilised condition which is obtained after a uniform current has been flowing for a sufficiently long time; however, in practice, this seldom if ever occurs. Load currents vary in magnitude and duration, and fuses vary as regards temperature time constant. Moreover, the fuse experiences mechanical and other stresses in varying degrees owing to temperature variations. One of the problems in practice is to equate the effects of variable loadings with those due to steady-state conditions so as to judge whether a particular fuse will be proof against deterioration in service. Whereas it is relatively simple to establish the short-time rating or the no-damage limit of a fuse corresponding to the shorter times, it is not always so simple to know the service loads in like terms unless a directly measured record of the load is available. Even so, a simple r.m.s. evaluation of load current with respect to time does not necessarily take into account the stresses resulting from transient peaks of short duration (Jacks, 1967).

The application of current-limiting fuses is covered by IEC Recommendation 282-1. In addition to the 'general purpose' duty of a fuse, this guide recognises an application area of growing importance, namely that of 'backup'. The latter duty is exemplified by the onerous nature of the protection required for h.v. motors, particularly against terminal box faults. In motor applications, the use of an associated current-interrupting device, e.g. a contactor, is standard practice up to ratings comparable with the full-load current of the motor. The fuse must, therefore, be able to interrupt satisfactorily all fault current beyond this level up to the maximum fault level of the system with minimum values of cutoff current and let-through energy, but it must not operate under the repetitive surge currents due to motor starting. The characteristics of general purpose fuses are no longer able to meet this specification economically, and this has led to the development of two distinct types of h.r.c. fuse for general purpose and backup applications, respectively (Feenan, 1971).

Some of the parameters available to the fuse designer to achieve the required operating characteristics are: variation of the element geometry with, for example, fragile elements wound on a central ceramic core for general purpose applications and fewer self-supporting elements of a larger cross-section and incorporating expansion bends for the backup application, variation of the element cross-section by tapering or necking to create selective multiple interruptions with localised pressure areas and/or cool zones, use of different amounts of filler and perhaps fillers of different properties.

Voltage is a vital parameter because the fuse is a voltage sensitive device, as indeed are most circuit interrupters. Voltage rating is chosen as that at which a fuse will interrupt safely all currents within its breaking-capacity rating within specified limits of transient behaviour of the circuit and with a margin for contingency. The minimum voltage at which a given fuse may be applied in a given system may be prescribed by the arc voltage which the system will stand. The maximum voltage at which the same fuse may be applied may also be decided by the maximum fault energy let-through which the system will permit; thus the voltage ratings chosen may be well within the inherent voltage capability of the fuse where no such limits are imposed. Arc voltage as such may be varied according to the method of arc control employed, and is a function of design; it may be controlled by variation of basically the same design parameters enumerated above, and the result is inevitably a compromise.

Load ratings of up to 450A in air and 140A in oil are available from single fuses up to 6·6kV and 11kV, respectively. Recent designs have concentrated on improved performance and protective characteristics, but trends towards the use of glass-fibre bodies would permit the generation of higher-quenching pressures to further extend ratings.

9.2.2 Expulsion fuses

The application of expulsion fuses is covered by IEC Recommendation 282-2. These fuses achieve current interruption by mechanical separation of their contacts due to the generation of arc-product pressure, and they may also be spring-assisted. Expulsion fuses are mostly applied out of doors, e.g. for rural distribution incorporating a dropout and/or display facility. However, they may also be made virtually nonventing, and thus suitable for indoor or in-cabinet operation, by the addition of a filtering device. Expulsion fuses built with a boric acid lined interrupting chamber evolve water vapour which serves to deionise the arc, and, in some indoor applications, this vapour is condensed and vented via a high-thermal-conductivity device to provide a display facility (Cameron and Smith, 1971).

There is almost no upper voltage limit for the design of expulsion fuses, and they find application up to at least 138 kV. In addition, because of their low resistance, they may be used for high load current levels, e.g. 200A in single units. The operating characteristic of these fuses is not as inverse as that of the h.r.c. fuse, and interruption does not take place until at least several cycles after the fault occurs. Therefore, although expulsion fuses are very economical, they are not as widely used as h.r.c. fuses because of ever increasing system co-ordination requirements.

9.2.3 Explosive fuses

The load-carrying capacity of the conventional current-limiting fuse requires an inherent thermal inertia corresponding to a minimum level of cutoff current which may be unacceptably high for large rates of rise of current, i.e. high prospective fault current. For such applications, there has been a development by the German firm Calor-Emag of a high-power triggered fuse the operating action of which is assisted by an explosive charge. This equipment has been called the 'Is-limiter' (Brueckner, 1959 and 1960; Keders, 1976).

Construction of the Is-limiter is shown in Fig. 9.1*a*, and Fig. 9.1*b* is a diagrammatic section to show the main components before and after operation. The post insulators are mounted on a base plate and carry the pole heads with contact blocks to which the interchangeable inserts are fitted. The pole heads are provided with cooling ribs and one of the post insulators (or both for higher-voltage ratings) contains an encapsulated pulse transformer connecting the tripping device to the detonator cap. The transformer also prevents direct electrical contact between the tripping network and the detonator cap which is at system voltage.

Through current is metered by specially designed current transformers located in series with the limiter, and for predetermined values of *di/dt*, semiconductor

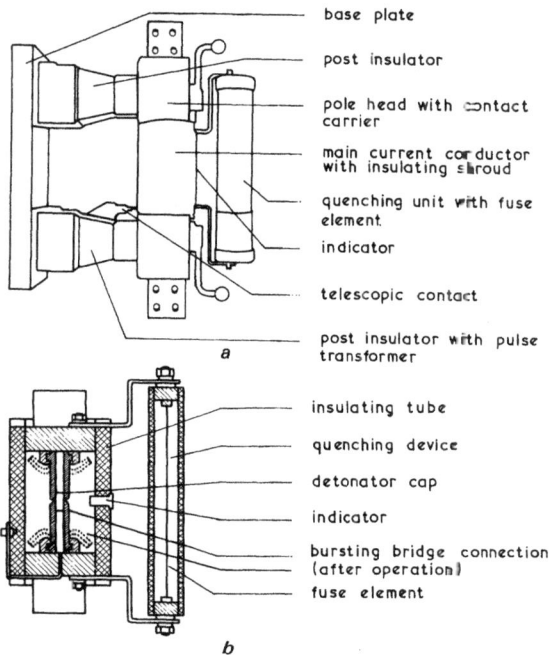

base plate

post insulator

pole head with contact carrier

main current conductor with insulating shroud

quenching unit with fuse element

indicator

telescopic contact

post insulator with pulse transformer

a

insulating tube

quenching device

detonator cap

indicator

bursting bridge connection (after operation)

fuse element

b

Fig. 9.1 Is-limiter

(*a*) construction
(*b*) operating principle

circuitry in a separate cubicle provides a tripping signal to trigger the charge of a power capacitor as a current pulse which explodes the detonator. The main current path is formed by the so-called 'bursting bridge', a short tubular conductor which is surrounded by a pressure-resistant insulating shroud containing the detonating cap. The bursting of the bridge is aided by four slots in an axial direction and a soldered joint at its midsection. On ignition of the detonator cap, the bridge opens after about 0·1 ms, the bars of the bridge being deflected as shown. The main current is then diverted to a parallel self-quenching fuse element of the conventional current-limiting h.c.r. type where it is interrupted. The fuse wire of this element is very lightly dimensioned so that the fault current is limited after about a further 0·5 ms.

Standard ratings of Is-limiter are available and have been installed in single units for system voltages up to 12 kV with current ratings up to 3000 A. For system voltages above 12 kV the inserts have a number of bursting chambers connected in series with one common parallel fuse element assembly; e.g. for 36 kV, 2000 A duty three chambers are arranged in series. Thus the device finds application in the power range of about 25-120 MVA (3 phase) where continuous supply is not essential.

9.3 Current-variation systems

9.3.1 Scope for applications

The density of power demand is continually increasing, and so therefore is the required firm supply at all voltage levels. This affects existing equipment, but even new circuit-breaker ratings cannot be increased indefinitely without severe economic penalties. Conventional solutions to the problem of increasing short-circuit levels include operating restrictions, complex sequential tripping schemes, segregation of supplies and the superposition of a higher-voltage system. The latter two methods are expensive and utilise rights-of-way which are becoming scarce.

At public-utility-system level, as well as the problem of increasing short-circuit ratings, there is a growing awareness of the need to contain system disturbances to prevent their escalation from causing a complete system shutdown owing to a cascading effect.

In industrial applications, there are increasing economic advantages for private generation schemes, but a practical limitation is often the short-circuit rating of existing switchgear, which may not be conveniently uprated or replaced, and/or the feedback of fault current to the local supply network. All of these requirements may be fulfilled by some form of automatic coupling, between system blocks, which presents minimum impedance to load current but which limits the through fault current. Since the mid 1960s, increasing attention has been given to the development of equipment to this end and it is known by such names as short-circuit limiter (s.l.c.) current-limiting device (c.l.d.), fault-current limiter (f.c.l.) etc.

Consideration has been given by a number of authors (Falcone *et al.*, 1974, Barkan, 1974 and 1980) to criteria for the performance of such equipment in order to define desirable characteristics and, in due course, to produce standard specifications. The following characteristics are agreed to be most desirable:

(i)   The first peak of fault current should be limited in order to restrict electromagnetic force on system components, particularly transformer windings and busbars. In addition the subsequent current should be reduced to predetermined values at the time of fault interruption. The device itself need not interrupt currents. The degree of current limitation required profoundly affects the design of the installation; e.g. high levels of limitation must be achieved by very rapid action which precludes electromechanical or thermally activated devices.

(ii) A device which presents a resistive impedance during operation is generally preferred because of the improvement in fault current power factor. However, in the normal state it should give rise to no more energy losses than, say, a normal circuit breaker:

(iii) The device should not give rise to unacceptable transient phenomena or harmonics. In principle, the operation of system protection should be unaffected.

(iv) The cost of the installation must be low enough to demonstrate the economic worth of the application. Fig. 9.2 indicates the potential market for fault-current limiters in the USA based on the results of a 1976 survey of Utilities (Barkan, 1980). This indicates that only a small multiple of the cost of a conventional circuit breaker is acceptable.

(v)   The installation should not occupy significantly more space than, say, a transformer of similar load rating.

(vi) The device should be at least as reliable as a conventional circuit breaker. This is a severe requirement in view of the additional components involved and one which is difficult to meet for all equipments which depend on sensing fault current in order to initiate control or switching actions. Moreover its mode of failure should not put at risk other system components which may be underrated due to its application.

Items (iv), (v) and (vi) together with the low-loss requirement of (ii) are the conflicting restraints of every engineering design and, in practice, trade-offs are necessary to meet particular circumstances.

The matter of reliability (vi) is extremely important and the need for confirmation of this feature will undoubtedly delay the introduction of many fault-current limiting devices for regular system use. However, an interesting parallel has been drawn with the application of surge diverters for system over-voltage protection. Following early difficulties these have gained acceptance and their use is now common for insulation co-ordination in system design.

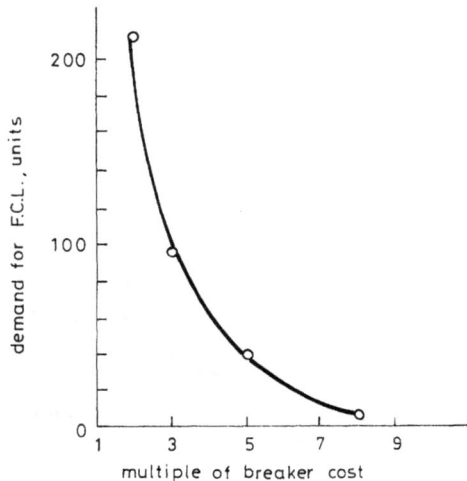

Fig. 9.2   Relationship between f.c.l. cost and utility acceptance (based on data from AEIC Survey, 1976)

A considerable variety of methods of fault-current limitation has been considered to date and four devices, representative of the range, which have been constructed are described in the following Sections. An important difference in principle between these is whether the change of circuit impedance is inherent or controlled in response to the operation of a sensing element.

9.3.2 Resonance links

The first published reference (Kalkner, 1966) to the use of an *LC* series resonance circuit for the restriction of fault current described the original work done by the late Dr. B. Kalkner of AEG Telefunken. Since then development has continued in Germany, and, on slightly different lines, by the GEC in the UK, the latter resulting in several installations under the name of s.l.c. — 'short-circuit limiting coupling) (John, 1968; McConnach, 1969; and Thanawala and Young, 1970). In its most simple form, as advocated by Kalkner, the link consists of identical circuits per phase of the schematic form shown in Fig. 9.3*a*. The reactance drop of a linear series reactor is eliminated, apart from that due to a small predetermined reactance which may be desirable for load-sharing purposes, by a series capacitor which is shunted by a nonlinear bypass circuit inoperative under normal-load conditions. When a fault occurs, the function of the bypass circuit is first to disrupt the series resonance condition, thus causing the series reactance to limit the through current and, secondly to restore the load current when the fault is removed from the system.

Alternative forms of bypass circuit were explored including triggered spark gap/ circuit breakers, transductors and nonlinear resistors, but the requirement for immediate and automatic action without fail was considered to preclude the use of

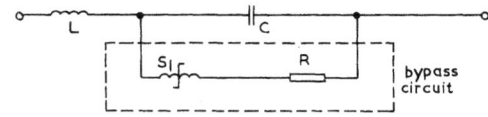

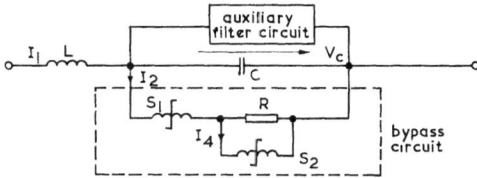

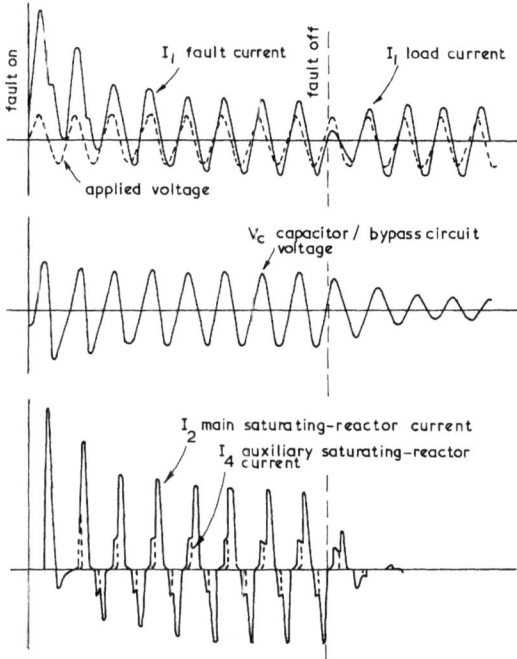

Fig. 9.3 Resonance links

    (a) schematic of basic type
    (b) schematic of double saturated reactor SLC type
    (c) typical performance of s.l.c.

any mechanical devices. The only fully successful solution was a self-saturating reactor with a sharp 'knee point' in the magnetic characteristic at a voltage just above its maximum operating voltage level which gave a reliable and automatic operation without the need for a control system. A sufficiently sharp knee point can be achieved by using ordinary transformer steels and normal winding techniques for the saturating reactors. A benefit of the saturating reactor auxiliary circuit is that it can be made to resonate partially with the series capacitor thereby contributing to the overall link impedance in its fault mode. The resistor shown in Fig. 9.3a is necessary to prevent self-excited oscillations and also to damp the currents in the closed circuit formed by the capacitor and the bypass circuit to enable postfault load operation to be quickly established.

The normal power loss in the damping resistor is negligible, since it carries only the magnetising current of the saturating reactor. The optimum value of damping resistance at fault onset is different from that required to damp the circulatory current. If chosen solely for the latter purpose, the losses and hence the resistor rating could be prohibitively high. This effect has been eliminated in the s.l.c. version of the auxiliary circuit (Fig. 9.3b), in which an auxiliary saturating reactor is connected in parallel with the resistor. Under fault conditions, this reactor saturates before the main saturating reactor, and effectively by-passes the resistor to a designed extent. When the fault is removed, the combined resistor-reactor losses are sufficient for the initial current damping, then the auxiliary reactor desaturates allowing the damping resistor to have its full designed effect in causing the main saturating reactor to desaturate rapidly.

A further feature of the bypass circuit resistor is that it can be designed to alter the phase of the fault-current contribution through the link which aids current interruption. Normal fault-current contributions are of low power factor, say 0·1 or 0·2, but that via the s.l.c. can be as high as 0·4 or 0·5, and this can give the unusual effect that the contribution of the link to the system fault level is less than its load rating.

The proportions of the bypass circuit components depend critically on the maximum through current for which operation is to be inhibited and the maximum fault current (and its duration) when operating. Standard available ratings of capacitors allow overvoltages of up to 2·7 times normal for 1 cycle falling to 1·6 times normal for 6 cycles. Design studies have enabled the voltages occurring across the link in operation to be matched to this characteristic, thus avoiding a special capacitor specification. In recent s.l.c. designs, advantage has been taken of capacitors using the plastic film-paper sandwich type of dielectric which enables 200–300 kVAr to be provided in one can which was previously rated at 100 kVAr and also reduces the load losses in similar proportions.

Fig. 9.3b also shows an auxiliary selective damping filter circuit the purpose of which is to eliminate the undesirable subsynchronous self excitations of large synchronous or induction machines which can be a feature of some series capacitor installations. (I.E.E.E. S.S.R. Task Force, 1976).

Fig. 9.3c shows the typical performance of an s.l.c. under test conditions. From this, it is seen that fault current is limited to about twice the peak load current even with the d.c. component.

After fault clearance, recovery to normal load current occurs in about three half cycles. Finally it is seen that, although the bypass circuit current is severely nonsinusoidal, the low impedance of the main capacitor effectively restricts harmonics to within the link circuit itself.

Several installations exist in industrial systems at voltages up to 11 kV and for normal load throughput ratings of up to 14 MVA and fault current limitation to about twice that value in situations where the prospective fault current corresponded to 500 MVA.

There are many advantages in the coupling of sections or blocks of a grid system by these links if fault levels are tending to exceed those of available switchgear. Costs have been quoted of the order of two to four times that of a transformer of equivalent load capacity. This may be compared with a cost for an asynchronous d.c. interconnection of the same load rating whose cost might be that of about 10—15 transformers, of equivalent capacity. The d.c. interconnection of course provides many control and transmission right-of-way utilisation features in addition to its inherent ability to reduce fault contributions.

Fig. 9.4 shows the installation of a 6.6kV 10 MVA s.l.c. in a UK oil refinery. The linear reactors in the foreground are the normal aircored type. The small tanks contain the reactors and resistors of the auxiliary filter circuits, and the large tanks

Fig. 9.4 6·6 kV 10 MVA s.l.c. installation at a UK oil refinery [Courtesy GEC]

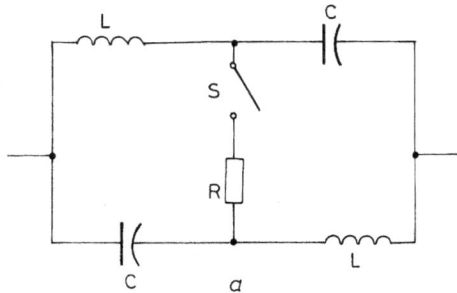

system rating 145kV 300MVA
source impedance: 0.062 + 1.86Ω
FCL reactance = 14Ω, damping R=7Ω
fault closing at 0

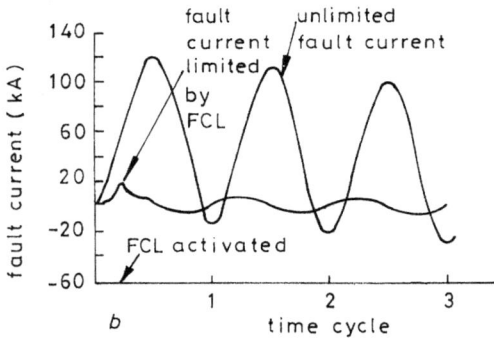

Fig. 9.5   Tuned-impedance current limiter
      (*a*)   circuit diagram
      (*b*)   fault currents

contain the main and auxiliary saturating reactors and the damping resistors, all natural oil cooled. The capacitors, mounted above the main tanks, are constructed and protected according to normal series capacitor bank practice with special attention to the rating for harmonics. The layout is quite flexible and capable of variation to suit the site available, e.g. in height etc. Typical areas required are of the order of 9 m² /MVA of load rating.

The resonance link has been shown to be a practical and economically justifiable installation, especially in the circumstances of growing industrial generation where fault levels and/or feedback to the local supply system cannot be contained by other means. The components are not new, but the design process requires extensive analytical studies, and each installation is virtually tailor-made.

Replacing the saturated reactor of the Kalkner tuned circuit by a fast-acting switch, which operates on the detection of fault current, allows greater freedom for the selection of component ratings and therefore leads to the possibility of lower costs – at the complexity of introducing a switching element.

Fig. 9.5a shows one of several early circuits which have been studied to explore these possibilities (Tahiliani, 1980). The switch S comprises a spark gap triggered by the potential across one of the capacitors. The L and C components are tuned to essentially zero impedance at standard system frequency and are chosen, with the resistance R, according to the required current limiting performance. With S open the device presents zero reactive impedance, but some losses, to through current. With S closed the circuit is parallel-tuned giving greatly increased impedance of resistive appearance. Fig. 9.5b gives the calculated performance of the circuit for the parameters shown. It is apparent that considerable limitation of the first peak of fault current can be achieved in principle. However, more detailed studies (EPRI, 1978) showed that the transient stresses on the components, particularly in the capacitors, resulted in increased component costs which narrowed the range over which the device was economically more attractive than the SLC. Typical figures quoted are costs equivalent to about two to three transformers and losses of 0.1% of normal rating. In consequence of the loss-cost penalty the ideal application of this type of limiter is considered to be at a tie position where normal current would be extremely low.

9.3.3 Controlled transformer-type interconnectors

With future high-voltage power-system applications in view, engineers of the Japanese Hitachi Co. have investigated some fault-current-limiting interconnectors with additional control features. Analytical and model studies have been carried out and 66 kV prototypes have been constructed (Hirayama *et al.*, 1971; Matsuda *et al.*, 1971).

The principle of one of these requirements is, with reference to the schematic diagram of Fig. 9.6a, as follows:

The major component is a series transformer with a leakage impedance of the necessary magnitude for fault-current limitation. The voltage drop across the series transformer due to normal power flow is compensated by a regulating transformer via a secondary winding, which causes reduction of the apparent impedance between the adjacent system blocks. A linear reactor can be inserted in series with the series transformer to increase further the interconnector impedance if preventive control of spread of disturbance is required.

A line diagram of the prototype voltage-compensation equipment is shown in Fig. 9.6b. The total effective area, allowing for a boundary fence, is about 400m^2, i.e. 10 m^2/MVA of load rating. The equipment was designed as single-phase units with SF$_6$ gas-insulated connections and SF$_6$ tap-change switches. The connections between the transformers and the reactors are oil-immersed via ducts, thereby leading to a compact arrangement.

Other types of interconnector investigated have different characteristics, and

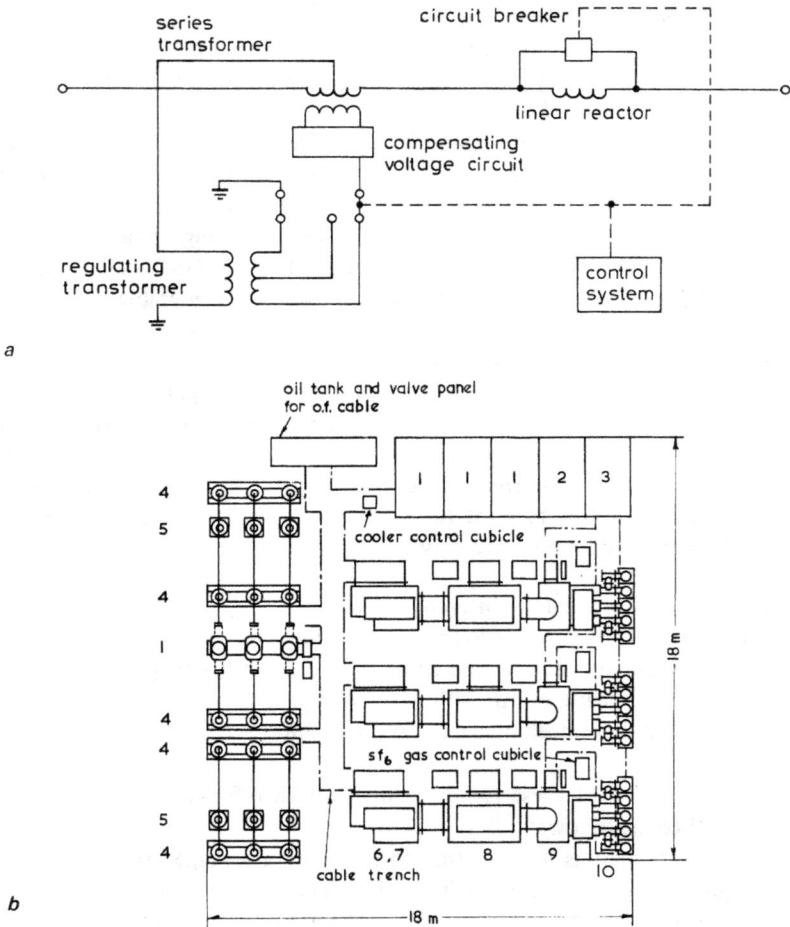

Fig. 9.6   66kV voltage-compensated system interconnection
(a)   schematic
(b)   arrangement

1 circuit breaker	6 tap-change reactor
2 potential transformer	7 linear reactor
3 lightning arrester	8 services transformer
4 cable head	9 earthing transformer
5 potential device	10 tap-change switch

only with the voltage-compensated equipment is the current-limiting feature independent of the control system. The control feature may be attractive for some system applications if the cost is not thereby increased significantly.

It is thus seen that the system interconnector carries out the same fault-current

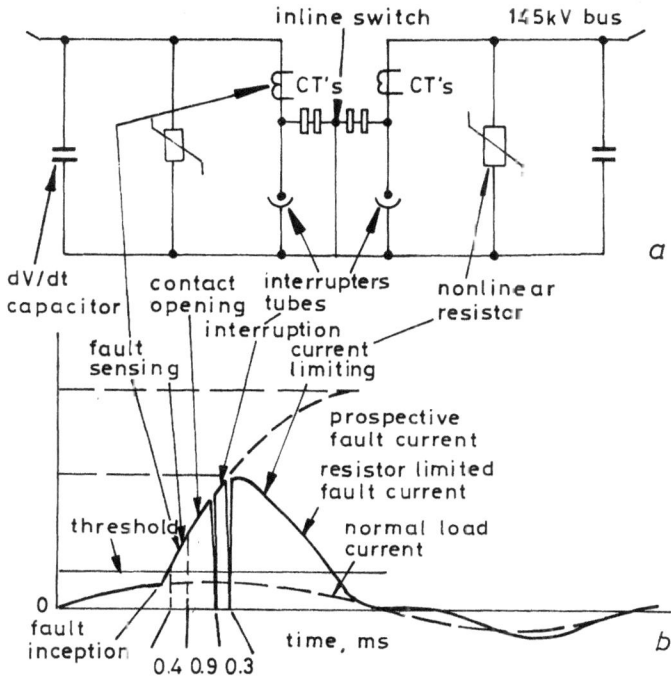

Fig. 9.7 Operating principle of the c.l.d.
(a) schematic
(b) normal load and fault currents

limiting functions as the resonance link, and provides some further features at the cost of a feedback-control system. Areas of the two devices are comparable. Costs are not available for the system interconnector, but it is reasonable to assume that the basic cost of the sensing and control equipment might require a minimum size of installation before it is justified. That little further application of this equipment has occurred indicates the difficulty of cost reduction for the design.

9.3.4 Switched-resistor current-limiting device (c.l.d.)

The operating principle of the c.l.d. is simply the ultra-fast detection of fault current to initiate the opening of a high-speed mechanical d.c. switch in order to divert or commutate the current into a nonlinear resistor. This limits the fault current which is then interrupted by magnetic means. The implementation of the principle calls for significant engineering research and development, and this, together with field tests to prove the system, is described in a number of references (Grzan *et al.*, 1978 and 1980; King *et al.*, 1980).

The schematic diagram of the main components and the operating principles of the c.l.d. are shown in Fig. 9.7. There are five essential elements:

(i)   A high-speed sensing and control unit detects and assesses faults in order to initiate current limiting action within 400µs if the fault current is expected to exceed a given peak value. The sensing system employs waveform comparison techniques and data transmission is by optical links. Particular care is necessary to avoid nuisance tripping due to electromagnetic interference.

(ii) A mechanical in-line switch, the characteristics of which include a contact opening speed in excess of 20 m/s, an arc-voltage rate of rise of about 1000 V/ms, and a dielectric recovery capability of 10 kV/µs after a dI/dt of 50 A/µs and a deionisation margin of 100 µs. These are stringent requirements and have been achieved by the use of an electrodynamic impulse motor drive to obtain the desired high contact opening speed and a fast-acting gas-blast system in combination with an arc chute for a high rate of arc voltage rise. The drive currents which reach 40 kA are provided from ignitron-switched capacitor banks.

(iii) Crossed-field interrupters are triggered to commutate rapidly the current out of the in-line switch. The current is held in the interrupters for 300 µs to give the in-line switch a chance to open fully and to deionise, after which time the interrupters cease to conduct and the current is then forced into the resistors. The interrupters, which were developed for h.v.d.c. switching applications, are basically cold-cathode glow-discharge devices using two concentric cylinders as electrodes. They operate at a gas pressure so low

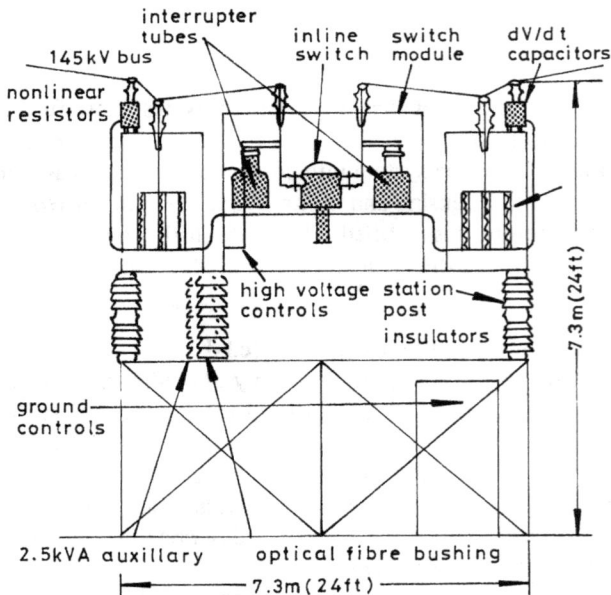

Fig. 9.8   Schematic assembly of one pole of a 145kV c.l.d.

that current can flow only when a magnetic field is applied. For this application the current-handling capability of the interrupter had to be increased from 2 kA to about 7 kA and it had to be made bipolar to withstand voltages of ± 100 kV.

(iv) A silicon-carbide limiting resistance must be rated to carry the limited fault current until a circuit breaker undertakes the final interruption. It has been calculated that for the prototype device 1·4 MJ is absorbed per phase during a 6·5 kA insertion spike and an average of 1·5 MJ per cycle until the breaker opens.

(v) A shunt capacitor restrains the rate of voltage rise during current interruption and prevents surges across the device after interruption.

Fig. 9.8 shows in a schematic assembly the major components comprising one phase of the 145 kV c.l.d. prototype and indicates its dimensions. In field tests on this prototype (Gallagher *et al.*, 1980) prospective symmetrical fault current of 5·2 kA peak was limited to 3·7 kA within 1·6ms from fault inception.

Additional development of the c.l.d. is necessary for higher ratings, and it may be some time before realistic costs and information on reliability become available.

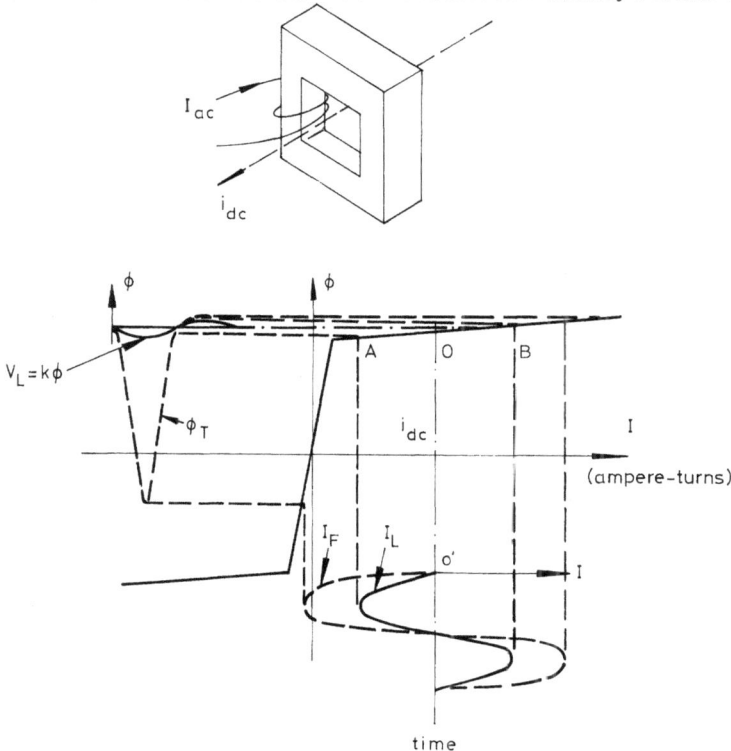

F g. 9.9 Principle of current limitation in a d.c. saturated iron core (Courtesy, K.C.Parton)

9.3.5 Superconducting short-circuit limiter

The possibility has always existed to control fault current by varying the impedance of a reactor as a function of the system current. In normal operation, high flux levels corresponding to saturation in a closed iron core could be set by a direct-current bias and modified as a result of increasing fault current. However, such a system using normal components would be uneconomic because of its size and the copper losses associated with the high d.c. levels required.

Experience with superconducting technology has enabled high d.c. bias levels to be maintained with virtually no copper losses although the cryogenic process itself requires a small energy input. Using this technology, one phase of a prototype superconducting short-circuit-limiting equipment has been developed by the NEI company based on work described by Parton (1978). Its load rating is 3 kV, 556A, corresponding to a 3-phase rating of 5 MVA and losses in normal operation are less than 0·1% of rating.

The operating principle is shown in Fig. 9.9 where normal a.c. operations with voltage V_L and current I_L is at a low fixed impedance corresponding to the saturated

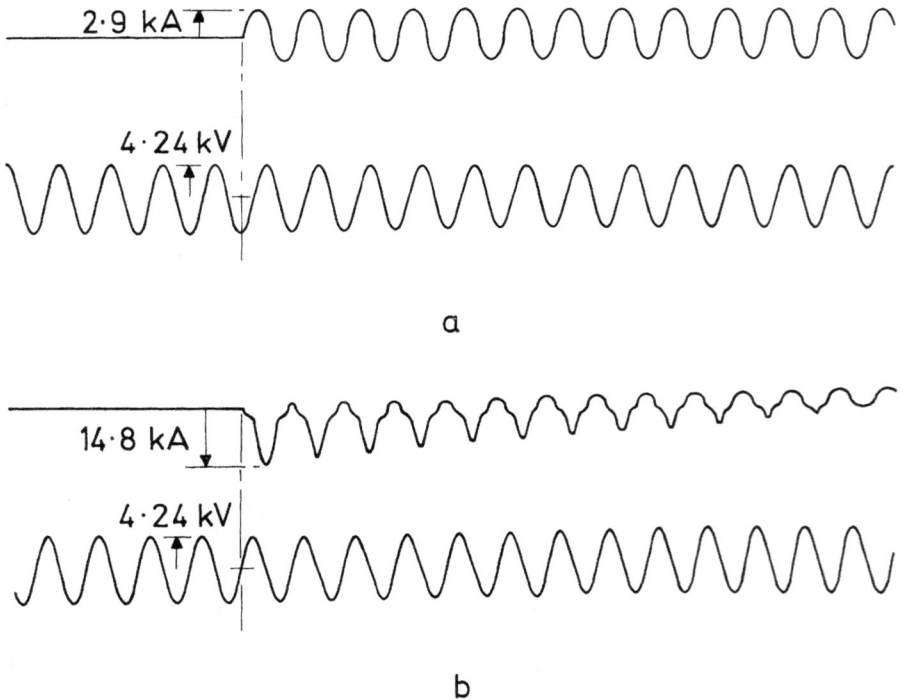

a

b

Fig. 9.10   Test records for prototype superconducting short-circuit limiter (Courtesy, NEI Parsons-Peebles)
(a)   fault at voltage peak
(b)   fault at voltage zero

slope AOB. On the occurrence of a fault the increasing alternating current tends to reduce the effective ampere-turns and hence the flux ϕ_T in the core, transferring operation during part of each cycle to the more steeply sloping unsaturated part of the characteristic. The greater 'back voltage' which results corresponds to a higher impedance, which in turn limits the first peak of fault current I_F by an amount determined by the physical arrangement of the core and coils.

Fig. 9.10 comprises oscillographic records of current and voltage during tests of the prototype which was designed to reduce prospective current peaks of 14 kA for faults occurring at voltage peak to less than four times rated load current peak, i.e. 3·15 kA. Fig. 9.10a shows that in the design case the first peak of fault current is limited to 2·9 kA and Fig. 9.10b shows that in the same circuit the first peak of asymmetric fault current, for faults which occur at voltage zero, is reduced to 14·8 kA from the prospective 28 kA. In normal applications the equipment would be designed to limit the most severe potential faults, i.e. at voltage zero, to required levels. An important feature is that current decrement is controlled by the design parameters of the equipment itself and is independent of the X/R ratio of the

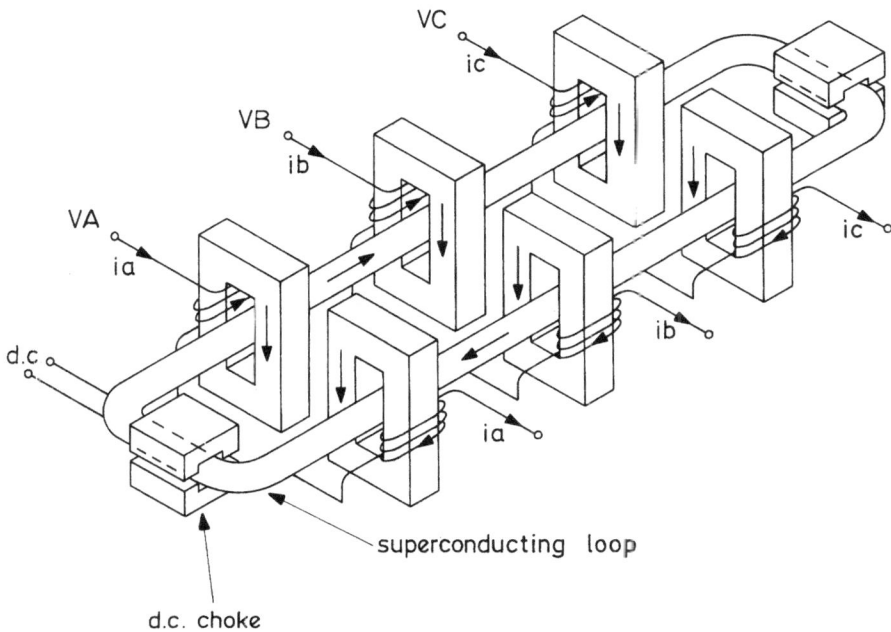

Fig. 9.11   Physical arrangement of superconducting short-circuit current limiter

circuit in which it is located. The inherent action of the superconducting limiter, its immediate and continuous response and after-fault recovery are well suited to those applications where system stability is an important consideration.

In order to provide for both positive and negative half-cycles of fault current, realisation of the principle requires two cores per phase threaded by the same bias current with windings of opposite polarity connected in series as shown in Fig. 9.11.

The technology of the device is based on conventional transformer techniques and proven superconducting developments. The installation incorporates a helium cycle via a compressor to supply the cryostat and a local helium storage tank to provide several days' standby capacity. Loss of superconductivity would lead to a fail-safe condition since removal of the d.c. bias causes reversion to the high impedance mode.

Early estimates of the cost of a 200 MVA superconducting limiter correspond to less than that of an equivalent double wound transformer, but costs are strongly dependent on scale and there is obviously a smallest size for which the cryogenic installation would tend to make the application uneconomic. Nevertheless, despite the present lack of application experience the superconducting short-circuit limiter must be considered among those systems which could contribute to solving the problem of fault-current limitation.

9.4 Synchronised switching

9.4.1 Short-circuit testing
There is a long experience of point-on-wave switching techniques, latterly called 'synchronised switching', in short-circuit testing-station service. It is necessary to control both the station master closing switch and the opening and closing of the circuit breaker under test to meet the various requirements of BS116 and IEC 56 as described in Chapter 10 and also to reduce the number of random tests which would otherwise be required to prove a circuit breaker.

Testing station closing switches are specially designed for such duties, early versions operating at 6–11 kV with current ratings of hundreds of kiloamperes, with contacts housed in compressed air to minimise pre-arcing and mechanically designed for short, consistent closing times. Even in older testing stations, timing control components have now been converted to solid state, using many of the techniques developed for synthetic testing.

The extreme duty required of modern testing-station breakers is such that it is desirable to exploit the advantages of synchronised switching to interrupt at the first current zero. The potential benefit of this is demonstrated in a recent paper (Kriechbaum, 1971) which shows that an arcing time of 2 ms corresponds to an arc energy of only about 15% of that of a symmetrical current loop or 6% of that of a total asymmetrical current loop. The same paper describes a generator test circuit breaker developed by the AEG Co. in which contact wear during interruption is greatly reduced by synchronised switching. The breaker has been tested to interrupt

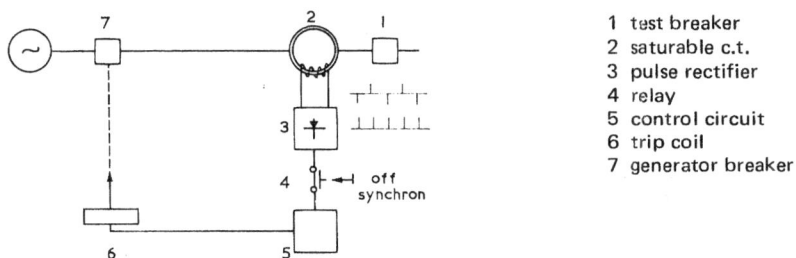

1 test breaker
2 saturable c.t.
3 pulse rectifier
4 relay
5 control circuit
6 trip coil
7 generator breaker

a

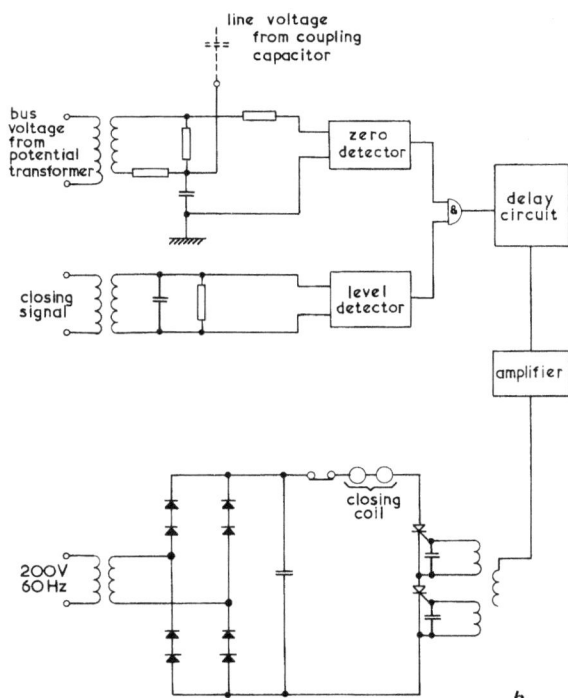

b

Fig. 9.12 Control circuits for synchronised operation

 (*a*) tripping circuit for AEG half-cycle testing station breaker
 (*b*) detecting circuit for synchronous closing of Fuji 550 kV circuit breaker

currents up to 170 kA, at recovery voltages of 12 kV, after one halfcycle with deviations in breaking time of less than 0·2 ms.

Features of the design of this breaker included: dimensioning of the nozzle cross-sections so that the arc itself throttles the blast at high instantaneous values of current thus reducing the arc energy, increase of the air pressure at a predetermined time ahead of current zero, storage of compressed air at 80 atm directly round about the nozzles with increase of the exhaust cross-sections behind the nozzles so that the greatest pressure drop occurs in the nozzles, shortening the conduits between the nozzles and the exhausts, a high relative speed of the contacts at the moment of separation so that dielectric recovery ensures withstand after only 1 ms, and finally the contacts themselves being designed as solid pressure contacts. The time elapsing from tripping impulse to contact separation is 6·8 ms, and appropriately earlier action is initiated so that the arc voltage develops to the necessary extinguishing value at current zero. Typical arcing times of less than 2·5 ms are achieved with the circuit breaker described. The tripping scheme (Fig. 9.12*a*) required to obtain the performance was developed from synthetic testing work. It utilises pulse signals from a saturated current transformer 2 at each current zero which are applied to the control circuit 5 to fire a thyristor which discharges a capacitor into the tripping coil 6. The armature of this coil directly activates the high-speed latch device of the generator breaker.

9.4.2 System applications

A circuit breaker rated at 3000 A, 550 kV, 35 000 MVA has been developed by the Fuji Electric Co. of Japan (Nitta and Kiyokuni, 1970; *Electrical World*, 1971), which employs synchronous detection techniques (Fig. 9.12*b*). Voltage signals from the busbar side and the line side are compared in a zero detector after the former have been phase shifted by a predetermined amount depending on the line length. A pulse is generated when the resultant voltage is zero. When the closing control voltage is applied to the circuit breaker, this pulse is fed to the following delay circuit through an AND circuit. The delay circuit is used to control the excitation time of the closing coil of the breaker, to close the contacts when the voltage becomes zero. Since the circuit-breaker closing time varies with temperature and pressure changes, this delay circuit is used to compensate for these changes as it also does for system-frequency variations. This circuit breaker also achieves one cycle interruption, light guides being used to transmit the tripping signal from ground level to photoelectric detectors at the operating heads. The time from tripping impulse to contact separation is of the order of 100 μs.

It is important to note that, to allow for failure of the tripping-circuit components, the design of the testing circuit breaker (Section 9.4.1) is such that its rated breaking performance can be achieved without the assistance of synchronised switching. It is difficult to see how this philosophy could be set aside for general system circuit-breaking duties.

It is therefore postulated that, for the foreseeable future, synchronised opening may be adopted where it will reduce contact wear in onerous situations (Beehler, 1973) or where it is necessary in terms of fast fault clearance to maintain system stability margins but not to enable circuit breakers to be applied where potential fault levels exceed their design ratings.

The case for synchronised closing is different. The economics of ultra-high-voltage systems depend significantly on their required impulse levels, which in turn are particularly affected by the overvoltages generated during transmission-line energisation (Chap. 3). For instance, it has been suggested that transmission voltages in excess of 1000 kV require switching surge overvoltages to be limited to no more than 1·7-1·8 times rated voltage while factors of less than 1·5 have been discussed for 1500 kV transmission. Thus synchronised closing applications must be expected to increase as indicated in Russian investigations for 750 kV (Beliakov, 1978).

When a circuit breaker is equipped for synchronised switching, this may be used either to be the sole means of controlling overvoltages as in the Fuji Electric breaker, or in conjunction with closing resistors to reduce the duty on the latter as in a 550 kV breaker developed by the American Westinghouse Co. (Yeckley *et al.*, 1971).

The adoption of synchronised switching will thus reduce to an acceptable level the number of occasions on which surge diverters would be required to operate. Meanwhile the presence of the diverters guards against the failure of the synchronised system due to auxiliary components or for any other reason that still allows the circuit breaker to close normally.

9.5 H.V.D.C. switching systems

9.5.1 The need for switchgear
The majority of h.v.d.c. schemes now in operation or under construction are point-to-point transmission schemes in which powers of up to 6000 MW are transmitted at voltages of up to ± 750 kV (Rumpf and Jarret, 1980). H.V.D.C. circuit breakers are not needed for this application since the control of the convertors of such a scheme permits rapid blocking and unblocking to enable transient line faults to be cleared and, in conjunction with the fast-acting isolators, permanent faults can also be dealt with. Ultimate protection is provided by circuit breakers on the a.c. side of each terminal.

An exception to the point-to-point scheme is the 460 MW, ± 266 kV Kingsnorth three terminal scheme (Calverley *et al.*, 1968) which incorporates line isolators for no-load switching only. However, this is not a true multi-terminal scheme and extensions beyond this number of terminals would involve unacceptable complexities in control and/or operating conditions (Ekstrom and Liss, 1972). Multiterminal schemes, which could be designed if appropriate d.c. switchgear were available, have advantages over 'nonswitchgear' schemes (Eisenak *et al.*, 1972: Kangiesser *et al.*, 1972: Long and Lian, 1974: Reeve, 1980). The most important of these advantages are summarised as: much reduced terminal station costs (up to half), lower losses and greater possibility to utilise the inherent overload capacity of the lines since this is not now inhibited by the terminal equipment ratings. Such circuit breakers should be capable of switching normal load currents and, with assistance from the control system, of clearing line faults.

There is thus a real case for the development of h.v.d.c. circuit breakers, and it is not unfair to say that h.v.d.c. transmission has so far developed in spite of the fact that no suitable circuit breaker has been available.

Consideration has also been given (Bowles *et al.*, 1976) to using an uncontrolled rectifier based on diodes at the sending end of transmission from remote generating plants. Control would be achieved by the thyristor-based invertor at the receiving end and the whole scheme would be less complex and probably less costly than a full h.v.d.c. system. However, protection of the line would require a sending-end d.c. circuit breaker capable of interrupting, unaided, full potential fault current — a severe duty which is not considered further here.

9.5.2 Circuit-breaker duty

As in the a.c. case the duties of a d.c. circuit breaker, according to its location, are: to connect and disconnect transmission lines and/or cables, to switch load, and to make and interrupt near or remote faults. There is also one additional duty, namely to connect and disconnect the large filter banks associated with each terminal. However, the major problems are those associated with the switching of load current and fault current.

Because of the time taken to detect, transmit and evaluate fault information and for the circuit breaker to act on this, short-circuit currents will probably have risen to values unacceptable to other equipment unless the system control scheme intervenes to limit current. Such action would assist the operation under fault conditions of the load switches which will be required for multiterminal installations and the development of h.v.d.c. circuit breakers with interrupting capability beyond $1.25 \times$ full-load current is considered unlikely to be cost-effective (Kangiesser *et al.*, 1972). Ratings for the foreseeable future will involve levels of load current up to 2500 A as standard and possibly 5000 A if short-term line overload capabilities are utilised. System voltage levels are now up to ± 750 kV d.c. Unlike a.c. operations, the voltage appearing across a d.c. breaker during and immediately after interruption is largely determined by the breaker because the breaker must itself bring the current to zero. For reasons of insulation co-ordination it is considered that this breaker-generated voltage should not exceed 0·6 times the rated direct voltage.

The smoothing inductance at terminal stations and the transmission-line inductances represent very large amounts of stored energy, of the order of many megajoules which must be dissipated by the breaker or by some other means of interruption.

Other less onerous switching duties associated with h.v.d.c. installations have been investigated (Kangiesser, 1972; Bowles, 1976; Housler, 1980) and relatively simple means found for their solution.

9.5.3 Methods of interruption

Many theoretical studies have been carried out to examine possibilities for the interruption of large direct currents (Electra, 1971). However, only methods using the commutation principle seem to be practical. The 'commutation principle' describes a process whereby artificial current zeros are produced by superimposed oscillatory current and there is a considerable body of experience of the techniques required for this purpose in connection with synthetic testing of a.c. circuit breakers.

A current zero may be achieved very simply by a circuit such as that shown in Fig. 9.13a. When it is required to open the circuit breaker the switch S is first closed causing the precharged capacitor C to discharge through the breaker. This causes the line current to be commutated through the capacitor, opposing the current through the breaker and enabling that current to be interrupted. A relatively small capacitor will satisfy the impedance requirements, e.g. typically 0·5 μF to interrupt 5000 A at 100 kV the first current zero being created within about 5 μs as shown by Greenwood and Lee (1972).

In practice, however, there are two serious problems to be considered. First, this value of capacitance is too low by many orders to absorb the energy which was stored in the system inductance without serious overvoltage. Secondly the rate of change of current represented by 5000 A in 5 μs is about 70 times that experienced by an a.c. breaker when interrupting a symmetrical 60 Hz current of 40 000 A. Moreover, the rate of rise of recovery voltage is correspondingly high. These are compelling reasons why the circuit must be modified to reduce di/dt before current zero and dv/dt afterwards.

A development of the American General Electric Co. and described in a paper (Greenwood and Lee, 1972) meets these requirements and is shown in Fig. 9.13b.

The self-saturated reactor L_4 has low inductance under load conditions but this rises on commutation as the circulatory current opposes the line current. In consequence, the rate at which the current declines is sharply reduced. Following interruption the rate of rise of recovery voltage is limited by the large capacitance C_1 which must be charged through the unsaturated value of L_4. Fig. 9.13c shows the relative performance of the simple and modified circuits. In the latter it is seen that a period of low voltage occurs and, by design, this may be 20-30 μs, which is sufficient for high-speed vacuum-breaker interruption.

A further modification adds a shunt surge suppressor, comprising nonlinear resistors connected by a triggered vacuum gap, to reduce the rating required of the main components by absorbing the higher surge energies.

9.5.4 H.V.D.C. vacuum circuit breakers

Using the above principles a number of experimental circuit breakers have been

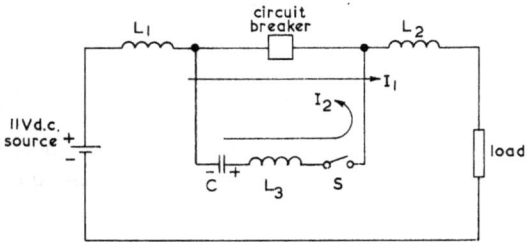

a

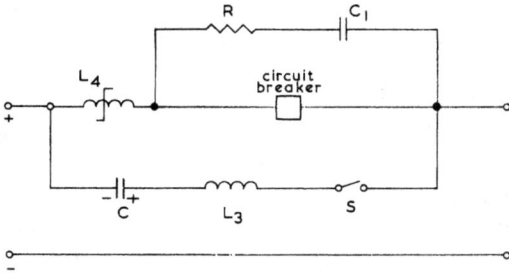

b

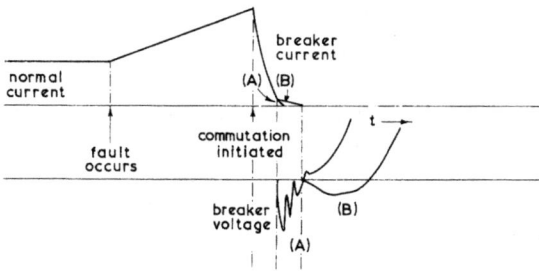

c

Fig. 9.13 d.c. interruption by the commutation process

    (*a*) basic circuit
    (*b*) modified circuit
    (*c*) performance of basic (A) and modified (B) circuits

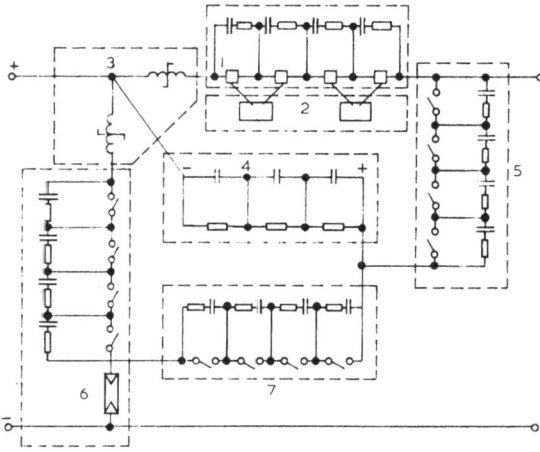

a

b

Fig. 9.14 An experimental h.v.d.c. vacuum circuit breaker

(*a*) circuit diagram for the 80 kV experimental h.v.d.c. circuit breaker

1 vacuum interrupters
2 mechanism
3 saturable reactors
4 commutating capacitors

5 commutating t.v.g.
6 surge suppressor
7 surge suppressor turn-off circuit

(*b*) 80 kV experimental h.v.d.c. vacuum circuit breaker (identification the same as (*a*)) [Courtesy GE (USA)]

constructed (Greenwood *et al.*, 1972) including an 80 kV, 5 kA 4-break version shown schematically in Fig. 9.14*a* and illustrated in Fig. 9.14*b*. The symbols in these Figures refer to the same components.

A particular feature of this equipment is the high-pressure hydraulic system which enables contact velocities of 9m/s to be attained. The vacuum circuit breaker requires a critical contact separation of only 5-15 mm to achieve full dielectric integrity, but, since extremely high accelerations (greater than 500 g) are involved, the mechanism must be of minimum effective mass. Since stress waves in common structural insulating materials propagate at speeds of the order of 3000 m/s it is evident that even a 300 cm insulated rod coupling the contacts and the power source would result in a minimum delay of 1 ms. Exceptionally rapid response is achieved in this design by locating the hydraulic prime mover at the contacts with solenoid operated valves controlled over light guides. During the experiments, 5 kA was interrupted with a system inductance of 60 mH and, by the action of the surge suppressors, the recovery voltage was restricted to 40 kV.

Later work (Anderson, 1978) using linear reactors and a standard vacuum interruptor unit rated at 45 kV improved this performance and demonstrated by synthetic testing methods the interruption of 8 kA against recovery voltages of more than 100 kV.

9.5.5 High-speed oil-cooled h.v.d.c. circuit breaker

Early work by Kind *et al.* (1968) includes descriptions of tests of German experimental low-oil-volume devices, to demonstrate two principles. These were; first, that forced oil cooling with very high relative speeds between oil and arc could generate the required arc voltages for interruption of large direct currents and, secondly, that series connection of switching elements could be used, as in a.c. interruption, to increase the voltage rating. These principles were extended successfully in later collaborative work by the AEG, ASEA, BBC and Siemens Companies (Ekstrom *et al.*, 1976). In this work a standard minimum oil circuit-breaker unit rated at 72·5 kV, 3150 A was used as the central element of a system which employed another variation of commutation principle as shown in Fig. 9.15.

In the closed position the isolating switch IS and the commutating switch CS are closed. Upon fault detection CS opens creating an arc voltage of 15-20 kV which causes the gap G_c to spark over connecting the uncharged capacitance C in parallel with the commutation switch. The inrush of current to C is in the opposite direction to that in CS and the arc path in CS is deionised. Thus the fault current i is commutated through the capacitance C, the voltage across which increases rapidly at several kV/μs. This voltage rise in turn causes sparkover of the specially designed d.c. surge arrestor A which gradually reduces i to zero while absorbing the energy of the interruption process. The surge arrestor reseals against a d.c. voltage of about two-thirds of its sparkover volage, thus eliminating the need for a residual current breaker.

Fig. 9.16 shows the interruption process schematically. In synthetic tests of a

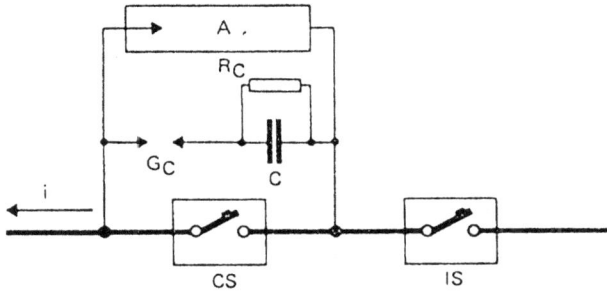

Fig. 9.15 Principal diagram of h.v.d.c. circuit breaker
    CS = commutation switch
    IS = fast acting isolation switch
    C = ccommutation capacitor
    G_C = spark gap
    R_C = discharge resistor
    A = h.v.d.c. surge arrester (energy absorber)
    i = current to be interrupted

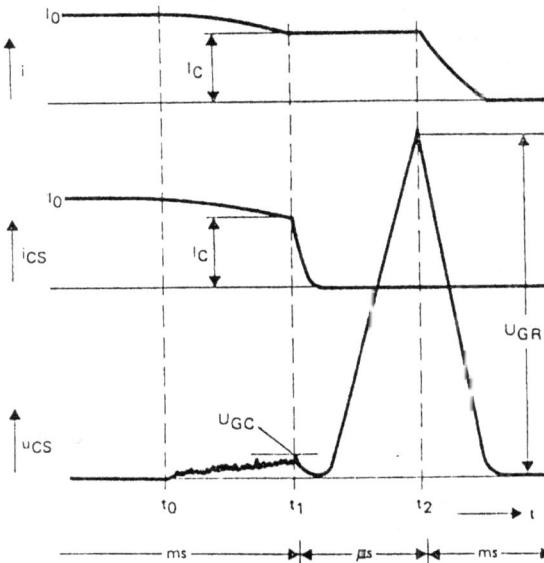

Fig. 9.16 Schematic oscillogram of a commutation of the current I_C into the energy absorbing resistor R
    i = total current
    I_0 = total current at the moment of contact sepa-ation t_0 of the commutation switch
    I_C = total current at the moment of the ir itiation of the commutation process (t_1)
    i_{CS} = current through the commutation switch
    u_{CS} = voltage across the commutation swi+ch
    U_{GC}= sparking voltage of gap G_C
    U_{GR}= spark ng voltage of gap G_R
    t_0 = moment of contact separation of the commut=tion switch
    t_1 = mom∋nt of sparking of gap G_C
    t_2 = mom∋nt of sparking of gap G_R

unit designed for 200 kV system voltage, currents up to 2500 A were successfully switched in circuits incorporating inductances of up to 1·9H. These tests were considered to prove that suitable h.v.d.c. circuit breakers based on this principle can be successfully developed for multiterminal systems.

9.6 Semiconductor switching

9.6.1 Static circuit breakers

Electronic applications have always involved the effective switching of both alternating and direct currents, albeit at milliamp levels in most cases. The introduction of semiconductor rectifiers and then thyristors (controlled rectifiers) in the late 1950s had seen power electronic applications grow from initial low ratings to standard equipments for up to about 5 kV and thousands of amperes by the late 1960s. At that time single thyristor cells rated at about 1·25 kV (peak inverse voltage) and 150 A (mean forward current) were in standard use for industrial control schemes, and thyristor 'strings' i.e. series/parallel arrangements of a relatively small number of these cells were employed to obtain convertor units of the required current and voltage ratings for industrial and traction applications.

A small number of all-thyristor circuit breaker equivalents have been developed for special applications and are in service in the USA. These equipments are of two basic types, i.e. the 'half-cycle' static circuit breaker and the 'current-limiting' static circuit breaker as described by Lee *et al.* (1971) based on work done in the American General Electric Co.

Fig. 9.17*a* shows the principal elements in a 'half-cycle' breaker and the form of current interruption. This is simply the reverse/parallel connection of a thyristor cell or string of cells with a signal from the system protective devices applied to the gate to block the turn-on of the appropriate cell at the next current zero. Overvoltages are small because there is no current chopping. In practice, of course, the overcurrent detecting system must be all electronic to achieve the necessary speed. Typical applications are 3-phase 240 V 700 A equipment to provide uninterruptible power to computer installations by rapidly inserting buffer impedance between the computer supply and other faulted busbars. The largest rated equipments in service are single phase installations for induction heating loads rated at 3 kV, 10 000 A continuous current with 100 000 A symmetrical r.m.s. interrupting capacity which is equivalent to a 5 kV, 900 MVA fault rating.

Fig. 9.17*b* shows the principal elements in a 'current-limiting' breaker. In this a thyristor cell or string T_1 is used in conjunction with four sets of bridge connected diodes D to carry both directions of alternating current flow. The other main difference from the half-cycle breaker consists of the addition of a 'turn-off' circuit, i.e. another thyristor T_2, a precharged capacitor C and a small reactor L. When an abnormal condition is detected, not only is the gate signal to T_1 blocked out but T_2 is fired so that the high-frequency capacitor discharge current opposes the fault current forcing it to zero as shown. All current ceases when the 'turn-off' capacitor

power thyristor

sensing circuit

a.c. input

a.c. output

voltage

current

control circuit for gate signals

a

'turnoff' circuit for current limiting action

sensing circuit

L

D   D   C

T₁

a.c. input   D   D   T₂   a.c. output

voltage

potential fault current

current

control circuit for gate signals

b

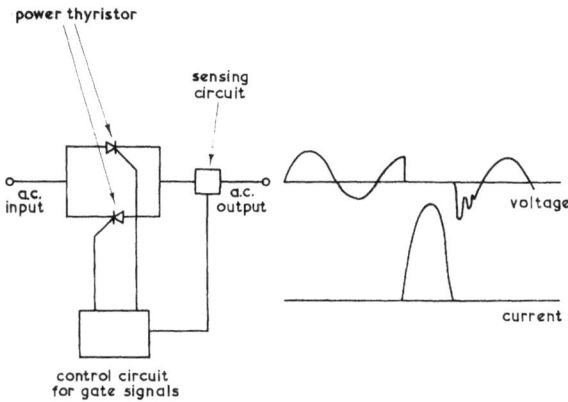

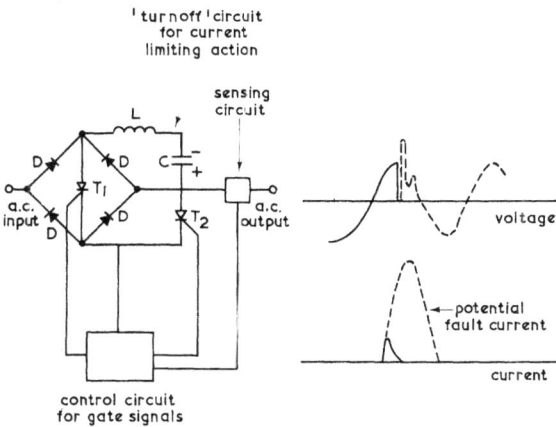

Fig. 9.17   Static circuit breaker principles

(*a*) half cycle static circuit breaker
(*b*) current-limiting static circuit breaker

is fully charged in the reverse direction.

Ratings of current-limiting breakers in service are 600 V, 1600 A and 3-phase for busbar-tie duty where the available short-circuit current on each busbar is 75 000 A symmetrical. The loads supplied are large spot welders requiring a very low system impedance to reduce voltage fluctuations during normal operation. By these means the system impedance is reduced to that corresponding to 150 000 A.

Other applications of thyristor controllers have been in the field of reactive power control where they compete with synchronous compensators and conventionally switched capacitors for the prevention of, for example, arc furnace flicker. In such cases, the duty of switchgear could obviously be combined with that of the controller, only an isolator being necessary for safety purposes. Another application foreseen for static breakers is the control of braking resistors during system transient stability disturbances since high-speed insertion and removal of such resistors is a prerequisite of their utilisation.

As a result of the development work on thyristor convertors for h.v.d.c. transmission, described later in this section, it is considered technically feasible to build static circuit breakers for almost any standard transmission voltage. It would be expected that such circuit breakers would take one of the favoured forms of the new thyristor convertors, namely outdoor equipments with all the components for one phase contained in one oil cooled tank with appropriate h.v. and control bushings (Hangsberger *et al*, 1970) or indoor type with air insulation and water cooling (Beriger *et al.*, 1976). Both types have been constructed for h.v.d.c. schemes rated at ± 500 kV and about 2000 MW. Unfortunately losses are at present unacceptably high, being of the order of 2% of load rating. Among the benefits which could result from the use of the static breaker are the elimination of transient overvoltages by the selective closing and opening of phases according to system conditions and the absence of mechanical maintenance requirements.

9.6.2 Power-thyristor string design
It was not until h.v.d.c. transmission provided the necessary stimulus that serious attention was given to the design of high-voltage high-power thyristor equivalents to the mercury-arc valves already in service at the end of the 1960s with ratings up to about 100-150 kV (peak) and 1800 A (mean d.c.) for multianode valves. Although thyristor convertors were not expected to be economic initially, either in terms of first cost or in terms of losses, their development was encouraged for several reasons. Among these were:

(*a*)   the reasonable prospect of decreasing costs in the relatively new thyristor cell technology due also to the increasing industrial demand for high-power devices (IEE Conference Publication 53, 1969)
(*b*)   the reduced maintenance costs as compared with high-vacuum mercury-arc valves
(*c*)   the reduced filter requirements, and hence a smaller terminal area requirement.

During the 1970s thyristor-controlled reactors and thyristor-switched capacitors have been increasingly used on a.c. systems for compensation and power-factor-correction applications. (Engberg *et al.*, 1979).

As yet there has been no corresponding high-power development for a.c. circuit-breaker purposes, but since the design problems involved are essentially the same as for the d.c. convertor they are reviewed here.

The combination of thyristor cells in series requires consideration of their voltage-sharing performance and, in parallel, their current-sharing performance. In early designs of convertor it was found that only about one-quarter of the voltage rating and less than one-half of the current rating of available cells could be utilised without exceeding required safety margins (Calverley, 1969). Presently available thyristors have typical ratings of 4 kV blocking and 2 kA average. For a 150 kV, 1800 A rating a single string of approximately 100 thyristors in series per arm could be utilised. A larger number of units than necessary is normally used to allow for failure, defective units being replaced during routine maintenance. Thyristors have now completely replaced the mercury-arc valve for all new installations.

A voltage factor of about one-third is now permissible following the introduction of gapless metal-oxide arrestors and due to improvements in semiconductor production techniques.

Part of the overvoltage factor is to allow for system and transient overvoltages outside the control of the convertor designer, but the remaining factor requires much study. Closer control of the manufacturing process and the use of neutron doping has contributed towards minimising variations in thyristor characteristics (Bergsjö *et al.*, 1976). For voltage sharing a protective network of resistors and capacitors is required as for normal grading of interrupter heads of conventional switchgear. Thyristor string grading networks are, however, far more complex than in the switchgear case.

The following factors are important in the design of grading strings:

(*a*) *Cell characteristics:* Variations are present in the particular characteristics of each cell which influence performance during various parts of its operating cycle. Delays occur before full forward conduction is achieved and are also encountered during turn-off. The effects of junction capacitance, rapid changes in applied voltage, reverse leakage currents and the necessity to remove charged carriers from the semiconductor material are all important factors to be considered (Brisby, 1979).

(*b*) *Grading networks:* The tolerances of components in the RC networks must be maintained and should be confirmed by short time overload and temperature cycling tests.

(*c*) *Stray capacitance:* The capacitance to earth of each module differs due to its particular position in the valve assembly, and consequently will influence the voltage distribution.

(*d*) *Firing system:* The firing pulses should have consistent, fast rise-times and appropriately be timed to minimise any effect on voltage distribution. Optical coupling is now almost exclusively used for thyristor firing systems; simplification in auxiliary requirements in the future could result from direct light triggering of the thyristor element.

The greater part of the overcurrent factor is taken up by the margin required for fault performance since excessive rise in the temperature of the junction reduces cell performance and can lead to breakdown under reverse voltage. When blocking has been unsuccessful on the first cycle the tolerable surge-current rating corresponds to only about six times the continuous-current rating for a cell with single-sided cooling, but, whereas double-sided cooling can significantly improve load performance, there is little improvement in fault performance. It has been proposed (Kriechbaum, 1971) that very fast acting shunt switchgear should be used

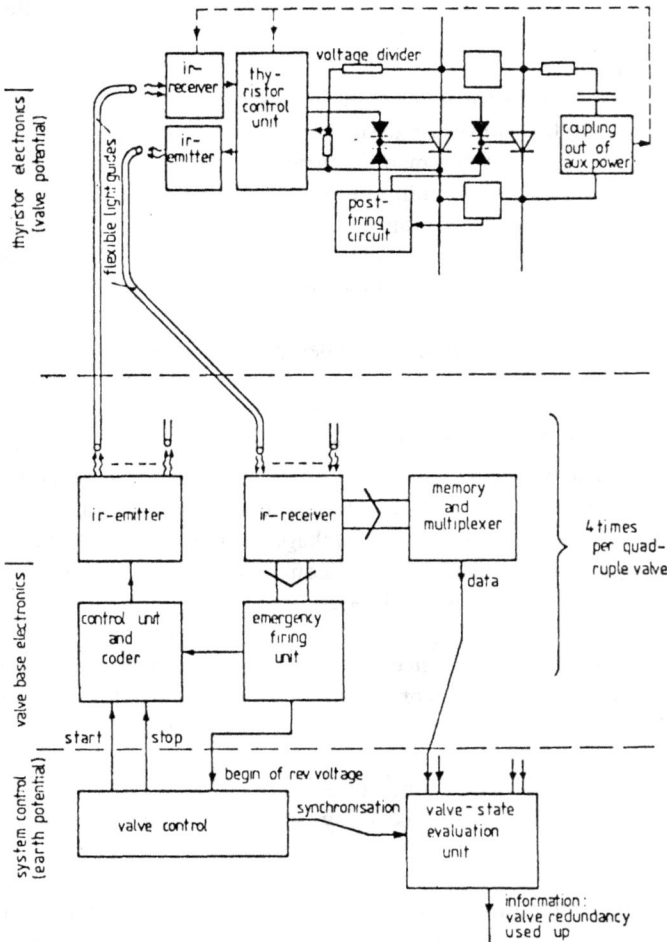

Fig. 9.18   Block diagram of BBC/AEG/Siemens Firing System

for the protection of semiconductor installations under fault conditions. This method is not favoured by the utilities and would, however, hardly be applicable to a static circuit-breaker installation.

The rate of change of thyristor current is an important consideration and non-linear reactors distributed down the string are commonly used to limit the rate of rise of current at turn-on. These, in turn, affect the voltage distribution of the network under transient conditions.

That the complete network of thyristors, auxiliary components and firing systems is very complex is shown by the block diagram of Fig. 9.18 from Beriger *et al.* (1976) which is based on the work of the Continental consortium of firms, AEG, BBC and Siemens. The overall design problem is extremely difficult and it is often necessary to resort to model studies to verify the performance of the chosen configuration.

Modern production techniques are significantly reducing the spread of cell characteristics and permitting higher cell-current ratings. Effort is now being aimed at improving the voltage characteristics, particularly in achieving a higher degree of cell self-protection when subject to overvoltages and high rates of voltage rise. A reduction in the complex electronic circuits at present required could then follow. Success in these directions would reduce both costs and operational losses.

9.7 Acknowledgments

The preparation of this chapter has required, in addition to the references listed, information which was freely given by the following organisations:

Aberdare Holdings Limited,
AEG-Telefunken,
ASEA Ltd.,
Brown Boveri & Co. Ltd.,
CALOR-EMAG Elektricitals Aktiengesellschaft
The Fuji Electric Co. (via Siemens Ltd.),
General Electric Co. (USA),
GEC (UK),
Hitachi Ltd.,
Northern Engineering Industries Ltd.,
Westinghouse Electric Corporation

The assistance of colleagues in Kennedy & Donkin is also gratefully acknowledged.

9.8 References

ANDERSON, J.M., and CARROLL, J.J. (1978): 'Appliability of a vacuum interrupter as the basic switch element in h.v. d.c. breakers,' *IEEE Trans.*, **PAS-97**, pp.1893–1900

BARKAN, P., and WILSON, D.D. (1974): 'Current-limiting devices for transmission and distribution applications,' Proceedings American Power Conference, pp.1105–1113

BARKAN, P. (1980): 'Reliability implications in the design of fault current limiters,' *IEEE Trans.*, **PAS-99**, pp.1734–1740

BEEHLER, J., and McCONNELL, L.D. (1973): 'A new synchronous circuit breaker for machine protection,' *ibid.*, **PAS-92**, pp.668–681

BELIAKOV, N.N., MOMAROV, A.N., and RASHKES, V.S. (1978): 'Results of internal overvoltages and electrical equipment characteristics measurements in the Soviet 750 kV network,' CIGRE Report 33–08

BERGSJO, N.J., and STUTESSON, S. (1976): 'Quality control of high power thyristors,' *ASEA J.*, **1**, pp.13–21

BERIGER, C., ETTER, P., HENGSBERGER, J., and THIELE, G. (1976): 'Design of water-cooled thyristor valve groups for extension of Manitoba Hydro h.v. d.c. scheme,' CIGRE Report 14–05

BOWLES, J.P., VAUGHAN, L., and HINGORANI, N.G. (1976): 'Specification of h.v.d.c. circuit breakers for different system applications.' CIGRE Report 13–09

BRISBY, K. (1978): 'Thyristors for h.v. d.c. applications.' *ASEA J.*, **3**, pp.68–71

CALVERLEY, T.E. (1969): 'Mercury arc and thyristor valves for h.v.d.c. transmission,' *Direct current*, **1**, pp.5–14

CALVERLEY, T.E., GAVRILOVIC, A., LAST, F.H., and MOTT, C.W. (1968): 'The Kingsnorth-Beddington-Willesden d.c. link.' CIGRE Report 43–04

CAMERON, F.L., and SMITH, D.R. (1971): 'The philosophy of power fuse design and application'. Paper presented to the Pacific Coast Electrical Association Engineering and Operating Section, California

EISENACK, H., KANNGIEBER, K.W., KIND, D., DOETZOLD, B., SCHULTZ, W., PUCHER, W., and JOSS, P. (1972): 'The h.v.d.c. circuit-breaker in intermeshed multi-terminal h.v.d.c. systems.' CIGRE Report 31–08

EKSTROM, A., HARTELL, H., LIPS, H.P., and SCHULTZ, W. (1976): 'Design and testing of an h.v.d.c. circuit-breaker.' CIGRE Report 13–06

EKSTROM, A., and LISS, G. (1972): 'Basic problems of control and protection of multi-terminal h.v.d.c. schemes with and without d.c. switching devices'. CIGRE Report 14–08

Electra (1971): 'H.V.D.C. switching devices and arrangements'. Report of Working Group 03 of CIGRE Study Committee 13 (Switching Equipment), pp.9–65

Electrical World (1971): '550 kV circuit breaker offers one-cycle tripping', p.62

ENGBERG, K., FRANK, H., and TORSENG, S. (1979): 'Reactors and capacitors controlled by thyristors for optimum power system VAR control.' Technical Seminar on 'Reactive Compensation in Power Systems' September, 1979. IEE Digest 1979/52

EPRI (1978): 'Controlled impedance short circuit limiter.' Report EL–857

FALCONE, C.A., BEEHLER, J.E., MEKOLITES, W.E., and GRZAN, J. (1974): 'Current limiting device – a utility's need,' *IEEE Trans.*, **PAS-93**, pp.1768-1775

FEENAN, J. (1971): 'Trends in the design of high-voltage high-rupturing capacity fuses', *Proc. IEE*, **118**, pp.216–221

FEENAN, J. (1980): 'New standard for fuses for h.v. motor circuits,' *Elect. Times*, 15th Aug. 1980, pp.9, 12

GALLAGHER, H.E., CARBERRY, R., KING, H.J., GRZAN, J., BEEHLER, J.E., and FAKHERI, A.J. (1980): '145 kV current limiting device-field tests,' *IEEE Trans.*, **PAS-99**, pp.69–77

GREENWOOD, A.N., BARKAR, P., and KRACHT, W.C. (1972): 'H.V.D.C. vacuum circuit breakers', *IEEE Trans.*, **PAS-91**, pp.1575–88

GREENWOOD, A.N., and LEE, T.H. (1972): 'Theory and application of the commutation principle for H.V.D.C. circuit breakers', *ibid.*, **PAS-91**, pp.1570–74

GRZAN, J., BEEHLER, J.E., KNAUER, W., and KING, H. (1978): 'Application and development of a fault current limiting device.' CIGRE Report 23–05

HANGSBERGER, J., SCHERBAUM, F., and WÜTHRICH, R. (1970): 'An oil-cooled H.V.D.C. thyristor valve for outdoor installation'. CIGRE Report 14–04

HAUSLER, M.P., KOLSCH, H., and PAULI, M. (9180): 'Application and switching duties of high speed switches in the d.c. switchyard of h.v. d.c. stations.' CIGRE Report 14–02

HIRAYAMA, T., HIROYOSHI, H., NAKAYAMA, K., MURAI, K., CKUDA, K., KURITA, K., and KATO, Y. (1971): 'Development of system interconnector cf dynamic characteristics', *Hitachi Rev.*, **20**, pp.373–381

HIRAYAMA, T., KATO, Y., MUKAL, H., and OYA, L. (1971): 'Developments of A.C. type interconnectors'. IEEE Paper 71 TP 609-PWR presented at the the International Symposium or. High-Power Testing, Portland, Oregon

IEE Conference Publication 53 (1969): 'Power thyristors and their applications'

IEEE SSR Task Force (1976): 'Analysis and control of subsynchronous resonance.' IEEE Publication 76 CH 1066–0–PWR

JACKS, E. (1967): 'Fuse performance data for modern applications'. Paper presented at the ERA Distribution Conference, Edinburgh

JOHN, M.N. (1968): 'S.L.C. – An automatic resonance link', *Electr. Times,* 29th August

JONES, K.M., WHALLEY, H., and CURE, J.R. (1969): 'The prospects of using thyristor strings in distribution system development'. *IEE Conf. Publ.* **53**, pp.422–432

KAIKNER, B. (1966): 'Short-circuit current-limiter for coupled high power systems'. CIGRE Paper 301

KANGIESSER, K.W., KOETZOLD, B., SCHULTZ, W., and PUCHER, W. (1972): 'H.V.D.C. circuit-breaker in intermeshed multi-terminal h.v. d.c. systems.' CIGRE Report 31–08

KEDERS, T., and LEIBOLT, A.A. (1976): 'A current limiting device for service voltages up to 34.5 kV.' IEEE Summer Meeting Paper No. A76 436–6

KIND, D., MARX, E., MOELLENHOFF, K., and SALGE, J. (1968): 'Circuit breaker for h.v.d.c. transmission'. CIGRE Report 13–08

KING, H.J., GALLACHER, H.E., and KNAUER, W. (1980): '145 kV current limiting device – construction and factory test,' *IEEE Trans.*, **PAS-99**, pp.911-918

KREICHBAUM, K. (1971): 'A half-cycle air blast generator breaker for high power testing fields', IEEE Paper 71 C57-PWR presented at the International Symposium on High-Power Testing, Portland, Oregon

KRIECHBAUM, K., and KRONDORFER, W. (1971): 'The overcurrent diverter. Control gear for the protection of installations'. Proceedings of the International Congress on Distribution Power Systems (CIRED), Liege, Paper 14

LEE, T.H., SMITH, J.A., and TITUS, C.H. (1971): 'Static power circuit breakers – a tool for the future'. Paper presented to the American Power Conference, Chicago, Ill, April 21

LONG, W.F., and LIAN, K.T. (1974): 'A design study for d.c. circuit breaker on a three-terminal system.' IEEE Conference Paper C74 131–9

MATSUDA, K., ARAI, S., HIRAYAMA, T., KURITA, K., HYUGA, S., SUDO, Y., NAKANO, S., MURAI, K., and OKUYAMA, K. (1971): '66 kV system interconnector', *Hitachi Rev.,* **20**, pp.382–385

McCONNACH, J.S. (1969): 'Short-circuit current limitation in industrial power systems'. Proceedings of the AIM International Congress on Distribution Power Systems, Liege, Paper 10

NITTA, Y., and KIYOKUNI, N. (1970): 'Synchronous closing circuit breaker for one-cycle interruption', *Fuji Elec. Rev.,* **16**, pp.38–44

PARTON, K.C. (1978): 'A new power system fault limiter.' *Elect. Rev. Int.,* **202**(5), pp.63–65

REEVE, J. (1980): 'Multiterminal h.v.d.c. power systems,' *IEEE Trans.,* **PAS-99**, pp.729-737

RUMPF, E., and JARRETT, G.S.H. (1980): 'A survey of the performance of h.v.d.c. systems throughout the world during 1975–78.' CIGRE Report 14–08

TAHLIANI, V.H., and PORTER, J.W. (1980): 'Fault current limiters an overview of EPRI research,' *IEEE Trans.,* **PAS-99**, pp.1964–1969

TAM, C.C. (1970): 'General survey of D.C. projects', *Direct Current,* **1**, pp. 128–121

THANAWALA, H.L., and YOUNG, D.J. (1970): 'Saturated reactors – Some recent applications in power systems', *Energy Internat.,* **7**, pp.20–25

YECKLEY, R.N., FRIEDRICH, R.E., and THUOT, M.E. (1971): 'E.H.V. breaker rated for control of closing voltage switching surges to 1·5 per unit'. IEEE Paper 71 TP 571-PWR presented at the International Symposium on High-Power Testing, Portland Oregon

Chapter 10

Circuit-breaker specification and testing

J. A. Sullivan, C.Eng., F.I.E.E.

10.1 Specification

The performance specification for circuit breakers has been subject to almost constant review since the end of the second decade of this century. The specifications produced by national committees have been legion. They include British Standards, the Verband Deutscher Elektrotechniker (VDE) and the American National Standards Institute (ANSI), to mention just a few. Also, the International Electro-Technical Commission (IEC), with its representation from almost all countries in the world, has been producing and revising its recommendations. This has meant that there has had to be a great deal of repetition of testing to satisfy the idiosyncrasies of the various specifications. In 1969, there was general agreement to the adoption of the recommendations of IEC as completely as possible in national standards. At the present time, therefore, the IEC is more active than ever in updating and revising its recommendations, and the national committees are in the process of changing their specifications, in most cases to comply word for word with the IEC recommendations, which, for circuit breakers below 1000 V, is IEC157; for circuit breakers above 1000 V, the relevant recommendation is IEC56. Hopefully, therefore, wherever possible, users will, in the future, be issuing specifications requiring tests in accordance with the recommendations of the IEC. Although this may not happen, and it would be a utopia for the designers, it will be generally true, and thus a great degree of standardisation of test requirements will profit all, and, not least, assist the circuit breaker designer to know what performance he is required to provide.

10.2 Environment

Before considering tests, we should consider the environment in which the circuit breakers will be used, as this forms the basis for test requirements.

In considering the effects that the environment may have on the performance of an item of electrical plant, one is confronted with a multitude of different aspects that must be considered at some stage in the design, manufacture or application of the plant. The immediate environment of the circuit breaker will be influenced, to some extent, by the housing or protective covering, or its absence, that is employed in the construction, and whether the installation is indoor or outdoor. It is possible to envisage complete encapsulation of a circuit breaker and its allied equipment, but it can be shown that, even under these conditions, the equipment may not be considered free of all external influences.

In considering the natural environment, it is found that this is, to a greater or lesser extent, dependent on the geographical location of the intended installation and the prevailing climatic conditions. It is possible to lay down minimum standards that cater for the conditions applicable to the majority of cases, and to draw attention to additional safeguards necessary for extreme locations and climates.

Pollution of the environment requires special safeguards to be taken. The more obvious instances are those involving the presence of salt, dust and chemical waste products. The extreme case of inflammable environments, such as may exist in mines and some industrial process plants, has caused the development of a complete range of electrical equipment adapted for these conditions. Man, together with lower forms of animal life, must also be considered part of the environment, and thus their adverse affect on the functioning of a circuit breaker cannot be overlooked. At least, for man, it is essential to protect not only the equipment from man, but to protect man from the equipment. The obvious requirement is to ensure safety from electric shock, but there may be other electrical or mechanical disturbances to consider. The engineer must therefore take into account the effects of an environment on the behaviour of a switchgear installation, and also the effects of that equipment on the environment. Adequate safeguards must be employed during the design and manufacturing stages, backed by sufficient tests to ensure compliance with recognised standards of performance.

10.3 Normal service conditions

The International Electrotechnical Commission have laid down atmospheric and climatic conditions that shall be withstood by standard circuit breakers. Special conditions are also referred to, and provide a basis for assessing the performance of a standard piece of equipment and the need for considering special applications. Service conditions considered as normal can be summarised as follows:

10.3.1 Ambient air temperature

The minimum ambient air temperature is specified as $-5°C$ for indoor installations and $-25°C$ for outdoor installations in temperate climates. For arctic climates, temperatures of $-20°C$ and $-50°C$ are recognised for indoor and outdoor installations, respectively.

Much equipment, particularly where designed for outdoor installation, is suitable for arctic conditions in its standard form. Other apparatus can often be supplied for low-temperature environments, when this is specified by employing special safeguards, such as low-temperature insulating and lubricating oils and strategically placed thermostatically controlled heaters.

Type tests to prove the correct functioning of equipment at subzero temperatures would take the form of an extension to the mechanical tests in a cold chamber simulating the service environment. Particular attention must be paid to seals subject to flexing during circuit breaker operation. All types of rubber become progressively harder and stiffer with reduction in temperature, and, in addition to the loss of flexibility, recovery from deformation becomes slower. Additional dielectric tests under low-temperature conditions may be indicated where gas insulation is employed.

10.3.2 Altitude

The altitude of the site of installation is considered for normal purposes not to exceed 1000 m above sea level. For installation at higher altitudes, it is necessary to take into account the reduction of the dielectric strength and the cooling effect of the air. The rated insulation levels specified for circuit breakers apply to installations not exceeding 1000 m, and the specified correction factors must be applied to the test voltages employed.

10.3.3 Atmospheric conditions

The ambient air is considered not to be materially polluted by dust, smoke, corrosive or flammable gases and vapours or salt. Where adverse conditions are known to exist, special consideration should be given to the design of those parts of a circuit-breaker, especially the insulation, normally exposed to the atmosphere.

For outdoor installations, the design should take into account the presence of condensation, rain, snow, a layer of ice or hoar frost up to 5 kg/m^2, rapid temperature changes, wind pressures of 700 N/m^2 and the effects of solar radiation.

For indoor installations, clean air is assumed. Humidity conditions are difficult to agree on, and, for high-voltage circuit breakers, are still subject to agreement. However, for low-voltage switchgear, a relative humidity not exceeding 50% at a maximum temperature of $40°C$ is recognised with higher relative humidity permissible at lower temperatures, e.g. 90% at $20°C$. Care should be taken of moderate condensation that occasionally occurs, owing to variations in temperature.

For indoor installations not employing additional enclosures, the ideal clean-air conditions are implied, and no tests for atmospheric conditions are required. For totally unenclosed circuit breakers installed outdoors, the normal service conditions apply, as previously stated. No general specification for tests under these conditions has been published, but the need to test in some instances is recognised.

For circuit breakers employing pressurised gas as an insulating medium, the rain tightness of the circuit breaker itself is assured, and additional precautions would only be necessary to prevent the ingress of water into any external mechanisms and cabling enclosures. For circuit breakers having enclosed volumes that may breathe atmospheric air, e.g. oil circuit breakers, a more stringent test may be indicated. Such a typical test of rain tightness should specify the intensity and direction of the rainfall and the duration for which it should be applied, possibly up to 24 h. Tests with a coating of ice or hoar frost would not normally be necessary unless there were moving parts, the correct functioning of which may be affected by icing. In such cases, e.g. circuit breakers with external isolating gaps or employing external drive linkages, tests should be performed in a cold chamber capable of producing the specified ice coating, after which correct operation of the circuit breaker should be demonstrated. Tests for the maximum wind pressure specified are not necessary, since the structure would normally be subjected to more severe conditions during short-circuit testing.

10.4 Mechanical testing

The circuit breaker is, without exception, entirely dependent on the correct behaviour of its mechanical parts to perform its function. The mechanical integrity of each unit is therefore of the utmost importance and must be adequately considered from the initial design stages, through type and routine tests in the factory, to site installation and commissioning checks.

The results of type tests, which must include a mechanical-endurance test and a verification of the operational limits, may be applied to all other circuit breakers of the same design and having the same operating characteristics. Adequate quality control is, of course, a prerequisite. Mechanical routine tests thus comprise a repeat of the tests for the determination of the circuit breaker operating characteristics, sufficient no-load duty cycles being included to check that adjustable parts have been set correctly and that the circuit breaker, its controls and interlocks function consistently.

10.4.1 Mechanical-endurance type test

Throughout its useful life, a circuit breaker may be called on to perform some thousands of operations without mechanical failure and in accordance with its predetermined characteristics. Typical requirements are for a maintenance-free life of 1000 close-open cycles, whereas the design life with maintenance may be nearer 20 000 operations. The circuit breaker is therefore subjected to a mechanical-endurance type test comprising a specified number of no-load closing and opening and close-open operations equal to its rated maintenance-free life. Following such a test, which is performed on a breaker complete with its operating mechanism, all components, including the contacts, must be in a satisfactory condition and should show no signs of excessive wear or deformation.

For design information, it is often necessary to exceed the specified number of mechanical-endurance operations during the development, although some lubrication or adjustments may be made at intervals. These latter tests will then provide the designer with additional data on maintenance requirements and may give advance indication of the possible causes of failure after extended service. The effect of long periods of static loading of mechanisms, corrosion and the deterioration of lubricants must also be taken into account.

10.4.2 Operating limits

A purchasing specification should quote a preferred value or values for the auxiliary operation supplies to a circuit breaker for any actuating coils, relays or motors employed. However, it is recognised that the voltage of such supply sources cannot be held constant, and acceptable operating limits must be defined.

Since many d.c. supplies are derived from batteries, it is to be expected that components relying on such supplies should be capable of functioning over the range of battery voltages from fully charged down to a partially discharged state. This operating range is typically from 85 to 110% of the rated voltage for closing devices, but is extended to 70 to 110% for shunt-trip devices. Similarly, if operation involves the use of stored energy in the form of compressed air or hydraulic pressure, correct operation over the range of pressures between cutout and cut-in of compressor or pump controls must be ensured.

Testing therefore involves the operation of a circuit breaker over the range of the various auxiliary supplies to check for correct behaviour. The main criterion to be met during such tests is the operation of the circuit breaker at the lower limit, without undue hesitation in performing the duties of short-circuit and load-current switching. Tests are made at the upper limit to demonstrate that energy input is not damaging. Also, it should be confirmed that no damage results from the inadvertent operation of the unloaded mechanism.

10.4.3 Component testing

Components and subassemblies may often be subjected to separate testing. This may be for reasons of convenience during early development work, when the overall design is not finalised, or may be due to the requirements for certain components to comply with separate standards. Examples of components requiring separate type tests are air receivers and other such containers subject to fluid pressure, be it transient or sustained, circuit-breaker bushings and some insulation components.

10.5 Withstand-voltage tests

Whilst there may be a considerable variation in the mechanical duty imposed upon circuit breakers in service according to their application, and their operating temperature may be variable and may not approach the maximum, the normal service voltage will be sensibly constant. They will periodically be subjected to overvoltages due to lightning, switching surges and in some cases to high temporary power-frequency overvoltages. The dielectric tests are therefore extremely important.

The test values mostly used are those given in IEC Publication 71-1 Table, III and IV, although for historical reasons many users require higher lightning impulse test values than are given in Table IV. The test requirements of IEC Publication 56-4 together with the amendment define the type and routine dielectric test requirements. IEC Publication 60 covers the test techniques and conditions for these tests.

The type tests shown in Section 1 of IEC 56-4 include power-frequency and lightning impulse system voltages below 300kV, together with switching-surge tests for system voltages of 300kV and above. These are all withstand tests and some users' specifications also include requirements for critical-flashover testing. In this section attention is drawn to some of the methods used to establish a better knowledge of the withstand capability of the circuit breakers. The IEC test requirements specify a 15-impulse withstand test which is considered to be adequate to demonstrate that the circuit breaker has a rated impulse-voltage withstand probability of 90% in a Gaussian distribution, if there are not more than two flashovers.

The waveform specified for the lightning impulse tests has a voltage risetime of $1·2 \mu s$ and a decay time to 50% of the peak voltage of $50 \mu s$. A voltage rise time of $1·2 \mu s$ will only occur when the lightning strike is to an overhead line shield-wire or tower within about 0·5km of the circuit breaker. This is because the steep voltage rise occurring due to the line-insulator flashover is rapidly attenuated. The voltage risetime for lightning strikes direct to a phase conductor has less than 5% probability of being as severe as the test risetime. For switching-surge tests the specified waveform in IEC 60 is 250/2500, with wide tolerances for both the front and tail times. Investigations have shown that high-voltage insulation structures with air gaps up to 6m have the lowest withstand-voltage level with this specified waveform. Above 6m, the most onerous front time increases and is $450 \mu s$ at 10m.

The typical variation of withstand voltage, expressed in terms of standard deviation in Gaussian distribution, is 2 or 3% for the $1·2/50$ waveform and in the range of 4-12% for the 250/2500 waveform. IEC 71, Guide to Insulation Coordination, uses 3% (lightning) and 6% (switching-surge), respectively, as a basis of determining withstand probabilities.

IEC 56-4 Modification No. 1 (1975) requires that for 300kV and above outdoor circuit breakers, the switching surge tests be made with positive polarity only dry and with both polarities wet. The power-frequency tests are wet and dry for circuit breakers lower than 300kV and dry only above 300kV.

The lightning and switching impulse tests, together with the 1 min power-frequency test, provide three points that give, with satisfactory approximation, a significant part of the voltage/time characteristic of the circuit-breaker insulation.

10.5.1 Design tests

Apart from meeting the aforementioned switching-surge-test requirements, which provide a 90% withstand probability, a more precise knowledge of the withstand voltage at higher probabilities is needed by the designer, together with the level of confidence inherent in the results of the tests.

All methods that are in general use assume the distribution to follow the Gaussian law. This is sufficiently accurate, but it should not be overlooked that, at low failure probabilities, that is at two or three standard deviations below the 50% withstand level, corresponding to a probability of 2·3 and 0·14%, respectively, this assumption is in some doubt. At this low-probability level, other distributions, such as the binomial distribution, provide different statistical information, and, in this case, a better result.

The methods used to determine the statistical information are described in Sections 10.5.1.1–10.5.1.4, with some guidance as to their efficiency.

10.5.1.1 Up-and-down method

In this method, the circuit breaker has applied to it test voltages that increase in discrete steps until breakdown occurs; thereafter the voltage is lowered after each breakdown and raised after each withstand. Such an iterative procedure enables the 50% breakdown level to be established with an acceptable degree of accuracy, but provides no information about the standard deviation of the distribution. A low-probability withstand level can be calculated, but this involves the assumption of a value for the standard deviation, which can involve a significant error. The method is suitable when comparison of performances is required.

10.5.1.2 Fixed-voltage multi-shot method

This is the most popular method and entails choosing a starting-voltage level by raising the test-voltage steps until a flashover occurs. This voltage level is then used as the first fixed voltage level, and 20 or 40 shots are made. The ratio of flashover to withstand is noted, and from this a decision can be made to either raise or lower the voltage by approximately 6%. Further test-voltage levels are chosen, and by this means four or five or more voltage levels, each with twenty or forty shots, are carried out, which can then be plotted on probability paper to establish by means of the best straight line, the standard deviation and the probability of flashover over the range of 50% ± two standard deviations. If there is difficulty in drawing a straight line between the fixed-voltage-levels results, a linear-regression method may have to be used.

10.5.1.3 Cumulative-frequency method

Examination of the point of breakdown on the impulse voltage during tests, particularly under positive-polarity conditions, indicates that virtually all breakdowns occur before the impulse crest is reached and at various voltage levels before the crest. Ouyang (1966) has shown how to make use of this phenomenon and to produce a cumulative frequency of breakdown characteristic by applying a number of impulses of identical magnitude, and recording, from film, the voltage at each breakdown. A typical characteristic is shown in Fig. 10.1. Only one voltage level is strictly necessary, and the analysis is claimed to provide significantly more information in the low-probability region by eliminating the need for extrapolation. Experience of this method is, however, limited.

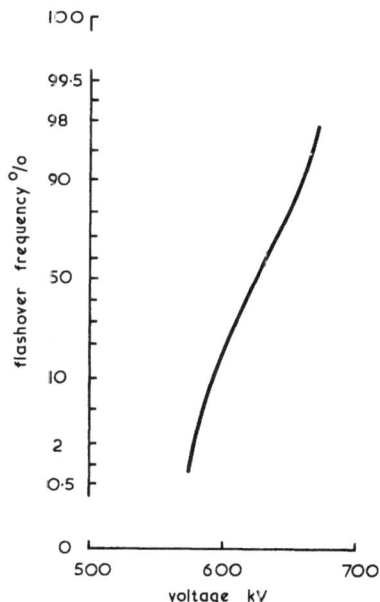

Fig. 10.1 Typical cummulative frequency of breakdown characteristic

10.5.1.4 Contour-lines method

This is an extension by Ouyang *et al.* (1970) of the cumulative-frequency method of analysis, rather than a different test procedure. It requires the recording of the time to breakdown, as well as the voltage at breakdown, and is only valid if breakdown occurs before the crest of the impulse voltage is reached. This method, as does the cumulative-frequency method, makes use of the quantitative analysis of the breakdown to provide superior information from a smaller number of tests.

10.6 Power-factor tests

Power-factor and capacitance measurements are frequently carried out on high-voltage switchgear insulation and, in particular, on high-voltage bushings, capacitors, current transformers etc. The high-voltage Schering bridge shown in Fig. 10.2 is being used to determine the variation of capacitance C_x and loss angle tan δ with voltage for a high-voltage bushing. The shape of the loss-angle/voltage curve gives an indication of the quality of the insulation, and an important characteristic is the 'knee point', which is defined as the voltage at which the tan δ/voltage curve exhibits its first marked change of slope (BS223:1956). Limits are sometimes set for the permissible increase in tan δ between two voltages expressed in terms of per-unit rated voltage U_n. Departures from the normal shape of these curves also enable the presence of moisture and other defects in the insulation to be detected. The Schering bridge is also used during thermal-stability tests at service temperatures on high-voltage bushings and capacitors.

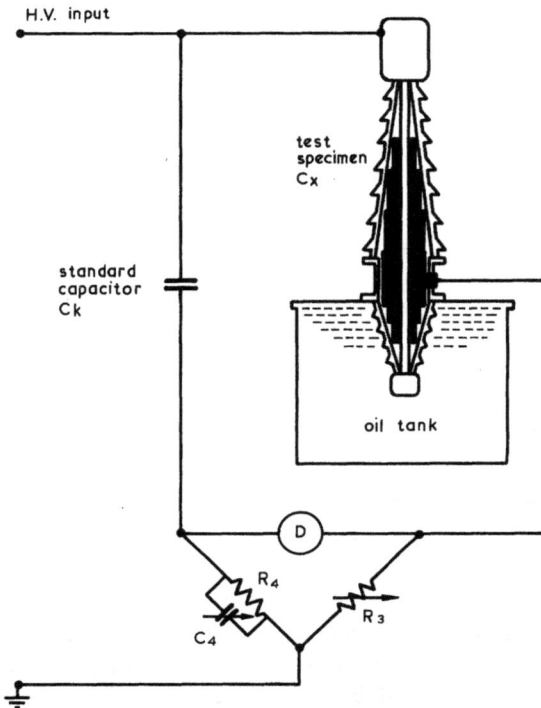

Fig. 10.2 High-voltage Schering bridge

Precautions must be taken to ensure that connections are reasonably discharge free and that the bridge is effectively screened and earthed. The detector D may be a vibration galvanometer or an electronic device that enables the balance condition of the bridge to be determined.

10.7 Partial-discharge testing

Localised electrical discharges may occur, owing to ionisation in cavities or voids within the solid insulation or within gas bubbles in insulating liquids; discharges may also occur along dielectric surfaces. Such internal discharges only partially bridge the insulation between the conductors, and, although they involve only small amounts of energy, they can lead to the progressive deterioration of the dielectric properties of the insulating materials. This is particularly true for the oil-impregnated-paper insulation of high-voltage bushings and current transformers.

The detection of partial discharges constitutes a useful nondestructive high-voltage test, and, although different circuits are available, the fundamental processes taking place can be explained by reference to the circuit of Fig. 10.3, which is in general use. As the alternating voltage U applied to the test specimen C_x increases with time, a corresponding increasing voltage ΔU appears across the small cavity c. ΔU depends on the capacitance c of the cavity, in series with the capacitance of the remainder of the specimen. The size of c has been exaggerated for the purpose of illustration. A point in the voltage wave is reached at which ΔU may be sufficient to cause ionisation and a discharge within the cavity. This, in turn, causes a sudden collapse of ΔU in a time of the order of 10^{-7} s or less. The resulting electrical charge $c\Delta U$ dissipated within the cavity causes a rapid transient current to flow round the closed circuit formed by the test specimen C_x, the discharge-free coupling capacitor C_k and the measuring impedance Z_m, so that the charge $c\Delta U$ is redistributed proportionally between the various capacitances in the circuit. The resulting current pulse is measured by a discharge detector, comprising an amplifier and an oscilloscope, which is connected across the tuned measuring impedance Z_m. Since the dimensions and capacitance of the cavity are unknown, the magnitude of the discharge in the cavity cannot be calculated. This discharge must therefore be related to the equivalent loss of charge, expressed in picocoulombs, at the terminals of the specimen, and is called the 'apparent' discharge magnitude q (IEC 270:1968). q can be deduced theoretically from the magnitude of the voltage pulse that appears across the impedance Z_m and from the equivalent capacitance network of the circuit (Kreuger *et al*,., 1964). In practice, of course, there may be discharges in many cavities or voids in the sample, all of which will contribute to the erosion, progressive deterioration and possible ultimate breakdown of the sample. The presence of multiple voids and the recurrent discharges that occur over the various parts of the voltage cycle can be seen on an oscilloscope, using an elliptical time base, as in Fig. 10.3. Other methods of

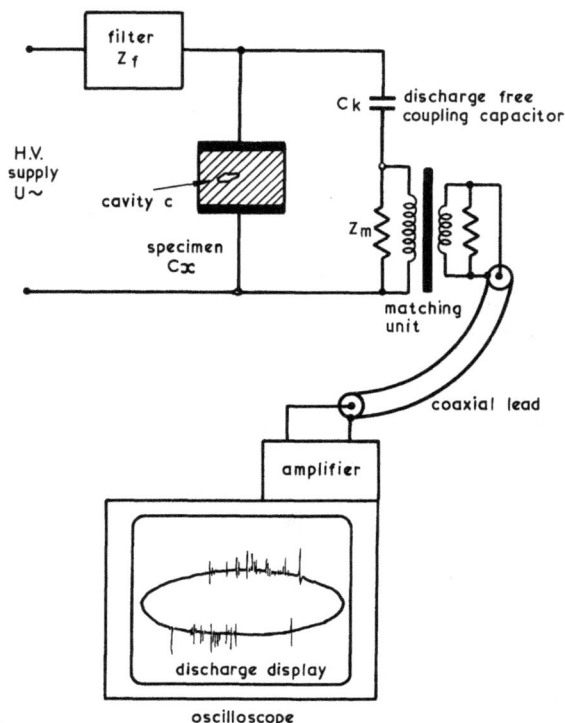

Fig. 10.3 Discharge test arrangement

presenting the discharge information are available, depending on whether the magnitude of the individual discharges, or of the integrated discharges, is required (BS223:1956).

Discharge circuits are usually calibrated by applying a known pulse of charge across the measuring impedance or, preferably, directly across the terminals of the test specimen C_x. Discharges of 1 pC or less can be recorded on test specimens by sensitive detectors, and, on specimens of small capacitance, discharges as small as 0·005 pC can be measured. With such small discharges, it is essential that the measurements be carried out in a screened room. The blocking capacitor must be free from discharge, and it is usual to employ a standard compressed-gas capacitor for this purpose. These are available for voltages up to 800 kV, at a capacitance of 50 to 100 pF. The transformer that provides the high-voltage supply must be discharge free at the test voltage. As a rough guide, testing transformers are usually discharge free up to about one-half their rated voltage. The addition of a filter Z_f helps to eliminate spurious discharges from the high-voltage supply.

It is believed that discharges below a certain magnitude cause only a slight deterioration of the insulation, so that an infinite life can be expected. This magnitude is called the 'permissible discharge magnitude', and it depends greatly on the voltage stress, since the life tends to decrease as the seventh to ninth power of the stress. For high-voltage bushings, the discharge levels are often measured at a voltage of 0·67 times the 3-phase rated voltage immediately following the 1 min power-frequency test. Practical experience shows that the following approximate values of the level of partial discharge seem to be acceptable (BS223: 1956):

Oil-impregnated-paper bushings	10 pC
Synthetic-resin-bonded bushings	100 pC
Synthetic-resin-impregnated bushings	20 pC
Cast-resin bushings	20 pC

Although it is accepted that, at this voltage, oil-impregnated-paper bushings should preferably be discharge free, purchasing specifications that require this test generally recognise the impracticability of obtaining complete freedom from background noise, and the limit they impose therefore takes this into account. They also generally require the test to be made prior to the other high-voltage tests, and that there should be no significant change in discharge level when the final test is made.

10.8 Radio-interference tests

External corona discharges may occur at the highly stressed points of high-voltage equipment. This, together with internal discharges, can give rise to interference with radio and television reception. Observations and measurements can be made in a laboratory to determine how this radio-interference (r.i.) varies with voltage. The equipment may be energised in the darkened laboratory, and the voltage at which the visible corona is extinguished is recorded. A more precise method is to use a circuit similar to that shown in Figure 10.3, except that a radio-noise meter (GPO measuring set) is used instead of the amplifier and oscilloscope of the discharge detector. By this means, radio-noise measurements can be made, usually at a frequency of 1 MHz. The reading of the microvoltmeter gives the radio-noise level in decibels above 1 μV, but as this gives the integrated discharge magnitude it cannot give the magnitude of individual discharges.

10.9 Temperature-rise tests

The current-carrying capacity of any item of plant is limited largely by the temperature rises in its component parts. One effect of excessive temperature rise is to cause deterioration of the electrical insulation, owing to ageing and to the differential expansion between the conductors and the surrounding insulation. This deterioration eventually results in failure, usually when the equipment is subject to undue stress of some kind. In a circuit breaker, the current-carrying parts are subjected to thermal and mechanical stresses, which can be due to either the normal current operation or to short-circuit conditions. The extent to which normal or short-circuit current influences the size and construction of the current-carrying parts depends on the rated current and breaking capacity. The mechanical strength required has a great influence on the size of the current-carrying parts, because of the need for rigidity and speed of movement. It is possible therefore that, in certain units, the sizes of the current-carrying parts would permit larger than the rated currents, especially for limited periods. The permissible temperature rises for circuit breakers are given in the standards appropriate to the type of switchgear. Typical figures, as agreed internationally for high-voltage a.c. circuit breakers, are given in Table IV of IEC 56-2: 1971.

Verification of the temperature-rise limits is by a type test that is mandatory for each design of circuit breaker. The tests should be made on an arrangement of the breaker, representative of its type, carrying a current not less than its rated normal current, and without exceeding the limits of temperature rise specified.

10.9.1 Test arrangement
Temperature-rise tests should be made on a new and complete assembly or subassembly with clean contact parts. The tests should preferably be made with the rated number of phases. However, where a multiphase circuit breaker is comprised of separate poles one single-pole unit may, for convenience or owing to limited test facilities, be tested, provided that it is equivalent to or not in a more favourable condition than the complete multipole circuit breaker in respect of heat generation and dissipation. Where a design provides alternative components or arrangements, the tests should be made with those components or arrangements for which the most severe conditions are obtained.

The circuit breaker, including current transformers and auxiliary devices, should be mounted to simulate as close as practicable the usual service conditions, including any covers and enclosures, and should be protected against undue external heating or cooling. Temporary connections to the main circuit should be such that no abnormal amount of heat is conducted away from, or conveyed to, the circuit breaker during the test. In case of doubt, the temperature rise at the terminals of the main circuit, and at the temporary connections at a distance of 1 m from the terminals is measured. The difference of temperature rise should not exceed 5°C.

The test should be made over a period of time sufficient for the temperature rise to reach a constant value. For practical purposes this condition is obtained when the variation does not exceed 1°C/h. The time for the whole test may be shortened by preheating the circuit with a higher value of current. Tests should be made at rated frequency and when circuit-breakers rated for 50 Hz are tested at 60 Hz and vice versa, care should be exercised in the interpretation of results.

Since the temperature rise of a circuit breaker can be attributed almost entirely to the losses caused by the passage of its rated normal current, then the test can be made with the correct value of current which may be derived from a source at any convenient voltage regardless of the breaker rated voltage. A very-low-voltage supply is used not only for economic reasons but also to enable temperature sensing devices to be applied directly to conductors normally at prohibitive voltage levels. A stepdown transformer with suitable means of adjusting the value of current can usually be fed from the manufacturers main supply, enabling tests to be carried out without the need for a special test laboratory.

10.9.2 Temperature measurement

For coils, the method of measuring the temperature rise by variation of resistance is generally used. Other methods being precluded unless it is impracticable to use the resistance method. The temperature of the winding, as measured by a thermometer before commencement of the test, should not differ from that of the surrounding medium (air, oil etc) by more than $3°C$.

For copper conductors, the ratio of the hot temperature T_2 to the cold temperature T_1 may be obtained from the ratio of the hot resistance R_2 to the cold resistance R_1 by the following formula

$$\frac{R_2}{R_1} = \frac{T_2 + 234 \cdot 5}{T_1 + 234 \cdot 5}$$

where T_1 and T_2 are expressed in $°C$. A simpler method giving results only slightly less accurate may be used for most tests by calculating the temperature rise on the assumption that $0 \cdot 4\%$ increase in resistance represents a $1°C$ increase in temperature. For conductors other than coils, the temperature of the different parts should be measured with thermometers, or thermocouples of any suitable type, placed at the hottest accessible spots. The temperature of oil in oil circuit breakers should be measured in the upper layer of the oil.

The points at which the highest temperatures will be produced cannot always be easily determined in advance of testing. Consideration of likely thermal sources and a study of the paths for electric current is necessary, therefore, in order to assist in the choice of points for measurement. It is to be expected that the insulation directly surrounding a conductor would be at a high temperature, although it would be very difficult to make direct measurements. However, in most cases, such conductors being of uniform cross-section can be considered as uniformly distributed heat sources, as with bare conductors, and a thermocouple reading taken at the exposed ends of the conductor will give a good indication of temperature rise of the buried portion. Contact assemblies form a local heat source and measurements should be taken at both fixed and moving contacts. Springs or other heat-treated components should be given special consideration and, especially where they may form a current path, direct temperature measurements should be made. Where current transformers form an integral part of the design these components also act as a heat source. Although the winding current in the current transformer is small, the large number of turns makes the I^2R loss appreciable, and also the core loss requires consideration. For truly representative conditions current transformers should be connected during the test to an external load equivalent to the rated burden of the transformer. Eddy-current heating in an appreciable form can occur in some circuit breakers, and although this is especially true of metal enclosed switchgear it is by no means limited to this type. First, those components which are particularly prone to thermal damage are those between phases such as bushing flanges and support panels. Secondly, those components which are in a current loop, such as the panels between cable and busbar compartments, or those which are in a sharp bend in the current path, could become dangerously hot.

For measurements with thermometers or thermocouples, certain precautions should be taken. Good heat conductivity between the thermometer, or thermocouple, and the surface of the part under test must be ensured. This necessitates firm attachment of the measuring device assisted by a wrapping of the bulb or junction with aluminium foil or another pliable conducting material. Thermocouples or the bulbs of the thermometers must be suitably protected against cooling from outside. The protected area, however, should be negligible compared with the cooling surface area of the apparatus under test. This can best be achieved by applying a pad of suitable insulating material such as plasticine or other plastic compounds or dry cotton wool over the thermocouple or thermometer bulb after attaching it to the item whose temperature is being measured. When bulb thermometers are employed in places where there is a varying magnetic field, alcohol thermometers should be used in preference to the mercury type as the latter are unreliable under these conditions.

The ambient air temperature should be measured by means of at least three thermometers or thermocouples. The measuring points should be equally distributed around the circuit breaker at about the average height of its current carrying parts and at a distance of about one metre from the circuit breaker. The thermometers must be protected against air currents and heat radiation. In order to avoid indication errors because of rapid temperature changes the measuring device may be put into small oil-filled containers with an oil content of about half a litre.

During the last quarter of the test period, the change in ambient temperature must not exceed $1°C/h$. If this is not possible because of unfavourable conditions at the place of test, the temperature of an identical circuit breaker under the same ambient conditions, but without current, can be taken as a substitute for the ambient air temperature. This additional circuit breaker should not be subjected to undue heat radiation.

10.9.3 Control and auxiliary circuits

Tests for temperature rise of coils and other control devices should be made with the rated current and frequency. Closing and tripping devices should be tested at their rated supply voltage. Circuits intended for continuous operation should be tested for a sufficient time for the temperature to reach a constant value.

For closing and tripping circuits that are energised only during the selected operation, the test conditions should be modified as appropriate. If the circuit breaker is provided with a device which automatically opens the circuit at the end of the operation, the device must be energised ten times in succession, the time interval between the instants of each energising being 2 s. Where the circuit breaker has no automatic device for opening the control circuit at the end of an operation the circuit should be energised and de-energised in a predetermined cycle appropriate to the application. This would normally take the form of a test energising the circuit ten times; the interval between the instants of each energising being 2 s, and the duration of each energising being 1 s. After cooling down, the circuit should then be energised once for a duration of 10–15 s. This latter requirement is the controlling feature.

10.9.4 Main circuit resistance

The resistance of each pole of the main circuit of a breaker should be measured for comparison between the circuit-breaker type tested for temperature rise and all circuit breakers of the same type subsequently routine tested. The resistance of any circuit breaker should not exceed the resistance of the type tested breaker by more than 20%.

The measurement of main circuit resistance should be made with direct current, recording the resistance, or direct voltage drop, across the terminals of each pole. The current chosen for the test should have any convenient value up to the rated normal current, but should preferably be at least 100 A. The measurements should be made before the temperature rise test and at the ambient air temperature, and should be recorded in the type test report.

10.9.5 Interpretation of results

The temperature rises of the different components should be referred to the ambient air temperature. If the ambient temperature during the test is between 10–40°C, no corrections are necessary to take account of the ambient temperature during the test, and the values of temperature rise specified are limiting values. Where an assembly or subassembly exceeds the specified values, it shall be considered to have failed the test. If the ambient temperature exceeds 40°C or is less than 10°C, standard conditions do not apply, and the results are subject to special agreement between the manufacturer and user.

It is known that the contact resistance of unplated contacts is a rather variable quantity for any given design, and is susceptible to change with length of service for each particular circuit breaker. Increase in contact resistance is more pronounced in circuit breakers that may remain in the closed position for a long time without being operated at a current near to the rated normal current and at a high ambient temperature. For this reason many modern circuit breakers have silver-plated main contacts, since for these the contacts resistance is not only lower but less likely to fluctuate. The result of this is that higher temperature rises can be permitted for silver-faced contacts, but the problem of durability of the silver facing can be open to incorrect interpretation. The quality of silver facing must be such that after any of the short-circuit test duties or after the mechanical endurance test, there is still a continuous layer of silver at the contact points. If this is not the case the contact must be regarded as not silver faced and the lower value of temperature rise applied. This aspect should be borne in mind when carrying out the examination of contacts following the short-circuit and mechanical-endurance-type tests.

10.10 Short-circuit testing

In this Section some of the important factors affecting the short-circuit capability of circuit-breakers and some of the characeristics of the various types of circuit-breakers are discussed. The requirements of IEC recommendations and national standards are referred to and discussed where they are of particular importance. Mainly, the emphasis is on the testing to assist the designer in assessing the ability of his equipment to perform short-circuit duties, rather than the ability to pass the tests required by the standards.

Circuit-breaker design is a very specialised form of engineering requiring mainly mechanical engineering knowledge. However, the circuit breaker does not always see things this way, and the designer must in addition know how to provide for the dielectric appetite of its structural arrangement and arc extinguishing medium. This usually means that he must rely on another specialist engineer, the 'short-circuit test engineer' to supply him feedback information on the capabilities of his protégé. The interchange between these specialists will be assisted by the circuit-breaker designer's understanding of the testing station equipment which the test engineer has at his disposal. For this reason a simple description of the testing station equipment and characteristics is included.

10.10.1 Testing standards

Co-operation within the International Electrotechnical Commission has resulted in agreed duty cycles which are now commonly specified. The duty cycle is defined in IEC 56-2 as either 0–3min – CO–3min – CO or 0–0·3s – CO–3min – CO and alternatively CO–15s – CO although the latter is not usually called for. The 0·3s interval represents the typical dead time for auto-reclose. The alternative duty cycle which includes the 15s interval represents the minimum time employed for delayed auto-reclose. Test duties at the full rated short-circuit current as well as at lower currents are specified. The American Standard also uses a duty cycle of CO–15s – CO and also allows 0–15s–0 followed by 0–15s – CO. The 15s time interval may be reduced at the discretion of the manufacturer. The circuit-breaker performance can be far from consistent owing to many factors such as the vagrancies of airflow or oil gas pressures, contact burning and the variable characteristics of the arc. In order to ensure that the circuit breaker has an adequate margin of safety to allow for such random behaviour, it is necessary during the development to carry out a considerable number of tests. For certification purposes, however, the number of tests is restricted.

A significant factor now included in the standards is the specification of the transient-recovery-voltage conditions. Although 'standard' levels have been proposed, the appropriate values remain a topic of considerable discussion, and there are significant differences existing between European and American standards in this respect and, indeed, other factors.

10.10.2 Transient recovery voltages

Many studies of our network recovery-voltage conditions have shown that the traditional method of representing the transient recovery voltage (t.r.v.) by means of a 1-cos waveform is not valid in that the initial rise of the voltage of this waveform does not represent service conditions. The 1-cos waveform results from the existence of a parallel damping resistor in the damping circuit of the t.r.v., where in fact a more realistic simulation is achieved by the use of a series damped circuit. The effect of a series connected resistor is to cause a much steeper voltage rise. In addition to this, local capacitance on the source side of the circuit breaker, e.g. the capacitance of busbars, limits the rate of rise for the first few microseconds. These factors have been taken into account by introducing a t.r.v. comprising an initial time delay of a few microseconds and a fast rate of rise of the voltage followed by a slower rate of rise. The assigned quantitative values and the specified shape of the t.r.v. differ significantly between European and American specifications. In the former instance, a single or double frequency transient with series damping is permitted, while the latter requires a double frequency transient, the initial section of which is simulated by an exponential transient, the latter being of the 1-cos form. These two methods are compared in Fig. 10.4, which illustrates the four-parameter method of specifying a t.r.v. together with a delay line to take account of series damping effect. When tests are to be performed in accordance with this method the test t.r.v. waveform must conform to, or exceed, the specified envelope. A two-parameter envelope is used to describe a single frequency transient, while a four-parameter envelope is required for the double-frequency case.

Although both of these methods provide a more accurate method of describing the t.r.v., there are considerable problems in simulating these transients in a testing station, particularly the four-parameter and exponential 1-cos types. Various methods of simulating these t.r.v. requirements have been investigated (Lageman *et al.*, 1972), and although the representation is feasible, a significant increase in the size of t.r.v. control capacitors installed in a testing station is necessary and can only be achieved by considerable capital expenditure. The two-parameter t.r.v. can readily be reproduced and is therefore widely used to represent the other t.r.v. waveforms (Fig. 10.5).

In the author's opinion, the necessity to reproduce such complex waveforms is

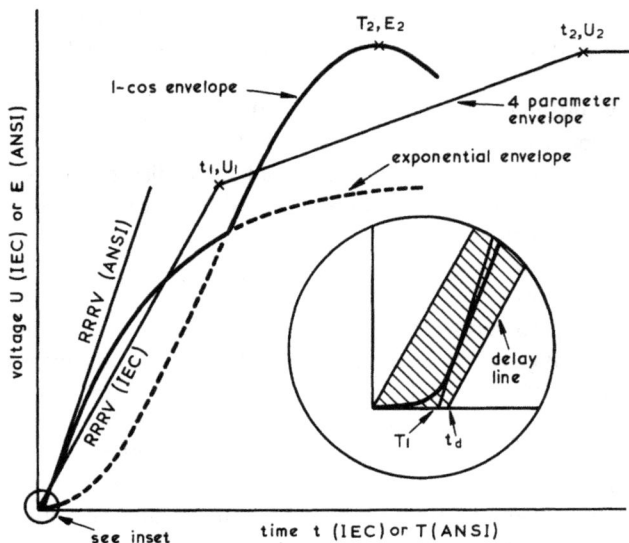

Fig. 10.4 TRV specification — comparison IEC and ANSI methods

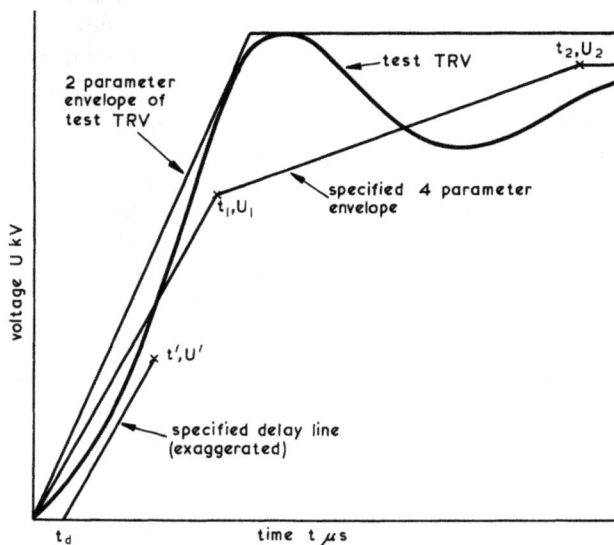

Fig. 10.5 Two parameter TRV used to represent other TRV waveforms

questionable. In practice, many designs of circuit breakers, particularly gas blast, are insensitive to the difference in time of occurrence of the t.r.v. peak as represented by the two- and four-parameter methods. If the specified intial t.r.v. can be withstood the success or otherwise of withstanding the full peak voltage is independent of the time of application within the limits imposed between the two- and four-parameter methods. The important thing is that standards now recognise that the increase in voltage during the first few microseconds after current clearance is extremely important and that the 1-cos waveform with parallel resistance damping cannot stress the circuit breaker in a satisfactory manner. Fig. 10.6 indicates how even a 1-cos waveform with a higher overall r.r.v. is significantly less onerous in the first few critical microseconds.

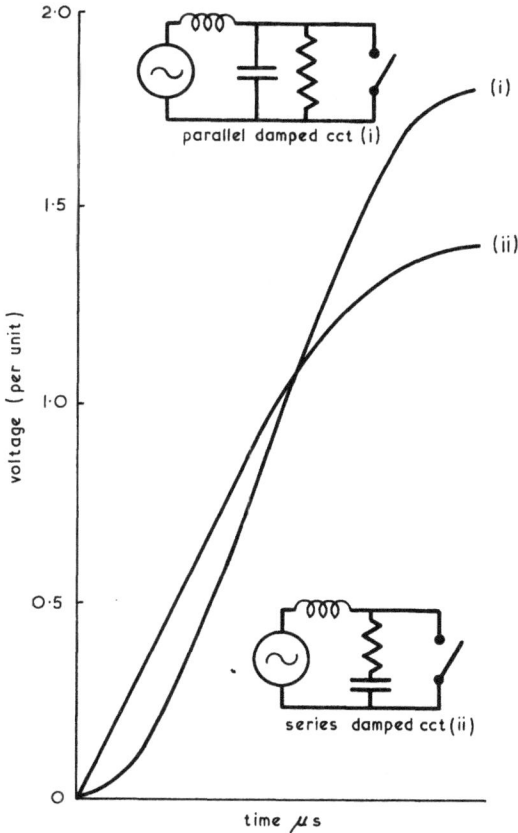

Fig. 10.6 Initial severity of TRV — comparison of waveforms

10.10.3 Inherent circuit parameters

The calculations shown in Section 10.10.5 are used to determine the inherrent parameters for the circuit, but these will be modified on test by the characteristics of the circuit breaker. The test parameters specified are, therefore, those inherrent in the circuit, assuming the circuit breaker has no modifying effect.

Theoretically, when current is interrupted at a zero, the full crest value of power frequency voltage appropriate to the power factor of the circuit is impressed across the open circuit breaker, but in practice there is a voltage drop in the generator, and this and the time constants of the generators, transformers, etc. has the effect of depressing the peak of the t.r.v. and r.r.r.v. This depression, particularly in the case of a generator, can result in a significant easement, as illustrated in Fig. 10.7. Allowance for this should therefore be made when assessing the inherrent characteristic of the test circuit.

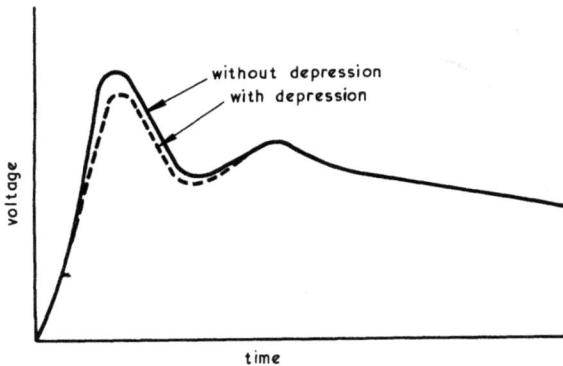

Fig. 10.7 Effects of voltage depression

The following methods are available for determining the inherrent t.r.v. waveform:

(*a*) direct short-circuit switching
(*b*) power frequency current injection
(*c*) capacitor current injection
(*d*) model networks
(*e*) calculation from system parameters

Methods (*a*)–(*c*), are most suited to use in short-circuit testing stations, and (*d*) and (*e*) are more appropriate to power systems.

Method (*a*) yields the best results when a circuit breaker is available which has a low arc voltage and negligible post arc current, e.g. vacuum circuit-breaker, and although compensation is necessary for the effects of arc voltage, the recorded t.r.v. takes into account the factors contributing to voltage depression, and is, therefore, often used as a check on other methods.

Method (*b*) involves the injection of a small current into the test circuit using a low-voltage source and recording the circuit response when the current is interrupted by means of a semiconductor diode. It can only be utilised when the test circuit is dead, and is thus mainly applicable to test plants, but has the disadvantage of making no allowance for the effects of corona or magnetic saturation.

Method (*c*) is similar to method (*b*), but uses a capacitor as the voltage source instead of a power frequency supply. It is more appropriate for measuring high-frequency t.r.v.s, such as those present on artificial transmission lines. The use of model networks is limited by the need to know fully the characteristics of the plant to be simulated, as well as producing an accurate model, particularly when characteristics vary with frequency. Calculation from system parameters is convenient using digital-computer facilities, provided the full parameters of all equipment are known. This method does not in general take account of depression effects.

10.10.4 Methods of test
Two conditions must first be considered:

(*a*) whether the testing station output is greater than the rating of the circuit breaker
(*b*) whether the rating of a single unit or phase of the circuit-breaker is greater than the maximum direct output of the testing station.

If condition (*a*) is met then a complete circuit breaker should normally be tested. If condition (*a*) is not met a unit test method must be employed, and this is possible only if the circuit breaker has been designed to permit the use of this method. Should condition (*b*) prevail then synthetic or as a last resort two-part test methods must be employed. In the event of condition (*a*) applying, the testing station parameters would be adjusted to provide the current, voltages, t.r.v. etc. appropriate to each of the duties stated in that programme, but in the case of condition (*b*) it must be modified to allow for the fact that unit tests are being made.

10.10.5 Voltage distribution
The methods used to modify the unequal distribution of voltage between several series-connected interruptors, which occurs due to inherrent capacitances, are:

(*a*) high or low ohmic, linear or nonlinear resistors across each interruptor
(*b*) capacitors across each interruptor
(*c*) a combination of (*a*) and (*b*).

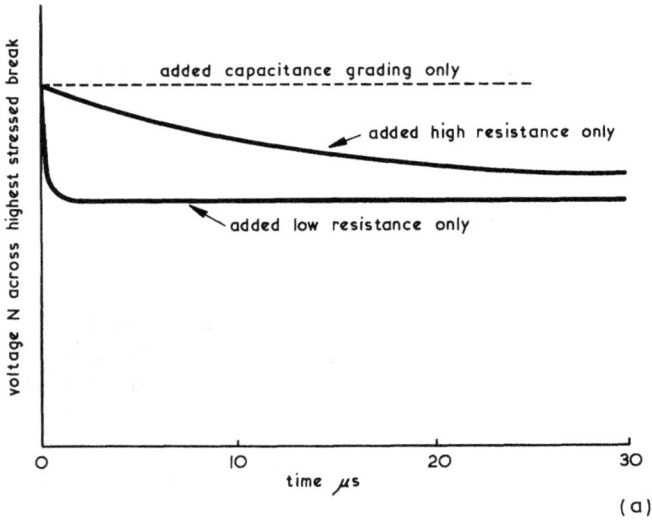

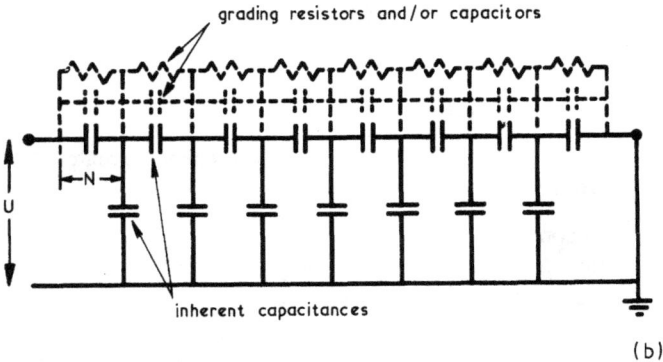

Fig. 10.8 Voltage distribution on multi-break circuit breakers

(a) Effectiveness as voltage stress against time
(b) Equivalent circuit for multi-break circuit breaker

Fig. 10.8 illustrates the effectiveness of the various forms of voltage control with respect to time for a circuit breaker with eight interruptors, the equivalent ladder network also being indicated. Capacitance grading, results in a distribution independant of time, and a very low ohmic resistor is required to ensure equal distribution across all interruptors.

The time scale shown suggests that the same voltage grading factor should be allowed when unit testing under short-line-fault conditions as for terminal-fault conditions, particularly when using capacitive grading. However, it is claimed that during the period immediately following current zero on circuit breakers exhibiting significant post arc conductivity, the post arc resistance path actually controls the distribution in a manner similar to that shown for low ohmic resistors, resulting in linear distribution. This aspect has not, however, been conclusively demonstrated.

For terminal-fault conditions, the distribution is determined for the condition of one terminal earthed and the other energised, but under phase-opposition conditions an easement is obtained as both terminals are energised. To a lesser extent this is true for 'short line' faults.

Measurements under power frequency conditions are acceptable when capacitive grading is used, but when resistors are utilised the measurement must be made at the required t.r.v. frequency as they are not frequency dependent. However, the distribution is usually calculated by solving the ladder network.

Consider a multiunit circuit breaker having n interrupters per phase, which is to be tested for a terminal fault of rated symmetrical breaking capacity I_t kA at a rated 3-phase voltage of V_n kV; the supply r.r.v. is (dV_s/dt)V/μs, and the first-phase-to-clear factor is $1 \cdot 5$;

$$\text{peak t.r.v.} = \hat{V} = \frac{V_n \sqrt{(2)}}{\sqrt{(3)}} \times 1 \cdot 5 \times k,$$

where k = amplitude factor, and peak

$$\text{frequency of t.r.v.} = \frac{\hat{a}V_s/dt}{2\hat{U}}$$

The most highly stressed interrupter of a circuit-breaker pole can be determined by calculation or by the test methods previously mentioned.

If this stress is $N\%$:

$$\text{the power frequency unit test voltage} = \frac{V_n \times 1 \cdot 5}{\sqrt{(3)}} \times \frac{N}{100}$$

$$\text{the transient peak voltage per unit} = \hat{V} \times \frac{N}{100}$$

$$\text{the r.r.r.v. per unit} = dV_s/dt \times \frac{N}{100}$$

$$\text{supply reactance} \quad Xs = \frac{V_n \times 1 \cdot 5}{\sqrt{(3)} \times I_t} \times \frac{N}{100}$$

$$\text{supply inductance} \quad Ls = \frac{X_s}{\omega}$$

and the supply capacitance $\quad Cs = \dfrac{1}{\omega_s{}^2 L_s}$

where ω = system radian frequency
and ω_s = input-circuit natural radian frequency

The parameters of the test supply can now be simulated in the testing station, appropriate series or parallel resistance being added to provide the degree of damping required. For 'short line' faults at any given distance, the artificial line parameters are determined as follows:

line surge impedance = Z_0

short-circuit current at a distant fault = I

line side peak voltage = $\hat{V}_l$

line side r.r.v. = dV_l/dt

Then the line side frequency = $\dfrac{dV_l/dt}{2\hat{V}_l}$

If for this frequency the most highly stressed interrupter is, from calculation or test, $N\%$ of the total pole voltage, then

peak voltage per interrupter = $\hat{V}_l \times \dfrac{N}{100}$

and the r.r.v.

$$= dV_l/dt \times \dfrac{N}{100}$$

For an amplitude factor k, the r.m.s. line voltage drop

$$= V \text{ r.m.s.} = \dfrac{\hat{V}_l}{k\sqrt{(2)}} \times \dfrac{N}{100}$$

Therefore the line inductance

$$L = \dfrac{V \text{ r.m.s.}}{\omega_0 I},$$

where ω_0 = line natural radian frequency.

The voltage wave travels along the line at a speed νm/s, which can be assumed equal to the speed of propagation of light waves. Thus the line capacitance

$$C = \dfrac{1}{\nu^2 L}$$

10.10.6 Unit testing

The national standards allow that each independent unit interrupter may be separately tested with its appropriate voltage and current. It is necessary first of all to define a unit interrupter.

A unit on a circuit breaker, whether for making or breaking purposes, is defined as that part of a circuit breaker which in itself acts as a circuit breaker and which in series with one or more identical and simultaneously operated making or breaking units forms the complete circuit breaker.

The distribution of voltage between units is determined by the grid of shunt and series capacitances inherent in the circuit breaker construction. This division can be modified by adding a permanently connected capacitor in parallel with each interrupter unit and/or by parallel connected resistors, which are in circuit until normal 50 Hz voltage conditions across the interrupter unit have been reached. The auxiliary interrupter units controlling the shunt resistors will, however, have a different voltage distribution pattern as this is controlled by the capacitance grid only. The voltage division is discussed in greater detail in Section 10.10.5.

The basic requirements that must be fulfilled in order to satisfy the validity of unit testing are as follows:

(*a*) The opening or closing of the interrupters of a complete pole must be practically simultaneous. In practice a tolerance of up to a quarter of a cycle of rated frequency is acceptable between the operation of contacts within a circuit breaker pole or between poles, and to achieve this the tolerance on each interrupting unit must be less, of the order of 3 ms.

(*b*) Where the supply of arc extinguishing medium is obtained from a source external to the units, the supply to each should be independent of that to all other units, the system being arranged to ensure that all are supplied virtually simultaneously.

(*c*) The exhaust arrangements of each interrupter of airblast circuit breaker and the magnetic forces associated with the current through an oil circuit breaker interrupter should in no way influence the operation of other interrupters.

Although these are the basic requirements necessary for the validity of unit testing, they involve other requirements that are sometimes overlooked.

A mechanism designed to operate three single-phase poles of a circuit breaker each with a number of interrupter units in series, is generally much too powerful for the operation of an individual interrupter unit, and if so arranged, for example on an oil circuit breaker, the contact opening speed and closing speed will be markedly increased. On the other hand, it is uneconomical and sometimes impracticable to design the mechanism in such a manner as to produce the limited energy required for the operation of one unit. Because of this, when a mechanically coupled 3-pole circuit breaker operated by a single mechanism is to be tested with only one unit operating under short-circuit conditions, and where under this condition the operating characteristics are significantly different to those of the fully operative circuit breaker, simulated loading must be added to correct the operating characteristics. When testing a single interrupter unit of the minimum-oil-content type it is often convenient to mount the interrupter on a shorter support insulator to that used on the complete circuit breaker. Here again it may be found necessary to simulate additional loading of the mechanism in order to achieve sensibly the same operating characteristic as that on the full weight circuit breaker. When operation under full short-circuit current produces an operating speed during the time of arcing which is significantly different to that at no load by say 25%, the loading on the mechanism can be quite different with all interrupters under short-circuit conditions compared with only one, and is invalid.

Single pressure type SF_6 circuit breakers generate their own pressure by a piston acting in a cylinder attached to the moving contact. Designers should recognise that the operating mechanism is correctly loaded only when an arc is present in all interrupter nozzles.

Proof of performance, however, can be demonstrated by performing tests on each unit with the operating mechanism loaded so that the moving contact characteristic is sensibly identical to that of a complete pole, together with supplementary tests on the complete pole to show that the mechanism has sufficient energy to operate all the interrupters simultaneously.

The limitations that exist in the testing station may prevent this being done at full rating on a multi-interrupter circuit breaker, but it is essential to simulate the correct arc duration. This may be achieved by carrying out a test at rated current and maximum available voltage, and with the initial t.r.v. severity increased in order to increase the arc duration to that obtained during unit tests. Although a synthetic test could be applied to a complete pole, it may well be invalid owing to the distortion of the test current caused by the summated arc voltages of the multiple interrupters of the test and auxiliary circuit breakers.

As well as the need to ensure correct operating characteristics during opening, it is necessary to be satisfied that the load imposed on the mechanism, due to the peak making current in all the interrupters, can be sustained. With the knowledge of prearcing time obtained from a full-voltage low-current breaking test, it is possible to obtain the same mechanism loading by ensuring the same prearcing time when a full current, but a lower voltage, is applied. On airblast circuit breakers and any other type of circuit breaker which has a small contact travel, this is unnecessary, but on oil circuit breakers this is achieved by including at the fixed contact a piece of fuse wire of sufficient length to initiate the required prearcing.

The magnetic forces on an arc in an interrupter unit caused by the current passing through an adjacent pole or other interrupters in the pole where they are not mounted in a straight line, can be simulated by arranging the connections to the interrupter to be physically in the same position as that occupied by the adjacent pole or interrupters.

The unit testing of bulk oil circuit breakers requires special consideration. Designs have been produced up to 132 kV with the three phases in one tank. These have up to two interrupters in each pole. Above 132 kV bulk oil circuit breakers have a tank per phase, but these may contain up to six interrupters in series. These all have some form of shunting resistor which is inserted during the opening stroke, which, besides limiting overvoltages due to current chopping or restriking on-line dropping, provides a control of the voltage division between the interrupters.

The gas bubble forced out of the interrupter during and immediately subsequent to the arcing can reduce the insulation strength of the oil between the interrupter and the tank and also between the interrupter units. For this reason ideally all these oil gaps should be stressed with the appropriate voltage during the unit testing. Various circuit-testing arrangements have been derived to achieve this, but none of them completely satisfies all conditions without overstressing one or more parts of the insulation system. For example, a three-phase-in-one tank circuit-breaker with two interrupters in each pole can be arranged with one of each pair of interrupters shorted out so that the interrupter units being tested in a 3-phase circuit can be tested with half the full voltage of the breaker and with the required currents. However, the oil gap to earth and between phases is then only stressed at half voltage. To overcome this the tank can be insulated from ground and energised at half voltage of opposite polarity to one of the phases. The other two phases, however, do not have the correct voltage to tank applied and the oil gap between the poles is still stressed only at half voltage. Another test arrangement, which is attributed to Dr. Biermanns of Germany, can be used on single-pole bulk oil circuit-breakers with two or more breaks per pole and enables a 3-phase connection of the testing station to be used to test two groups of interrupters. When interrupting a short circuit between three phases which is clear of earth, the second and last phases to clear do so at the same instant, and in fact one of these two can be considered as redundant. Biermanns therefore arranged to introduce the third phase into the tank to the cross-bar via a through bushing in the tank wall. Although the current and the recovery voltages are correct and thus the amount of gas produced is substantially correct, the voltage to tank is still only half the full value in each phase.

Unit testing of bulk oil circuit breakers therefore cannot fully represent all service conditions. However, this type of circuit breaker tested as indicated above has been extensively used and has proved eminently satisfactory in service. Therefore, in considering unit testing as a means of proving the rating of bulk oil circuit breakers sensible judgement must be exercised.

The duty cycle, which is the basis of the IEC Recommendations and national standards, includes only three break operations for each required short-circuit current level. There are, however, many variables to contend with, in the speed of contact movement, variations in pressure of the extinguishing medium and not the least, the random characteristics of power arcs, consequently, three operations is quite inadequate to indicate to the circuit-breaker designer with sufficient confidence the capability of the circuit-breaker design. It is now widely accepted that the statistical approach which has been applied to switching impulse testing is equally applicable to short-circuit testing. Unit testing is one of the useful tools available to the circuit-breaker designer which allows him, though not inexpensively, to obtain a sufficient number of test results to enable a better understanding of the probability of success or failure of a design at a particular required short circuit rating. The statistical performance obtained in this way is, however, applicable only to a single interrupter unit and poses the question of how it may be extrapolated to indicate the performance of the multiunit circuit breaker.

This aspect has been studied in detail by Braun (1972) by investigating the performance of a single and two interrupter units of a circuit breaker in the probable failure area. This work included the investigation of the effect of the various modes of breakdown, i.e. thermal and dielectric breakdown of the interrupters. It was concluded that the distribution functions relating to the short-circuit-current performance differ according to the breakdown mechanism. In the case of thermal breakdown, i.e. post-arc current effect, an ideal voltage distribution exists between interrupters which results in the performance of two interrupters in series being greater than twice that of a single interrupter unit. In the case of dielectric breakdown, however, it was concluded that unit testing is only valid at low levels of failure probability; in the latter case the voltage on each interrupter is controlled by the external impedances including any added parallel capacitance or resistance and in the former case by the post arc current condition in each interrupter unit. As the post-arc current in each series connected interrupter is the same the voltage division is equal, whereas the external impedances may not ensure that this is so. Thus, although this work confirms the validity of unit testing, it demonstrates the need to ensure that the rating is assigned at a low probability of failure level and emphasises the need to differentiate between the two modes of failure.

Unit testing does not, however, completely demonstrate the performance of a 3-phase circuit-breaker under actual service conditions. When tested in a short-circuit testing station, the unit has only two opportunities per cycle to interrupt the current, whereas the 3-phase unit has six opportunities, enabling one phase to assist the other in interruption. Wilson *et al.* (1971) have presented a comprehensive analysis of the improved reliability obtained by assessing performance on a 3-phase basis.

10.10.7 Point on wave
As the ability of a contact gap to sustain the impressed recovery voltage depends on its dielectric strength, which in turn is dependent on the pollution of the dielectric, it follows that the energy released between the point-on-wave of contact separation and the natural current zero is a decisive factor.

Circuit breakers, such as the air-blast type, which utilise an external supply of arc extinguishing medium have a definite range of arcing times. A typical air-blast circuit breaker may have a minimum arcing time of 3 ms and a maximum of 22 ms. In fact, the most probable maximum would be 12 ms, but owing to random conditions the probability of 22 ms may be very small. The maximum amount of information can be obtained by grouping the tests at, or close to, this minimum time to current zero. Separation of the contacts at say between 6—8 ms before a current zero can represent a relatively easy test condition. For this reason the single-phase test procedure for circuit breakers with arcing times not exceeding one cycle for the first pole to clear is specified by IEC in the following manner:

For the test duty at full rated symmetrical short-circuit breaking current, a minimum of three operations are made for the rated operating sequence. During the first valid breaking operation, contact separation must occur sufficiently close to one current zero to ensure that arc extinction does not occur at this current zero. It may be necessary to make more than one test to achieve this. During the second breaking operation, contact separation is then advanced approximately 60° electrical from that of the first valid breaking operation.

If during the second breaking operation arc extinction occurs at the first current zero, then the third breaking operation is made with the same timing of contact separation as the first valid breaking operation. If during the second breaking operation arc extinction does not occur at the first current zero, then the third breaking operation is made with contact separation advanced approximately 60° electrical from that of the second breaking operation.

For the test duty with asymmetrical breaking current, the sequence of testing must additionally take account of the asymmetrical form of the current wave and the corresponding energy dissipation in major and minor loops of arcing. The first valid breaking operation is arranged such that contact separation occurs during the minor loop of current and sufficiently close to the current zero to ensure that arc extinction does not occur at this current zero, i.e. to ensure that the circuit breaker is subjected to a full major loop of arcing. For one of the other two breaking operations, contact separation is advanced, and for the other retarded, by approximately 60° electrical from that required for the first valid breaking operation.

The arc duration of circuit breakers which generate their own arc extinguishing energy, such as oil circuit breakers, is also influenced by the point-on-wave of contact separation, but they are necessarily more random in their behaviour. It is therefore essential to investigate more fully the performance throughout the complete cycle of current to ensure a proper understanding of the capabilities of this type of circuit breaker is obtained. This is particularly relevant for asymmetrical current tests where although the correct d.c. component may be obtained, the duty will be considerably eased unless a final major loop of current is obtained.

10.10.8 Arrangements for test

Reference has already been made earlier in Section 10 to some of the provisions for short-circuit tests, particularly those relating to unit testing. The following gives an insight into further miscellaneous arrangements which may be made, depending somewhat on the type of circuit breaker to be tested, and whether the tests to be made are for development purposes or to obtain certification to a given standard. The number of tests which may be made on one particular design of circuit breaker are limited for practical and economic reasons. It is therefore essential that in order to assess with sufficient confidence the capabilities of a design the most onerous conditions, commensurate with those possible in service, should be applied for each test duty.

For a 3-pole circuit breaker, all short-circuit making and breaking requirements can, in principle, be proved with a 3-pole circuit breaker with all poles operating together. A circuit breaker having all its arcing contacts supported within a common enclosure can only be tested as a complete unit to be representative of its type. Where each circuit breaker pole is a separate unit it may be tested for convenience, ór owing to limitation of available testing facilities, as a single-pole unit provided that it is representative of the complete 3-pole circuit breaker in respect of:

(*a*) speed of make
(*b*) speed of break
(*c*) availability of arc-extinguishing medium
(*d*) power and strength of closing and opening devices
(*e*) rigidity of structure

The circuit-breaker operating devices are normally set during the tests at the minimum voltage or pressure as applicable, and in the case of gas-blast circuit breakers the supply of arc extinguishing medium should also be set at minimum pressure. Exceptions may be made in those cases where operation at minimum levels may make accurate control of the test sequence difficult, e.g. inconsistent breaker opening times, or where operation at maximum pressure could be more onerous, e.g. current chopping at low current levels.

Where the physical arrangement of one side of a circuit breaker differs from that of the other side, the live side of the test circuit should be connected for test to that side of the circuit breaker which gives the more onerous conditions with respect to voltage to earth, unless the circuit breaker is especially designed for feeding from one side only.

When a single break air circuit breaker, such as that of Fig. 5.3 of chapter 5, clears short-circuit currents the largest quantity of gas is generated at the fixed contact side of the chute because there is no arc transfer to the runner at that side, and the arc moves more quickly up and burns for a longer time there. Earthed metal in metalclad constructions is approximately the same distance from all points at the top of the chute, and therefore with a low-impedance cooler the possibility of earth faults when the circuit breaker is opening is greater when the fixed contact is connected to the supply side, i.e. a 'dead-blade' connection.

With high-impedance coolers, gases cannot escape so easily and there is a greater tendency for the ionised gases to move down the arc chute and to stress the insulation between the fixed and moving contacts. This may cause further breakdowns across the contact gaps until failure to earth below the moving contact blade takes place. With high-impedance coolers, therefore, the circuit-breaker is generally more highly stressed during 'live-blade' tests.

Single break bulk oil circuit breakers have the interrupter supported from the bushing on one side, the other bushing supporting the hinge of the moving contact. The emission of gas within the tank is therefore asymmetrical and the position of any earthed metal within the tank, such as a dashpot, can determine the most onerous arrangement in the test circuit. For example, the gas rising from the interrupter may bridge the oil gap between it and the dashpot; with the h.v. test lead connected to the interrupter bushing this gas column will be electrically stressed, but not so if the interrupter bushing is connected to the ground potential lead.

Similar conditions may apply to other types of circuit breaker which are characterised by the lack of symmetry of the contact and/or interrupter arrangement. Examples of circuit breakers falling into this category must include single-break small-oil-volume circuit breakers, dead tank SF_6 or vacuum units or any circuit breaker having an odd number of interrupters per pole. In cases of doubt as to which test condition would be most onerous, specified groups of test duties in the test series may be performed with opposite test connections.

When setting up a circuit-breaker test arrangement adequate facilities must be included for monitoring of the behaviour during test. Essential for meaningful analysis of circuit-breaker performance are oscillographic records of test quantities, such as applied and recovery voltage, current and relative timing of operating device initiation, contact separation etc. In addition, the above minimum recordings may be supplemented by oscillograph traces of contact travel, oil or gas pressures and auxiliary devices. During the test for rated short-circuit making current it is necessary to confirm that the circuit breaker closes fully and latches closed without any undue hesitation. It is sometimes difficult to establish whether or not a circuit breaker has latched and at what instant of time latching occurred unless this operation is suitably recorded. The attachment of a simple sensing device, e.g. microswitch, to the latching mechanism will enable this event to be recorded on the oscillograph together with the other phenomena.

10.10.9 Failure diagnosis

Failure of a circuit breaker to interrupt may be due to a number of factors and often it is possible to determine the reason for failure from an examination of the t.r.v. up to the point of breakdown. The modes of failure can be categorised as (*a*) thermal breakdown due to energy balance conditions and (*b*) dielectric breakdown of the contact gap. Dielectric breakdown can either occur at a relatively low voltage, such as short-line fault conditions and is mainly confined to air-blast circuit breakers and usually is a function of the electrode configuration, or may occur at or near the maximum value of the t.r.v. generally for terminal fault conditions and is common to all circuit-breaker designs. Failure in this manner is often indicative of the need to increase the rate of separation of the contacts (for example, by increasing the speed or travel) or by modification of the venting area of the interrupter.

The ability to recognise these various types of breakdown is important, particularly during development testing, as it provides an indication of the modifications which may be necessary to improve the performance. Thermal reignition, such as the breakdown shown in Fig. 10.9 is characterised by a smooth transition in the voltage trace through current zero and may occur at more than one current zero. This phenomenon is present when the rate of dissipation of energy from the arc is less than the rate of energy input to the arc.

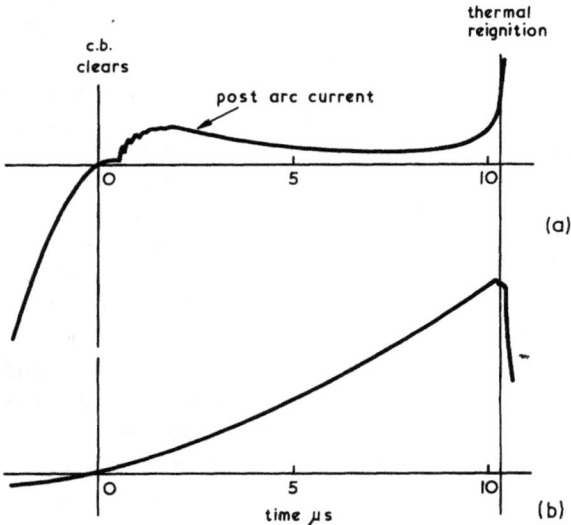

Fig. 10.9 Thermal reignition of a circuit breaker

(*a*) current trace
(*b*) voltage across circuit-breaker contacts

A breakdown at the peak of the transient recovery voltage on an air-blast circuit breaker may be due to turbulent air flow at the nozzle. Particular attention has to be paid to the profile of the nozzle in order to obtain the optimum short-circuit performance. There are sophisticated techniques available, such as Schlieren photography which are aids to the development of the optimum nozzle profile but the presence of the intensely hot arc between the contacts affects the air flow

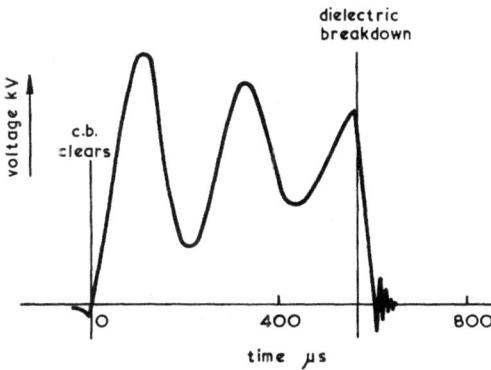

Fig. 10.10 Interruption failure due to dielectric breakdown

conditions and to obtain effective information in this way can be as complicated and expensive as a programme of tests on the nozzle profiles.

Occasionally another type of failure can occur, that of failure after the peak of the transient recovery voltage as illustrated in Fig. 10.10. This is generally owing to particles released into the contact zone after melting of the metal has occurred. It may also occur owing to turbulent air-flow conditions on an air-blast circuit breaker, particularly those with very high operating air pressures. Measures which may be required to obviate this type of failure include changing to a more arc resistant material or applying a coating of high-temperature material such as zircon on upstream non-current-transfer faces of contacts or again by attention to the air-flow conditions. On oil circuit breakers it may be necessary to improve the pressure of the oil in the contact gap which can be reduced to a very low value by the displacement of the moving contact.

Although the interpretation of the mechanism of the breakdown is by no means as simple as the foregoing may suggest it is, nevertheless, essential to be able to identify which mechanism of failure exists in order that the appropriate corrective measures can be taken.

10.11 Synthetic testing
Market conditions necessitate the minimal number of interrupters per phase. However, the rating of the interrupter unit has historically been dictated by the maximum output available from direct testing stations. This, plus the exorbitant cost of significantly increasing the direct output of test plants, has in the last 10—15 years stimulated interest in synthetic test methods first postulated in the early 1930s.

Two distinct phases exist during fault interruption, the arcing period comprising a high current and low voltage, and the post zero period comprising no current and high voltage. It is thus possible, theoretically, to utilise two separate energy sources, one to provide the current and the second to provide the recovery voltage, but the problem lies in the transition from one condition to another around the period of current zero, while permitting the same interaction between circuit breaker and test circuit, as exists in a direct circuit. The degree to which this is achieved by synthetic circuits is a measure of their validity, unfortunately this expression cannot be quantified, primarily because the phenomenon existing at the current zero period is still not fully understood, despite the extensive research that has been made.

Synthetic test methods have been employed since the early 1930s, one of the earliest being that attributable to Biermanns, which has been referred to in Section 10.10.6. In this and most other instances significant distortion of the current and/or voltage occurs leaving the validity of the test open to considerable doubt.

There are many methods of combining the two energy sources, but two basic systems are at present utilised, namely current injection and voltage injection, the difference between the two being that in the former case the circuit breaker is conducted through the zero period by the voltage source and in the latter by the current source.

Extensive researches initiated in the early 1960s aimed primarily at comparing direct and synthetic test methods have resulted in the synthetic test methods achieving a high degree of respectability for interruption tests in that they are considered suitable for type and acceptance testing as well as being an ideal developmental tool. The reader is referred to Anderson *et al* (1966) and (1968).

Research into methods of synthetic testing for making capacity and interruption for circuit breakers with low ohmic shunt resistors is still proceeding.

10.11.1 Voltage injection

This test is primarily attributable to the work of Slamecka. A simplified schematic diagram is shown in the Fig. 10.11. The method is as follows:

The current circuit supplies the transient voltage across the circuit breaker up to a point some microseconds after current zero, the slope of the initial t.r.v. so provided being controlled by the series connected resistor capacitor combination. The high-voltage circuit is arranged to be switched in at a particular time after current zero by controlling the setting of the sphere gaps located in the current source. Operation of these gaps connects in the high-voltage circuit, at which time the circuit breaker is stressed by a t.r.v. comprised of the sum of both energy sources. The high-voltage source may be connected either across the test or auxiliary circuit breaker, the latter being more usual, resulting in circuits designated either parallel or series voltage injection.

When 'short-line' fault tests are performed the artificial line is connected between the circuit breaker and earth in the high-current circuit, with the effect

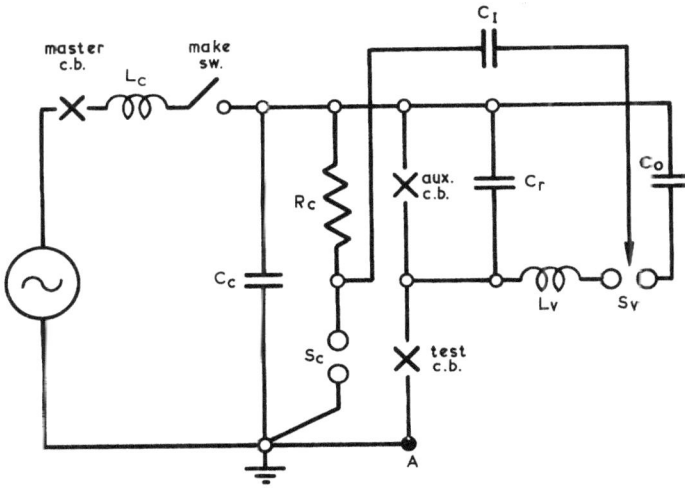

Fig. 10.11 Simplified diagram of voltage injection circuit

that the initial stressing after current zero is still provided by the high-current circuit. Development testing of t.r.v. sensitive gas-blast circuit-breakers may thus adequately be performed without the high-voltage circuit. A major disadvantage, however, is that the first line component, inductance L_c, must be capable of carrying the full short-circuit current and this thus limits the test current. The LC units for the remainder of the line are rated only for the line voltage drop. To overcome the limitation introduced by L_c a further circuit has been developed which treats the artificial line as a further source of voltage to be connected in at the appropriate instant, unfortunately introducing still more complications. Fig. 10.12 illustrates this additional circuit which is connected between point A and earth in place of the solid connection of the terminal-fault circuit Fig. 10.11.

Use of the voltage injection method requires the close matching of the initial t.r.v. in the current and voltage sources so as to ensure a smooth transition from one to the other. A useful, if not essential aid, therefore, is a transient-network analyser.

A basic problem of all methods of synthetic testing is that of obtaining consistency of operation of the triggering system, and one of the attributes of the voltage injection method is that once the heavy current gaps are set to the particular spacing, consistent operation can be achieved and the speed of testing approaches that obtainable in a direct test station.

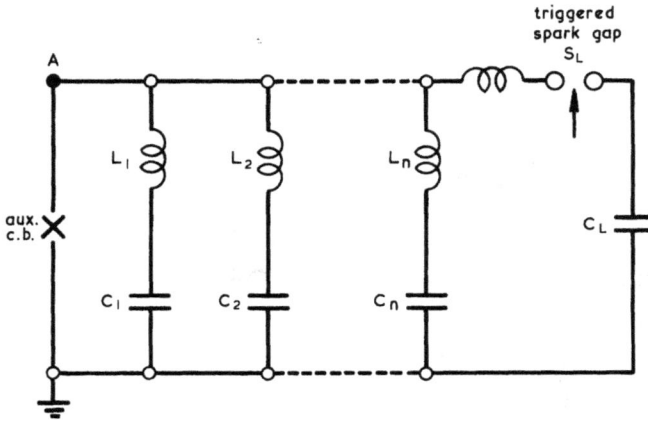

Fig. 10.12 Additional circuitry for short-line fault tests

10.11.2 Current injection
The current injection method relies upon the superposition of current and not voltage; a typical method is shown schematically in Fig. 10.13*a*. The method was developed quite independantly, and yet virtually at the same time, in Germany by Weil and Dobke and in the UK by the Electrical Research Association. It resulted from development work on the use of a capacitor voltage source and high-frequency test current. This indicated that the rate of change of current through zero significantly affects the performance of a gas-blast circuit breaker, and that an equivalent performance is obtained if the di/dt of the high-frequency current is identical to that of a normal power frequency current at zero. A typical current injection circuit operates in the following manner:

Closure of the make switch initiates the flow of current I_c through the current circuit, the master, auxiliary and test circuit breakers being already closed. Tripping of the auxiliary and test circuit breakers is adjusted such that contact separation of both occurs simultaneously. This frequently requires initiation of an opening operation prior to current flow in the test circuit. At time t_1 of Fig. 10.13*b* a high-frequency current is injected into the test circuit breaker by triggering the spark gap S in the voltage circuit, the current through the test breaker being increased by an amount I_v. The auxiliary circuit breaker carries only the power frequency current and readily interrupts at current zero (time t_2) as it is subjected only to the relatively low voltage of the generator. Arcing in the test circuit breaker

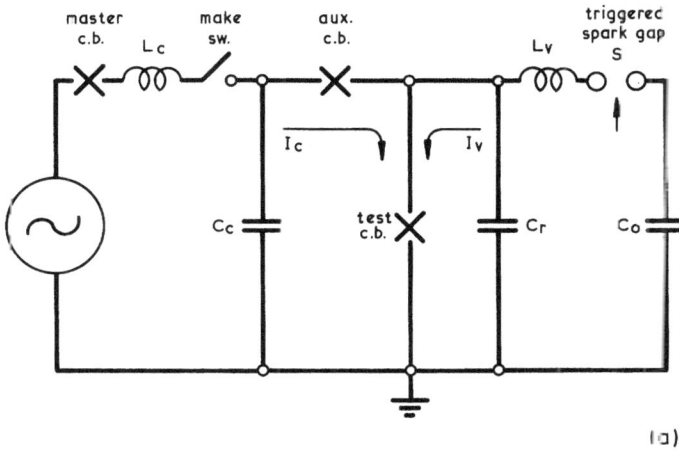

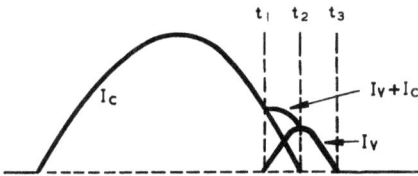

(b)

Fig. 10.13 Current injection method

    (a) simplified circuit diagram
    (b) Injection time sequence

is prolonged to time t_3, i.e. the zero of the injected current thus subjecting the circuit breaker to a direct recovery voltage transient from the capacitor source. The t.r.v. of this circuit is controlled by the parameters C_r and L_v. Should the circuit breaker fail to interrupt at this point, subsequent current flow comprises only that of the voltage source, thereby limiting the possibility of damaging the circuit breaker, a significant aspect when performing development tests.

As in the voltage injection method, the voltage source may be connected across either the test or auxiliary circuit breakers, providing either parallel or series current injection circuits. The latter is more suitable when low t.r.v. frequencies are utilised, the distortion being lower than in the parallel case.

10.11.3 Validity of synthetic testing

It is essential during synthetic tests that circuit breakers shall be subject to conditions as severe as those of full direct test methods in order that they are similarly stressed. This entails meeting the following requirements:

10.11.4 Current interrupted

The form and magnitude of the power-frequency current flowing through the test circuit breaker, particularly during the final loop prior to arc extinction, should not be less than that of the equivalent direct test. A measure of the probable degree of current distortion is given by the ratio of current source voltage to test circuit-breaker arc voltage. In the synthetic test, the auxiliary circuit-breaker arc voltage is also present, resulting in the further reduction of this ratio to a level at which significant distortion may occur. An easement of the duty may result due to reduction of the peak and the duration of the current loop by the effects of distortion. This can be compensated by introducing a small degree of asymmetry into the final current loop or by reducing the inductance in the current circuit. The effects of current distortion and its compensation are illustrated in Fig. 10.14. It is necessary, because of this distortion, to permit a tolerance on the amplitude and duration of the final current loop; the practice usually adopted is that the peak value should not be less than 100% of the rated value and the duration of the final current loop should be within ±10% of the time for nominal test frequency.

10.11.5 Rate of change of current

The rate of change of current through the test circuit breaker di/dt at the instant of current zero should not be less than that of the equivalent direct test. In the current injection case this entails selecting the frequency and amplitude of the current from the voltage source (i.e. injected current) so as to achieve the same rate of change as would be imposed by the sinusoidal power frequency current.

The voltage injection method has an advantage in this respect, the correct value being readily achieved because the test current is the actual power frequency current.

10.11.6 Prezero current period

In a direct test circuit, the arc voltage of the circuit breaker increases as the current approaches zero, and this may cause the current to reach zero prematurely. Synthetic test circuits should enable a similar reaction to occur. This the voltage injection method permits but in the current injection scheme the transition from current source to voltage source should be complete before there is a significant change in the arc voltage of the circuit breaker, in order to ensure validity. This requirement together with the rate of change of current, determines the frequency and magnitude of the injected current. A frequency in the range 300–1000 Hz is normally used for parallel current injection circuits, the precise value being dependent on the arc voltage characteristic of the circuit breaker being tested. The

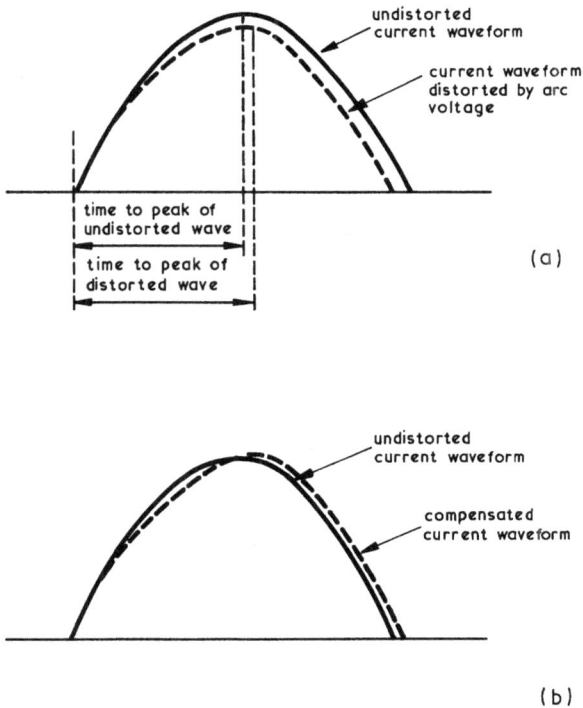

Fig. 10.14 Current distortion

    (*a*) Effect on waveform
    (*b*) Compensated waveform

series current injection circuit causes less distortion of the test current permitting the use of lower injection frequencies.

10.11.7 Post-arc duration

It is necessary to ensure the voltage circuit of the current-injection scheme reproduces the correct conditions in the post-arc current period, and that the form and magnitude of the inherent t.r.v. correspond to the specified test requirements.

    This will be achieved only when the losses in the voltage circuit are small, e.g. C_r being much smaller than C_0 in Fig. 10.13. A ratio of 10:1 is generally regarded as the minimum acceptable. A larger value of C_r has the two-fold effect of modifying the t.r.v. frequency and reducing the level of the direct recovery voltage due to charge sharing between the two capacitances.

Compensation can be provided by increasing the value of inductance L_v and the initial charge on the voltage source capacitor.

In the case of the voltage injection system, application of the t.r.v. should not occur during the post-arc period, yet shall be sufficiently early to satisfy the rated time delay (t_d) requirements of the t.r.v. The current source must reproduce the correct post-arc current conditions and the initial portion of the t.r.v.

10.11.8 Recovery voltage

The capacitance used for the voltage source gives a recovery voltage, which is direct, with an exponential decay. The acceptability of this voltage form is limited by the presence of low ohmic resistors on the test circuit breaker. Where these are greater than 1 kΩ a direct recovery voltage is permissible provided the time constant is greater than 8 ms, and the initial peak of the t.r.v. is of correct value, but for resistors of lower value an alternating recovery voltage method is in practice necessary.

The recovery voltage produced in the voltage scheme is primarily direct, but has an alternating component superimposed from the current source. However, similar limits of time constant are applicable.

10.11.9 Injection time

In order to ensure a smooth transition from power-frequency current to the injected current ideally the peak of the latter should coincide with the zero of the power frequency current, but in practice a tolerance of ±20 electrical degrees (referred to the injection frequency) is permissible.

In the series current injection method the voltage circuit should be applied to the test circuit breaker in such a way that the current in the auxiliary breaker reaches zero between 200 and 1000 μs before zero of the current in the test circuit breaker.

The injection time appropriate to the voltage injection scheme is not critical, being simply that required to ensure a smooth transition from one voltage source to another while satisfying the rated time delay.

10.11.10 Synthetic test procedure

Test procedure with synthetic circuits differs somewhat from that for direct tests primarily because of the limitation imposed by the preselection of the current zero at which the circuit breaker has to attempt interruption. A step-by-step procedure is usually adopted, requiring testing of a circuit breaker at various points on wave of contact separation. The procedure is to apply the recovery voltage transient at the first current zero for each point-on-wave condition. If the circuit breaker interrupts on all tests, further tests are not required and this becomes a valid test. In the event of a reignition, the tests are repeated with the application of the high-voltage circuit delayed until the second current zero, with forced reignition at the first to ensure continuation of arcing. Should reignition again occur it may be necessary to carry out a further series with reignition at two current zeros if this extent of arcing is not a disqualification. The forced reignition circuit is not always necessary with the voltage injection method as the initial voltage transient of the heavy-current source is often sufficient itself to ensure reignition.

10.11.11 Multi-loop testing

Both the voltage and current injection methods that have been described are applicable only at the ultimate current zero of the circuit-breaker arcing time, but obviously the total arcing energy from the initiation of the arc has an influence on the final clearance. Preliminary testing is necessary to establish how many current zeros as a maximum the circuit breaker may pass before clearing, and the synthetic test circuit must include facilities which enable the ultimate current zero to be reached.

An obvious, but expensive, method of multiloop testing is to provide the number of voltage source equipments equal to the total number of current loops desired. A further method is to delay the opening of the auxiliary circuit breaker until the final loop of arcing predicted in the main circuit breaker, and utilise the high t.r.v. of the current circuit to ensure the test circuit breaker reignites at the current zero prior to that at which interruption is predicted. This is usually limited to the first zero after contact separation when testing gas-blast circuit-breakers. A more reliable method is that known as forced reignition and shown schematically in Fig. 10.15a. The operating principle is that a current pulse having a steep front is injected into the test circuit from a charged capacitor, at $5-10\ \mu s$ prior to current zero, the polarity of the pulse being opposite to that of the power frequency current. Provided the pulse magnitude, time constant and energy are of sufficient magnitude, the current through the circuit breaker will be rapidly reversed and the current flow maintained until the next power frequency half period is established. This phenomenon is illustrated in Fig. 10.15b. One set of reignition equipment is required for each forced reignition.

In an alternating-current system the circuit breaker, after fault interruption, is subjected to an alternating recovery voltage of power frequency. In a synthetic circuit using a capacitor for the voltage source the initial t r.v. can be arranged to be closely similar to this. but thereafter is simply the direct voltage of the charged capacitor; this subjects the circuit breaker to a more severe condition if this is maintained at the peak a.c. level.

10.11.12 Resistance-switched circuit breakers

Resistors that shunt the arcing gap of an interrupter form a parallel path through which the voltage source energy can be discharged, which, depending on the ohmic value of the resistor, may result in the rapid decrease of the recovery voltage to an unacceptable degree. Test circuits have, therefore, been developed to provide an alternating recovery voltage source.

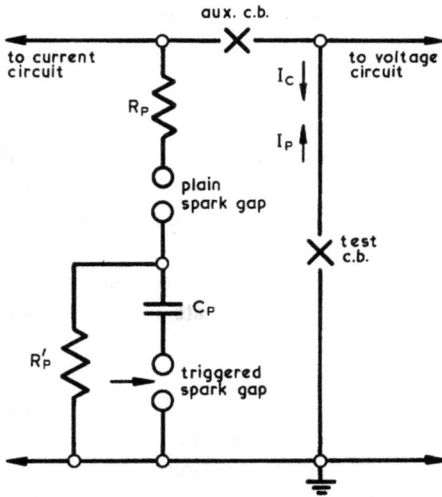

(a)

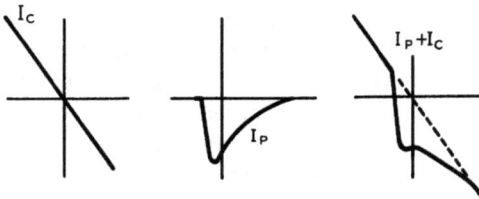

(b)

Fig. 10.15 Multi-loop synthetic tests

(a) Additional reignition circuitry
(b) Current reversal by pulse addition

Appropriate circuits are available, as shown in Fig. 10.16*a*, which require a second voltage source, comprising a generator/transformer unit, connected across the test circuit breaker. This voltage source contributing a small but significant portion of the total 50 Hz test current. However, as this does not flow through the auxiliary circuit breaker, it combines with the injected current loop and produces an alternative current zero, at a short time before the natural zero of the injected loop. The test circuit breaker will attempt to interrupt at this alternative current zero. It is possible to adjust the circuit parameters so that the required rate of

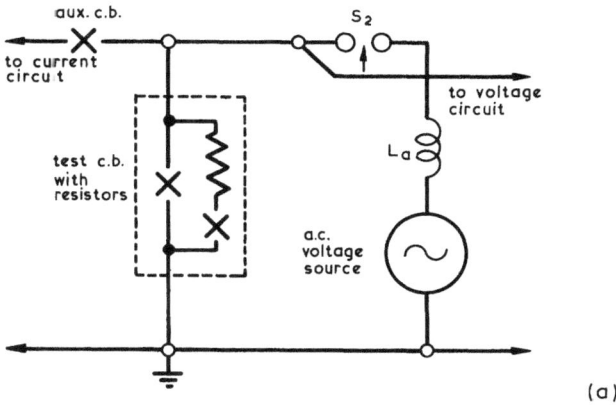

(a)

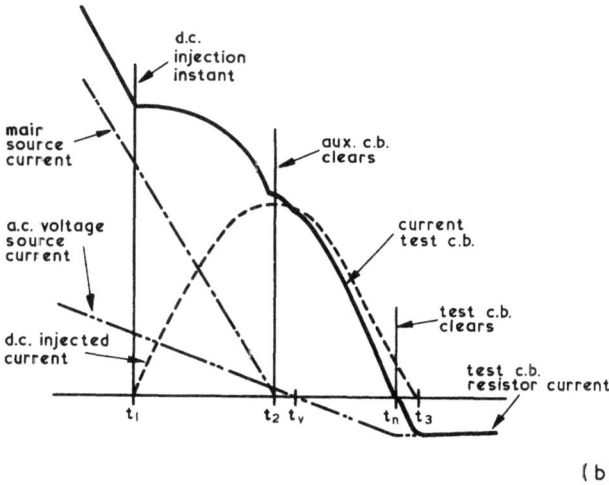

(b)

Fig. 10.16 Synthetic testing of resistance switched circuit breakers

    (a) Additional circuitry for alternating recovery voltage

    (b) Injection time sequence

change of current is still achieved. To achieve the correct recovery voltage conditions, the instantaneous current in the alternating voltage source must, at the instant of the current zero, be equal to that appertaining to the circuit-breaker shunt resistance, under steady-state recovery voltage conditions. This can conveniently be achieved by varying the position of the current zero (t_v) of the alternating recovery voltage source with respect to the injected current. This is most easily achieved by controlling the instant at which this current is initiated by means of a spark gap S_2. The use of this circuit is not in practice simple and requires very accurate control and triggering of the spark gaps. A further disadvantage is that the generator transformer unit will most likely be of the same rating as those used for the voltage source and this will reduce the power available for the current source, thereby conflicting with one of the objects of the synthetic test method, which is to increase the overall output of a testing station.

10.11.13 Short line fault: current injection
For this method the artificial line is included in the voltage test circuit, thus both the supply and line transient voltages are applied to one terminal of the circuit breaker under test, unlike the voltage injection method which more closely simulates the system conditions. It has the advantage, however, that the line sections need only be rated for the current of the voltage source rather than that of the main current source. The location of the artificial line in the test circuit is the same irrespective of whether the parallel or series current injection method is used.

10.11.14 Making capacity
A typical synthetic test method for the making capacity of a circuit breaker is a circuit arranged with the primaries of the transformers in the voltage and current circuits in parallel. The secondary side of the voltage-circuit transformer is connected across the test circuit breaker, and the secondary side of the current source transformer is connected through a spark gap of adequate rating, resulting in an arrangement which is the inverse of that used for circuit-breaker opening tests with an alternating voltage source.

The test procedure is as follows: Initially the test circuit breaker is open and the rated voltage is applied from the voltage circuit transformer, thus during the closing stroke the circuit breaker is subjected to the correct applied voltage resulting in full prearcing. When prearcing commences, the small current flowing energises the sensing device which then operates the triggered spark gap permitting full short-circuit current to flow through the test circuit breaker. It is, however, necessary that the spark gap is capable of handling the full making current. Where it is desired to perform a close/open operation this is achieved simply by adding the basic voltage capacitor source together with the auxiliary breaker to the test arrangement. It does, however, require the introduction of capacitance in both the voltage and current sources, which on current initiation can cause oscillations in the local circuits formed by these two capacitors. These oscillations can affect the behaviour of the sensing device in the spark gap, but this can be overcome by incorporating a damping resistor in series with the current source capacitor.

10.11.15 Composite test duties
Close/open duties and even autoreclose duty cycles are possible by using a combination of some of the described test circuits, but the complicated circuitry

which results is fraught with difficulties, for practical reasons a compromise between precise representations of service conditions and a close representation involving simpler test circuits has to be agreed. A complete synthetic circuit for performing short-line fault tests with multiloop arcing and 50 Hz recovery voltage is extremely complex and the numerous triggering devices used must be reliable in order that a successful test can be made.

10.12 Test facilities

The first commercial short-circuit testing station in the UK was opened in 1929. This comprised a single generator together with step-up power transformer, current-limiting reactors and the appropriate control equipment. Since that date the number of such installations has increased considerably and their capacity grown to as many as five generators, in an attempt to match the growth in power-system fault levels. Two basic test methods are at present utilised which are defined as follows:

Direct testing: A method whereby the current and voltage are obtained from the same energy source.
Synthetic testing: A method whereby the current and voltage are wholly or in part obtained from separate sources.

Either of these methods may be applied to:

(*a*) the complete 3-phase circuit breaker
(*b*) a single phase of the circuit breaker
(*c*)) a unit of one phase of the circuit breaker.

Direct testing laboratories can provide substantially the majority of operating conditions in an actual network, but they are costly. Synthetic testing laboratories extend the scope of the machines by providing the recovery voltage from a capacitor bank, but as they are as yet unable to satisfactorily reproduce all the requirements of short-circuit testing and because of insufficient confidence on the part of the users this form of testing has not obtained complete acceptance.

The largest direct test facility is at KEMA, Arnhem, Holland. No. 2 Laboratory is capable of up to 2500 MVA 3-phase making and breaking capacity at voltages from 11 to 173kV and 1800 MVA single-phase from 11 to 300kV. No. 4 and 5 Laboratories share four machines suitable for 50 and 60 Hz operation. The output is 9000 MVA 3-phase from 9 to 272kV and with single-phase has various outputs ranging from 305kA at 9kV to 11kA at 475kV. There is also a synthetic test laboratory using No. 4 Laboratory machines as the current source. No. 3 Laboratory has d.c. test facilities.

These testing stations must be able to reproduce as nearly as possible all of the conditions previously outlined, i.e. to vary the fault current between zero and 100%

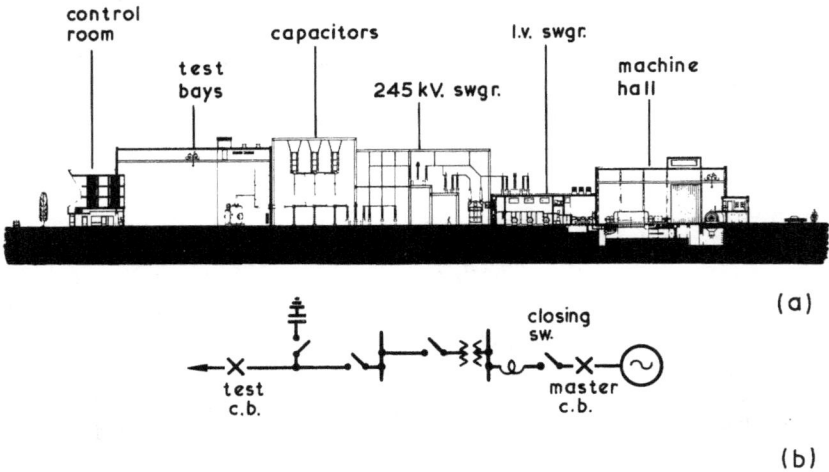

Fig. 10.17 Typical high-power test station (K.E.M.A.)

    (a) Sectional view
    (b) Simplified line diagram of connections

rating with the various recovery voltages and transient recovery voltages imposed by the actual system. This in turn requires not only special equipment, but imposes peculiar design requirements even on those items that appear to be conventional.

Fig. 10.17 is a sectional view of a typical high-power testing station comprising four generator units all of which can be operated in parallel. The rather specialised equipment in such test layouts is described in some detail in the following Sections.

10.12.1 Short circuit generators

The generators provide the energy used during the test. It is essential to obtain maximum return for the capital invested, i.e. maximum possible output from a particular size of machine thus that transient and subtransient reactance must be as low as possible to provide the maximum current. This in turn requires that the leakage flux is minimal and individual windings as close coupled as possible. The machine may be subjected to single- or multiphase faults, which necessitates special precautions being taken in the reinforcement of the end windings of the machine.

During the period the machine is short circuited the torque developed must be absorbed by the foundations on which it is mounted, giving rise to the need for particular attention to the design of the foundations. One solution is to mount the machine on shock absorbing springs.

The generator is driven by an induction motor directly coupled to the shaft and driven from the external supply system. It is disconnected from the supply immediately prior to the application of short circuit, the inertia of the rotor being maintained by means of its large fly-wheel effect. To provide flexibility of machine output many are arranged with grouped windings which can be connected in parallel or series.

10.12.2 Short-circuit making switch

This switch is used to connect the generator to the test circuit at a precise point on the voltage wave, high-speed operation being essential to enable the precision in closing instant to be obtained. Prearcing during closure of the switch must be minimal, as this would affect the consistency of point-on-wave control. These switches are frequently of the compressed air type, closure taking place in a chamber pressurised at up to 60 atm in order to minimise the degree of prearcing.

10.12.3 Master circuit breaker

The purpose of this circuit breaker is to protect the equipment within the testing station against the consequences of a failure of the test circuit breaker to interrupt the fault current. It is located between the generator and transformer, and because of its position in the electrical circuit it is subjected to an extremely high r.r.v. because of the high natural frequency of the circuit, and must interrupt the full output of the generator. On many of the older test stations oil circuit breakers are used for this duty, but some recent generators are of increased size and fault ratings of 150–200 kA are required. A recent design of air-blast circuit breaker developed specifically for this duty (by AEG) is claimed to have a tripping time of 7 ms and maximum arcing time of 10 ms.

The master circuit breaker must also be of high mechanical integrity, and must be capable of many short-circuit operations without attention so that operation of the testing station is not restricted by the need for frequent maintenance.

10.12.4 Transformers

The testing station transformers are subjected to particularly onerous conditions compared with a normal design in that:

(*a*) They must withstand repeatedly the axial and radial forces imposed on the windings by short circuits on their terminals.

(*b*) The insulation level of the windings must be capable of withstanding the overvoltages generated within the station, particularly when the test circuit breakers are interrupting magnetising or capacitive current. In many instances surge arresters are installed to prevent damage to the windings.

(*c*) Their impedance must be as small as possible to ensure that the testing station output is not significantly reduced. This requires a transformer having an impedance of the order of a quarter of that of a normal transformer.

(*d*) The nominal rating of the transformer is unimportant.

10.12.5 Current-limiting reactors and resistors

The short-circuit test current must be controlled to within very close limits, together with the circuit power factor. For this purpose, air-cored reactors are installed between the generator and transformer together with resistors, the reactors and their terminals being arranged such as to enable the rapid changing of connections to provide an extensive range of inductance and for series or parallel grouping, permitting a large range of the impedances and hence current values, to be selected. Air-cored reactors are also used in the voltage recovery circuit of synthetic test plant and these are of lighter construction as they do not have to carry the short-circuit current. Resistors are usually of the metal grid type and are also provided with tapping facilities.

10.12.6 Connections

The connections of the plant, particularly those at generator voltage, increase the inductance and capacitance of the test circuit. The physical arrangement of these and their length therefore must be carefully considered and has an important influence on the layout of the station.

10.12.7 Artificial transmission lines

A fault located some distance along an overhead line from a switching station (a kilometre or so from the circuit-breaker terminals) is a special test arrangement requirement. A high-frequency voltage transient is impressed on the 'line' terminal of the circuit breaker, in the form of a saw tooth wave. This condition must be reproduced in a short-circuit testing station, and in the first attempts to do this a short length of transmission line was constructed.

This line design had to simulate the varied conditions presented by single and bundle conductors, i.e. surge impedance in the range 300–450 Ω. Simulation of such varied conditions, particularly when testing individual units and of a circuit breaker, was too cumbersome with actual lines, and the majority of testing stations now use an artificial line comprising various series-parallel combinations of inductance and capacitance. In practice, an actual line comprises an infinite number of inductance and capacitance elements but the line transient can be adequately reproduced by six to eight 'artificial line' elements. Typical artificial lines may comprise cast-in-resin reactors together with capacitors and changeover links for rapid adjustment arranged in the equivalent T or π formation of the line being represented. During the opening of the circuit breaker, a voltage greater than that of the insulation withstand level of the components of the line may arise, thus, in some test stations, surge arresters are provided at the line terminals, and serve not only to protect the line but avoid applying an 'evolving fault' to the circuit breaker in the event of the artificial line insulation breaking down.

10.12.8 Capacitors

Banks of capacitors of up to hundreds of MVAR are required for the testing of circuit breakers for the switching of shunt capacitor installations and are also used to test for overhead line and cable no-load switching. Capacitors are also required for the control of the circuit natural frequency for normal short-circuit tests.

10.12.9 Measuring and control equipment

Apart from the main circuit equipment described above it is necessary to co-ordinate the operation of all these main circuit elements and accurately measure the voltage and current conditions at the circuit breaker under test. Both the instant of operation and the sequence of operation of the main circuit elements has to be controlled and the measurement of changes in voltage and current over a few microseconds near to a current zero as well as over several seconds is necessary. Referring to Fig. 10.18, a typical test sequence is as follows:

(*a*) The generator is run up to speed, unexcited, with its driving motor connected to the external supply. For an opening test, the test circuit-breaker will be closed and the master circuit-breaker and closing switch will be open.

(*b*) The master circuit-breaker is closed.

(*c*) Field exciter is switched on raising the generator voltage to the appropriate value. Then in a rapid automatic sequence:

(*d*) The generator driving motor is tripped.

(*e*) The closing switch is closed.

(*f*) The circuit breaker under test is tripped.

(*g*) The master circuit breaker is tripped.

The automatic sequence controller must of necessity be accurate and consistently reliable. Pendulum operated switches set at the appropriate position on the arc of

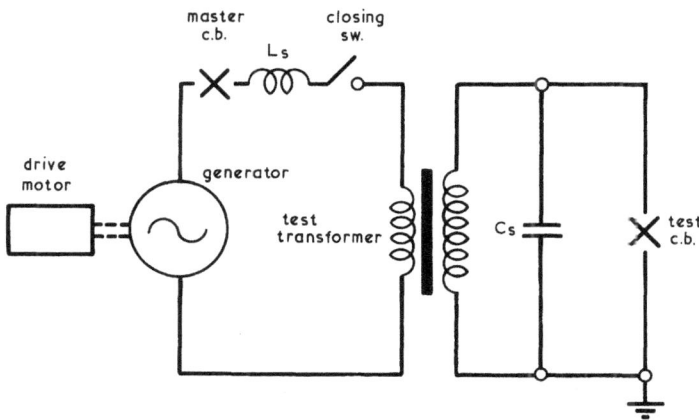

Fig. 10.18 Simplified diagram for direct s.c. tests

the pendulum travel, or contacts around a constant speed rotating drum or electronic black boxes are methods of providing the sequence of operations.

The recording equipment used includes electromagnetic oscillographs for voltages and currents with frequencies up to 1 kHz, rotating drum recording cathode ray oscilloscopes for relatively low t.r.v. frequencies (up to 5 kHz), together with high-speed single sweep c.r.o.s when frequencies as high as 200 kHz are present. During development testing it may be necessary to measure within a period of a few microseconds of a current zero which requires extremely sophisticated equipment such as field effect transistors.

10.12.10 Auxiliary plant
The direct voltage for trip and closing coils may be available either from a battery or a machine, usually the latter, and is variable over the range of voltage required for testing. As a point of interest, the higher winding resistance of a machine affects the time constant, particularly of large solenoid closing mechanism coils. In rare cases this difference can enable a circuit breaker to close against short-circuit throw-off forces with the minimum operating voltage derived from a machine and yet fail to do so from a battery. The compressed air, oil and, in some testing stations, the SF_6, supplies and regulating facilities are available.

10.12.11 Synthetic recovery voltage source
A direct recovery voltage for any method of synthetic testing requires a large capacitor bank. It is necessary for the voltage to be varied over a wide range, and there must be adequate energy stored to provide the high-frequency injected current without distortion. The battery must be arranged into several segments so as to provide a large selection of series parallel connections.

A typical battery installed in one UK testing station has four racks each of six rows, and each rack has eighteen $4 \cdot 4$ μF 25 kV d.c. capacitors. The storage capacity of the complete bank when fully charged is 600 kJ. Rapid selection of the appropriate capacitance is achieved with a bank of pneumatically operated switches, remotely controlled, permitting the capacitance to be varied from $1 \cdot 6$ μF at any charge voltage up to 600 kV d.c. to 480 μF at voltages up to 50 kV and with many intermediate settings. The d.c. charging equipment in this installation has a 150 kV charging transformer rated 100 kVA, and a four stage multiplier rectifier tank using silicon diodes. After the bank has been charged it is isolated from the charging unit by means of a remotely controlled, pneumatically operated switch.

A battery for a voltage injection test circuit is much smaller than that described above as the energy required is much smaller.

10.12.12 Spark gaps for synthetic circuits
Probably the most important item of equipment in the synthetic test circuit is the triggered spark gap which controls the instant of injection in both current and voltage schemes. They must be capable of performing this duty to within ±100 μs of a preselected instant, while permitting adjustment of the breakdown voltage level of

the gap over a wide range. The electrodes must be capable of repeatedly carrying injection currents of up to the order of 5 kA. A number of test installations utilise a 'Trigatron' gap for this purpose. These gaps have fast and consistent operating times of the order of one microsecond, but after a small amount of electrode burning the operation becomes inconsistent. Another type in use is a plasmajet injection gap, the principle of operation of which is as follows: A charged capacitor is caused to discharge into a small chamber in one of the gap electrodes. The ionised gas produced emerges as a jet through a small orifice, causing breakdown at levels as low as 10% of the untriggered voltage breakdown level; an adequate margin of safety is provided against the uncontrolled operation during test. The gaps can take the form of hemispherical electrodes operating at normal atmospheric air pressure with adjustment to vary their withstand level, or as in one case they are housed in a pressurised vessel. Variation of withstand level is achieved by adjusting the air pressure.

The spark gap is situated at the test potential, a common method of triggering the gap being by means of an optical light link between ground and test potential.

10.12.13 Reignition equipment for synthetic circuits
A typical circuit is that shown schematically in Fig. 10.15 and comprises a capacitor which can be charged to a voltage of approximately 50 kV d.c. and is isolated from the test circuit by the sphere gaps at each end. A few microseconds before current zero the ground potential gap is triggered and discharges the capacitor through the limiting resistor, thus forcing the current to a premature zero and reversing the arc voltage in the test breaker. Hence the test circuit breaker continues to arc until the next zero and if required a further reignition can be applied using a similar circuit. Triggering of the reignition process is critical to within microseconds, and to obtain this precision a trigger pulse is obtained by a gating relay from a ferrite cored peaking transformer connected to the ground potential connection of the test circuit breaker.

10.12.14 Auxiliary circuit breaker for synthetic circuits
This circuit breaker is used to isolate the current source from the high-voltage source and must, therefore, have a performance superior to that of the circuit breaker being tested. Frequently, it may constitute an identical design to the test breaker, but using an increased number of interrupter units in order to reduce the voltage per interrupter. There is a limitation to the number of interrupter units used by the need to minimise the total arc voltage to within the capability of the current source machine as described in Section 10.11.4.

10.13 Capacitive current switching

The switching of capacitive currents is involved in the disconnection of capacitor banks and cables on no-load. It also occurs during the interruption of the charging currents of long unloaded lines, and under all these conditions the changes in power frequency voltage of the source due to a change in power factor and regulation, may result in severe over voltages being experienced by the circuit breaker.

If after a period of a quarter of a cycle after capacitive current has been interrupted, the dielectric between the fixed and moving contact breaks down, this is termed a restrike. It is the restrike that can, depending on the electrical characteristics of the source and the load, cause severe electrical and/or mechanical stresses on the circuit breaker.

It is sometimes mistakenly understood that if a circuit breaker is tested in accordance with the national standards to a duty requiring a maximum of 12 opening operations during which no restrike occurs, that the circuit breaker can be deemed altogether to be restrike free. However, even if no restrike occurs during a test series the probability of restrike is, at best, 1 in 13, i.e. 7·8%. This probability is too high for development testing and a more realistic probability is 2·3%, i.e. two standard deviations below the 50% restrike value.

One important criterion is the voltage drop in the source impedance which results in a high-frequency transient component impressed on the circuit breaker interrupting gap (Fig. 10.19).

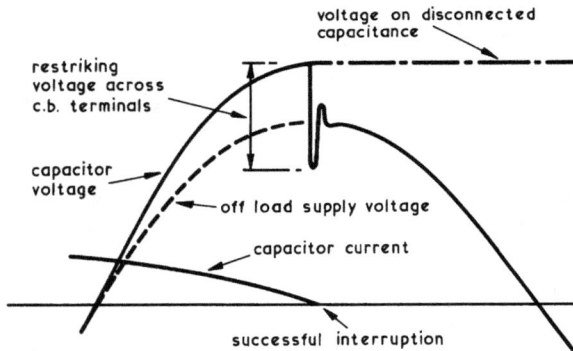

Fig. 10.19 Restriking voltage with capacitive load and high source impedance

Prior to current interruption, the source side voltage consists of the sum of the main source voltage and the voltage drop across the inductive components. After current interruption only the main source voltage exists, thus the circuit-breaker source side terminal voltage is reduced by the voltage drop across the source inductive elements. It is essential, therefore, to ensure that the circuit-breaker can operate satisfactorily in a circuit having a high source impedance.

When the capacitance of the source side approaches that of the load, on restriking there will be a very high transient current flow. This extremely high transient current occurring as an arc in an oil circuit-breaker arc control device, which has very little gas within it, can create extremely high hydraulic pressures. It is therefore necessary to ensure that the circuit breaker will withstand these high currents in a circuit with approximately equal capacitance on each side of the circuit breaker and with a low source impedance.

The testing of circuit breakers under all the above conditions is covered in IEC Recommendation 56-4 (1971). The rated value of the capacitive current in amperes of the circuit-breaker is usually assigned by the manufacturer and the tests specified are as follows:

10.13.1 Line charging current

Line-charging current breaking tests are intended to prove the ability of a circuit-breaker to switch overhead lines on no load and also overhead lines in series with short lengths of cable. The tests are applicable to circuit breakers having rated voltages of 72·5 kV and above, as below this voltage tests are considered unnecessary. The tests may be made as field or as laboratory tests and they may be made single or 3-phase with the limitation that single-phase tests may only be used for circuit breakers that are restrike free and which are intended for use on a system with an effectively earthed neutral. Also 3-phase laboratory tests with concentrated capacitor banks are valid only for circuit breakers which are restrike free.

In the equivalent supply circuit (Fig. 10.20) the value of the source side impedance must initially be high enough so that its short-circuit current does not exceed 10% of the rated short-circuit current of the breaker, but an overriding requirement is that the voltages variation caused by switching the capacitive current must not exceed 10%. The value of the source side capacitance C_2 must be as low as possible subject to its prospective transient recovery voltage not exceeding the prescribed limits. With this high impedance circuit, the following 3-phase make—break test duties are carried out:

Test duty 1: Ten tests with a current of 20—40% of the rated line-charging breaking current.

Test duty 2: Ten tests with a current of 100—110% of the rated line-charging breaking current.

Test duty 3: This is performed only if restrikes have occurred on the previous duties. The circuit as Fig. 10.20 is used, but with as low a value as possible of the source side impedance consistent with the short-circuit current not exceeding the rated short-circuit current of the circuit breaker. The value of C_2 must be at least equal to the line or load capacitance C_1.

For all these tests a 3-phase overhead line may be used, or alternatively concentrated capacitors with a series noninductive resistor of about 250 Ω may be used to represent an overhead line and its surge impedance.

For 3-phase tests, and for single-phase field tests the test voltage between phases at the breaker immediately prior to opening must be equal to the rated voltage of the circuit breaker. In the case of single-phase laboratory tests using capacitor banks such as would be carried out on single-phase single-unit h.v. circuit breakers, the test voltage must be as near as possible equal to 1·2 times the system highest voltage divided by root three.

10.13.2 Cable charging current

The test circuit is the same as that depicted in Fig. 10.20 except that the capacitance C_1 is now that of the actual cable or alternatively a capacitor with a series connected 25 Ω noninductive resistor to simulate the surge impedance of the cable. The test voltages and the earthing requirements for the circuit breaker and load capacitance C_1 differ somewhat from those associated with the line-charging current breaking tests.

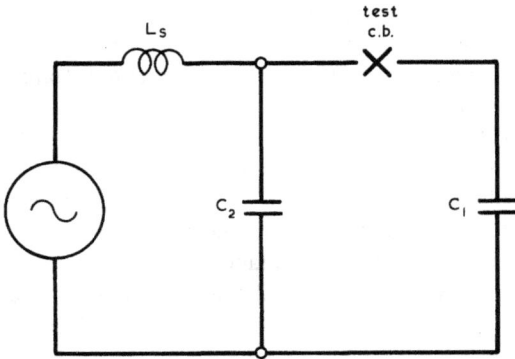

Fig. 10.20 Capacitive current switching circuit

The test duties 1 and 2 consist of the same percentage values of the rated cable-charging breaking current as those required for overhead lines, and are carried out using the high impedance version of the circuit in Fig. 10.20. Test duties 3 and 4 are a repeat of duties 1 and 2, respectively, but using the low impedance version of the circuit.

10.13.3 Capacitor bank current

The test circuit is again similar to Fig. 10.20 in its high and low impedance versions, and the capacitance C_1 is now that of the actual or equivalent capacitor bank. The test duties 1—4 are similar and consist of the same percentage values of the rated single capacitor bank breaking current as those itemised for overhead lines and cables.

The above capacitive current switching tests have several features in common. Tests may be made single phase or 3-phase in which case a test duty consists of ten shots. In the event of single-phase testing using point-on-wave control, each test duty will consist of 12 tests distributed at intervals of approximately 30 electrical degrees. When point-on-wave control is not used the number of tests in each test duty will be 30.

The characteristics of the capacitance circuit must, with all necessary test measuring devices, such as voltage dividers, included, be such that the voltage decay does not exceed 10% at the end of an interval of 10 ms. after final arc extinction. The decay of voltage is much influenced by voltage transformers so that voltage dividers are preferred from the measurement point of view. Voltage transformers also give rise to ferroresonance phenomena during breaking operations unless special precautions are taken.

10.14 Low inductive current switching

The switching of low inductive currents is necessary when items of equipment such as shunt reactors, unloaded transformers, reactor loaded transformers and motors are installed on a system. The origin and magnitude of the overvoltages produced under such conditions have been considered in chapter 3.

Large power transformers, having cores of cold rolled grain oriented steel have low values of steady state magnetising current, so that even if these small currents are 'chopped' during switching it has been found that the overvoltages produced are not serious. Dangerous overvoltages could, however, be produced if the switching operation occurs during the period when heavy magnetising inrush currents are flowing.

The danger exists from the unstable characteristics of the arc between the contacts of the circuit breaker and the overvoltages produced by the suppression of the current before the natural current zero. The overvoltage is related to the instantaneous value of the current which can be chopped by a circuit breaker and the time before current zero that the contacts open. This current varies between 20 and 80 A peak for gas-blast, oil and SF_6 circuit breakers. It is not a problem with free-air circuit breakers as the post arc resistance is low and this suppresses serious overvoltages. The maximum overvoltages are therefore produced when transformer magnetising inrush current or shunt reactors are switched. The latter duty is particularly onerous owing to the high natural frequency of shunt reactors.

The selection of an appropriate test circuit, the method of test and the representation of the parameters of the system have been considered for some years, but international agreement has not been achieved. Some test circuits

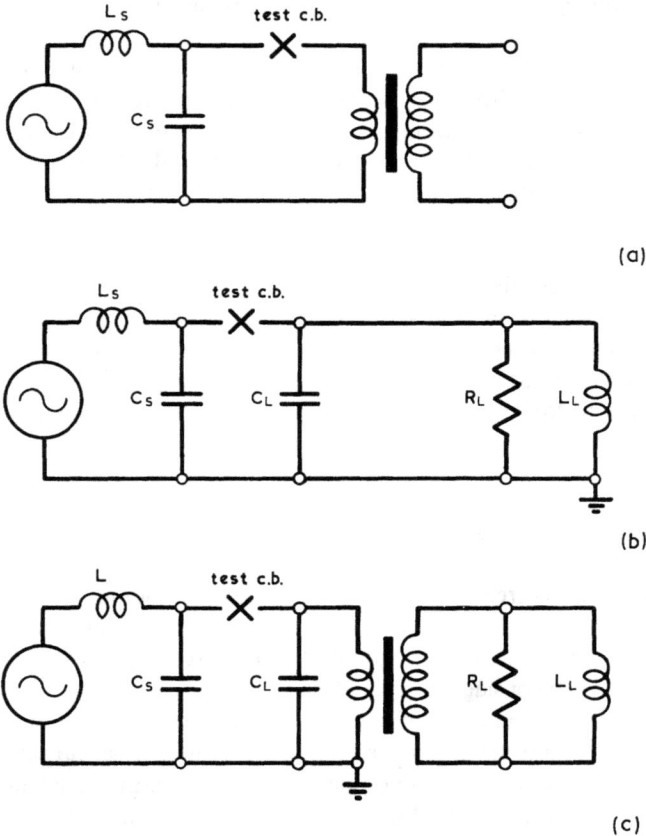

Fig. 10.21 Inductive current switching circuits

(a) Open-circuit transformer load
(b) and (c) simulated loads

representing these conditions which have been proposed are shown in Fig. 110.21.
Another circuit which has been used is shown in Fig. 10.22. This is recognised as
being severe and has not been accepted as an alternative, but it is strongly favoured
by some users and is at least an easy circuit to produce and is useful for
comparative purposes.

The tests may be made on a complete 3-phase or single-phase unit, but where
this is not possible because of limitations in the testing capacity, tests are
sometimes made on the maximum possible number of interrupter units in the
circuit breaker.

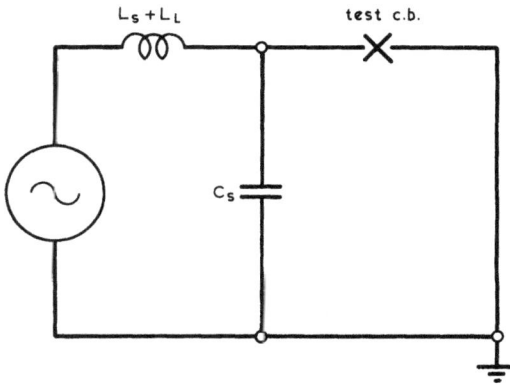

Fig. 10.22 Simple inductive current test circuit

In the absence of international agreement specifications covering low inductive current switching are varied and vague. Some purchasing specifications ask for tests with the circuit shown in Fig. 10.22 at currents of 10 to 15, 20 to 30, 50 and 100 A.

These four levels give a reasonably clear indication of the performance of the circuit breaker. The number of tests specified ranges from six with random point of wave to ten at 18° intervals around a half cycle of current. The overvoltages are measured on the supply and load side of the circuit breaker. The tests should be carried out at the maximum operating pressure where the arc stability is affected by increase in pressure, such as in SF_6 and air-blast circuit breakers.

The maximum allowable overvoltages are subject to agreement between the manufacturer and purchaser but as a guide an overvoltage not greater than 0·6 times the equipment BIL can be tolerated. On high-voltage systems this may be too high and in these cases some purchasers are prepared to take advantage of the transformer or shunt reactor arresters to limit the overvoltage rather than complicate the performance of the circuit breaker by requiring breaking resistors and other associated switches.

10.15 Evolving faults

It may at times be necessary to demonstrate that a circuit breaker can withstand the effects of, and successfully interrupt, evolving faults. The term 'evolving fault' is used to describe a power arc which is initiated by a voltage surge and differs from the normally considered short circuit in that it does not commence until the circuit breaker contacts are partially open. Depending on the instant of initiation the evolving fault may seriously affect the circuit breaker or its performance.

An evolving fault may occur between phases or from the phase to earth and in particular may be provoked by overvoltages generated during the interruption of low inductive or capacitive currents. However, tests are normally performed only for the case of a single phase to earth fault and for convenience employ a circuit representing an unloaded transformer. Such a circuit will probably have already been employed for the normal type test for transformer magnetising current interruption. The fault level of the input circuit should be equal to the rated breaking capacity of the circuit breaker under test.

For each series of tests to be undertaken, the evolving fault is initiated by a rod gap connected between the high-voltage terminal of the transformer to be switched and earth. The gap is set initially to a power frequency breakdown voltage peak equal to 90% of the maximum overvoltage recorded during normal magnetising current tests. If after about five tests no evolving fault occurs the gap is reduced and further tests performed. A total of three evolving faults are considered adequate to verify the behaviour of the circuit breaker, but they should preferably have been initiated at three different levels of magnetising current.

10.16 Circuit breaking in parallel

It is sometimes necessary to arrange breakers to operate in parallel. This is inherent in a system where circuits are controlled by circuit breakers arranged in mesh formation. If trip impulses are then given to the paralleled circuit breakers it can be postulated that commutation of the arc from one circuit breaker to the other can occur with resulting high stresses in the interrupters and possible failure to interrupt successfully. It is possible to avoid such a condition if the trip impulse to one circuit breaker is delayed for one or more cycles, but this may not be practical due to system stability requirements.

Testing for this condition may be considered necessary, particularly where circuit breakers are to be used which employ current dependant interrupting devices. There is no recognised standard of testing for the case of parallel switching. Influencing factors in addition to the obvious ones of system fault level and t.r.v. requirements are:

(a) characteristics of the two circuit breakers connected in parallel
(b) relative impedances of the two main current paths.

For the general case it can be assumed that the two circuit breakers are identical and that there would be no significant difference in characteristics beyond a possible small difference in opening time of the main contacts from trip initiation. The test is therefore performed with two parallel connected circuit breakers or breaking units, depending on available test facilities. Tests should be conducted with both simultaneous contact separation and with contact separation staggered by not more than one cycle. Since the disturbance created within the interrupters will be proportional to the fault current the tests can normally be limited to those in a circuit with full rated breaking capacity, except where it is considered the longer arc durations resulting from low fault level interruption are valid and may be more onerous. The impedance of the two current paths around a mesh will not be equal, and thus the current sharing between circuit breakers must be determined, the most onerous case being for the second circuit breaker to open to have the greater share.

10.17 Special service conditions

Circuit breakers may be required to operate under service conditions outside the scope of generally accepted standards for several reasons. It may be that the climatic conditions are more severe than those mentioned earlier in this text, e.g. abnormal temperature, humidity, wind loading etc. A second factor is that of pollution of the environment due to natural or artificial causes, e.g. fog, salt spray, fungi and bacteria, dust, chemical vapours and flammable gases. A third and very large group of abnormal conditions may result from the location of installation. The siting of apparatus may cause abnormal conditions of heating, ventilation and humidity, such as may be experienced in boiler rooms and some industrial plants. Location of circuit breakers on board ships gives rise to a number of problems requiring special attention as do some other modes of transport. The location of circuit breakers of some types in residential areas may also impose special conditions with regard to low noise operation. The installation of switchgear in areas of the world where the probability of earthquakes is high produces abnormal conditions for which special design and testing is becoming increasingly demanded.

Because of the very abnormality of many of these service conditions it is very difficult to agree on a standard of testing to show the ability of a specific piece of switchgear to function satisfactorily in all respects in service. However, considerable attention has been paid to some of the special service requirements resulting in the publication of standards or methods of testing by manufacturers and other interested bodies. More general agreement has been reached on the standards of performance expected of circuit breakers for operation in flammable gas environments and national and international specifications have been issued. The following text makes reference to some of the more commonly recognised special service conditions and the methods employed for ensuring safe and reliable operation under these conditions.

10.17.1 Flammable atmospheres
The use of electricity as an acceptable source of power in coal mining went ahead in the last decade of the 19th century, the need to provide some degree of protection

for flammable atmospheres following shortly afterwards. More recently there has been a rapid expansion in other hazardous industries, e.g. petrochemical industry, resulting in a more worldwide demand for suitable equipment for these conditions. In all hazardous industries the first defence against explosion of flammable gas is adequate ventilation. For many reasons this is not a perfect answer and other steps must be taken to guard against or eliminate potential sources of ignition. It is found impracticable to prevent external gas from entering an enclosure due to the breathing which results from changes in temperature and the consequent expansion and contraction of the enclosed atmosphere when the apparatus is taken on and off load. Gas that enters an enclosure is therefore subject to ignition, especially when arcing occurs, such as at the contact of a circuit-breaker. The development of flameproof enclosures for apparatus is on the basis that when gas within the enclosure is ignited, the flame will not be communicated to the external atmosphere.

Tests to determine whether a circuit breaker enclosure is flameproof are not normally carried out with the apparatus energised. Gas ignition tests are carried out by a recognised test authority and these tests are conducted with all the component parts assembled as for use but without being operated. The equipment is installed in a test cell which is equipped with facilities for filling the circuit breaker enclosure with gas and also ensuring that the atmosphere immediately surrounding the test piece is also flammable. One of a number of different gases may be used, dependent on the environment being represented. The tests are conducted from a remote point where observations can be made of any visible disturbances, emission of incandescent particles etc., the gas mixture within the enclosure being ignited by a suitable device specially installed for the test.

In the UK the test authority for all flameproof equipment is the Safety in Mines Research Establishment. This organisation has in the past been responsible only for the flameproofing test, it being the separate responsibility of the manufacturer to ensure adequate testing for the other rated capabilities of the switchgear. However, the rating and performance tests do not generally include tests in flammable gas mixtures. Although flameproof equipment has given satisfactory performance in the field, it is believed desirable to examine in detail any effects that may occur when apparatus is operating electrically in an enclosure containing a flammable gas. For example, it is recognised that flammable gas may originate within the enclosure, such as gases derived from the cracking of hydrocarbon oil in oil circuit breakers. Research is now being undertaken to evaluate the hazards involved under such conditions by performing short-circuit and other switching tests in flammable gas atmospheres. Subsequently, it may be necessary to give consideration to more extensive use of this test procedure.

10.17.2 Pollution

Although certain forms of pollution are found to give rise to erosion or oxidisation of the metallic components of circuit breakers, e.g. chemical vapours, the greater cause of failure of switchgear in service is pollution of insulation. For reasons of economy the provision of protective enclosures is normally limited to particularly severe cases and those not of prohibitive size or complexity such as in the lower range of voltages.

There has been some considerable activity internationally in attempting to determine the validity, repeatability and reproducibility of pollution tests. Research has covered the possibility of conducting tests with both natural and artificial pollution of the insulator surfaces with comparison of results between the two forms of pollution wherever possible. It has been indicated that it may be necessary to consider dividing the tests into two distinct classes to cover the pollution which can occur under coastal conditions, i.e. high salt content, and industrial atmospheres, i.e. heavy dust deposits.

One salt fog test is proposed by Lambeth *et al.* (1968) to simulate the conditions to which outdoor insulators in coastal areas may be subjected. Briefly, the test consists of subjecting a cleaned insulator to a salt water fog while energised at the maximum working voltage of the system. The fog is produced by arrays of nozzles, on opposite sides of the insulator, which direct a fog of droplets at the insulator by means of compressed air. The highest salinity at which three or more tests out of four are withstands is called the withstand salinity, and is regarded as the criterion of performance. The salinity values are chosen from recommended values which increase in geometric progression, from 2·5 g/1 to 226 g/1, because of the shape of the curve of flashover voltage against salinity.

The saline fog test should preferably be carried out in a plastic tent or in a separate small high-voltage laboratory since the equipment must be specially designed to withstand the corrosive effects of the salt spray. A small chemical laboratory is also required where the salt solution can be mixed and fed through plastic pipes to the compressed air jets in the adjacent main high voltage laboratory. Although the test appears to be complicated, it is claimed that good correlation has been achieved between these tests and the results of long term natural pollution testing.

Dust deposit tests have been submitted with several variants to simulate conditions of industrial pollution. In one such test the insulator is sprayed with a uniform coating of a suspension of diatomatious earth in distilled water containing enough calcium chloride to give the desired layer conductivity. The coated insulator is then erected for test and wetted by steam. A voltage of 2·5 kV/m length of the insulator is applied, and the leakage current measured to obtain the layer conductance. The measurements are continued until the maximum conductivity is obtained in about 20 minutes. The full test voltage is then applied for up to 15 minutes to determine a withstand or a flashover value. The insulator is then dried and wetted again, and another withstand or flashover measurement made. Five such measurements are taken for the layer conductivity in question. The series of measurements are then repeated for other values of the layer conductivity, and finally the characteristic of withstand voltage in terms of layer conductivity is determined. One variant of the test employs a slurry of methylcellulose, floated chalk and salt with which the insulator is coated to give a conducting viscous layer containing water, which is then energised without further wetting. An objection to this method is that the pollution coating is uniform whereas in service it can be distinctly nonuniform.

At the Plenary Session of CIGRE in 1968 the position had been reached that, as a result of a joint international investigation, the salt fog test method had been accepted as a valid and suitable method of measuring the pollution performance of insulators. However, it was agreed to carry out a similar investigation on the dust deposit tests to assess them in the same manner. It is noted that tests with thick layers of low conductivity (industrial pollution) influence the development of discharges very differently from thin layers of high conductivity (salt pollution).

10.17.3 Earthquakes

The increasing use of high-voltage installations in areas of the world recognised as earthquake zones has created a need for an international standard for a test of earthquake proofing. One problem in devising such a test is to establish a 'standard earthquake' up to and including which electrical equipment can be made earthquake proof. For the design of circuit breakers and other electrical apparatus the earthquake recorded at El Centro in California in 1940 is used as a standard reference. On this particular occasion a ground acceleration of up to 0·33 g was measured, being most pronounced in the horizontal direction. See Hitchcock (1969) and Gallant *et al.* (1971).

Calculations have shown that the number of earthquakes having an intensity 1·5 times as much as the proposed 'standard' would probably occur only twice in 100 years. Since the generally accepted life of electrical equipment does not exceed 50 years, a figure of 1·5 times the ground motion at El Centro, i.e. 0·5 g is considered practical.

Although there is no internationally agreed specification for earthquake tests, many tests are now made on high-voltage circuit breakers which are mounted on a platform accelerated by a vibrating machine. The vibratory drive can generally be adjusted to produce the required waveform and frequency and in some instances actual earthquake recordings can be used to stimulate the vibrations. The circuit breaker is first assembled as nearly as possible as in service on the test device in order to subject it to a horizontal vibrating acceleration applied at the base. When tests are specified the waveform used is of the El Centro standard with acceleration raised to 0·5 g. Alternatively, tests are required with a sinusoidal vibration at the resonence frequency of the equipment with 2 cycles of vibration giving 0·5 g or 3 cycles giving 0·39 g or with 5 cycles at 0·25 g. Stress measurements are made where practicable and a detailed examination after test, of the condition of cemented joints and other nonelastic components. It has been found beneficial to loosely sling those units mounted at the top of porcelain columns from an overhead beam or tripod in order that failure of the column does not cause unnecessary damage to other components. Further tests to simulate vertical earthquake accelerations may be necessary in those cases where large weights are supported unequally, e.g. horizontally mounted interrupter heads. These tests can often be made by mounting the test piece on the platform at right angles to the normal mode.

10.17.4 Shipboard installations

Electrical installations on board ship are, because of the very nature and functions of the vessel, subject to more severe conditions of service than most land based equivalents. For this reason it has been found necessary over the years to specify special requirements for electrical equipment, and to develop special testing techniques to meet these needs. System voltages, and therefore the circuit breakers protecting them, do not normally exceed 5:5 kV, but the security and reliability required of the installed equipment must be compared with that of a major transmission scheme. Circuit breakers are required in the first instance to have characteristics not less than those specified in the recognised national and international standards. In addition, the following are among the special tests to which equipment for use on board ships must be submitted. It should be noted that circuit breakers would normally be specified for enclosure to give a minimum protection against accidental contact and droplets of water. The special tests would therefore be performed with the circuit breaker mounted in any such enclosure used in service. A hygroscopic test is made by placing the unit in an enclosure containing air with a relative degree of humidity of between 91% and 95% and at a temperature from 20° to 30°C. The duration of the test is 48 h after which the circuit breaker and associated equipment should show no damage and in addition should be capable of withstanding the normal dielectric test at system frequency. In addition, the insulation resistance between all incoming and outgoing terminals measured at a direct voltage of at least 300 V is not to be lower than 20 MΩ.

A test to check the insensitiveness to the motion of the ship should be made. The circuit breaker unit is inclined from its normal position to any angle up to $22°30'$ in all directions to ensure that, in particular, the open and closed positions are not affected. Where the circuit breaker is solenoid operated, an additional check is also made to ensure that correct operation at the minimum operating voltage of the coils is not affected under these conditions.

A vibration test is specified. The circuit breaker, placed in its normal working position, is submitted to vibrations of a frequency between 0–25 Hz with a total amplitude of 2 mm. The direction of application of the vibration should be the most onerous and would normally be parallel to the axis of motion of sliding parts and plugging contacts. The vibrations are first applied progressively from 0 to 25 Hz and returning to zero over a period of approximately 5 minutes, noting the most unfavourable frequency. The vibration test is then maintained at the selected frequency for a period of 1 h. If no critical frequency is apparent the test is made at 25 Hz. During the test period the circuit breaker is required to withstand the passage of a current not less than 80% of its rated normal current. The test is considered as satisfactory if no signs of deterioration or damage are apparent and if no malfunction of the protective or interlock devices occurs. In addition, no slackening should have taken place at any bolts or connections and the apparatus should still be able to withstand the normal dielectric tests.

10.17.5 Noise

The increasing public intolerance of the harmful effects of industry and commerce on the environment has caused many repercussions. Not the least of these is the consideration which must be given to the generation of noise by machines. The air blast circuit breaker has long been recognised as a potential source of noise nuisance due to the explosive nature of the short blasts of compressed air released during switching operations. Where such noise problems can be anticipated, other types of circuit breaker may be installed or adequate noise reducing enclosures employed. However, it is not always possible to find an acceptable alternative to the airblast circuit breaker, and as a result much development work has been done in order to quantify and reduce where possible the noise levels. Work by Fawdrey *et al.* (1970) has been particularly aimed at the silencing of the airblast circuit breaker, but has also caused a more critical appraisal of the noise performance of all circuit breaker types.

Noise assessment is difficult, both as regards the generator and the listener. Work is being done internationally to rationalise the measurement of radiated noise, and some agreement has been reached on this as applied to sustained or even intermittent sounds, e.g. rotating electric machines and transformers. No such agreement has yet been reached on the evaluation of short duration noise, such as that produced by circuit-breaker operation. It is generally accepted that the characteristics required of a circuit breaker should be specified in terms of sound pressure levels throughout the frequency spectrum. Investigations have also shown that it is essential that the low-frequency components should be limited, even to the extent of transferring the energy available in these components to higher-frequency components by appropriate silencer design. This is both because the low-frequency components are the most objectionable and because these components lose considerably less energy in transmission through the atmosphere than the higher-frequency components.

In the UK, noise level tests are specified where considered necessary as type tests. Measurements are required to be made at a height of $1 \cdot 2$ m above ground level at a distance of 25 m from the centre of the circuit breaker and through an arc of $360°$. A pressure transducer is used, associated with a precision sound level meter set to its fast response and linear scale for direct determination of the overall sound pressure level. Because of its short duration the noise is also recorded on tape to enable a detailed harmonic analysis and a check on the instrument reading to be made in the laboratory. The noise recording is played back successively through a 1/3 octave band analyser to determine maximum r.m.s. sound pressure levels for each 1/3 octave wave band. On this basis, present requirements prescribe that for high-voltage circuit breakers the no load sound pressure level must not be greater than 95 dB at the most disadvantageous point 25 m from the circuit breaker. It is appreciated that under the conditions of interruption of a heavy short circuit the additional release of energy could significantly increase the noise level emitted. However, because of the extreme rarity of high fault operation no account is at present taken of such operations in the specification of noise levels.

10.18 References

ANDERSON, J. G. P., ELLIS, N. S., MASON, F. O., NOBLE, R. G., ORTON, L. H., REECE, M. P., and STEEL, J. G. (1966): 'Synthetic testing of a.c. circuit breakers. Pt. 1 – Methods of testing and relative severity', *Proc. IEE*, 113, (4), pp. 611–621

ANDERSON. J. G. P., ELLIS, N. S., HICK,M. A. S., MASON, F. C., ORTON, L. H., REECE, M. P. and STEEL, J. G. (1968): 'Synthetic testing of a.c. circuit breakers. Pt. 2 – Requirements for circuit breaker proving'. *Proc. IEE*, 115, (7), pp. 996–1007

BRAUN, A. (1972): 'Validity of unit testing on circuit breakers exhibiting statistically varying breaking capacity', *Proc. Inst. Electr. Electron. Eng.*, PAS–91, pp. 791–796

BS223 (1956): 'High-voltage bushings'

FAWDRY, C. A., GIFFORD, R. A., McCULLOCH, D. G., and MORRIS, C. (1970): 'Suppression of noise in air-blast circuit breakers', *Proc. IEE*, 117, (5), pp. 968–978

GALLANT, C. R., HAMMOND, G. W., BYBEL, D., and STAHL, D. P. (1971): 'A testing program for qualification of switchgear to be subjected to an earthquake experience', *Proc. Inst. Electr. Electron. Eng.*, PAS–90, pp. 339–349

HITCHCOCK, H. C. (1969): 'Electrical equipment and earthquake', *N.Z. Eng.*, 15th Jan. pp. 3–14

IEC56 (1971): Pt. 2, 'Rating'. Pt. 4, 'Type tests and routine tests'

IEC270 (1968): 'Partial discharge measurements'

KREUGER, F. H. (1964): 'Discharge detection in high voltage equipment' (Heywood Temple)

LAGEMAN, B., and DE VRIES, M. N. D. (1972): 'High voltage circuit breaker testing in accordance with new USA Standard for transient recovery voltage', *Proc. Electr. Electron. Eng.*, PAS–91, pp. 812–818

LAMBETH, P. J., LOOMS, L. S. T., LEROY, G., PORCHERON, Y., CARRAR, G., and SFORZINI, M. (1968): 'The salt fog artificial pollution test'. CIGRE, Paris, Report 25–08

OUYANG, M. (1966): 'New method for the assessment of switching impulse insulation strength', *Proc. IEE*, 113, (11), pp. 1835–1841

OUYANG, M., and CARRARA, G. (1970): 'Evaluation and application of impulse test results', *Electra*, No. 13

WILSON, R. W., HARDERS, C. F. (1971): 'Improved reliability from statistical redundancy of 3-phase operation of high-voltage circuit breakers', *Proc. Inst. Electr. Electron. Eng.*, PAS–90, pp. 670–681

Chapter 11

Design criteria for reliability, maintenance and safety

J. A. Sullivan, C.Eng., F.I.E.E.

11.1 Introduction

The success of any design of a circuit breaker is clearly dependent on its commercial viability, but no circuit breaker that is consistently unreliable in service can be commercially viable. It is therefore necessary for the design engineer to ensure that circuit breakers provide an acceptable level of reliability, and this entails in its turn a compromise between failure rate and the cost of development and production. Attention to detail can reduce the statistical levels of failure very considerably without affecting cost of production, but the development cost to achieve this may be increased and this must be accepted as equivalent to an insurance premium against repairs, replacements and even loss of orders.

Other chapters show that circuit breakers derive from a knowledge of many disciplines, among which may be mentioned plasma physics, insulation technology, electromagnetic theory, materials and structures, mechanics, fluid dynamics, control techniques and production technology. Added to this is the fact that the designed life of circuit breakers is some 30 or so years, thus ageing, electrolytic corrosion, and chemical degradation of the components become important considerations.

Chapter 10 covers circuit breaker specifications, type testing and routine testing. Such testing is done mainly on new equipments and shows that set qualities and standards are being maintained. This chapter examines sources of service experience, and emphasises that attention to detail design can avoid the pitfalls which occur when the equipment is no longer new.

Experience indicates that where such attention is given during the design stages, the level of reliability is improved not only in service but in reproducibility; thus production costs as well as site costs are reduced.

Observance of the factors highlighted in this chapter will help to ensure that reliability, simple maintenance and safety are designed into the full life of the equipment.

11.2 Operating statistics

In order to gauge properly where attention should be concentrated to achieve improved reliability the designer must have available extensive information on the qualities of materials including the effects of ageing and chemical degradation, particularly of gasket and insulating materials, and carry out life tests on subassemblies and components. He must also have available statistics of the types of troubles encountered in past service. Unfortunately for this reliance must be placed on the user to record properly and provide the information, but experience shows that such information is very scarce, as users tend to regard such records as of low priority.

For some classes of failure, for example where there has been danger to life, there is usually a statutory requirement to report occurrences. In the UK this is the concern of the Factories Inspectorate, and hence some statistics are available from them although not always of a detailed nature.

The CIGRÉ Study Committee 13 is attempting to assess the failure rate of circuit breakers in service. A report of 1972 (13–07) to the Congress concludes that only by a standardised method of reporting failures can a useful dossier of information be provided, and it is to be hoped that this will lead to a clearer understanding of the areas of design which need more careful attention.

There are other sources of information such as the German 'Verband Deutscher Elektrotechniker', but again with no clear definition of 'failures', and hence little confidence can be attached to these statistics. One fairly reliable but unpublished source of information shows the following types of failures for high voltage air blast circuit breakers.

Table 11.1

Type of failure	Percentage of all failures
electrical	10
air leaks and air conditioning	20
pipes and joints	3
control circuits	10
air valves	15
auxiliary switches and interlocks	12
mechanical	25
unknown	5

Although these failure rates apply to outdoor circuit breakers of the air-blast type and although oil circuit breakers have some fundamentally different components, the evidence suggests that for all types of circuit breakers, electrical breakdown represents only a small proportion of the total number of failures. If high reliability is sought therefore, attention must be directed especially towards the mechanical design of components such as valves, gaskets and auxiliary components, mechanisms and trip gear.

11.3 Reliability

Unless the designer has a clear understanding of the mechanical and electrical duties and the forces they represent on the circuit breaker he cannot be in a position to produce equipment which is reliable and predictable in its operation. Methods of calculating forces, particularly accelerating and decelerating forces of moving parts and the characteristic strengths and weaknesses of materials must be understood. Much of this information is inevitably empirical, but nevertheless useful and necessary in predicting performance. Mistakes may be made and failure of components will occur when unaccounted for forces are beyond the capability of the materials. These can only be minimised by the disciplined collation of information and design techniques and the routine use of same. This data bank must be continually updated and extended as experience is gained. Each design should be recorded giving the loading conditions, the methods (and their origin) used in calculating the forces and the subsequent dimensions of materials selected to withstand them. Unless this is set out in a satisfactory and consistent manner and recorded systematically for each and every design, reliability and reputation will suffer. A designer is only as good as his data bank and the use he makes of it.

In the following Section some guidance is given on the approach which the design engineer may usefully use to ensure reliability for some typical examples where unreliability is commonly experienced.

11.3.1 Toggles and bearings

The bearings of circuit breakers present special problems owing to the fact that they remain static for long periods. This is particularly so in toggle linkages, the design criteria for which were discussed in chapter 4. The pressure on the designer to increase the mechanical advantage in order to reduce the trip force is considerable, and the finer the toggle angles used the more critical this trip action becomes, especially in service due to increased friction at the pivot points.

To ensure consistent operation even with a gross increase in friction at pivots, due to corrosion and dirt or slight misalignment, it is good practice to assume that plain bearings will reach a coefficient of friction of the order of 0·25, and if this results in unsatisfactory stability conditions, ball or roller bearings must be used, which can then be assumed to have a coefficient of friction of about 0·1. In effect, the assumed coefficients should be an order of magnitude greater than the coefficients under ideal conditions.

The operating frequency requirements of circuit breakers varies considerably. Low-voltage breakers, particularly in process plants, may have a frequency of operation akin to that of contactors. Even high-voltage circuit breakers used to control generators which are switched twice a day to meet load requirements may be required to operate several hundred times a year. Generally, however, the problem is that the circuit breaker is not operated sufficiently frequently. It may remain closed for days, weeks or even months on end. It is well known that static loading of bearings causes the lubricant to be displaced so that the bearing reaches ultimately a state of zero lubrication and that friction or stiction is then very high, resisting initial movement. Greases and oils tend to increase in viscosity with temperature reduction and some solidify with time and lack of movement.

As a general rule, therefore, the design should minimise the number of bearings so far as is possible consistent with the need to keep toggle angles open sufficiently to include the built in allowances for increases in friction that can be expected.

The bearings that are required should preferably be designed not to need lubrication in service throughout their life, for example by using self lubricating sintered bronze bearings, impregnated with lubricant of the colloidal graphite type, or roller bearings or ball bearings with permanent grease packing, so that the effects of deteriorating lubrication on initial friction are of less consequence, or p.t.f.e. impregnated bearings which are now widely used and generally speaking have a good service record.

Nylon bearings can be used to meet a wide range of low stiction requirements in mechanisms, but the inherent advantages are offset by dimensional instabilities associated with some grades of this material. This requires careful selection, and allowance for changes in bearing clearances that may be due for example to moisture absorption or thermal effects.

Where lubrication is necessary, then proper provision for it is essential, e.g. grease or oil nipples. It is then necessary to give clear instructions in the maintenance handbook of the type of lubricant to be used and intervals at which the lubrication is necessary in service, a selection requiring detailed consultation with the lubricant manufacturer.

Another aspect of bearing design is to ensure that bearing effort is not increased by corrosion. Pins should therefore be made out of rustless materials, e.g. stainless steel. The bushes should for preference be made of nonrusting materials, e.g. bronze. Yet the cost of special bushes for the bearings cannot always be justified, since rustless pins can give satisfactory service in practice working in mild steel bearings machined in mechanism plates or linkages, more especially where the friction angles are not critical.

Another form of corrosion deterioration is associated with duralumin type alloys sometimes used for linkages, for which purpose they are very suitable, more especially as it is also a good bearing material. One word of warning is necessary, however, since some aluminium alloys are susceptible to corrosion growth, and, if exposed to outdoor conditions, this growth can fill the bearing clearance with disastrous results.

11.3.2 Running fits

Clearances for pins in links must be given particular attention. It is good practice to use the maximum clearance consistent with reliability of movement. For mechanism bearings a minimum clearance including all tolerances of 0·1 mm is a useful guide. Where closer tolerances are essential, ball or roller bearings should be provided and where the need for precision is minimal 0·4 mm nominal clearance is adequate. Variation in tolerances between different assemblies causing alignment errors also require consideration, and it may well be essential to build prototype mechanisms to the extremes of bearing clearances and alignment errors in order to check that the operation is satisfactory.

11.3.3 Auxiliary switches

A significant proportion of failures can be attributed to auxiliary switches and limit switches. One estimate is that these are responsible for 10% of circuit breaker failures. Auxiliary switches are generally manufactured by specialist firms and unfortunately some are derived designs from control switches and are inherently unsatisfactory for high speed movement. The control that can be exercised by the circuit breaker designer is thus limited to the selection of the type purchased, the mounting of it, the method of coupling it to the mechanism, and the angular acceleration and deceleration to which the auxiliary switch is then subjected by the circuit breaker. In the selection, cognisance should be taken of the terminal type and arrangement, the torsional rigidity of the moving contact system, the type of contact material and contact springing arrangement, the accelerations and shock the assembly will stand, and finally the insulation between adjacent contacts. The circuit breaker design engineer should ensure that the contacts are suitable for the breaking duty and that the plating will ensure good electrical contacts for many years without attention. Silver plating not less than 0·07 mm thick is a suitable coating. Beryllium copper has good elastic properties, and, although its conductivity is only 20% of that of copper, it has an adequate current carrying capacity and is one of the best materials for contacts as it can be used without the need for separate springs.

There are a wide variety of satisfactory terminal arrangements available although in this area many users have their own opinions as to which should be used. It is, however, generally agreed that separate pressure plates that do not rotate on the conductor are superior, together with a screw of adequate size to provide a sufficiently high pressure, without damage when a torque of 10 Nm is applied. For accessibility, terminals arranged in one plane rather than distributed around the periphery of the auxiliary switch are superior. The electrical security of the auxiliary switch requires that adjacent terminals are separated by insulation with good antitracking properties, such as melamine, and providing a creepage distance of 1 cm. This is particularly important with d.c. circuits, especially where the auxiliary switch is mounted in a damp environment. This insulation barrier between adjacent pairs of terminals and contacts should project above the electrically conducting parts sufficiently to ensure that they cannot be bridged by a flat metal object such as a steel rule.

Fixing the auxiliary switch firmly in position is a common-sense requirement, but there have been many instances where sheet steel cubicles have been used as the base, and it is not then surprising that deflections have resulted in inconsistent operation. The drive must also be as direct and rigid as possible because the angular accelerations and the shock loading applied to auxiliary switches by high speed circuit breakers can be considerable. The auxiliary switch itself must also be satisfactory for these shock loadings, which can be minimised by appropriate design of the driving linkage angles.

11.3.4 Springs

Tension springs have an inherent fundamental weakness in the increased stressing at the end loop, aggravated by the notch effect created by the sharp bends required to obtain the necessary end fixing. Wherever possible it is preferable for the designer to use compression springs, but where this is unacceptable, due to other features of the design, an improvement is to use a plate for the end fixing which is located within the last few turns of the spring.

All springs fatigue and, in the design, the stress should be arranged for one million operations wherever this is possible, but in special cases where this cannot be accommodated at least one hundred thousand operations should be provided for. To obtain one million operations, the stress has to be approximately 90% of the permissible stress for one hundred thousand operations, 82% of the stress for ten thousand operations. These differences are minimal and the improved reliability obtained by the use of the lower stresses is appreciable.

11.3.5   Gas and high pressure oil leaks

Capillary types of pipe connectors, which entail cleaning the ends of the pipes and the application of heat, provide a neat joint, but conditions on site sometimes in open trenches are not conducive to efficient jointing by this method. Experience has shown that a higher proportion of leaking joints occur with this method of jointing, and it should therefore be restricted to factory made joints. Mechanical pressure type joints give less trouble, but great care in following the manufacturer's recommendations of materials for the nut, body and ring for the fluid pressure and fixing instructions is necessary. Where the type having swaged on collars is used, the pipe line should be dismantled after initial assembly to inspect the position of the collar on the pipe.

Castings always present a possibility of leakage due to porosity, and machining small components from solid bar or the impregnation of castings are ways of reducing this liability.

Hydraulic mechanisms are also liable to leakage problems and at the high pressures normally used even the oil pipes themselves may leak owing to porosity or minute cracks, so that it is good practice to over-pressure test all such components. Further, if oil leakage develops in service this can be in the form of a very fine oil spray which can present a risk of explosion in confined spaces. This requires then that high pressure oil components should be segregated from electrical switching components and be separately ventilated.

11.3.6 Gas valves

Valves are responsible for a high proportion of faults on air-blast and SF_6 circuit breakers and to a smaller extent on compressed air closed oil circuit breakers. The types of valves involved are numerous and include stop valves, reducing valves, servo valves and blast valves. The troubles include the effects of foreign matter and water which is carried over the air supply system, the design of which is not within the scope of this chapter. It is, however, very necessary that particular attention should be paid to the design of the gas supply system by the circuit breaker designer.

The choice of valve seat material entails technical assistance from the material supplier, if it is of resilient material, and the circuit-breaker designer must ensure that the optimum loading is adhered to and that the seal is always made on the designed diameter. All seating materials rely on the elastic distortion caused by pressure for a seal, and trouble has been experienced where the movement of the seat material under pressure enables a seal to occur on either the inside or the outside of the design diameter due to bad contouring of the piston face. Indeed this has led to the serious condition of complete loss of air from pressurised interrupters in which the contacts are maintained open by air pressure. There is also some risk of the valve seat material adhering to the metal valve when it has been held closed for long periods resulting in mechanical damage to the valve seat on opening. Metal-to-metal line contact valves offer security from these troubles, but their high cost restricts the extent to which they can be used and they are not suitable for valves swept by exhaust gas.

Piston movement within the cylinder can be affected by friction particularly where scouring or metal 'pickup' occurs. Lubrication can prevent this but it introduces additional risks, for it must be applied sparingly, and with the gas flow it is difficult to ensure that the lubricant will remain where it is required. A range of materials which have self lubricating properties or minimal coefficients of friction are available, e.g. p.t.f.e. plastic coatings or complete articles, and are preferable to liquid lubricants, but here again attention must be paid to dimensional stability. Improvements can also be obtained by chrome plating and polishing of cylinder bores. Silver plating of one component and chrome finish on the other has been found satisfactory. To avoid scouring, edges should be chamfered, and this also assists assembly.

In general, it is important to note that the moving force on the valve should have a high factor of safety over the resisting gas pressure force, adequate to swamp all stiction which may arise from sticking of valve to valve seats, from piston to cylinder walls, or valve stems to valve guides, after prolonged periods without movement. Such sticking can be caused by corrosion, dirt, the use of incompatible materials, and can be aggravated by low temperature. The actual ratio of opening force to sealing force required will of course vary with the conditions, but will in general need to be increased the smaller the valve diameter. The ratio may be expected to fall in the range of perhaps 2–10 fold, and the greater the ratio used to ensure opening reliability, the greater will be the consequential need to provide adequate damping at completion of movement.

11.3.7 Gaskets

It is not difficult to see that gaskets can sometimes fail to do their job of forming a gas or liquid seal when one considers the difficult task they have of compensating for all the various loading on them in service and during transportation. Bushings, to choose an example, have relatively massive porcelain insulation cantilevered from the usual lifting point, and the accelerating forces due to 'snatch' imposed by a careless crane driver can cause very great variations of load to be applied to the gaskets at each end of the porcelain insulators.

The most effective gasket seal is obtained when it is recessed into one of the metal faces with the compression limited to the optimum amount by the clamping bolts. 'O' ring seals are the most usual form and application of this principle, but flat gaskets can be used in the same way, providing the groove is of sufficient size (with all tolerances) to accommodate the compressed gasket. Working to the instructions given by the gasket manufacturer will provide a good seal, but care must be exercised against excessive or unevenly applied greasing of the gasket.

The gasket manufacturer must be informed about factory processes as well as the service environment to which the gasket will be subjected. Rubber bonded cork material for example loses much of its resilience and takes a permanent set when subjected to combined high temperature and low vacuum as employed for the processing of insulation materials.

Positioning of the gasket is important, for with internal pressure the relieving load is a function of the diameter over which the pressure is exerted. In general the seal should be made on the smallest possible diameter, but due allowance must be made for eccentricity and manufacturing tolerances on the diameter of porcelains. These tolerances are obtainable from the manufacturers, but an effective inspection system is necessary to ensure the tolerances are not exceeded.

In an extremely cold environment the majority of gasket materials become harder and less resilient. Where the circuit breaker is to be subjected to temperatures of the order of $-40°C$, appropriate materials must be selected.

The location of gaskets should be such that the risk of displacement during assembly is minimal, and such that if displacement occurs it can be seen during inspection.

11.3.8 Shock loading

Circuit breakers that are essentially high speed mechanisms have considerable acceleration and deceleration forces, and the design should control these to known values. This requires that the decelerations at the end of the opening movement can be measured and the forces therefore established. If the control is by means of oil dampers, then the design should ensure that these dampers are sealed and do not require refilling, or that they are automatically recharged from oil in the circuit breaker, and that this is not lost in two or three operations when the circuit breaker has its oil tanks removed. If the deceleration is provided by resilience in the stops, e.g. by washers and general elasticity, then very accurate recording will be necessary to check that the limiting decelerations are in fact acceptable in combination with extensive life testing.

11.3.9 Finishes

The finish of components in circuit breakers can have an important bearing on reliability and maintenance. Comment has been made on this in various chapters including this one. While finishes constitute a large subject in itself a short resume may be useful in this chapter.

Most common metals have a tendency to corrode or oxidise in normal environment, e.g. iron and steel rust, metals in combination tend to corrode electrolytically, the surfaces of copper, brass, bronze, etc., tend to oxidise and produce bad electrical contact. It is therefore normal to provide a protective finish to minimise deterioration and to arrange that exposed metals in combination should be closely related in the electrochemical series.

A decorative finish, which is important for other reasons, is also provided by protective coatings.

Compatible finishes between engaging metals may also be required to prevent 'seizing' or corrosion growth.

The chemical and physical nature of metallic surfaces are important in the choice and life of protective coatings. It is for example vital that the first coating effectively adheres or keys to the metal. The second coating, if needed, must be compatible with the first.

In coating technology, cleanliness of the surfaces to be coated is of prime importance. For example, the presence of grease, oil, corrosion products, dirt and millscale, all of which affect adherence, must be eliminated.

In the preparation of metal surfaces two steps are usually taken. First, the removal of organic substances, such as oil and grease, which can be done by immersion in trichlorethylene or a similar degreasing agent. Secondly, the removal of inorganic material, such as scale, corrosion products and mineral matter. This is preferably done by shot blasting or alternatively by immersion in a solution of hydrochloric acid or sulphuric acid. then rinsed and dried. It is important that the cleaned metal surfaces should be covered with the protective coating as soon as possible and certainly within a period of four hours, to prevent further contamination of the surface.

Protective coatings include deposition of metals by electroplating, hot-dipped galvanising, metal spraying and painting.

Cadmium and zinc are commonly used for plating iron and steel components, zinc being more widely used because it is cheaper. Cadmium is a better finish for marine conditions, but there is little difference between cadmium and zinc for most industrial applications. Tin is also used for the protection of steel components, and hot galvanising for cubicles and heavy components. Tin, silver and chrome plating are used for current carrying parts of which silver has the advantage that silver oxides formed on exposed surfaces are generally good conductors. Also included among protective coatings are phosphating treatments, i.e. immersion in manganese or zinc phosphates, which are then readily sealed by oiling or painting. This is normally applied to aluminium components.

Paints used for general purposes, e.g. enclosures, are based mainly on alkyd resin and are air dried. Modified alkyd resin paints are used for stoving. Paints based on epoxy and polyurethene resins are used mainly where chemical resistance is of importance.

All of these finishes add to the dimensions of the components. This must be allowed for in the design. It should also be noted that where it is necessary the 'finishes' may have to be 'stopped off' bearing faces for dimensional reasons and to reduce seizing.

11.4 Electrical failure

One of the principal causes of electrical failure in circuit breakers is contamination of the insulation surfaces. Contamination can be reduced by controlling the field stress over the insulation which will reduce the concentration of contamination in the high stress areas.

Nevertheless, contamination cannot be prevented in for example oil circuit breakers or in circuit breakers situated in contaminated atmospheres, such as in cement mills. It is then essential that all insulation surfaces that can be contaminated shall be visible in the normal maintenance procedure operation. Concealed stressed insulation can lead to serious failures because it will not receive routine cleaning.

One of the major causes of failures in air-blast circuit breakers has been due to condensation in the blast supply pipes when these operate normally at atmospheric pressure. Adequate and continuous air conditioning can prevent this but in addition the prevention of any free moisture generated at the upper end of the tube from flowing into the pipe by suitable mechanical design can be helpful. Provision for cleaning the pipe can also aid in ensuring reliability.

The electrical failure of circuit breaker bushings in outdoor equipment is owing to ingress of moisture through failure in the gasketing systems far more frequently than causes associated with failure of the insulation material itself. This then requires careful detail attention to the gasket flange design and adequate tightening instructions from the point of view of maintaining continuous pressure on the gasket material and ensuring correct assembly in the first instance. Built in spring loading is clearly of assistance in achieving the former aim.

11.5 Maintenance

Mechanisms should be designed so that they are easy to adjust and that the final adjustment is not critical. If adjustment is required in more than one place, then the adjustments in series should as far as possible not be interdependent but should be capable of being set finally on their own. Adjustments should also be easy to lock and easy to inspect.

If maintenance requires slow moving of the contact system, then the design should accept a tool which permits slow closing and slow opening of the circuit breaker under manual control without risk of the trip gear releasing.

Parts that require replacement owing to wear, such as contacts, should be easy to replace and should be easy to lock in position at the correct adjustment.

Means should be provided or at least provision made for connecting a travel recorder and pressure recording equipment and a timing device for accurate checking of these important aspects of the operation of the circuit breaker. The plus and minus tolerances for acceptable performance should be given in the maintenance manual.

There are no other hard and fast rules that can be applied to the design of circuit breakers for safety, much of it is in the application of good common sense. Perhaps the only other possible rule can be said to be, design for the convenience of maintenance. The inconvenience of requiring special tools to dismantle and reassemble components which have restricted access may be frustrating to the maintenance personnel. Many accidents can be attributed to the use of inappropriate tools used rather than troubling to obtain the correct ones. If the use of a special tool is unavoidable it is better to design arrangements in such a way that there is no alternative. For example an interrupter fixing bolt can be countersunk with the head sunk beyond the reach of an open spanner.

For convenience, if for no other reason, it is preferable to arrange for the simple withdrawal of components, for example interrupter units or indeed the complete circuit breaker where it is sufficiently small for handling, to facilitate regular maintenance. Conversely, it is preferable to deter by design interference with components which do not require regular maintenance, but to ensure that when necessary the complete component can be replaced conveniently.

Physiologists constantly remind us that man was not originally designed to stand erect and particularly not designed to lift heavy weights. A mass of 35 kg should be regarded as the optimum maximum lift and therefore any component of greater mass requires special lifting facilities. As it is preferable to avoid the use of such facilities, it is good practice to arrange subassemblies in such a way that they are well below this mass and, for safety, to avoid assemblies with a mass not sufficiently greater to deter maintenance personnel from attempting to lift them manually. Outdoor mounted circuit breakers, if they form an insulation structure as for example low-oil-content or air-blast circuit breakers, necessarily have to be mounted on high earthed structures or even if they are ground mounted much of the maintenance work has to be carried out at a high level. The design of the maintenance platforms is as much the responsibility of the designer, as is the circuit breaker. The platform must be capable of easy and safe erection and must be arranged to be dismantleable into short sections which can be carried easily about within a live switching station. Metal scaffold poles for example are not satisfactory as they are a danger during transport across the switching station. The construction of the platform, to ensure its easy and safe erection cannot be over stressed and for this reason is mentioned before the facilities it should provide. If it is difficult to transport and erect there is a danger that a substitute will be used at the expense of safety.

The height of the platform should be 1 m below the height of the main area of work to provide the optimum comfort and safety of personnel. The platform should be arranged to avoid as far as possible forward stretching and, most important, should have safety rails and a 'kick board' to prevent tools and components rolling or falling off.

Indoor circuit breakers of the 'plug-in' type can be maintained with the truck withdrawn to just outside the housing. It is preferable to avoid disconnection of the trip circuit by arranging the wiring to the coil in a flexible conduit of sufficient length to enable the circuit breaker to be withdrawn for maintenance, without the need for disconnection. Access to the contacts for inspection or changing should not require disconnection of essential interlocks or main current carrying connections. Heavy components such as arc chutes of air circuit breakers, which must be moved for access to contacts, should be arranged to be pivotted upwards and securely locked in this position if their mass is inconvenient for manual removal.

All that has been said on this subject is common sense and accidents attributable to poor design generally occur only where common sense has not been used. It is also necessary to take cognisance of the requirements of the Workshops and Factories Acts.

The compilation of clear maintenance instructions is again a responsibility of the design function, since this will affect reliability in service. Maintenance instructions should be presented in a form which will be clear to maintenance crews, and should establish erection instructions and both routine and rectification procedures.

This includes routine cleaning and lubrication, contact inspection and replacement, mechanism adjustment, checking of mechanical performance characteristics, checking of insulation characteristis, tightening procedures for oil and pressure gas seals, and for porcelain to metal joints. It should include not only the procedures but identification of when these are required, and of areas in the circuit breaker where the adjustment can compensate for wear, and for areas where it is preferable to replace complete assembly. This is particularly important with interrupting components where the permissible degree of wear can be ascertained only by short circuit test experience.

It should also include procedures necessary for cleaning oil, detail procedures necessary for removal and replacement of SF_6 gas with special reference to control of cleanliness and moisture levels, procedures for maintaining cleanliness when dismantling of high pressure oil and gas lines, and safety precautions in dealing with stored energy devices.

11.6 Simplicity

Perhaps the most important issue in the design of circuit breakers is to maintain simplicity in the overall construction and concept. The more components there are in a circuit breaker the more likely electrical or mechanical failure is, since the failure of any one will normally cause the failure of the complete circuit breaker. Apart from duplicated trip coils or any duplicated equipment which is paralleled, the only components in a circuit breaker that can be considered to be fail safe in character are the interrupters when mounted in series with one another. In this case to some extent they are in support of each other by reducing the duty on each interrupter, and if one fails the others will have some chance of performing the overall duty. Further if more interrupters are used in series, the smaller is the duty on each interrupter and the less is the chance of any one interrupter failing to carry its share of the duty. Nevertheless, the overall essential need for reliability requires that the number of interrupters shall be kept to the minimum that will accept the rating, because the increase in risk of mechanical failure or electrical failure of the assembly, through having more interrupters than are needed, is likely to outweigh considerably any increase in reliability achieved by reducing the rating on an interrupter below the level which it can normally accept.

Reduction in the number of interrupters in series can however fall short of the optimum requirements for reliability at the point when increasing the duty per interrupter involves undue increase in unit complication, thus increasing the total failure risk.

11.7 Reference

CIGRÉ (1972): Interim report (13–07) on the study of the reliability of high-voltage circuit breakers

Chapter 12

Insulation applied to
circuit breakers

D. Legg, B.Sc., C.Eng., M.I.E.E.

12.1 Technique of insulation design

In the design of a circuit breaker it is necessary to estimate the electrical and mechanical stresses to which the circuit breaker components will be subjected and to relate these to the known electrical and mechanical strengths of the materials to be used. A designer must be aware of the insulating materials that are available together with their basic electrical and mechanical properties.

The electric stress E imposed on an insulating (dielectric) material is the force acting on a unit charge located in the dielectric and is related to the kinetic energy acquired by charged particles in the dielectric. The electric stress E is numerically equal to the voltage gradient

$$E = \frac{-dV}{dx}$$

and is usually measured in kilovolts per millimetre or megavolts per metre. Voltage gradients can be calculated directly for simple electrode geometries (e.g. concentric cylinders), but for many of the geometries encountered in circuit breaker design it is necessary to estimate the electric field gradients by field sketching, by electrolytic tank plots or preferably by using a digital computer. The important stresses are (*a*) the maximum voltage gradients which usually occur at the surface of electrodes, and (*b*) the voltage gradients along solid insulation surfaces.

The electrical strength of insulating materials, (solid, gaseous and liquid) is the maximum stress a material can withstand and is found by experiment. A designer must, therefore, refer to the available technical literature for data on electric strength, and it is an unfortunate fact of engineering life that designers are faced with a considerable number of factors which affect, and generally reduce, the electric strength of an insulating material. Electric strength can be influenced by temperature, voltage waveform, pressure, electrode geometry, electrode material, the volume of insulation involved, mechanical stresses, the previous history of the material, and the presence of impurities. Although the variability in electric strength may present a problem, a designer allows for this by adjustments to safety factors. When the properties of an insulating material can be specified precisely, then safety factors can be small and vice-versa.

At the early stages in the design of a new circuit breaker it is customary for the designer to prepare a rough layout showing the principal concept. The optimum dimensions of solid insulating components and clearances in gaseous or liquid dielectrics are unknown at this stage and the designer is faced with the task of evaluating the shape and size of the circuit breaker components to meet the required voltage levels. The electrical stresses throughout the circuit breaker can be determined with any desired accuracy when the design is finalised, but in the initial stages it is possible to obtain approximate dimensions by relating the design to simple electrode geometries (such as concentric cylinders, sphere plane etc.) for which the electrical stress can be calculated.

A useful concept in the evaluation of the electrostatic voltage gradient in a contact gap is the ratio of average to maximum gradient (E_{av}/E_{max}) in the region of the minimum distance between conductors. The average gradient E_{av} is simply the applied voltage divided by the distance between electrodes. The maximum voltage gradient in the gap E_{max} usually occurs at an electrode surface, and it is the estimation of this gradient that is important to the designer because electrical breakdown generally depends on the maximum gradient not the average gradient.

The variation of voltage gradient in a much simplified circuit breaker geometry is illustrated by way of example in Fig. 12.1. The field plot of the voltage distribution for a pair of hemispherical ended contacts in oil (Fig. 12.1a) was derived by a digital computer. More important, however, are the potential gradients shown in Figs. 12.1b and c. The maximum gradient in this example occurs at the tip of the high-voltage contact and is approximately 2.4 times the average gradient in the contact gap V/g. (More precise voltage gradients can be derived by difference techniques proposed by Mattingley and Ryan (1971), and the maximum gradient is more accurately 2.7 times the average gradient.) At the point X on the outer surface of the insulating tube the gradient is only $0.15 V/g$, but is 1.2 times the average surface gradient V/l. The maximum surface gradient occurs at Y having a value of $1.5 V/l$ or $0.19 V/g$. The breakdown gradient E_s for the dielectric to be used is subsequently found from published experimental data and is compared with the gradients derived from the field plot. If the calculated gradient at any point approaches E_s then the designer has the choice of modifying his dimensions or changing the insulating materials.

Field plots, similar to that of Fig. 12.1, produced by analytical computer techniques have the advantage that the voltage gradients can be obtained on direct print out. The process, however, can be expensive and it may be expedient to carry out preliminary calculations using the concept of relating the design to simple

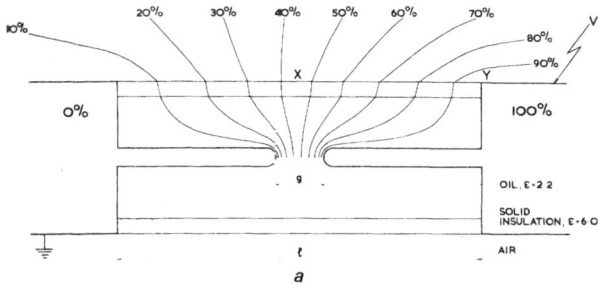

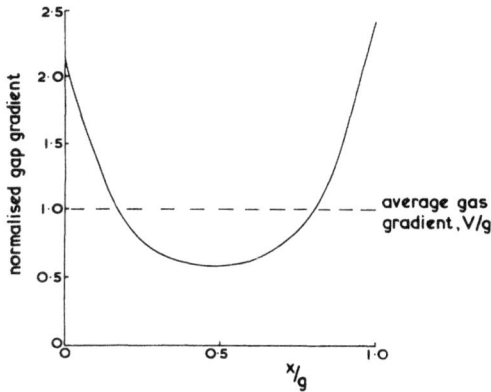

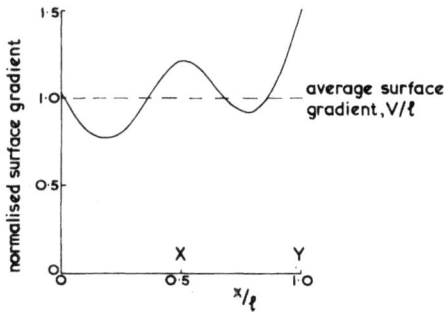

Fig. 12.1 Voltage distribution for hemispherical ended contacts in oil

    (*a*) field plot showing equipotential lines
    (*b*) variation of voltage-gradient in the contact gap
    (*c*) variation of voltage-gradient along outer surface of insulating tube

geometries. The ratio E_{av}/E_{max} can be calculated precisely for simple electrode geometries. The ratio, sometimes called the utilisation factor η of the electrode geometry, measures the 'inferiority' of the field system in comparison with that between infinite-plane parallel electrodes (uniform-field electrodes)*. For uniform field electrodes $E_{av} = E_{max}$ and hence $E_{av}/E_{max} = 1$, while for nonuniform field electrodes $E_{av}/E_{max} < 1$.

From the definition of η, one can write the simple relationship

$$V = E_{max}\eta g \tag{12.1}$$

where g is the electrode gap and V is the voltage applied to the gap.

If the voltage across the gap is raised, sparkover (breakdown) or corona threshold will eventually occur when the maximum gradient reaches a particular sparking gradient E_s at a sparking voltage or corona threshold voltage V_s. Ryan (1967) has shown that the power-frequency breakdown voltage can often be predicted for gas insulating systems by substituting V_s and E_s in eq. 12.1 to give

$$V_s = E_s\eta g \tag{12.2}$$

This simple relationship can be used for simple and perturbed electrode geometries. For perturbed systems, η is evaluated from numerical field studies and the appropriate value of E_s is obtained from experimental data for a similar unperturbed field system.

E_s is not a constant even for a gas at a given density, but varies dependent upon the shape, disposition and spacing of the electrodes. For engineering purposes E_s does, however, fall within reasonable limits for a particular dielectric as will be shown later.

For design purposes it is useful to collate data on the breakdown *gradients* of dielectric gases, for use together with values of η, when calculating the breakdown voltage in simple geometries. The alternative is to refer to an almost unlimited number of breakdown voltage versus gap curves quoted in the literature. Some values of η are tabulated for a few simple geometries in Table 12.1. It should be noted that η is greater for a geometrically symmetrical electrode configuration than for an asymmetrical arrangement. Utilisation factors for eccentric arrangements are given by Ryan and Walley (1967) and values of potential gradient across the electrode gap are given in a paper by Mattingley and Ryan (1971). As a simple example of using 'utilisation factors', consider a concentric-cylinder electrode system of radii 50 mm and 200 mm. From the Tables, η the ratio E_{av}/E_{max} is 0·462 (i.e. the maximum gradient at the surface of the inner electrode is 2·17 times the average gradient).

The breakdown voltage from eq. 12.2 is $V_s = E_s \times 0·462 \times 150$. Suppose the gap insulation is air at one atmosphere then the sparking gradient E_s at power

---

*The reciprocal relationship $f = 1/\eta$, termed the field enhancement factor, is sometimes quoted in the literature

Table 12.1   Values of E_{av}/E_{max} for some simple geometries

(a) *Concentric cylinders and spheres*

$\dfrac{R}{r}$	$\dfrac{E_{av}}{E_{max}}$	
	Cylinders	Spheres
1·05	0·976	0·952
1·10	0·953	0·909
1·20	0·912	0·833
1·30	0·875	0·769
1·40	0·841	0·714
1·50	0·811	0·667
1·60	0·783	0·625
1·70	0·758	0·588
1·80	0·735	0·556
1·90	0·713	0·526
2·00	0·693	0·500
2·20	0·657	0·455
2·40	0·625	0·417
2·50	0·611	0·400
3·00	0·549	0·333
4·00	0·462	0·250
5·00	0·402	0·200
6·00	0·358	0·167
7·00	0·324	0·143
8·00	0·297	0·125
9·00	0·275	0·111
10·00	0·256	0·100
50·00	0·080	0·020
100·00	0·047	0·010

r = radius of inner cylinder or sphere
R = radius of outer cylinder or sphere

(b) *Parallel cylinders and sphere gaps*

$\dfrac{r+g}{r}$	$\dfrac{E_{av}}{E_{max}}$	
	Cylinders	Spheres
1·10	0·9837	0·9675
1·20	0·9679	0·9361
1·30	0·9528	0·9038
1·40	0·9382	0·8697
1·50	0·9242	0·8340
1·60	0·9106	0·7976
1·70	0·8975	0·7613
1·80	0·8849	0·7258
1·90	0·8727	0·6917
2·00	0·8608	0·6592
2·20	0·8382	0·5997
2·40	0·8170	0·5474
2·60	0·7970	0·5019
2·80	0·7782	0·4622
3·00	0·7603	0·4276
4·00	0·6838	0·3075
5·00	0·6232	0·2380
6·00	0·5739	0·1935
7·00	0·5328	0·1628
8·00	0·4979	0·1403
9·00	0·4679	0·1233
10·00	0·4418	0·1099
25·00	0·2513	0·0416
50·00	0·1543	0·0204
100·00	0·0914	0·0101

r = radius of each cylinder or sphere
g = spacing between cylinders or spheres

Table 12.1   continued

(c) *Cylinder plane and sphere plane*

$r + g \over r$	Cylinder plane	Sphere plane
	$E_{av} \over E_{max}$	
1·05	0·976	0·952
1·10	0·953	0·909
1·20	0·912	0·833
1·30	0·875	0·769
1·40	0·841	0·714
1·50	0·811	0·667
1·60	0·783	0·625
1·70	0·758	0·588
1·80	0·735	0·556
1·90	0·713	0·526
2·00	0·693	0·500
2·20	0·657	0·455
2·40	0·625	0·417
2·50	0·611	0·400
3·00	0·549	0·333
4·00	0·462	0·250
5·00	0·402	0·200
6·00	0·358	0·167
7·00	0·324	0·143
8·00	0·297	0·125
9·00	0.275	0·111
10·00	0·256	0·100
50·00	0·080	0·020
100·00	0·047	0·010

r = radius of cylinder or sphere
g = gap between cylinder (or sphere) and plane

(d) *Hemispherical ended rod-rod*

$g \over r$	$E_{av} \over E_{max}$
3	0·44
5	0·31
7	0·24
9	0·20

r = radius of each hemispherical ended rod
g = spacing between tips of rods lying on a common axis

Note: The figures given in Table 12.1*d* are derived for infinitely long rods arranged as a rod-rod electrode gap. Lower values of E_{av}/E_{max} are obtained if one rod is connected to an earthed plane (Steinbigler, 1969). In Fig. 12.1, for which g/r = 5, the value of E_{av}/E_{max} in the contact gap is 0·414 (Fig. 12.1*b*). The difference between 0·414 and the value 0·31 given in Table 12.1*d* is due mainly to the large terminal electrodes in the circuit-breaker model

frequency would be appox 3·3 kV(peak)/mm (eq. 12.8) and the power frequency value of V_s would be

$$V_s = 3·3 \times 0·462 \times 150 = 229 \text{ kV(peak)}$$

By comparison, a uniform-field electrode system having a gap spacing of 150 mm would have a breakdown voltage of approximately 393 kV(peak) (eq. 12.4).

12.2 Gaseous insulation

12.2.1 Gases used for circuit breakers

Gases are used in circuit breakers both for arc extinction and to provide electrical insulation between contacts when the circuit breaker is in the open position. The insulating properties of a gas can be assessed in two principal modes; first, during arc interruption, the transient recovery of electric strength of flowing gas can be measured as it deionises after current zero, and, secondly, the longer term dielectric strength can be measured when static gas is used purely as an insulator at a constant density. The 'insulating' gas in a circuit breaker during arc interruption in fact must be able to change from a good conductor to a good insulator in a few microseconds.

In some circuit breakers a gas is used for both arc extinction and for long term insulation (e.g. in a gas-blast pressurised circuit breaker) while in an oil circuit breaker the evolved gas (hydrogen) is only used for arc extinction and oil provides the long term insulation across the gap.

Air and SF_6 are by far the most common gases used in circuit breakers. Apart from hydrogen which plays an important role in oil circuit breakers other gases like carbon dioxide, nitrogen, oxygen and the hydrocarbon gases have only been used in experimental circuit breakers. A comprehensive survey of the electrical properties of these gases is recorded by Clark (1962).

Both the circuit-interrupting ability and the electric strength of a gas usually increases as the gas density is increased. The breakdown voltage of a gas in a uniform electric field is a function of the product of the relative gas density δ and the electrode separation g, i.e. $V_s = f(\delta g)$. If the gas temperature is constant the sparking voltage is a function of the product of pressure and gap, i.e. $V_s = f(pg)$. This relationship is popularly known as Paschen's law and the form of the variation between breakdown voltage, pressure and gap spacing is sketched in Fig. 12.2.

For circuit breakers other than vacuum circuit breakers the region of application of gas insulation lies to the right of the minimum shown in Fig. 12.2. The minimum of the curve is typically at 350 V and $pg = 0·007$ bar mm for air, and at 500 V, $pg = 0·0035$ bar mm for SF_6 according to Oppermann (1972). For the right hand part of the curve, Paschen's law can be written up to the limits of its validity in the form

$$V_s = Ax + B\sqrt{x} \qquad (12.3)$$

where V_s = sparking voltage, $x = \delta g$, A and B are constants. All the compressed

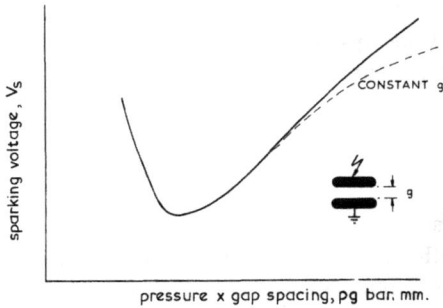

Fig. 12.2 Variation between breakdown voltage, gas density pressure and gap spacing for uniform-field electrodes. The broken line shows the kind of deviation from the Paschen characteristic which occurs in practice

gases fail to obey Paschen's law at some stage, i.e. the sparking voltage becomes significantly lower than expected from eqn. 12.3 for a given value of δg. Failure of the law is detected in practice, for example, when at a fixed electrode spacing increase in gas pressure above a certain level produces a less than proportional increase in the sparking voltage (Fig. 12.2 and 12.3).

Although the general concept of Paschen's law is useful to a designer it must be emphasised that for many practical combinations of electrode geometry and gas pressure the deviations can be considerable. The deviations, which usually occur at stresses of the order of $E = 10-20$ MV/m and at gas pressures above about 10 atm, are caused by prebreakdown currents (field emission) which distort the electric field and Paschen's law applies to uniform fields only. The value of stress at which deviations from the law start is also dependent on the electrode material, electrode separation, the area of the electrodes and the dust content of the gas.

12.2.2 Electric strength of air
12.2.2.1 Uniform fields
In a uniform field, the breakdown strength is virtually independent of the waveshape of the applied voltage, therefore the breakdown voltage level at a given value δg is virtually the same for switching impulse, lightning impulse and peak power-frequency voltage. At high stress levels (>10 MV/m), however, the long term power frequency hold off levels (after say 1 hour duration) can be significantly lower than the 1 min hold-off level.

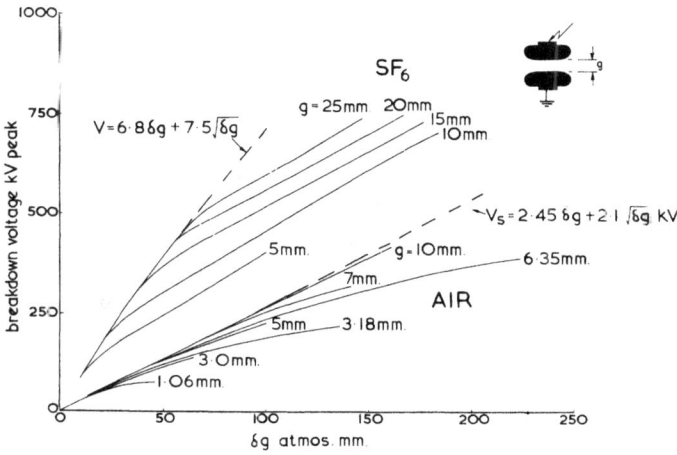

Fig. 12.3 Breakdown voltage of air and sulphur hexafluoride as a function of the product gas density times gap spacing for uniform-field electrodes

Results in **air:**
1·06, 3·18 and 6·35 mm gaps, d.c. breakdown (Howell, 1939)
3 mm gap, d.c. breakdown (Holt, 1956)
5, 7 and 10 mm gaps, 50 Hz breakdown (Finkelmann, 1937)
Results in SF₆: (Opperman, 1972)

The breakdown voltage of air between 'conditioned'* uniform-field electrodes is shown in Fig. 12.3. Numerous workers have derived empirical relationships covering the Paschen characteristic, for example Bruce (1947) and Boyd *et al.* (1966). The maximum value of breakdown voltage attained for a given value of δg before deviations from Paschen's law occur is indicated by the broken line, which for the data given in Fig. 12.3 can be represented by the equation

$$V_s = 2{\cdot}45\ \delta g + 2{\cdot}1\ \sqrt{\delta g}\ \text{kV(peak)} \tag{12.4}$$

where V_s is the sparking voltage, δ is the relative gas density† and g is the gap spacing in millimetres.

*See Section 12.2.4
†In eq. 12.3 − 14, δ the gas density relative to s.t.p. is given by

$$\delta = \frac{P}{1{\cdot}013} \times \frac{293}{T}$$

where P is gas pressure in bars absolute and T is gas temperature in deg K. For practical engineering accuracy at normal operating temperature δ can be taken simply as the gas pressure in atmospheres absolute (1 atm = 1·013 bar and 1 bar = 10^5 N/m²). For this reason in the associated graphs of Figs. 12.3, 12.5 and 12.6, δ is shown in atmospheres although in fact it is dimensionless

It will be noted that the curves drawn for constant value of g gradually fall below the broken line indicating a deviation from the Paschen characteristic. The derived values of V_s obtained from eqn. 12.4 must be used with great caution because, as can be seen, the curves start to flatten at pressures of about 10 atm.

From eqn. 12.4, the equivalent breakdown gradient E_s can be derived as

$$E_s = 2 \cdot 45\delta + \frac{2 \cdot 1 \sqrt{\delta}}{\sqrt{g}} \text{ kV(peak/mm)} \qquad (12.5)$$

For very large gaps at atmospheric pressure ($\delta = 1$) the electric strength of air tends towards $2 \cdot 45$ kV (peak)/mm while at 10 mm and atmospheric pressure E_s is approximately $3 \cdot 1$ kv(peak)/mm. It should be noted that the breakdown gradient is not constant even for a uniform field geometry but falls as the electrode spacing increases. Eqn. 12.5 is plotted for values of δ from one to three in Fig. 12.4. For air pressures above about 10 atm, field emission produces large deviations from the Paschen characteristic and the increase in E with pressure becomes progressively less proportionate and significant differences can exist between short term and long term power-frequency sparkover levels. It is obviously meaningless to quote the breakdown gradient for a compressed gas even in uniform fields without also quoting the electrode separation and gas density.

12.2.2.2 Nonuniform fields

In nonuniform fields, the breakdown voltage varies with voltage waveform and its polarity. Of the test voltages applied to circuit breakers, lightning impulse voltages

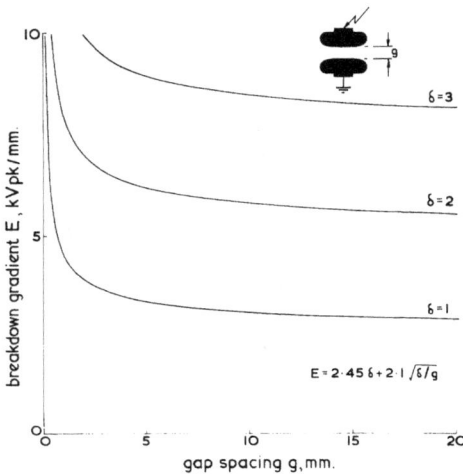

Fig. 12.4 Variation of breakdown gradient with gap spacing for uniform-field electrodes in air

usually give the highest breakdown levels, peak power-frequency voltage the lowest with switching impulse in between. Some practical geometries, e.g. coaxial cylinder arrangements, can be considered to be only slightly nonuniform (quasiuniform) if the gap spacing is small relative to the electrode dimensions and the sparkover levels of air in these fields is just below the uniform field strength. The divergence of the field can be assessed by the value of the utilisation factor η. In more divergent fields, ionisation occurs in the region of highest stress at voltages below the sparkover level. This ionisation is called corona and it becomes necessary to arrange design stresses below the corona inception stress rather than below the higher breakdown stress because corona should be avoided in most circuit-breaker applications. The stress required for the onset of corona in a concentric-cylinder electrode geometry depends more on the radius of the inner cylinder than the radial electrode spacing. Peek (1929) measured the a.c. corona inception stress in air at relative gas density values close to unity and showed that for concentric cylinders the corona inception stress E_i was given by

$$E_i = 3 \cdot 1 \delta + 3 \cdot 0 \ \sqrt{\frac{\delta}{r}} \ \ \text{kV(peak)/mm} \tag{12.6}$$

where δ is the relative gas density, r is radius of inner cylinder in millimetres; for $\delta \leqslant 1$ and r up to about 30 mm. (This expression is similar in form to that of eqn. 12.5, but the radius of the inner cylinder is involved rather than the gap spacing as in eqn. 12.5.) Expressions similar to eqn. 12.6 are obtained for corona-inception stresses for parallel cylinders and sphere-plane gaps.

It will be seen that in eqn. 12.6 the inception stress increases as r decreases. The smaller r the more rapidly does the stress fall with distance, and because electrons must be accelerated over a finite distance to cause avalanches then higher stresses are required at the electrode to produce corona.

For quasiuniform fields the corona inception stress E_i is not significantly different from the breakdown stress E_s. The minimum breakdown stress for concentric-cylinder, sphere/plane and concentric-sphere geometries is found in each case to depend on the radius of the most highly stressed electrode and it is convenient to give breakdown data for these arrangements by plotting rE_s against $r\delta$. Fig. 12.5 shows the relationship between rE_s and $r\delta$ for concentric cylinders and concentric spheres under power frequency conditions. For air the relationship can be represented by

$$50 \ \text{Hz} \quad r\delta < 40 \quad rE_s = 3 \cdot 17 \, r\delta + 8 \cdot 7 \tag{12.7}$$

$$r\delta > 40 \quad rE_s = 2 \cdot 55 \, r\delta + 38 \tag{12.8}$$

while under negative impulse conditions (Fig. 12.6) the relationship is

$$-\text{ve} \ 1/50 \quad rE_s = 2 \cdot 55 \, r\delta + 41 \tag{12.9}$$

In the above equations E_s is measured in kilovolts(peak) per millimetre, δ is the relative air density and r is the radius of the smaller electrode in millimetres. The equations can also be used for the approximate minimum values of E_s experienced

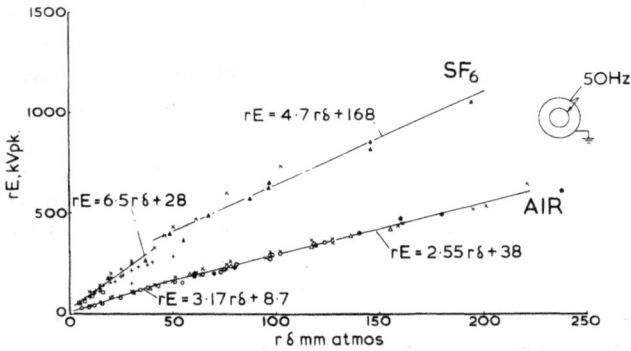

Fig. 12.5 Relationship between rE and rδ for concentric cylinders and concentric spheres in air and sulphur hexafluoride for power frequency voltages

     o   concentric cylinders in air
     x   concentric spheres in air and SF_6
     △   concentric cylinders in air (Gossen, 1948)
     •   concentric cylinders in air (Finkelmann, 1937)
     □   concentric cylinders in SF_6 (Howard, 1957)
     +   concentric cylinders in SF_6 (Zaleskii, 1970)
     ▲   concentric cylinders in SF_6 (Kawaguchi, 1970)

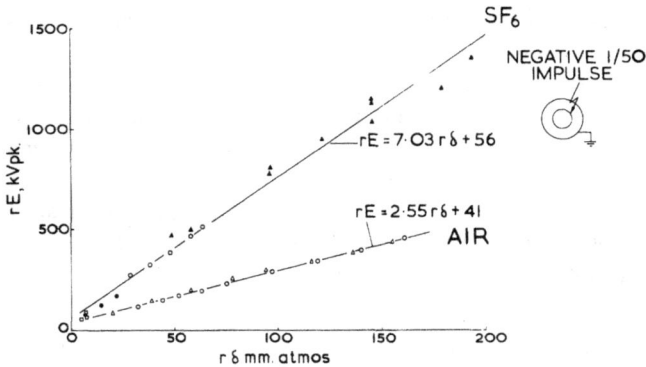

Fig. 12.6 Relationship between rE and rδ for concentric cylinders in air and sulphur hexafluoride for negative lightning impulse voltages

     o   values in air and SF_6
     △   values in air (Gossens, 1948)
     •   values in SF_6 (Mulcahy, 1964)
     □   values in SF_6 (Howard, 1957)
     ▲   values in SF_6 (Kawaguchi, 1970)

with sphere-plane electrodes. It is worth noting that for large values of r, $E_s = 2 \cdot 55 \, \delta$ kV(peak)/mm which is similar to the uniform-field gradient for large gaps.

For large open-type circuit breakers, data is required on the flashover voltage to ground and between phases in air at atmospheric pressure. The fields involved are usually highly divergent and a rough estimate of breakdown levels can be obtained using data from rod—rod and rod—plane geometries shown in Fig. 12.7. The electric

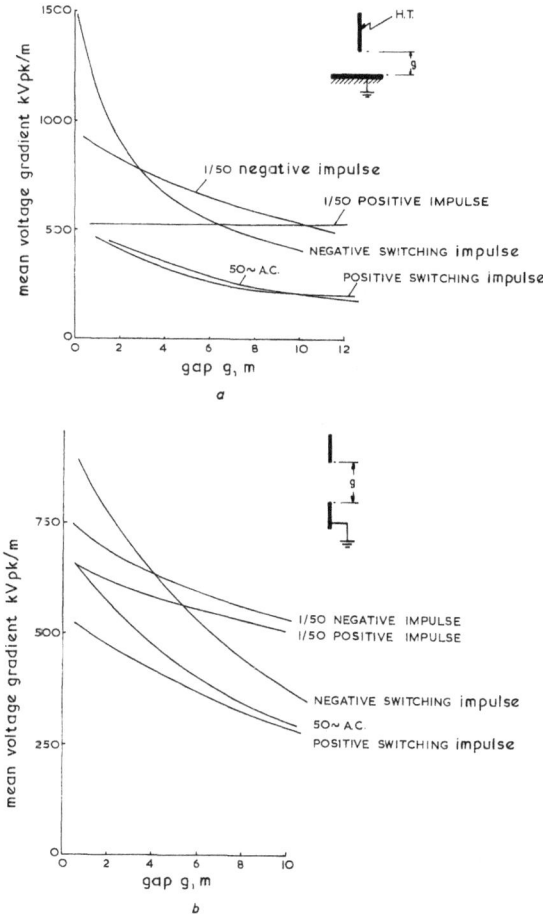

Fig. 12.7 Mean voltage gradients corresponding to the 50% flashover voltage for rod gaps in air at atmospheric pressure

    (*a*) rod-plane gaps
    (*b*) rod-rod gaps

field of a practical open terminal high-voltage circuit breaker is however not so divergent as that of a rod—plane gap, and for phase-to-phase clearances or phase-to-ground clearances, open high-voltage breakers can be adequately represented by sphere—plane or ring—plane gaps. Degli Esposti *et al.* (1971) have recorded breakdown data for sphere—plane and ring—plane gaps in air spacings up to 7 mm and have derived the maximum permissible surface gradients in order to avoid corona for spherical electrodes (Table 12.2). The surface gradient at corona inception was found to depend on the electrode radius of curvature and be largely independent of the gap spacing.

Table 12.2   Maximum acceptable surface stress for spherical electrodes in sphere-plane gaps of the order of 1—7 m (Degli Esposte *et al.*, (1971)

	Maximum surface stress kV(peak)/mm		
Radius of curvature m	Switching impulse	Peak 50 Hz	Direct voltage
0·25	4·0	3·5	3·0
0·7	2·7	2·0	1·8
1·0	2·2	1·5	1·3

12.2.3 Electric strength of SF_6

SF_6 has an electric strength of about 2·5 times that of air at the same pressure in uniform fields. The general properties of SF_6 are described in chapter 7 and the purpose of this Section is to show that the breakdown characteristics are similar to those described for air in Section 12.2.2, with the important difference that the breakdown levels for SF_6 are higher at corresponding pressures. The data for SF_6 have been plotted on the same graphs used for air (Figs. 12.3, 12.5, and 12.6) to facilitate this comparison.

From Fig. 12.3 the values of V_s and E_s, over the range of density and gap spacing in which the Paschen characteristic holds (in a uniform field), can be expressed as

$$V_s = 6 \cdot 8\delta g + 7 \cdot 5\sqrt{(\delta g)} \text{ kV(peak)} \tag{12.10}$$

and

$$E_s = 6 \cdot 8\delta + \frac{7 \cdot 5\sqrt{\delta}}{\sqrt{g}} \text{ kV(peak)/mm} \tag{12.11}$$

in which δ is the relative gas density and g is the gap spacing in millimetres.

Expressions similar to those previously obtained for air can also be derived for concentric—cylinder and concentric—sphere geometries. From Figs. 12.5 and 12.6 the following approximate relationships are derived for SF_6 in which E_s is measured in kilovolts(peak) per millimetre and r is radius of smaller electrode in millimetres. Again the equations can be used for the approximate

50 Hz $r\delta < 40$ $rE_s = 6 \cdot 54 r\delta + 28$ (12.12)

$r\delta > 40$ $rE_s = 4 \cdot 70 r\delta + 168$ (12.13)

−ve 1/50 impulse $rE_s = 7 \cdot 03 r\delta + 56$ (12.14)

minimum values of E_s experienced with sphere—plane electrodes. The breakdown gradients for switching impulse voltages of various waveforms have been measured for coaxial cylinder geometrics by Ryan and Watson (1978) and they found that the breakdown gradient was lower with waveforms of 230/1700 μs than for lightning impulses of 1/50 μs. Gaseous insulation takes a finite time to breakdown and a knowledge of the breakdown voltage V as a function of breakdown time t is of practical importance for insulation coordination in SF_6 switchgear. Ryan and Watson (1978) have derived voltage-time ($V \sim t$) characteristics from their work on impulse breakdown levels of coaxial-cylinder electrode arrangement in SF_6 within the pressure range 0·1-0·6 MPa. The time to breakdown covered the range 0·3-500 μs and the $V \sim t$ curves show an inverse characteristic with breakdown levels falling with larger values of t and an extremely wide scatter between overvoltage level and time lag. Taschner (1976) measured voltage-time characteristics in SF_6 for coaxial cylinders of diameters 11 cm and 33 cm and found a steep increase in breakdown strength for breakdown times below 1·5 μs. Taschner concluded that SF_6 switchgear should be protected against overvoltages with rise times in this range by surge diverters with flat $V \sim t$ characteristics. Voltage-time characteristics of concentric cylinders and sphere-plane electrodes in SF_6 subjected to lightning impulse voltages and linearly rising voltages up to 500kV/μs are given by Knorr *et al.* (1980). A characteristic of SF_6 is that it maintains its high dielectric strength even if diluted with another gas. Howard (1957) and Cohen (1956) have measured the electric strength SF_6—air and SF_6—nitrogen mixtures, and for moderate total pressures of about three to four atmospheres the strength of $SF_6 - N_2$ has been shown to fall by only one tenth as the percentage of SF_6 in the mixture is reduced from 100% to 30%. With SF_6—air mixtures the reduction of electric strength with reducing SF_6 content is more pronounced. Takuma *et al.* (1972) have given empirical relationships to illustrate this effect. Cookson and Wootton (1978) have measured breakdown characteristics of hydrogen—SF_6 mixtures and Cookson and Pedersen (1978) have reviewed the potential use of mixtures of SF_6 with nitrogen, air and carbon dioxide for 1200kV systems. Malik and Qureshi (1979) have reported that mixtures of SF_6 and other fluorocarbon and perfluorocarbon gases can have dielectric strength superior to that of pure SF_6. Farish *et al.* (1976) have demonstrated that the dielectric strength of mixtures of SF_6 and nitrogen can be less sensitive to electrode surface roughness than that of pure SF_6 for certain ranges of roughness.

12.2.4 Conditioning effects

At pressures of up to a few atmospheres the breakdown voltage V_s of a gas is virtually independent of the number of breakdowns across the gap, provided that

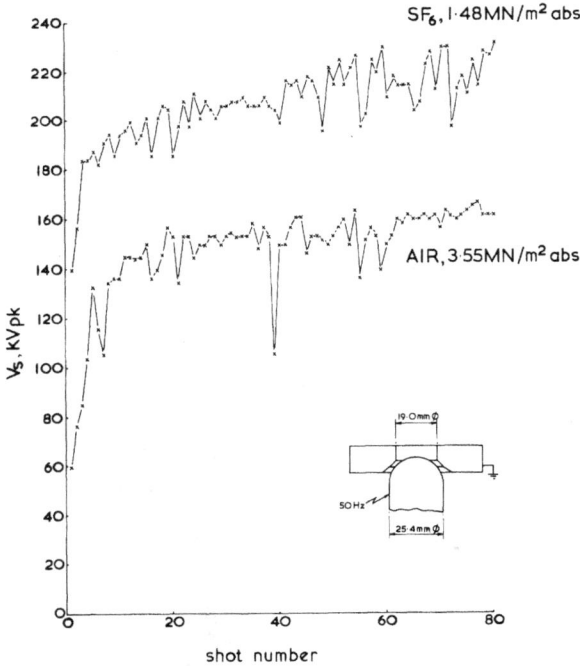

Fig. 12.8 Effect of spark conditioning on the power-frequency breakdown voltage in compressed air and sulphur hexafluoride. Copper nozzle and contact stem, minimum electrode gap 3 mm

the discharge energy is such that the electrode surfaces are not significantly damaged. At higher pressures, however, when the applied field E is of the order of 10 kV/mm, V_s can increase with successive sparkovers until a relatively constant value is reached (Fig. 12.8). This effect is called spark conditioning and the initial breakdown voltage of new 'virgin' electrodes can be less than half the spark conditioned breakdown-voltage level achieved after 100 or more sparks (Cookson, 1970). The number of sparkovers required to attain the conditioned state in a particular gas is dependent on the electrode materials and increases with increase in gas pressure and electrode area.

Electrode gaps are more readily spark conditioned if the electrodes are clean and smooth and their surface area is kept to a minimum.

Conditioning has been observed in air and SF_6 (Fig. 12.8), nitrogen, hydrogen and carbon dioxide. It is a complex electrode effect which is not fully understood and may also be influenced by the presence of dust in the gas and on the surface of the electrodes. Ikeda and Yoda (1971) have observed an appreciable conditioning effect for SF_6 under a.c. breakdown conditions when dust is present. They found a pronounced spark conditioning for stresses as low as 30 kV/mm at gas pressure below 9 atm.

From an engineering viewpoint the conditioning effect is of major importance because the values of breakdown voltage and stress quoted in the literature are invariably the conditioned values (as indeed are the data shown in Fig. 12.3), but obviously for some practical applications the first breakdown level should be allowed for.

The problem of conditioning is most likely to be significant in permanently pressurised gas-blast circuit breakers operating at pressure above say 30 atm in air and 10 atm in SF_6. Circuit-breaker electrode gaps can be spark conditioned using a low energy high-voltage source, but it is more common to base design stresses on the unconditioned state and assume that the initial breakdown level of the contact gap can be as low as half the fully conditioned level. After a number of circuit-breaker operations, breaking very small currents the sparkover level of the contact-nozzle gap can increase and the designer must allow for this when co-ordinating the breakdown voltage levels of the circuit-breaker contact gap and the solid insulation in parallel.

12.2.5   Effect of particles

Dust particles cause a reduction of the breakdown voltage in gases and the effect is more marked when E is greater than about 5kV peak/mm and gas pressures are above several atmospheres. The presence of dust can substantially reduce both the initial and the fully conditioned breakdown voltages obtained with clean electrodes, and it is therefore important to eliminate dust, fibres and particularly metal particles from circuit breakers using compressed gas insulation.

Under a.c. conditions both metallic and non-conducting particles can oscillate between the electrodes, and if the voltage is sufficiently high, particles can move across the electrode gap. Cooke *et al.* (1977) have measured the movement of spherical and wire particles between coaxial cylinders under d.c. and a.c. applied voltages. They have recorded the 'lift-off' fields for particles and shown that under a.c. conditions particles can bounce on the lower electrode and with increasing voltage eventually cross the entire gap to trigger flashover. Small spark discharges ('microdischarges') which occur when the charged particles impact on an electrode in uniform fields are thought to initiate gap breakdown causing reduced sparkover levels. Measurements by Cookson and Farish (1972) on the influence of conducting particles during a.c. breakdown of compressed SF_6 have shown that in a 150mm/250mm coaxial electrode system the breakdown voltage was reduced by about 40% by the introduction of fine copper wires 1·6mm long at an SF_6 pressure of 0·5 MN/m^2. Generally the longer the wires the lower the breakdown voltage. At the same pressure, aluminium spheres of 1·6mm diameter reduced the 'clean' breakdown level by approximately 30%.

An extensive and comprehensive study of the effect of conducting particles on the electric strength of compressed-gas systems is recorded in EPRI report EL-1007 (1979) in which it is demonstrated that breakdown occurs when wire particles are very near but not touching an electrode and that breakdown depends strongly on the number of particles present and on the time of application of voltage in the range of minutes to hours. Breakdown levels are shown to be dependent on the size of wire particles, and typically with coaxial electrodes in SF_6 at a pressure of 0·4 MN/m^2, the breakdown voltage was reduced by 30%, 48% and 60% from the clean level by the introduction of copper wires of 0·45mm in diameter and 3·2mm, 6·4mm and 12·7mm in length, respectively.

In practical circuit breakers dust and conducting particles would usually be very much smaller than the particles used by Cookson and Farish. With reasonable precautions during manufacture it is possible to eliminate particles having a dimension greater than, say, 0·5mm, and the reduction in breakdown voltage caused by particulate matter is then very much smaller. Dust can be generated even in a clean circuit-breaker during switching operations when metallic particles are formed by friction and from arcing in the contact assembly. Careful design is required to reduce the formation of metallic dust to a minimum.

Cellulose fibres in the gas are probably a greater hazard than nonconducting dust particles. Mulcahy (1972) has shown that, even at atmospheric pressure, the presence of cotton fibres a few millimetres in length can reduce the d.c. breakdown level of a 10mm uniform-field gap by as much as 40%. The percentage reduction in breakdown level was very roughly proportional to both the length and the number of the fibres. It is important, therefore, that even nonconducting dust and fibrous material be removed from any equipment using high-pressure gas insulation because particles as small as 10^{-2} mm can reduce the sparkover level and increase the scatter in V_s for a given value of δg.

12.2.6 Effect of electrode area

The breakdown gradient in a compressed gas decreases significantly with increasing electrode area if the pressure and field is high enough. It has been reported by Philip and Trump (1966) that in nitrogen—carbon dioxide mixtures at a pressure of 2 MN/m^2 an increase in anode area from 10^3 to 10^6 mm^2 resulted in a reduction of the d.c. breakdown gradient by about half, but at pressure of 0·5 MN/m^2 and fields below 10 kV/mm the effect of electrode area was negligible. The electrode area effect has also been noted in SF_6 and air. Ryan and Watson (1978) have shown that the withstand gradient in SF_6 decreases with increase in the stressed electrode area for negative lightning and switching impulse waveforms. An increase in electrode area of 10^3 produced a reduction in withstand gradient of 18% and 34% for negative lightning impulse and switching impulse voltages, respectively, for coaxial cylindrical electrodes in SF_6 at 0·4 MPa. Consequently breakdown values obtained in the laboratory with small-area electrodes must be used with caution in the design of large circuit breakers using compressed gas immersion. The number of conditioning sparks required to reach a final level of breakdown voltage in compressed gases becomes greater as the electrode area is increased at a given pressure. For example, the small electrode area shown in Fig. 12.8 only about 80 sparks were necessary.

12.2.7 Effect of temperature

In uniform fields the breakdown voltage is independent of the gas temperature if the density remains constant, within the limits of Paschen's law, because $V_s = f(\delta g)$. If electrodes are enclosed in a sealed chamber then if the temperature is raised, the gas pressure increases but the gas density remains constant and the breakdown voltage should be constant. Paschen's law has been verified for air at atmospheric pressure for temperatures up to 900°C by Frank (1928) and up to 1100°C for nimonic electrodes by Alston (1958), Powell and Ryan (1972). Eqn. 12.5 can be used with acceptable accuracy for air temperatures up to 900°C at 1 atm, 50 Hz. Above atmospheric pressure however, the relationship of eqn. 12·5 does not hold to such high temperatures; for example at pressures of about 0·5 MN/m² the relationship only holds to about 100°C.

Hot spots sometimes exist on electrodes, for example after an arc has burned in a circuit breaker. It has been shown by Alston (1958) that the effect of a hot spot on a uniform field electrode is almost the same as if the whole electrode gap was at the temperature of the hot spot. This is owing to the lowering of the electric strength of the gas close to the hot spot and once ionisation occurs in this region the field becomes so distorted that complete breakdown tends to occur. Hot spots on electrodes in nonuniform fields have a much smaller effect. A 1200°C hot spot on either electrode of a sphere-plane gap lowers the breakdown voltage by less than 20% in a gap consisting of a 25 mm sphere, 80 mm away from the plane (Alston 1968).

12.2.8. Surface flashover of insulating spacers in gases

The most simple spacer-electrode geometry is that of a solid dielectric right-cylindrical spacer held between uniform field electrodes. In theory, if the spacer is clean and has an electric strength greater than that of the gas, the sparkover voltage should remain unchanged on introducing the spacer because the equipotential field lines remain unchanged. In fact flashover will occur across the gas/solid interface of the spacer at a lower voltage than the gap breakdown level. This can be due to a number of imperfections, for example: corrugations on the surface of the cylinder which distort the field locally, imperfect contact between the ends of the spacer and the electrodes, and particulate contamination on the surface of the spacer and surface charges.

The ratio of surface-flashover voltage to the gas breakdown voltage (with spacer removed) is sometimes called the 'spacer efficiency' and typical values for short cylindrical spacers between uniform field electrodes lie between 0·5 and 0·9 at gas pressures of a few atmospheres. Generally the spacer efficiency decreases with increase of gas pressure, with increasing spacer permittivity and with decreasing surface resistivity. For spacers between uniform-field electrodes it is generally found that a clean cylindrical cylinder has a higher flashover voltage than that of a clean corrugated spacer and it has been reported that corrugations only help if the spacer surface is unclean. In uniform fields at gas pressures in the range $0 \cdot 1 - 5 \ MN/m^2$ spacer efficiencies as high as $0 \cdot 90$ have been measured for materials such as p.t.f.e., polypropylene and perspex in SF_6, and values in the range $0 \cdot 4 - 0 \cdot 7$ have been obtained for filled epoxy resin.

In practical circuit breakers the spacer shape is rarely as simple as the cylindrical spacer described for uniform field electrodes. In many cases the flashover gradient across an insulating surface is much less than the sparkover gradient between the terminal electrodes with the spacer removed. For example, in air at atmospheric pressure solid insulators are often designed for a maximum stress of about $60-100$ V (r.m.s.)/mm at the working voltage for indoor service while for outdoor service the equivalent stress could be as low as $10-20$ V (r.m.s.)/mm. These stresses should be compared with the mean sparkover gradient of a rod-plane gap in air at atmospheric pressure which is approximately 500 V (peak)/mm for short gaps under a.c. and positive impulse conditions (Fig. 12.7a). When spacers are used in circuit breakers using compressed gas insulation it is possible by careful design to obtain spacer efficiencies approaching unity for quasiuniform geometries. Typical of practical spacers are disc or conical spacers used with concentric-cylinder electrodes, as for example for gas-filled terminal bushings of dead tank SF_6 switchgear. The gradient along the surface of a conical spacer can be made less than the maximum gradient between concentric-cylinder electrodes so that sparkover will take place at a point remote from the spacer surface. Work on simple truncated spacers in concentric cylinder electrodes has shown that the spacer efficiency falls appreciably with increase in gas pressure. Experiments carried out by Greenwood and Whittington (1972) on conical epoxy resin spacers of $0 \cdot 45$ m outer diameter in a coaxial SF_6 electrode geometry, showed that for impulse voltages there was little to be gained in mean breakdown level by increasing the gas pressure above $0 \cdot 5 \ MN/m^2$. They found that the spacer efficiency fell from $1 \cdot 0$ at $0 \cdot 15 \ MN/m^2$ to $0 \cdot 4$ at $0 \cdot 5 \ MN/m^2$ and concluded that the insulating strength at higher pressures must be influenced more by the spacer than by the gas.

The electric strength of a spacer-gas interface as well as being dependent on gas pressure is markedly affected by particulate contamination and the relative humidity of the insulating gas. Banford (1976) has shown that conducting particles, particularly copper filings, placed on the surface of an insulating spacer between electrodes do not move so readily under the action of applied electric fields as they do between bare electrodes due to forces acting between the insulator surface and the particles. Banford and Hampton (1973) working with small disc spacers between concentric cylinder electrodes, in SF_6 at a pressure of $1 \cdot 5 \ MN/m^2$, found that conducting particles on the surface of a spacer reduced the impulse flashover level considerably. Positive flashover was most affected, the flashover voltage with $35-75$ μm particles being 65% of the clean flashover level and with $125-210$ μm particles only 45% of the clean level. Under a.c. conditions, with surface contamination by particles in the range $125-210$ μm Banford and Hampton measured the mean flashover stress of $15 \cdot 3$ kV/mm at $1 \cdot 5 \ MN/m^2$ SF_6

gas pressure. Nitta *et al.* (1977) showed that fine metal powder, which scarcely impairs breakdown voltage in a gas gap, reduces flashover levels along the surface of dielectric spacers. Fine copper powder of average grain size of less than 30 μm reduced the a.c. flashover voltage of a conical spacer in SF_6 by 50% of the clean level. Banford (1973) has devised a technique whereby contamination on the surface of spacers in compressed gas may be detected by d.c. measurements of surface leakage currents because the current/voltage characteristics of contaminated spacers differ from those of clean spacers.

Circuit breakers using compressed gas insulation are filled with gas which is sufficiently dry to avoid condensation of water on surfaces at the minimum working ambient temperature. Ushio *et al.* (1971) have shown that when the amount of moisture in SF_6 gas is enough to condense on the surface of an epoxy insulator as dew then the flashover voltage across the insulator can fall to 20–60% of the dry level; but below 0°C when moisture condenses as a form of frost the flashover voltage remains at virtually the dry flashover level. If frost does form on insulating surfaces then when the temperature of the equipment rises above 0°C the frost subliminates directly to vapour without reduction in flashover level. From an insulation standpoint, therefore, it is only necessary to keep the dew point of moisture below 0°C with a reasonable margin of safety. For SF_6 circuit breakers, other factors (e.g. decomposition of SF_6 in the presence of moisture) dictate that the dew point should in fact be much lower than 0°C at the working pressure, and the moisture level in SF_6 switchgear is usually restricted to an upper limit of about 20 p.p.m. wt/wt which corresponds to a dew point of −25°C at 4 atm. Nitta *et al.* (1977) have measured the degradation of epoxy-resin surfaces exposed to arced SF_5 and showed that the influence of the decomposition products is very dependent on the moisture content of the gas, and while the surface resistance of silica-filled epoxy decreased drastically in wet conditions after exposure to arced SF_6, alumina-filled epoxy remained relatively intact.

The general conclusion to be drawn is that both spacers and gas should be dry and clean, particularly at high pressures. At lower pressures (say up to about 4 bars absolute) dry insulating dust has no appreciable effect on surface flashover but metallic particles can reduce surface flashover levels significantly.

12.3 Vacuum insulation

An absolute vacuum should approach the notion of perfect insulation because of the absence of collisions between charged particles and residual gas. Vacuum insulation is said to exist when the mean free path of any electrons is greater than the spacing between electrodes. If the vacuum is better than 10^{-4} torr $(<10^{-2}$ N/m^2)* the length of the mean free path is of the order of metres, therefore if the electrodes are separated by about 10 mm (as they would be in an open vacuum circuit breaker) an electron is likely to cross the gap without

*1 atm = 760 torr = $1 \cdot 01325 \times 10^5$ N/m^2

collisions. The normal mechanism of breakdown in gases at atmospheric pressure, due initially to the formation of electron avalanches, is therefore not possible.

In practical vacuum insulation two factors destroy the notion of perfect insulation. First, the presence of gases adsorbed on the electrode and solid insulating surfaces within the vacuum, and, secondly, the emission of charged particles from these surfaces. The complex electrical breakdown mechanism in vacuum has been reviewed by Hawley *et al.* (1965).

12.3.1 Electric strength of a vacuum

A useful comparison of breakdown data for design purposes is almost impossible because breakdown levels are so very dependent on the 'purity' of the vacuum and electrodes. General guidelines are given in the following Sections on some parameters which influence the electric strength of vacuum devices.

2.3.1.1 Dependence on electrode separation

For vacuum gaps up to one or two millimetres the breakdown voltage is approximately proportional to the gap separation, reaching about 50–100 kV at 1 mm with breakdown stress levels approaching 100 kV (peak)/mm. With larger gaps (>2 mm) the breakdown stress decreases as the gap increases (as it does with compressed gas, solid and liquid insulation) and at a spacing of 10 mm the breakdown gradient is about 20–30 kV (peak)/mm. Fig. 12.9 shows the breakdown voltage as a function of electrode spacing and illustrates the large variation in results due to differences in the physical condition of the contacts and residual gas.

12.3.1.2 Pressure effect

For pressures below 10^{-2} N/m² the breakdown voltage is found to be virtually independent of pressure, at least for small gaps of one or two millimetres, down to below 10^{-5} N/m². At pressures above 10^{-1} N/m² the breakdown voltage falls sharply (Fig. 12.10).

For larger gaps, the breakdown voltage does vary with pressure and a pronounced peak in breakdown voltage has been recorded in the region $1-10^{-2}$ N/m² by Arnold *et al.* (1963), i.e. the breakdown voltage is actually increased by using a relatively poor vacuum. Arnold, using sphere-plane stainless steel electrodes and a gap of 20 cm, found that the breakdown voltage increased from a roughly constant level of 650 kV over the pressure range $10^{-4}-10^{-3}$ N/m² to a peak of 1400 kV at about 5×10^{-2} N/m² before falling markedly with further increase in pressure. A similar pressure effect has been noted by Kuffel *et al.* (1970) for flashover across insulating spacers in vacuum with spacer lengths as small as 5 mm (Fig. 12.10).

12.3.1.3 Conditioning

A conditioning effect is found in a vacuum gap which is similar to that found in high-pressure gases and liquids, that is the breakdown voltage increases with

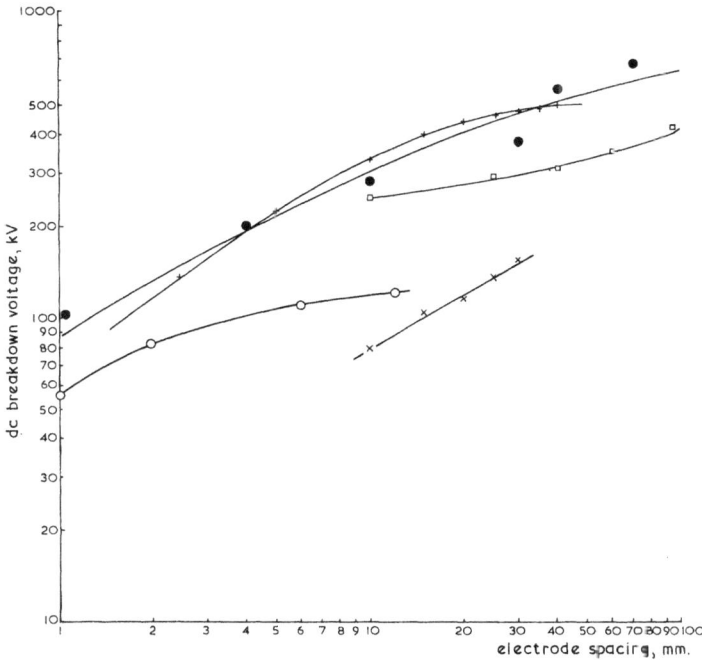

Fig. 12.9 Variation of d.c. and impulse breakdown voltage with electrode spacing in vacuum

+ sphere-plane steel electrodes, outgassed and partially conditioned, pressure 10^{-5} torr (Anderson, 1935)

• sphere-plane steel electrodes, outgassed and partially conditioned, pressure 10^{-6} torr (Trump *et al.*, 1947)

□ uniform-field stainless steel electrodes, outgassed but unconditioned, pressure 10^{-6} torr (Zanon, 1964)

× uniform-field o.f.h.c. copper electrodes outgassed but unconditioned, pressure 10^{-8} torr (Watson *et al.*, 1968)

○ negative 1/50 impulse voltage breakdown for a practical 11 kV vacuum interrupter (50% breakdown level)

repeated sparking until a steady level is reached when the electrodes become 'conditioned' (Hawley, 1968). The breakdown voltage usually rises steadily to a plateau in 10–100 sparkovers and the increase from first breakdown to final condition level is typically more than a factor of two. Conditioning can be also achieved by allowing prebreakdown currents in the gap to flow for a period or by heating the electrodes to a high temperature in the vacuum.

12.3.1.4 Electrode material

Electrode surfaces are obviously of major influence in vacuum breakdown and the material of both the cathode and anode has an influence on the breakdown level. Table 12.3 gives an indication of the influence of electrode material on breakdown.

Hawley (1968) gives the ranking of materials in order of decreasing electric strength (for direct voltage) as tungsten, molybdenum, tantalum, stainless steel, iron, nickel, aluminium, copper, lead, beryllium and carbon. Surface finish is not thought to be of major importance but the breakdown voltage is usually increased

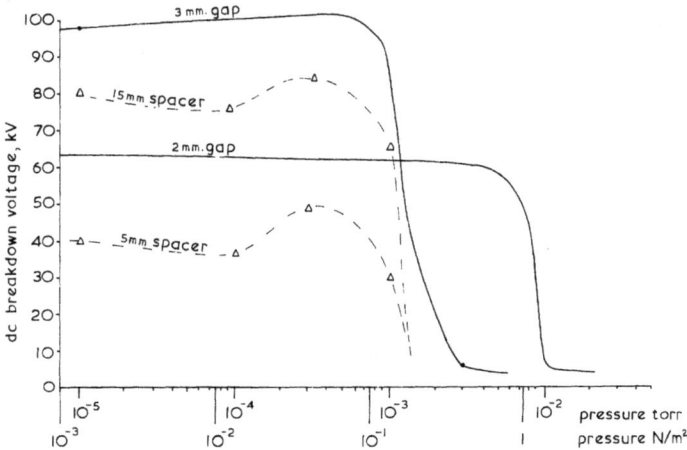

Fig. 12.10 Variation of d.c. breakdown voltage with pressure in vacuum
Solid lines — breakdown across uniform field o.f.h.c. copper electrodes

△ (broken lines) breakdown across cylindrical Plexiglass spacers of 25 mm diameter between nickel plated uniform field electrodes (Kuffel, 1970)

Table 12.3   Withstand voltage of a 1 mm gap for conditioned electrodes of various materials (Trump, 1954)

Electrode material	Voltage withstand across 1 mm gap
	kV
Steel	122
Stainless steel	120
Nickel	96
Aluminium	41
Copper	37

if the electrodes are clean and smooth. Contamination on electrode surfaces can lower the breakdown strength and for this reason sealed-off baked-out systems have improved insulation levels.

12.3.1.5 Electrode shape
Increasing electrode area can reduce the withstand voltage of a gap. Denholm *et al.* (1963*a*) have shown that with electrodes of 10 cm^2, 50 kV is held across a 1 mm

gap while the hold off voltage is reduced to 25 kV when the electrode area is increased to 1000 cm². The effect of electrode shape is not so critical in vacuum as it is for compressed gases in which uniform field geometries have the highest breakdown levels. In a vacuum the breakdown voltage can even increase in some cases as the electrode geometry becomes more nonuniform (Rabinowitz, 1964). Most gaps also withstand a higher impulse than a peak alternating voltage and a higher peak alternating voltage than direct voltage (Denholm, 1958).

12.3.1.6 Flashover across solid insulation in a vacuum

If a solid insulator is placed between electrodes in a vacuum the breakdown over the surface of the insulator will occur at a voltage lower than that of the original gap. Gleichauf (1951) has shown that the breakdown voltage is strongly dependent on the material of the insulator but independent of electrode materials. Tests on 19–25 mm long cylinders between uniform field electrodes by Gleichauf (1951), Shannon *et al.* (1965) and Strivastava (1968) have shown mean flashover gradients of the order of 1·6–9·0 kV/mm for Pyrex glass, porcelain and alumina, under d.c. conditions. Flashover voltages vary considerably, however, and are very dependent on the form of junction between the metal and insulator, the insulator geometry and the material porosity. Roughening the surface of the insulator, for example, in the region of the cathode can actually increase the breakdown voltage. Rohrbach (1968) has reported that the flashover voltage is twice as high for bare porcelain compared to glazed porcelain.

High-voltage bushings have been constructed for vacuum applications. Denholm *et al.* (1963b) constructed a 'graded' bushing using Pyrex glass and thin aluminium voltage-grading rings to hold off 1000 kV with a bushing 300 mm long, and Geramin *et al.* (1966) designed high-voltage vacuum bushings using high density alumina ceramics. Rohrbach (1968) has described the construction of bushings, for introducing voltages of 600 kV into vacuum vessels at CERN, which use high-voltage porcelain tubes approximately 400 mm long at the vacuum end of the bushing.

12.4 Insulating liquids

The only insulating liquid to be used extensively for high-voltage circuit breakers is mineral oil manufactured from petroleum. The properties of the mineral oil used for oil circuit breakers (often known as 'transformer oil') are specified in BS 148 (1959) and IEC 296 (1969) and typical values are shown in Table 12.4

Oil used in circuit breakers may have a slightly higher viscosity than that used for transformers in order to achieve a higher flashpoint. The function of oil in a circuit breaker is to insulate live exposed contacts from earth and from each other and to produce hydrogen gas for arc extinction.

Apart from hydrocarbon oils there has been little application of other insulating

Table 12.4   Typical properties of commercial mineral insulating oil for use in switchgear (For limiting values see BS 148, 1959)

Acidity	0:05 mg KOH/g
Breakdown strength (50 Hz)	10 kV (r.m.s.)/mm
Breakdown strength (1/50 μs)	25 kV (peak)/mm
Permittivity (50 Hz)	2·2
Specific gravity	0·86−0:89
Flash point	135°C
Pour point	−45°C
Coefficient of expansion (volume)	7×10^{-4} per°C
Thermal conductivity	0·39 cal/cm/sec/°C
Power factor	0·001

liquids to circuit breakers. Some research has been carried out on liquid SF_6 and on silicones. Fluorocarbons (fluorine compounds) have been developed which can stand repeated arcing and have good dielectric properties. In particular, fluoro-carbon insulating liquids have exceptionally high thermal conductivities, but they are expensive and as yet have not yet been used for circuit breakers.

The main disadvantages of the mineral oils used in circuit breakers is that carbonisation of the oil occurs (thereby reducing the insulation strength) and that there is also an inherent fire risk. Many 'noncombustible' oils have been tried (mostly chlorinated organic compounds) but they have found little application.

Water has been used for circuit breakers (Zajic, 1957) because the direct fire risk is eliminated. The problems of insulation are however formidable and water (expansion) circuit breakers are not used extensively. The dielectric strength of domestic water is only good when the stress is applied for very short periods, e.g. under impulse conditions. Distilled water, however, has reasonable electric strength; under impulse voltages of short duration (<20 μs) its electric strength is similar to that of transformer oil.

12.4.1 Electric strength of mineral oil

Insulating oil used for circuit breakers and transformers always contain impurities. Insulating oils readily absorb moisture, gases and particulate matter all of which lower the electric strength and cause considerable scatter in breakdown voltage measurements. The scatter in results makes it difficult for a designer to correlate the breakdown values derived by different authorities. It is possible to obtain electric strengths of the order of 100 kV (peak)/mm for highly purified transformer oil but oil so processed (degassed and filtered) is very sensitive to changes in the impurity content and the addition of even small amounts of impurities lowers the breakdown values appreciably. Practical values of breakdown strength are only of the order of 10 kV (peak)/mm under 50 Hz stresses and 20 kV (peak)/mm under

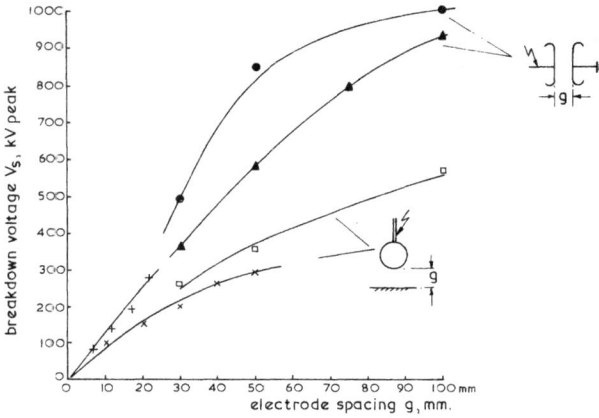

Fig. 12.11 Power-frequency breakdown voltage of good quality insulating oil for uniform-field and sphere-plane electrodes

- ●   mean breakdown levels for linearly rising 50 Hz voltage, uniform field electrodes (Brown Boveri Review, 1967)
- ▲   1 min withstand voltage, uniform field electrodes (Ganger, 1968)
- +   mean breakdown for linearly rising voltage, large area uniform field electrodes (Nelson *et al*., 1971)
- □   mean breakdown levels for linearly rising voltage for 500 mm diameter sphere to plane (Brown Boveri Review, 1967)
- x   1 min withstand voltage, sphere-plane electrodes (Hauschild, 1970)

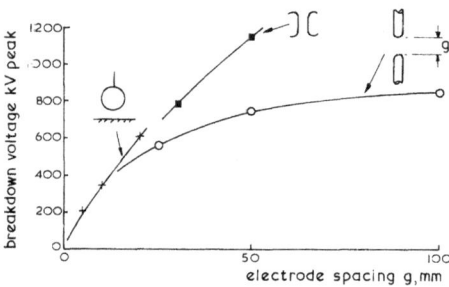

Fig. 12.12 Impulse breakdown voltage of good quality insulating oil

- +   sphere-plane electrodes (Hauschild, 1970)
- ■   50% breakdown levels, positive polarity, uniform field electrodes (Brown Boveri Review, 1967)
- ○   50% breakdown levels, 20 mm diameter rod-rod with hemispherical ends (Kratzenstein, 1970)

impulse stresses. The electric strength when impulses are applied is much less dependent on the purity of the oil than is the case with power-frequency voltage.

Some data on the electric strength of good quality transformer oil are given in Figs. 12.11 and 12.12.

12.4.1.1 Effect of impurities

The oil used in circuit breakers is not subjected to elaborate purification treatment because high purity cannot be sustained in service. The oil contains some impurities and the electric strength depends more on the nature of the impurities than on the nature of the oil itself; for this reason tests for electric strength are virtually tests for purity.

The main forms of impurity which concern the switchgear engineer are:

(a) *Gases:* Gas bubbles have a breakdown strength lower than that of the oil and breakdown in bubbles can trigger off total breakdown.

(b) *Moisture:* Globules of water can elongate in the stressed region between electrodes and form a low impedance bridge across the electrodes.

(c) *Fibres:* The combination of moisture and fibres produces a major hazard to the electrical strength of oil because moist fibres tend to line up across the contact gap.

(d) *Conducting particles (e.g. metallic inclusions and carbon particles):* These cause local enhancement of the electric field and are attracted into the electrode gap.

Mineral oils that are exposed to gases are capable of absorbing them into solution. The amount absorbed depends upon the chemical structure of the oil and is proportional to the volume and temperature of the oil and the pressure of the gas. Apart from normal gas absorption into solution, electrical discharges may induce either gas evolution or absorption depending upon the aromatic content of the oil. The gassing properties of oils are of major importance in oil-paper insulation (transformers and cables), but are of less importance in circuit-breaker applications where oil is used in bulk. Gas bubbles may be evolved from supersaturated oil by agitation, e.g. during the operation of a circuit breaker. The electric field E in a gas bubble immersed in oil of permittivity ϵ is approximately ϵ times the field in the surrounding oil. If the field in the bubble exceeds the discharge inception stress of the gas then breakdown will take place inside the bubble which causes further gas formation through decomposition of the oil. A breakdown of this kind is called a 'partial breakdown' because only the gas bubble, not the complete electrode gap, has broken down. The process can, however, under certain circumstances extend to cause complete breakdown. In general, gas bubbles in switchgear rise inside the circuit-breaking chamber and are dispelled into the gas region above the oil.

Moisture may be absorbed into bulk oil direct from the atmosphere or from condensation on tank walls owing to temperature changes. At normal temperatures oil is capable of holding about 50 p.p.m. in solution before saturation. At high temperatures the solubility of the water increases and it is possible on cooling for small water particles to exist in the oil giving a water-in-oil emulsion. Water, even in small quantities, can cause a reduction in the electric strength of oil. Zein Eldine and Tropper (1955), investigating the power-frequency strength of treated transformer oil to which water was added, prepared a water-in-oil emulsion with the smallest possible water particles and found the electric strength to be appreciably affected by moisture. For moisture contents of 50 p.p.m. the electric strength fell to about one half of that for moisture-free oil (Table 12.5).

Table 12.5   The effect of water content on the electric strength of transformer oil (Zein Eldine and Tropper, 1955)

Water content parts in 10^6	Breakdown strength, 50 Hz voltage kV (peak)/mm
0	50
30	30
50	23
100	14
150	11
200	9
500	8

Water globules are physically unstable in an electric field and they can elongate rapidly into filaments across electrode gaps to cause breakdown (Krasucki, 1962). Failure of switchgear in service, which may be ascribed to excessive water in the oil, is however extremely rare. The amount of water is checked at the time switchgear is commissioned, and it is usual to check for moisture periodically during routine maintenance. The methods used for testing oil samples taken from circuit breakers are given in BS 148 (1959).

The presence of cellulose fibres in insulating oil has a marked effect on the breakdown strength, particularly if the oil also contains moisture. The use of unsuitable cleaning cloths, such as cotton waste, can leave large quantities of minute fibres which contaminate the oil. Most fibres become induced electric dipoles and tend to drift towards a place of maximum stress where they may align 'head-to-tail' to form bridges across a contact gap in a manner analogous to the aligning of magnetic dipoles (iron filings) in a magnetic field. If the voltage stresses are sufficiently high, prebreakdown electrical discharges can take place at intervals along the bridges with the formation of gas leading to final breakdown in the gas.

Solid particles existing in oil experience a force in an electric field if the permittivity of the particle is different to that of the oil. It has been shown by Abrahm and Becker (1932) that spherical particles experience a force given by

$$F = r^3 \frac{\epsilon_2 - \epsilon_1}{2\epsilon_1 + \epsilon_2} E \operatorname{grad} E \qquad (12.15)$$

where r is the radius and ϵ_2 the permittivity of the particles, ϵ_1 is the permittivity of liquid and E is the electric field strength. (If the particle has a charge q the force is increased by an additional amount equal to qE). If $\epsilon_2 > \epsilon_1$ the force will urge the particle to the strongest part of the field, the effect being greater if the field is nonuniform. The force increases as ϵ_2 increases and for conducting particles ($\epsilon_2 \to \infty$) the force becomes equal to $r^3 E$ grad E. For uniform field electrodes the particles are attracted from outside into the uniform part of the field where the field is strongest. In the uniform field where E is constant and the gradient of E is zero, the force is zero and the particles are in equilibrium. The particle itself causes a degree of local field enhancement so that other particles are attracted to form a chain across the electrode gap. Particles of ionic size (radius about 1 Å $r = 10^{-4}$ μm) tend to remain dispersed in oil as do particles of low permittivity. Even in the absence of an electric field, however, particles of a few Angstroms in diameter tend to unite into larger clusters of diameter say 500–3000 Å (Kok, 1961). This phenomenon is called flocculation. Kok has demonstrated that a simple relationship exists between the size of particles of high permittivity and the breakdown strength (E_s is inversely proportional to $r^{3/2}$), hence the larger the particles the lower the breakdown strength. Flocculation can also occur with fibres, and it is thought that flocculation is a necessary process to cause particles and fibres to 'cement' together sufficiently in a bridge to lead to breakdown. The process of flocculation is very sensitive to the degree of acidity of the oil. The acidity of insulating oils generally increases during a period of service, this promotes flocculation and the increase in the size of particle clusters reduces the electric strength.

Of the various products of deterioration produced by arcing in oil circuit breakers the most prolific is carbon. The quantity of carbon produced is proportional to the magnitude of the arc energy and it increases progressively with the number of switching operations. The carbon produced reduces the electric strength of the oil, but owing to the low working stresses used for circuit breakers it is found that a considerable amount of carbon may be tolerated without significantly impairing the reliability of the circuit breaker. Carbon particles are attracted to the surfaces of bushings immersed in oil if the permittivity of the dielectric forming the bushing surface is greater than that of the oil. Carbon line up on oil circuit-breaker bushings can reduce the insulation security and must be removed during routine inspection. Carbon line up can be reduced by coating the bushing with a material which has a permittivity less than that of the oil so that carbon particles are repelled. Low permittivity coatings are difficult to apply in practice but reasonable results have been achieved with s.r.b.p. bushings using an epoxy varnish.

12.4.1.2 Effect of oil volume
The volume of oil lying in the stressed region between electrodes is said to influence the breakdown voltage, the breakdown voltage decreasing as the stressed volume

increases. This viewpoint is based on the fact that as breakdown stress decreases as the electrode area increases and also decreases as the electrode gap increases, then breakdown stress is inversely proportional to the stressed volume of oil lying between the electrodes. Nelson *et al.* (1971) working with large-diameter cylindrical geometry electrodes (substantially uniform field) have shown that the electric strength is dependent on the volume of oil between electrodes both for power-frequency and 1/50 μs impulse voltages (Table 12.6).

Table 12.6   The influence of stressed oil volume on the uniform field breakdown strength for linearly rising 50 Hz voltages and 1/50 μs impulse voltages (Nelson *et al.* 1971)

Stressed oil volume	Breakdown strength		
	50 Hz		Impulse
m^3	kV (r.m.s.)/mm	kV (peak)/mm	kV (peak)/mm
0·0010	12·0	17·0	32·0
0·0025	9·9	14·0	26·6
0·005	9·5	13·4	24·0
0·01	9·0	12·7	21·5
0·02	8·6	12·2	19·5
0·03	8·5	12·1	19·2

Designers should, according to Nelson *et al.*, attempt to minimise the stressed oil volume consistent with providing the necessary contact gap and withstand voltage. The volume effect, however, is dependent on the field geometry and care should be exercised in applying the 'volume theory' to nonuniform fields. The 'stressed oil volume' for nonuniform fields is usually taken as the oil volume in the highly stressed region above the 90% equipotential line.

12.4.1.3 Effect of frequency

The breakdown strength of most insulating media is dependent on the waveshape of the applied voltage and insulating oils are no exception. The breakdown strength of oil under 1 minute 50 Hz voltages is much less than that obtained under 1/50 μs impulse voltages. A finite time is required for flocculation of particles and for stable bridges to form in the region of maximum stress, hence it would be expected that very significant increases in breakdown strength should be achieved as the time of application of voltage is reduced to the microsecond region. The impulse ratio (impulse voltage breakdown level divided by the peak 50 Hz breakdown value) is generally of the order of two to four for oil and depends on the electrode geometry. Hauchild (1968) shows an impulse ratio of 3·6 for sphere–plane geometries using 1·2/50 μs impulse voltages and 1 minute 50 Hz voltages. For switching-impulse voltages of 100/3000 μs, Hauchild's results gave an impulse ratio of 2·3. Experiments described in Brown Boveri Review (1967) indicate impulse ratios of 1·5 for uniform field electrodes, 1·7—2·3 for cylinder-plate electrodes and 1·8—2·7 for coaxial cylinders. For switching impulse voltages it was found that the impulse ratio decreases as the rise time of the switching impulse voltage increases and with a rise time of 500 μs the impulse ratio varied from 1·3 for uniform field electrodes to 1·7 for coaxial cylinders.

12.4.1.4 Effect of hydrostatic pressure

The electric strength of oil increases with an increase in applied hydrostatic pressure even if the duration of the test voltage is as short as 1 μs. The electric strength is increased because gas bubbles (which play an important part in the breakdown mechanism) are compressed and also because of the greater solubility of gas at high pressures. The variation of breakdown strength with pressure depends on the amount of gas initially present in the oil. Kao and Highan (1956) found that the 50 Hz electrical strength of dry and filtered oil increased in strength by a factor of 1·3 and untreated oil containing moist dissolved air increased by a factor of 1·7, over the pressure range 0—250 lb/sq in gauge. Zein Eldine and Tropper (1955) showed that for carefully treated oil the electric strength was only slightly dependent on pressure for uniform fields, but for nonuniform fields breakdown values increased with increase in pressure. For oil containing some gas the breakdown voltage in nonuniform fields increased by about 15% over the pressure range 0—200 lbs/in^2. For practical purposes the increase in breakdown strength of oil with pressure is too small to be of real value. Some minimum oil circuit breakers, however, employ pressurised oil in the interruptor chambers to reduce the probability of restrikes across the small gas bubbles drawn at the contact tips during capacitive switching.

12.4.1.5 Flashover across solid insulation in oil

The flashover voltage gradient along the surface of clean solid insulation immersed in oil is usually lower than the breakdown strength of the oil itself. Ideally, it is the oil which breaks down along the oil/solid insulation interface, the influence of the solid insulation being the way in which its shape determines the electric field applied to the oil at its surface. In practice the surface of the solid insulation attracts some form of contamination and the flashover gradient is usually much lower than the inherent strength of the oil. Because of the complex nature of the surface flashover process, breakdown results can vary considerably and as a consequence there is little published data on this subject. As a very rough guide for practical circuit-breaker insulators which are reasonably clean the surface flashover gradient can be taken as about 1/5 that of the strength of the oil, i.e. about 1 kV (peak)/mm for 50 Hz voltages and 2·5 kV (peak)/mm for impulse voltages. As for any insulation design it is important to use field plots to find the maximum gradient in the circuit-breaker structure and to assume that breakdown is most likely to be initiated in this region.

12.5 Solid insulation

12.5.1 General

The reliability of a circuit breaker is very dependent on the quality of its structural insulating members. Solid insulation in a circuit breaker is nearly always stressed mechanically as well as electrically. The mechanical stresses may be imposed by internal gas pressure, by electromagnetic forces, by the weight of the circuit breaker itself, by forces imposed by the operating mechanism of the breaker and by external forces such as earthquakes.

The processes causing dielectric breakdown in solid insulation are more varied than those involved in gases and liquids. Apart from thermal breakdown (caused by cumulative heating and high loss tangent), solid insulation can fail internally as a result of partial discharges in voids and externally due to tracking and erosion caused by surface discharges. For these reasons the design of solid insulation in a circuit-breaker is both complex and of vital importance.

The type of solid insulation used in a circuit breaker is widely varied. Porcelain and glass fibre are commonly used for circuit-breaker chambers. Phenolic-paper, epoxy-paper and cast epoxide-resin materials are used for bushings and a wide range of plastics and plastic composites are used for component parts of the circuit breaker.

12.5.2 Plastics

The range of commercial polymeric materials is so extreme that the choice of the most suitable polymer for a circuit-breaker component presents a formidable problem to a designer. The final choice seldom results in a plastic which is positively superior to all others in the required environment.

The numerous factors which influence the choice of a plastic dielectric include:

electrical properties
cost and availability
size, shape and method of processing
mechanical and physical properties
temperature characteristics and flammability
effects of environment, e.g. oil, moisture, decomposition products

The electric strength is rarely of importance in selection, in any case nearly all plastics have an electric strength in excess of 15 kV (peak)/mm for samples a few millimetres in thickness. What is of greatest importance in the selection of materials is usually the interrelated factors, cost and the method of manufacture.

Plastics can be grouped into three main types depending on their molecular structure, as follows:

thermoplastics
thermosetting plastics
elastomers

and the method of processing plastics also depends largely on the molecular structure.

12.5.2.1 Thermoplastics

In thermoplastics, few or no chemical bonds exist between adjacent long polymeric chains. Their stiffness arises from the inherent stiffness of the polymer chain, a degree of mechanical interlocking and intermolecular forces (Van der Waal forces). The action of heat on thermoplastic polymers, therefore, is to weaken the intermolecular forces and allow the molecules to move more freely, i.e. to melt. Because of the different factors influencing the stiffness of the thermoplastics, a soft rubbery state is often reached before melting occurs. Also, because no intermolecular bonding occurs the melting process is largely reversible and thermoplastics can often be reprocessed several times without serious degradation.

These phenomena are made use of in the processing of thermoplastics, e.g. injection moulding, extrusion, vacuum forming and blow moulding. Each of these processes uses either high pressures to force the molten material into a moulding tool (injection moulding) or die (extrusion) or else applies a relatively low pressure to mould high viscosity rubbery material into a relatively simple shape (as in vacuum forming). Blow moulding is a combination of the two main techniques, namely extrusion followed by shaping under air pressure.

For thermoplastics, only injection moulding produces complex mouldings and because of the high pressures involved (tool-locking pressures of 2–5 tons/sq. in of projected moulding area are required) this process is usually only economical for large numbers of mouldings greater than say 5000, or for small simple articles.

The thermoplastics most commonly used in circuit breakers include polyamides (nylon 6 and 6·6), polycarbonate, polytetrafluoroethylene (p.t.f.e.), and poly-methylmethacrylate (p.m.m.a.). Nylon and polycarbonate are both resistant to oil, have heat distortion temperatures greater than 90°C and are used extensively in arc-control devices in oil circuit breakers. P.T.F.E. is highly resistant to chemicals and is finding application in SF_6 circuit breakers.

P.M.M.A. (Perspex) although perhaps not a true engineering thermoplastic, does find application in certain air circuit breakers. Perspex has the ability to depolymerise rather than decompose when heated and this effect has been used to advantage in the side plates of air circuit-breaker arc-chutes up to 11 kV.

Polyvinylchloride (p.v.c.) and polyethylene are used for insulation and air feed pipe applications, respectively, where mechanical properties and heat distortion temperatures are of secondary importance.

The properties of thermoplastics commonly used in circuit breakers are listed in Table 12.7.

12.5.2.2 Thermosetting plastics

Thermosetting plastics are produced by the cross-linking of sometimes quite small polymer chains to form large three-dimensional polymer networks.

Table 12.7   Properties of plastic materials used in circuit breakers

Properties	Thermoplastics							Thermosets			
	Nylon 6	P.M.M.A. (Perspex)	P.V.C. flexible, filled	Polyethylene high density	Polypropylene	P.T.F.E.	Polycarbonate	Epoxy casting resin silica filled	Phenolic resin bonded paper	Epoxy fine weave glass fabric laminate	Polyester dough moulding compound
Specific gravity	1·14	1·19	1·3	0·96	0·91	2·2	1·2	1·8	1·35	1·7	1·8
Electric strength,[1] kV/mm	15·7	15·3	11·8	19·2	26·0	19·0	14·2	18·0	15·0	14·0	10·0
Permittivity	3·8	3·4	5·5	2·3	2·5	2·0	3·0	4·0	5·2	4·0	5·7
Comparative tracking index[2]	**	>700	>700	**	>700	>700	90	240	100	290	>700
Heat distortion temperature,[3] °C	60	100	100*	55	60	100	140	110		110	200
Thermal conductivity, W/mK	0·23	0·18	0·17	0·53	0·12	0·25	0·19	0·65	0·21	0·65	0·85
Coefficient of expansion[4] per °C × 10⁻⁵	10	7·3		12	10	10	11	3	10	2·8	2·8
Modulus of elasticity,[5] GN/m²	2·2	3·0		1·2	1·4	0·4	2·1	9·0	7·0	2·1	7·0
Tensile strength,[5] MN/m²	70	84	25	35	38	35	56	85	70	210	70
Moisture absorption,[6] % in 24 hours	1·8	2·0	1·0	0·01	0·02	0·0	0·18	0·04	2·7	0·1	0·025
Flammability, cm/sec[7]	s.e.	0·06	s.e.	0·04	slow	n.f.	s.e.	s.e.	s.e.	s.e.	slow

*Maximum working temperature
**Heavy erosion
s.e. self extinguishing
n.f. nonflammable

Test methods:
1 ASTM D 149 (1964)
2 BS3781 (1964)
3 ASTM D 648 (1956)
4 ASTM D 696 (1970)
5 ASTM D 1708 (1966)
6 ASTM D 570 (1963)
7 ASTM D 635 (1968)

Because of their highly cross-linked structure, thermosetting plastics do not melt but only soften slightly with the application of heat, eventually decomposing by bond fracture at high temperatures. Thermosetting plastics are generally resistant to oil.

The polymers before crosslinking (resins) and the crosslinking agents (hardeners) are usually liquids or low melting-point solids and this, again, influences the method of processing. For example, resin, hardeners and fillers can be mixed and poured, possibly under vacuum, into relatively inexpensive moulds to produce large complex castings. Because only low pressures are involved these castings can contain conductors, windings and capacitor foils, thus producing a wide range of electrical equipment. Or again, the low viscosity resins and hardeners can be mixed to produce resin bonded glass fibre (r.b.g.f.) laminates for high strength circuit-breaker components or, with high grade electrical paper, synthetic resin bonded paper (s.r.b.p.) laminates for relatively inexpensive sheet insulation and condenser bushings. The resins and hardeners can also be combined with fillers to produce compounds able to be moulded under heat and pressure by relatively fast production techniques.

Thermosetting plastics include the polymers of urea, melamine, and phenol with formaldehyde, and also epoxide, polyester and silicone resins. Of these the phenol formaldehyde (phenolic) and epoxide resins are the most widely used in circuit-breaker components.

12.5.2.3 Elastomers

Elastomers are special types of either thermoplastic or thermosetting polymers characterised by their high elasticity when subjected to large strains (e.g. up to 100% elongation).

Most elastomers are of the thermosetting type with very long polymer chains (M wt $3 \times 10^5 - 3 \times 10^6$) and a low degree of crosslinking. The polymer molecules which are normally in a coiled formation can be extended under stress, resulting in the characteristic low modulus of elasticity of elastomers. The stiffness of elastomers is dependent on the crosslink density as seen in the formation of ebonite by the addition of large quantities of sulphur to natural rubber.

Because of their high molecular weight, uncured elastomers are generally of high viscosity and can only be processed by compression moulding, extrusion or similar high pressure processes. A further limitation to the size of component which can be produced is the rather poor thermal conductivity of elastomers which makes uniform curing of thick sections very difficult. Because of this the use of elastomers in circuit breakers is often restricted to relatively simple shapes. Certain elastomers particularly polyurethanes and silicones, are available in a castable form and have found use in certain specialised applications. Thermoplastic polyurethane elastomers are also available. Some properties of elastomers are given in Table 12.8.

Table 12.8   Properties of elastomers

Properties	Ethylene propylene rubber (e.p.r.)	Chloroprene (Neoprene)	Butadiene-Acrylonitrile (Nitrile)	Polyurethane	Chlorosulfonated polyethylene (c.s.p.)
Specific gravity	0·86	1·25	1·0	1·25	1·2
Permittivity	3·2	9	2·9	6	7—10
Maximum continuous operating temperature, °C	150	100	150	85	150
Resistance to: oil	P	G	E	E	G
ozone	E	F	P*	E	E
tear	F	G	G	E	F
abrasion	G	E	G	E	E
water absorption (swelling)	E	G	E	G	G
flame	P	G	P	F	G

*Addition of p.v.c. will improve the resistance to ozone appreciably

E — excellent   F — fair
G — good   F — poor

12.5.3 Cast epoxide resin

Cast epoxy insulation is used in circuit breakers for bushings, operating links, stand-off insulators, arcing chambers and interphase barriers.

The majority of simple castings used in high-voltage circuit breakers are made by the reaction of a bisphenol epoxy resin and an anhydride hardener and are cast under vacuum. A bisphenol resin (e.g. Araldite CT 200) cured with phthalic anhydride (Araldite HT 901) and using silica flour as a filler has a high plastic yield temperature, a low coefficient of thermal expansion and a low temperature rise due to exothermic reactions. Typical properties of this type of resin are shown in Table 12.7. For outdoor applications use has been made of cycloaliphatic epoxy resins which are significantly less prone to tracking than bisphenol resins.

The primary purpose of using silica flour in epoxy castings, apart from reducing the cost, is to reduce the coefficient of thermal expansion thereby reducing the chemical shrinkage on gelation and thermal shrinkage due to temperature changes.

Other fillers such as alumina are used for special applications, for example in epoxy resin arcing chambers for SF₆ single-pressure breakers. Some fillers also tend to improve the resistance of epoxy castings to wet tracking. The surface arc resistance of epoxy resins is enhanced by aluminium oxide trihydrate filler.

12.5.3.1 Stress control

The great advantage of epoxy resin to the switchgear designer is that complex conductor systems can be cast directly into insulators by virtue of the simplicity of the process and the low shrinkage of epoxy resins. At voltages below 10 kV, epoxy insulators are designed primarily with mechanical strength in mind and the problems of internal and external electrical stress control are relatively small. The radial thickness of resin required around a conductor is generally dictated by the mechanical strength required for the component rather than by the electric requirements.

Stress control in epoxy resin bushings is most readily achieved by the inclusion within the casting of one or more metallic foils. A single metal screen (Fig. 12.13) can be inserted in a casting to reduce the voltage gradient at the surface of a bushing in the proximity of the earthed flange. Multiple 'condenser' foils are difficult to include in a straight casting, but they can easily be incorporated in epoxy paper bushings. Epoxy paper bushings are made by winding on to a conductor mandrel from a continuous sheet of resin-coated paper in a manner similar to that used for phenolic-resin bonded paper bushings. Condenser foils can be introduced into the winding at predetermined intervals. Epoxy crepe-paper bushings can be made by winding crepe-paper tape on a conductor and impregnating the completed winding with a liquid epoxy resin. Epoxy corrugated-paper bushings are bushings in which a skeleton structure of corrugated paper is used to support the condenser stress-control foils and which is subsequently impregnated with silica-filled epoxy resin.

12.5.3.2 Electric strength

The 50 Hz a.c. strength of filled epoxy resin for uniform field conditions is of the order of 100 kV peak/mm for a gap spacing of one or two millimetres falling to about 20 kV peak/mm for gap spacings of 30 mm. Data by Goulsbra *et al.* (1970) on the electric strength of some epoxide resins are shown in Fig. 12.14. The electrical strengths shown in Fig. 12.14 are the average values E_{av} given by the ratio of the breakdown voltage V_s and electrode separation g. The maximum breakdown gradient E_{max} at the surface of the spheres is only a few per cent larger than E_{av} because the electrode spacings quoted in Fig. 12.14 give virtually uniform field conditions. Goulsbra *et al.* showed that the addition of 200 parts per hundred by weight of silica filler reduced the electric strength of bisphenol resin by approximately 25% for both 1/50 μsec and 50 Hz voltages. It was also demonstrated that the electric strength of epoxide resins is dependent on mechanical stress concentrations set up within the resin due to the encapsulation of materials having

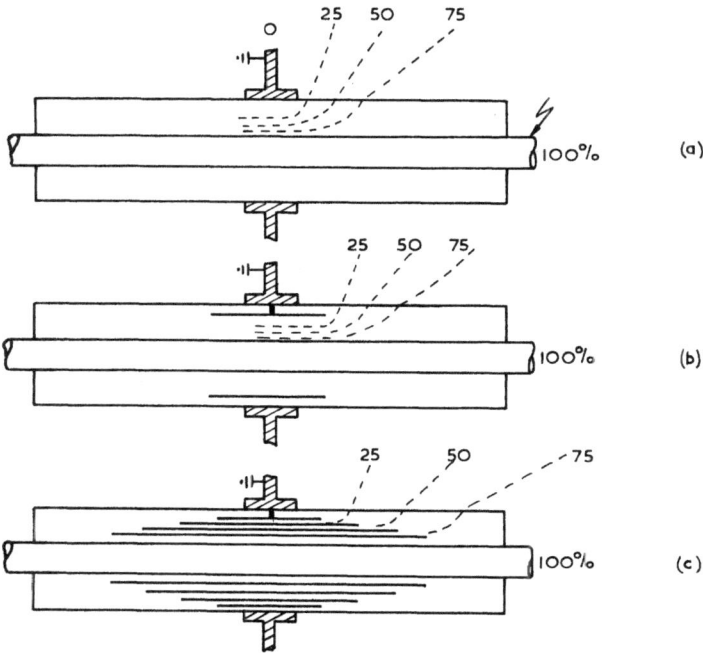

Fig. 12.13 Effect of metallic grading foils in a simple bushing

(a) without foils
(b) with a single foil at earth potential
(c) with several 'condenser' foils
The presence of grading foils improves the distribution of both the radial gradient and the surface gradient

different coefficients of expansion. Annealing improves the breakdown strength of unfilled bisphenol resin by about 17%. Fig. 12.15 shows some data obtained for breakdown of filled bisphenol epoxide resin in a concentric-cylinder electrode geometry. The average radial breakdown gradient in this practical test arrangement falls to approximately 11 kV peak/mm at the largest radial spacing of 38 mm corresponding to a gradient at the surface of the inner conductor of 23 kV (peak)/mm. For this type of geometry breakdown is most likely to occur at the ends of the foil where the stress may exceed that at the surface of the central conductor.

Breakdown stress of various resin systems as a function of the time to breakdown has been measured by Schmid (1967) under continuous alternating

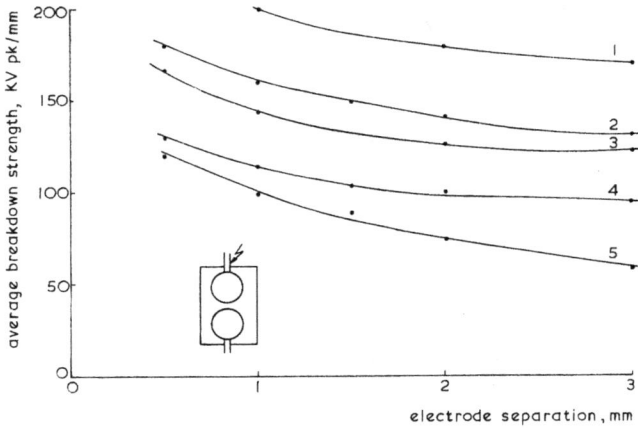

Fig. 12.14 Breakdown strength of expoxide resin for 20 mm diameter spherical electrodes (Goulsbra *et al.*, 1970)

    1 unfilled bisphenol resin CT200, 1/50 impulse
    2 silica filled bisphenol resin CT200, 1/50 impulse
    3 unfilled bisphenol resin CT200, 50 Hz
    4 silica filled bisphenol resin CT200, 50 Hz
    5 silica filled cycloaliphatic resin ERL 4289, 50 Hz

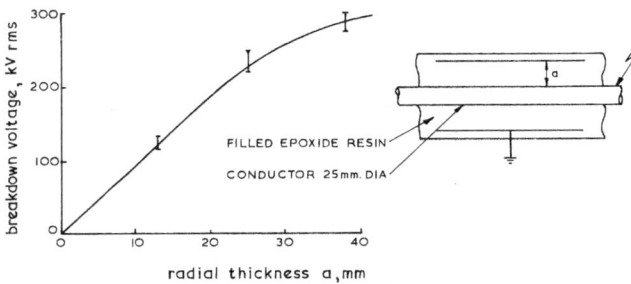

Fig. 12.15 Variation of 50 Hz breakdown voltage with radial thickness of epoxide resin for a concentric conductor-foil electrode arrangement

voltage conditions. Schmid working with 2 mm thick sheets of silica-filled Araldite CT 200 between 50 mm diameter spheres showed that the time to breakdown increases appreciably with a relatively small reduction in the continuously applied 50 Hz stress (Table 12.9) until at a certain stress, the stress versus time-to-breakdown curve changes to a constant stress line. Thus for filled bisphenol resins stressed with 50 Hz voltages either breakdown occurs within a short time (100 s) or no breakdown can be expected for a long period.

Table 12.9   Time to breakdown of filled bisphenol resin as a function of 50 Hz stress at $20°C$ (Schmid, 1967)

Applied stress	Time to breakdown
kV/mm	s
28	1
25	10
21	100
19	10^3
<19	$>10^4$

Schmid also showed that both mechanical stress and temperature produced a drastic reduction in the long term strength. Zolidziowski (1967) has measured the life of epoxy resin subjected to repeated $1/50 \mu s$ impulses. The breakdown stress for unfilled bisphenol resin nominally $0·5$ mm thick in a sphere-plane configuration was 170 to 150 kV (peak)/mm for lives of $1-10^2$ impulses and decreased to $130-115$ kV (peak)/mm for lives of 10^3-10^5 impulses. The decrease in the breakdown strength with the number of impulses is therefore relatively small.

Epoxy resin castings have small water absorptions of about $0·1-0·4\%$, and hence are not too adversely affected by ambient humidity.

12.5.3.3 Mechanical strength

The mechanical properties of typical epoxide resins are shown in Table 12.7. Cast epoxy resin of the type used in circuit breakers is a fairly brittle material, and there is a danger that badly designed components may crack due to transient mechanical loads or due to strains imposed by thermal shock. The flexural strength and modulus of elasticity (which should be as high as possible) are modified by the amount and type of filler used. For unfilled epoxy resins the coefficient of expansion is substantially higher than that of conductor materials (Table 12.10). The difference between the expansion coefficients of the resin and the conductor embedded in the resin introduces a stress in the resin if the insulator is operated over a wide temperature range. The effect of adding mineral filler, such as silica, is to reduce the coefficient of thermal expansion by a factor of two or more and also to increase the thermal conductivity by a factor of three of four. When a liquid

Table 12.10   Coefficients of linear expansion x 10^{-6} per °C

Copper	16
Brass	18
Aluminium	25
Filled CT 200	34
Unfilled Ct 200	63

resin converts to a solid (polymerisation) a small increase in density occurs which appears as a dimensional shrinkage. This is of some importance when casting resin around conductors or other metal parts because it can lead to stress cracking. Shrinkage is reduced to about 1% by the addition of inorganic fillers and by careful design of the casting tool. Despite the relatively low shrinkage of filled epoxy resin on curing, for some complex arrangements of conductors it is still possible for large stresses to be set up in the resin. Excessive strains can be eliminated by the use of cushioning materials applied around the conductors, for example a thin layer of p.v.c. tape.

12.5.4 Reinforced plastics

Many plastics, although having good dielectric properties, display relatively poor mechanical strength which rates them unsuitable for those applications where both high electrical and mechanical strength are required. The mechanical strength of some 'plastic' materials is greatly improved when reinforced with fibrous materials such as glass filaments. Glass-reinforced epoxy resin insulation is used extensively in circuit breakers in the following forms:

(a) *Filament wound structures:* glass-roving impregnated with resin, wound helically on a mandrel. When closed-end containers are wound the mandrel must be dismantled, deflated or melted when the winding is complete and the resin cured. The ratio of hoop strength to axial strength is determined by the winding helix angle.

(b) *Vacuum impregnated glass cloth:* dry glass-cloth wound on a mandrel and placed into a mould where the winding is impregnated with resin under vacuum. This process can produce very high strength tubes, the ratio of hoop to axial strength being determined by the weave of the glass cloth. Techniques (a) and (b) are used for circuit breaker pressure vessels, feed pipes and explosion pots.

(c) *Pulltruded rods:* impregnated glassfibre in roving form pulled through a heated die where gelation of the resin takes place; final cure being achieved in a tunnel oven in line with the die. Rods and bars of almost unlimited length and of very high tensile strength can be obtained. Operating rods for circuit breakers are often constructed using this technique.

(d) *Epoxy-glass laminates:* sheets of glass-cloth or unidirectional glass filaments bonded with epoxy resin to form a laminated plate. These are used for machining into circuit breaker operating links or phase barriers.

(e) *Dough mouldings:* chopped glassfibre mixed with resin and mineral filler to form a 'dough' which can be moulded under a low mechanical pressure and cured by heat. Dough mouldings are at present limited for use as primary insulation to switchgear having rated voltages of about 11 kV.

12.5.4.1 Electrical properties of glass-reinforced epoxide resin

The dielectric strength of glass-epoxy materials varies widely according to the type of glass filament reinforcement and on the degree of adhesion between resin and glass. For electrical purposes, the filaments are made from 'E' glass, a low alkaline borosilicate glass. The surface of the glass filament is usually coated with a chemical coupling agent to achieve good adhesion to the resin and gaseous cavities are eliminated either by vacuum impregnation or by careful winding. It is also important that the resin is cured uniformly throughout the insulator to avoid irregularities in the voltage gradient. The electric strength of glass-reinforced epoxy insulation varies considerably according to the form of construction. Typically the breakdown strength of vacuum impregnated glass-cloth epoxy material is in the range 15–20 kV (r.m.s.)/mm under 50 Hz alternating voltages for material thickness of 1–5 mm.

The electrical properties are dependent on the mechanical and thermal stresses imposed on the insulation. The dielectric strength of epoxy glass is usually evaluated when the test piece is unstressed mechanically. In circuit breaker structures, such as an arc-extinction chamber, high electric and mechanical transient stresses can occur simultaneously (e.g. during capacitor switching) and the electrical properties may be reduced if the glass-resin adhesion is impaired. The electrical properties also deteriorate if temperatures above 120–140°C are maintained for long periods.

12.5.4.2 Mechanical properties of glass-reinforced epoxide resin

A large proportion of the applications of glass-epoxy insulation in circuit breakers is in the form of tubular structures, which are designed to withstand steady and transient internal pressures. For a simple tube with metal end fittings the glass-epoxy cylinder wall is stressed both in the longitudinal and circumferential directions. The longitudinal stress is $S_L = pr/2t$, where p = internal pressure, r = inner radius and t is the wall thickness. The circumferential stress or hoop stress is twice the longitudinal stress. The axial and hoop strength of the cylinder can be modified by adjustments to the helix angle in filament wound tubes or in the type of weave for cloth wound tubes (e.g. more filaments in the circumferential direction to increase the hoop strength). Glass filaments have a tensile strength about 40 times that of epoxy resin so that high strengths can be obtained only by high glass contents. The mechanical strength of structures (cylinders and rods) made with glass fabrics or rovings increases proportionally with the glass content and tensile strengths up to 100 Kg/mm² can be obtained when the glass content approaches 80–90% in glass-epoxy rods.

Glass-epoxy rods manufactured by the pulltrusion technique have extremely high tensile strengths. For example a 25 mm diameter cylindrical rod can have a tensile breaking strength of 0·68 MN (tensile strength 1·4 GN/m²). The problem with such rods is to obtain an effective bond between the end fittings and the glass rod. For many constructions the end fittings will pull away from the rod before the rod fractures.

The long-term mechanical strength of glass-epoxy structural members under both continuous and cyclic load is of importance. The distortion of a member under a continuous static load is determined by a creep test in which samples of material or complete structures are stressed over a long period (at stresses which are a percentage of their short-term strength) and their deformation is recorded as a function of time. The creep test establishes the behaviour of the material under conditions where the period of stress represents a significant factor, the aim being to determine the stress which the material can maintain for an indefinite period of time without breaking. A curve of breaking stress against time-to-break (Fig. 12.16) can give the designer the necessary factors of safety.

During a stress-strain test, on say a cylindrical glassfibre-resin pressure vessel tested with hydrostatic pressure, the material will show little or no creep at the very low loads. At higher hydrostatic pressures the material will deform linearly and

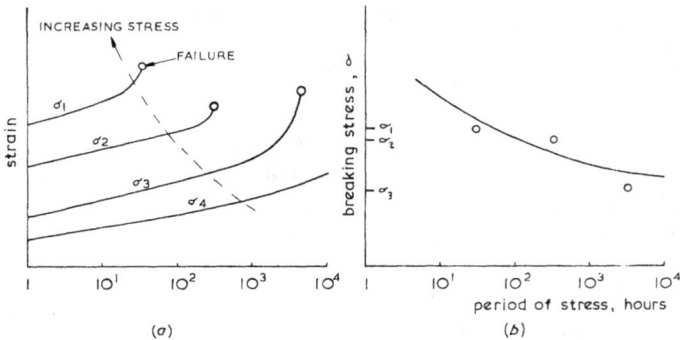

Fig. 12.16 Creep testing of epoxy-glass insulation

(a) strain plotted as a function of the time duration of stress for various levels of stress σ_1, σ_2 etc.

(b) curve of breaking stress against period of stress derived from (a) eventually debonding of the resin-glass interface starts, resulting in the formation of microcracks. With further increase in load more cracks will develop and finally the tube will fail completely as the individual glass filaments rupture. (On a hydraulic test the presence of cracks is indicated by weeping). For any given period of loading, 'weeping' occurs at a stress level well below the ultimate (rupture) stress, and is obviously important because the electrical properties are permanently impaired. Typically, debonding may occur at 20—40% of the ultimate stress. Practical design stresses are usually not more than about 10% or less of the ultimate stress. For example, if the working pressure in a glass-epoxy gas reservoir is 1 p.u., the hydraulic static test pressure could be 2.5 p.u. and the bursting pressure 5—10 p.u. Typically the ultimate hoop strength for a glass cloth-epoxy tubular balanced pressure vessel is about 3000 Kg/cm² with an axial tensile strength of about 1200 Kg/cm². Working hoop stresses would therefore be about 300 Kg/cm².

Bax (1970), working with filament wound epoxy-glass tubes, showed that the onset of weeping always occurs in the same region of strain (1—1·4%). independently of the time required to reach that level. Bax suggested that designs should be based on a strain limit below which deformation is completely elastic and reversible. For reinforced epoxy tubes this elastic limit according to Bax was about 0·5% tangential strain at a tangential stress of about 10·5 Kg/mm² for a winding angle of 30° and 0·35% at a tangential stress of 13 Kg/mm² for a winding angle of 54° 30'. Failure occurred at a strain of about 6%. The time to fracture under cycled stress indicates the fatigue behaviour of glassfibre reinforced plastics. Cyclic loading causes a pronounced reduction in time to failure during creep tests. The flexural (bending) strength of the material is measured by bending a sample as a beam between two simple supports. The test is sometimes called the crossbreak test (BS2782) and a flexural test as in ASTM D790.

The mechanical properties of glass epoxy are also dependent on temperature. The range of application of glass fibre reinforced epoxide resins lies in the range −20°C—+160°C. It is often stated that the mechanical properties of glass fibre epoxy are comparable to those of steel but one major exception is their loss of strength at elevated temperatures compared to steel. Designers must be aware of the temperature limitation which is usually specified as the temperature of deflection, i.e. the temperature at which mechanical and electrical properties deteriorate markedly (BS2782). Kidd (1967) has measured the deflection temperature of epoxide bonded glass-cloth laminates using a specimen stressed as a double supported beam (as for crossbreaking strength) with a high stress equivalent to 10% of the ultimate crossbreaking strength. A nominated critical level of deformation was taken as a measure of the temperature of deflection. Typical values of the deflection temperature for various resin systems range 105—150°C.

12.5.5 Paper insulation

Paper insulation is used in combination with other dielectric materials to make circuit breaker bushings and grading capacitors.

12.5.5.1 Synthetic resin-bonded-paper (s.r.b.p.) bushings

S.R.B.P. bushings are manufactured by winding continuous resin-coated paper sheet into a cylindrical form. The paper is first coated with epoxy or phenolic resin and then wound onto a cylindrical mandrel (which may be the conductor) under heat and pressure. If the bushing is to be a condenser bushing then conducting layers of metallic foil or metal-coated paper are inserted at appropriate intervals during the winding process, each foil forming a cylinder within the bushing.

The radial thickness of the s.r.b.p. bushings is limited in practice by thermal instability produced as a result of dielectric heating in the synthetic resin while the length of the s.r.b.p. bushings is limited to some extent by practical limitations in the length of paper-coating machines. Owing to the limit imposed by thermal instability s.r.b.p. bushings are rarely manufactured for voltage ratings greater than 420 kV.

12.5.5.2 Oil-impregnated-paper (o.i.p.) bushings

These bushings are made by winding paper onto a cylindrical conductor, metallic condenser foils being inserted in the winding. After vacuum drying, the paper winding is impregnated with highly processed oil. O.I.P. bushings can be made longer than s.r.b.p. bushings, because if necessary the winding can be spirally wound with paper tape. Dielectric losses are less than with s.r.b.p. hence o.i.p. bushings can be made for rated voltages at least up to 1500 kV.

12.5.5.3 Epoxy-resin paper bushings

Apart from bushings wound with epoxy-coated paper, (s.r.b.p.) bushings can also be made by winding with crepe-paper tape and impregnated under vacuum with epoxy resin (Legg and Rothwell, 1967). Aluminium grading foil can be inserted into such a bushing at predetermined radial intervals (usually at about 2 mm spacing). Using this technique, straight or cranked capacitor bushings can be made. Because crepe paper is impregnated (rather than coated) with resin the bushings are gas tight and hence are suitable for high pressure gas circuit breakers.

12.5.5.4 Stresses in circuit breaker bushings

The physical dimensions of graded resin-paper bushings are determined initially by the requirement to withstand type approval and routine tests in the factory and a satisfactory performance in these tests should subsequently ensure long trouble-free performance in service.

A number of electrical tests applied to a prototype bushing during type and routine tests are specified in BS223 (1956). Often the bushing design is dictated by one particular test and the remaining tests are readily accommodated. For example, condenser bushings designed to withstand the 50 Hz momentary dry withstand voltage test often have ample margin to withstand the impulse tests. Typically s.r.b.p. bushings for 11 kV circuit breakers may be designed to have a radial stress of 2 kV (r.m.s.)/mm at the discharge inception voltage. For rated voltages of 33 kV and below, bushings are usually required to have an apparent discharge level of not more than 100 pC at 0.69 E where E is the line voltage. If the radial stress is chosen to be 2·0 kV (r.m.s.)/mm at 7·6 kV the remaining 50 Hz stresses would be as follows for an 11 kV bushing:

Applied voltage	*Corresponding radial stress*
Discharge voltage 0·69E = 7·6 kV	2·0 kV(r.m.s.)/mm
Maximum working voltage $\dfrac{1·1E}{\sqrt{3}}$ = 7 kV	1·8 kV(r.m.s.)/mm
Routine overvoltage tests 27 kV	7·1 kV(r.m.s.)/mm

The axial stress over the graded part of a bushing is usually based on the type-approval 50 Hz test and the impulse type tests. For a bushing in air, typical design gradients are of the order of 0·3–0·45 kV (r.m.s.)/mm at the 50 Hz momentary withstand test voltage and 0·4–0·5 kV (peak)/mm for lightning impulse type tests. For bushings immersed in oil, the axial breakdown strengths of s.r.b.p. and o.i.p. bushings are less than that of the bulk oil when the bushing controls the surface gradient. For both types of bushings, the axial gradient under oil is designed on the basis of about 1·0 kV (r.m.s.)/mm for 50 Hz type tests and about 2·0 kV/mm for impulse tests, according to Barker (1968), i.e. about double the design axial gradients for the 'air end' of a bushing.

Bulk type bushings (i.e. not condenser graded) have been made of porcelain, cast resin and s.r.b.p. for circuit breakers rated up to 22 kV, but it is more usual to use condenser bushings for voltages down to 11 kV because they are shorter. The physical size of circuit breaker bushings has an important bearing on the overall size of the switchgear, for example, if the circuit breaker bushings are short then the isolating movement of a breaker in a metalclad panel is also short.

There are a number of ways metallic condenser layers can be arranged in a bushing. Common practice is to make the partial capacitances equal (which means the voltage across each capacitance is also equal) and the axial lengths of neighbouring capacitor layers differ by equal amounts. A fairly constant axial gradient is achieved while the radial gradient varies throughout the wall thickness of the bushing. The maximum radial gradients for s.r.b.p. should be about 2·0 kV/mm at the maximum working voltage and the corresponding figure for o.i.p. is about 3·5 kV (r.m.s.)/mm.

For bushings designed to have a constant axial gradient the radial gradient varies and will be a maximum either at the conductor or at the outer earth layer (flange). The stress at the inner conductor E_c at a voltage V for a bushing with an infinite number of layers is given by

$$E_c = \frac{V}{2}\frac{\alpha+1}{\alpha}\frac{1}{r_0 \log_e \beta} \tag{12.16}$$

where

$$\alpha = \frac{l_0}{l_n} \quad [l_0, l_n \text{ lengths of first condenser layer (adjacent to the conductor) and earthed layer, respectively}].$$

$$\beta = \frac{r_n}{r_0} \quad (r_0, r_n \text{ radii of conductor and earthed layer, respectively}).$$

The stress at the earthed condenser layer E_e is given by

$$E_e = \frac{V}{2} \alpha + 1 \frac{1}{r_n \log_e \beta} \qquad (12.17)$$

If $\beta > \alpha$ the maximum stress is at the conductor and if $\beta < \alpha$ the maximum stress is at the earthed layer. When $\alpha = \beta$ the radial gradients at the conductor and earthed foil are equal. The values of α and β chosen for any bushing depend on many factors, for example: the conductor diameter (current rating), the provision of current transformers and relation between 50 Hz and impulse ratings. As a guide, α and β usually lie in the range 3–4·5. A minimum bushing radius is obtained if $\alpha = \beta = 3·6$. If $\alpha = \beta = 3·15$, the bushing has minimum cross-section of insulating (minimum winding time). For convenience in manufacture the layer diameters are rationalised to about the nearest 0·25 mm, so partial capacitances are only approximately equal and axial layer clearances are adjusted to compensate for this.

In the simple formulas quoted, an infinite number of layers is assumed. To take into account a finite number of layers one must use $r_n - 1$ in place of r_n which gives rise to the difficulty that $r_n - 1$ is not known until the bushing has been fully calculated. For preliminary calculations this refinement is unnecessary because with a bushing having a large number of condenser foils the error in the derived stresses is generally less than 2%.

The number of condenser foils used in a bushing should be as large as possible to reduce the stress at the ends of the foils. The stress at foil ends can be about three times the radial gradient between foils, and it is probably more important to estimate the gradient at the end of the foil than the radial gradient. Fig. 12.17 illustrates the electrostatic field near the foil end, in which region the stress is very dependent on the spacing between layers. Bushings can in fact fail owing to the long term damage under 50 Hz voltages caused by discharges at the ends of capacitor foils (Barker, 1967).

The actual number of layers chosen is based on experience and typically seven partial capacitances are used for 11 kV s.r.b.p. bushings and 15 for 33 kV bushings, with foil spacings of about 1 mm.

12.5.6 Porcelain

12.5.6.1 General properties

Electrical porcelain is composed from clay substance, quartz and felspar. The approximate limits of composition for the three main components are given by

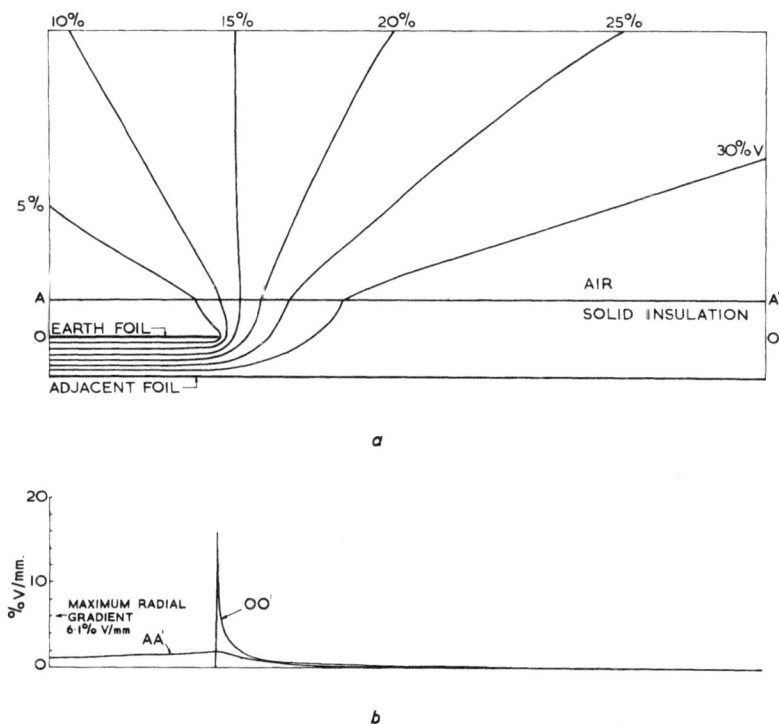

Fig. 12.17 Stress at the end of a condenser foil in a four layer graded bushing

(a) field plot in the region near the end of the shortest (earth) foil equipotential lines as a percentage of the total applied voltage V
(b) comparison of the field gradient along the radius of the earth foil (OO') and along the surface of the bushing (AA') with the maximum radial-stress between the foils

Ware (1967) as

Clay substance	40–60%
Felspar	25–35%
Quartz	20–40%

There are considerable variations in the type of porcelain manufactured in different parts of the world owing to the availability of raw materials in particular localities and the variation in manufacturing techniques. Electrical porcelain can, to some extent, be tailor made to meet specific requirements. Clay content can be reduced and silica (quartz) increased to give improved mechanical properties when fired. In general, higher felspar contents lead to higher puncture strength, high clay content improves the thermal shock resistance and high quartz content increases the mechanical strength. As the felspar content controls the vitrification of the porcelain body the percentage values are not varied significantly from the figures given by Ware. For the manufacture of high mechanical-strength porcelain the 20–40% quartz content is replaced wholly or in part by alumina.

The overall quality of porcelain is of vital importance in circuit breaker application, because porcelain components are usually subject to high mechanical and electrical stresses. Meticulous care is taken by the porcelain manufacturers to avoid inhomogeneities in the body such as pores and foreign particles, and laminations or folds which can occur in the plastic shaping stages of manufacture. Porcelain is extremely impermeable to gases and liquids owing to its low porosity and absence of open pores. Porcelain can be checked for porosity by immersing freshly broken pieces of a sample insulator in a 0·5% alcoholic solution of fuchsine dye under a pressure of 2000 lbs/in^2 for 24 hours (BS3297, 1960). When the test pieces are broken there should be no sign of penetration of fuchsine into the body.

Porcelain insulators are usually glazed to provide a smooth impermeable surface. Glazes are glasses of low fluidity of widely varied compositions, but mostly based on silica and various oxides (e.g. BaO, ZnO and CaO). The smooth surface obtained by glazing makes the insulator easier to clean both manually and by rain. Correctly applied glaze (i.e. in compression) improves the mechanical strength of an insulator by about 30%.

12.5.6.2 Design criteria

The puncture strength of porcelain is generally of the order of 15–25 kV (r.m.s.)/mm for slabs of porcelain up to 3 mm thick, but higher values are quoted by Salmang (1961). Puncture of porcelain insulators is rarely a problem for designers but surface flashover often is. Cylindrical shedded insulators of about 250 mm in length will flashover in dry conditions in those regions where the 50 Hz stress approaches about 500 V (r.m.s.)/mm, while under clean wet conditions the flashover gradient can be about 300 V (r.m.s.)/mm. If the insulator is both dirty and wet the flashover voltage can be greatly reduced. The flashover voltage of a polluted porcelain insulator is dependent on the dimensions and shape of the insulator and on the degree of pollution. The creepage length of the insulator is particularly important, and the risk of flashover is generally reduced when the creepage distance is increased. The creepage length of shaded porcelain insulators used outdoors in a clean area is of the order of 16 mm/kV (r.m.s.) of the system voltage (e.g. a total creepage length of about 4·8 m for a 300 kV system). In areas where porcelains are polluted (e.g. by air-borne salt in coastal regions) the creepage length used can be 25–40 mm/kV system voltage. The mechanism of pollution flashover has been extensively studied and is excellently reviewed by Lambeth (1971). Porcelain is not susceptible to tracking and erosion as is organic insulation.

One possible solution to the problem of pollution on porcelain insulators has been the use of a continuous layer of semiconducting glaze. A semiconducting glaze is designed to carry a current of the order of 1 mA at normal voltage and is intended to stabilise the voltage distribution along the length of the insulator thereby inhibiting the formation of the 'dry bands' which lead to pollution flashover. The heating effect also retards the deposition of dew. Stabilised insulators have been tried for many years but practical difficulties have been encountered owing to electrolyte corrosion of the ferrite glazes used. Recently, glazes based on a tin oxide/antimony oxide system have been developed which are more resistant to corrosion as well as having a smaller negative temperature coefficient.

The mechanical strength of normal electrical porcelain, as determined by glazed ceramic test pieces, is in the range $8\cdot4-12\cdot4$ kN/cm^2 (Robinson, 1972) and higher strength aluminous porcelains have recently become available with mechanical strengths about 50% greater. Solid core porcelain insulators are commonly used for circuit breakers, in sizes up to about 200 mm core diameter and up to 2 m in length. High-voltage stacks are constructed from solid core units usually about 750–1250 mm in length. Robinson (1972), quotes a maximum bending moment for solid core porcelains of about 51 KNm. Hollow porcelain insulators are also used for circuit breakers, e.g. as housings for compressed SF$_6$ insulation and as support insulators for gas-blast breakers. Tall slender insulators of this type can be made withh height to bore ratios up to of about 10:1 although a ratio not exceeding 6:1 is normally recommended.

Porcelain insulators may be designed to withstand a number of mechanical type and routine tests, for example as follows:

(*a*) a temperature-cycle test in which the insulator is immersed alternately in hot and cold fluid
(*b*) hydraulic tests to determine its ability to withstand internal pressures
(*c*) cantilever tests for hollow support insulators
(*d*) tensile tests to check both porcelain and metal end fittings
(*e*) oil-tightness and gas-tightness tests
(*f*) vibration tests to check the soundness of flange/porcelain cementing
(*g*) ultrasonic checks to detect internal faults (used as a back up to other tests).

Some typical properties of electrical porcelain are given in Table 12.11.

12.5.7 Electrical degradation of solid insulation
The principal mechanisms of long term failure of circuit breaker solid insulation are thermal instability, discharge erosion and tracking.

12.5.7.1 Thermal deterioration
Heat is generated continuously in electrically stressed insulation and the amount generated can increase as a circuit breaker becomes older due principally to

Table 12.11   Typical properties of electrical porcelain

(a) Values obtained from test pieces in accordance with BS1598 (1964)

	Normal body	High strength body
Specific gravity	2·4	2·6–2·7
Electric strength	15–25 kV(r.m.s.)/mm	15–25 kV(r.m.s.)/mm
Surface resistivity at 65% rh	$>10^{14}$ Ω	
Volume resistivity at 20°C	$>10^{14}$ Ω cm	
Permittivity	5·7–7·0	5·7–7·0
Power factor	0·017	0·017
Tensile strength	35–55 MN/m²	50–80 MN/m²
Compressive strength	440–540 MN/m²	660–810 MN/m²
Crossbreaking strength, unglazed	80 MN/m²	125 MN/m²
glazed	125 MN/m²	160 MN/m²
Young's modulus	70 GN/m²	83–96 GN/m²
Linear coefficient of expansion	$4–6·5 \times 10^{6}$/°C	$5–6 \times 10^{6}$/°C
Thermal conductivity	0·003–0·004	cal/s/cm/°C

(b) Typical practical stress values for design of insulators

Crossbreaking stress (bending loads)	17–48 MN/m²
Tensile stress (axial loads)	13–26 MN/m²
Hoop stress (internal pressure)	20–25 MN/m²
Torsional stress	13 MN/m²

moisture absorbed by the insulation. Under uniform a.c. stress the power generated in an insulator is given by

$$W = 2\pi f c V^2 \tan \delta \text{ watts}$$
$$= 2\pi f \epsilon V^2 \tan \delta \text{ watts per cm}^3 \qquad (12.18)$$

where V is applied r.m.s. voltage, c is capacitance and δ is the loss angle, and ϵ is the permittivity. Both ϵ and $\tan \delta$ increase with temperature for many dielectric materials, and hence the dielectric energy losses are some function of temperature. If the rate of dielectric heating ultimately exceeds the rate of cooling by thermal transfer then the temperature increases indefinitely and the insulation burns out or breaks down.

The condition of circuit breaker bushings in service can be assessed by measuring $\tan \delta$ at intervals of say every five or 10 years. The change in $\tan \delta$ relatively to its value at the time of manufacture gives an indication of any deterioration. For check tests on circuit breaker bushings in service it is usual to record the variation of $\tan \delta$

with voltage using a high-voltage Schering bridge which may reveal the nature of any measured deterioration (Barker, 1968).

12.5.7.2 Tracking

Tracking is the formation of a permanent, conducting path or 'track' across the surface of organic insulation caused by degradation of the insulating material. Day (1968) has shown that the three essentials of tracking are:

(*a*) the presence of a conducting film on the surface of the insulation
(*b*) a process whereby the leakage current through the conducting film is interrupted with the production of sparks
(*c*) the degradation of the insulation caused by the sparks.

A conducting film is usually formed when atmospheric moisture is absorbed by some form of contamination such as industrial deposits, carbonaceous dust or salt.

When moisture is deposited on the surface of a dirty insulator there are nearly always some constituents of the contamination that dissolve in the water to form an electrolytic solution. Leakage currents flow through the contaminant and the heating effect of these currents tends to evaporate the water thereby increasing the resistivity of the surface film. The nonuniformity of the conducting film will give rise to small areas having much higher resistivities than the remainder of the film with the result that the majority of the applied voltage appears across these areas and small discharges form as the areas 'dry out' and the surrounding air breaks down. In extreme cases the discharges formed may elongate to cause complete flashover of an insulator but more generally they are extinguished after a few cycles. The roots of the discharges cause high localised heating and may result in degradation of the insulator surface either by tracking or erosion. Materials, which when thermally degraded encourage carbon formation will track while those materials which decompose in the form of volatiles will erode. Bisphenol epoxy resin, polycarbonate and phenolic resin bonded paper are examples of tracking materials. Cycloaliphatic epoxy resins, nylon and polythene erode.

Various test methods have been devised to evalute the performance of insulating materials subjected to surface discharges. The more important methods are:

(*a*) ASTM D 2132 (1968): Dust-and-fog tracking and erosion resistance of electrical insulating material
(*b*) ASTM D 495 (1970): High-voltage, low-current arc resistances of solid electrical insulating materials
(*c*) ASTM D 2303 (1968): Liquid contaminant, inclined-plane tracking and erosion of insulating materials

(*d*) BS3781 (1964): Method of determining the comparative tracking index of solid insulating materials

Extreme care, however, must be taken when assessing the results of these tests in terms of practical applications. The majority of materials used for circuit breakers track (e.g. p.v.c. synthetic resin bonded paper, bisphenol epoxy resin, dough moulding materials, polycarbonate, etc.) and hence the dust fog and inclined-plane tests are of little use for evaluating such materials because these tests are particularly severe on materials which track easily. The dust fog and inclined-plane tests are more useful for determining the rate of erosion of materials such as cycloaliphatic resins and silicone rubbers. The tests more appropriate to tracking materials are arc-resistance and comparative tracking index test. Manufacturers usually quote these tests when specifying materials since they at least are reproducible and capable of differentiating between two tracking materials to a degree of accuracy not obtainable with the dust-fog test. The performance of various materials subjected to the comparative tracking index test, BS3781 (1964), are shown in Table 12.7.

12.5.7.3 Voids and partial discharge

The presence of voids in a circuit breaker insulator can have a serious influence on the life of the insulator. The electrical stress in thin gaseous voids lying parallel to the field is approximately ϵE_0 where ϵ is the permittivity and E_0 is the average stress in the dielectric surrounding the void. The breakdown strength of a gas near atmospheric pressure is less than that of solid dielectric materials, so it is possible for discharges to occur in the voids at relatively low values of E_0.

If the voltage applied to say a circuit breaker bushing is sufficient to ionise the gas inside a void, the bushing is said to be 'discharging' at that voltage. The voltage at which discharges are first detected is called the discharge inception voltage V_i. The magnitude of the discharges is usually quite small being limited by the impedance of the solid insulation, but under a.c. stressing these 'partial discharges' recur in successive cycles and can cause progressive erosion. The rate at which insulation erodes under the action of partial discharges increases with the amount by which the applied voltage exceeds the inception voltage. Mason (1960) has observed that the alternating voltage life of some forms of insulation which are subject to discharges varies roughly as $(V_i/V)^n$, where V_i is the discharge inception voltage, V is the applied voltage and n varies between five and nine. Thus doubling the applied voltage (assuming V exceeds V_i) will reduce the life of the insulation by a factor of $2^5 - 2^9$.

Probably the most common cause of failure in high-voltage apparatus is long term deterioration caused by discharges eventually followed by sudden dielectric puncture. The detection and measurement of partial discharges is therefore of some importance to manufacturers and users of switchgear, and a very powerful technique for making discharge tests was devised by Mole (1952). The ERA discharge detector designed by Mole can be used to detect partial discharges as small as 0·1 pC for samples having a very small capacitance, but sensitivities of 5 pC are possible for capacitances as large as 0·1 μF. Ideally there should be no detectable discharges (say <1 pC) in any component insulator of a circuit breaker either at working voltage or at the highest transient voltage experienced in service. The ideal of 'zero' discharge is unduly conservative, and experience has shown that discharge levels of over 100 pC can often be tolerated. Materials like glass and porcelain are virtually unaffected by internal discharges, some materials are fairly resistant to discharges (e.g. silica-filled epoxide resin) while some (e.g. oil-impregnated paper) deteriorate more rapidly. Discharges once started usually increase in magnitude with stressed time, but discharges can become short circuited by semiconducting films inside the void and discharging is terminated. The tolerable level of discharge for various kinds of insulation is found by experience and maximum levels are sometimes quoted in specifications. For example, new synthetic resin bonded paper bushings used in circuit breakers must have a maximum discharge magnitude of 100 pC, according to BEBS S17 (1968), measured at a voltage of 0·75 times rated voltage (for rated voltages 22—44 kV). For complete circuit breakers, similar levels are generally acceptable.

The variation of discharge level with applied voltage can be used to assess the condition of switchgear in service to augment data obtained from measurements of insulation resistance, capacitance and power factor. Hill and Cresswell (1973) have described how discharge testing can be used to evaluate the condition of 33 kV metalclad switchgear, both in a new condition and after a service life of over 40 years. They adopted a criterion that long-serving 33 kV switchgear exhibiting discharge levels in excess of 1000 pC at 20 kV test voltage would be removed from service for reinsulation.

12.6 Acknowledgments

The author is indebted to colleagues in the Research Department, A. Reyrolle & Co. Ltd; the Electrical Engineering Department, University of Strathclyde; and the Bushing Co. Ltd, for their help and advice; and to Allied Insulators Ltd. for data included in Table 12.11.

12.7 References

ABRAHM, M., and BECKER, R. (1932): 'Electricity and magnetism' (Blackie and Son), chap. 5

ALSTON, L. L. (1958): 'High-temperature effects on flashover in air', *Proc. IEE*, **105**, pp. 549—553

ALSTON, L. L. (1968): 'High-voltage technology' (Oxford University Press), chap. 3

ANDERSON, H. W. (1935): 'Effects of total voltage on breakdown in vacuum', *IEEE Trans.*, PAS-54, pp. 1315–1320

ARNOLD, K. W., BRITTON, R. B., ZANON, S. C., and DENHOLM, A. S. (1963): 'Electrical breakdown between a sphere and plane in vacuum'. Proceedings of the Sixth International Conference on Ionisation Phenomena in Gases. Paris, p. 101

ASTM D 570 (1963): 'Test for water absorption of plastics'

ASTM D 1708 (1966): 'Test for tensile properties of plastics by use of microtensile specimens'

ASTM 635 (1968): 'Tests for flammability of self-supporting plastics

ASTM D 696 (1970*a*): 'Test for coefficient of linear thermal expansion of plastics'

ASTM D 790 (1970*b*): 'Tests for flexural properties of plastics

BEBS S17 (1968): 'Partial discharge testing of demountable synthetic-resin bonded and oil impregnated paper bushings of rated voltage 22 kV and above

BANFORD, H. M. (1973): 'Current/voltage characteristics of gas/solid interfaces' in 'Diagnostic testing of high voltage power apparatus in service'. *IEE Conf. Publ.* 94, pp. 152–156

BANFORD, H.M. (1976): 'Particles and breakdown in SF_6-insulated apparatus', *Proc. IEE*, 128, pp.877-881

BANFORD, H. M., and HAMPTON, B. F. (1973): 'Surface flashover in compressed SF_6', 8th University Power Engineering Conference, Bath

BARKER, H. (1968): 'High-voltage bushings', in ALSTON, L. L. (Ed.): 'High-voltage technology' (Oxford University Press), chap. 12, pp. 228–253

BAX, J. (1970): 'Deformation behaviour and failure of glass fibre-reinforced resin material', *Plastics and Polymers*, February, pp. 27–30

BOYD, H. A., BRUCE, F. M., and TEDFORD, D. J. (1966): 'Sparkover in long uniform field gaps', *Nature*, May, pp. 719–720

BROWN BOVERI REVIEW (1967): 'The electric strength of insulating oil with power-frequency voltage, switching surges and standard impulses', *Brown Boveri Rev.*, 54, pp. 368–376

BRUCE, F. M. (1947): 'Calibration of uniform field spark-gaps for high-voltage measurements at power frequency', *J. IEE*, 94, Pt. III, pp. 138–149

BS148 (1959): 'Insulating oils for transformers and switchgear'

BS223 (1956): 'High voltage bushings'

BS2782 (1970): 'Methods of testing plastics'

CLARK, F. M. (1962): 'Insulating materials for design and engineering practice' (Wiley and Sons), chap. 7

COOKE, M.C., WOOTTON, R.E., and COOKSON, A.H. (1977): 'Influence of particles on AC and DC electrical performance of gas-insulated systems at extra-high-voltage', *IEEE Trans.*, PAS 96, pp.768-777

COOKSON, A. (1970): 'Electrical breakdown for uniform fields in compressed gases', *Proc. IEE*, 117, pp. 269–270

COOKSON, A. H., and FARISH, O. (1972): 'Partial-initiated breakdown between co-axial electrodes in compressed SF_6'. Proceedings of the International Symposium on High Voltage Technology, Munich University, pp. 343–349

COOKSON, A.H., and PEDERSEN, B.O. (1978): 'High voltage performance of mixtures of SF_6 with N_2, Air and CO_2 in compressed gas insulated equipment'. Fifth International Conference on Gas Discharges. pp.161-164

COOKSON, A.H., and WOOTTON, R.E. (1978): 'AC corona and breakdown characteristics for rod gaps in compressed hydrogen, SF_6 and hydrogen-SF_6 mixtures', *IEEE Trans.*, PAS-97, pp.415-423

DAY, A. G. (1968): 'Tracking and breakdown of solids due to discharges' in ALSTON, L. L. (Ed.): 'High-voltage technology' (Oxford University Press), pp. 158–169

DEGLI-ESPOSTI, G., ELLIS, E., MOSCA, W., and THIONE, L. (1972): 'Considerations on the maximum allowable surface gradients on the electrodes of high voltage apparatus'. Proceedings of the International Symposium on High Voltage Technology, Munich University, pp. 91–97

DENHOLM, A. S. (1958): 'The breakdown of small gaps in vacuum', *Can. J. Phys.*, **36** pp. 476–493

DENHOLM, A. S., BRITTON, R. B., and ARNOLD, K. W. (1963): 'The ability of a voltage graded surface to support a high voltage in vacuum and in a pressurised gas', *Rev. Sci. Instrum.*, **34**, p. 185

DENHOLM, A. S., McCOY, P. J., and COENRAADS, C. N. (1963*a*): 'The variable capacitance vacuum insulated generator – A progress report' Proceedings of the Symposium on Electrostatic Energy Conversion PIC–ELE 209/1

EPRI, USA (1979): 'Investigations of high-voltage particle initiated breakdown in gas-insulated systems'. Report EL-1007

FARISH, O., IBRAHIM, O. E., and CRICKTON, B. H. (1976): 'Effect of electrode surface roughness on breakdown in nitrogen-SF$_6$ mixtures, *Proc. IEE*, **123**, pp. 1047–1050

FINKLEMANN, E. (1937): 'Der elekrische Durchschlag verschiendener Gase unter hohem Druck', *Archiv fur Elektrotechnik*, **31**, pp. 282–286

FRANCK, S. (1928): 'Ansangspannung und gazdichte bei verschiedenen electrodenformen', *ibid.*, **21**, p. 318–374

GANGER, B. (1972): 'Electric strength of air with high alternating and impulse voltages and pressures up to 100AT',. Proceedings of the International Symposium on High Voltage Technology, Munich University, pp. 363–361

GANGER, B. E. (1968): 'The breakdown voltage of oil gaps with d.c. voltage', *IEEE Trans.*, PAS-87, pp. 1840–1843

GERMAIN, C., JEANNEROT, L., RORBACH, F., SIMON, D., and TINGUELY, R. (1966): 'Technological developments of the CERN Electrostatic Separator Programme'. Proceedings of the Second International Symposium on Insulation of High Voltages in Vacuum, pp. 279–287

GLEICHAUF, P. H. (1951): 'Electrical breakdown over insulators in high vacuum',*J. Appl. Phys.*, **22**, pp. 535–541 and 766–771

GOSSENS, R. F. (1948): 'Gas under pressure as an insulant', CIGRÉ Report 117

GOULSBRA, D., RYAN, H. M., SHAW, N., and WILKINS, R. (1970): 'Some electrical breakdown studies in epoxide resins' in 'Dielectric materials measurements and applications' *IEE Conf. Publ.* **67**, pp. 266–270

GREENWOOD, P., and WHITTINGTON, H. W. (1972): 'Surface flashover in compressed SF$_6$' in 'Gas discharges'. *IEE Conf. Publ.* **90**, pp. 233–235

HAUSCHILD, W. (1970): 'The dependence of the breakdown voltage of inhomogenous insulating gaps in oil on impulse voltage rise time and impulse voltage duration', *Elektrie*, **24**, pp. 244–246

HAWLEY, R., ZAKY, A., and ELDINE, M. E. (1965): 'Insulating properties of high vacuum', *Proc. IEE*. **112**, (6), pp. 1237–1248

HAWLEY, R. (1968): 'The electrical properties of high vacuum' in ALSTON, L. L. (Ed.): Chap. 4, 'High-voltage technology' (Oxford University Press), pp. 59–94

HOLT, E. H. (1956): 'The electric strength of highly compressed gases',*Proc. IEE*, **103**, *Pt. A,* (7), pp. 57–68

HOWARD, P. R. (1957): 'Insulation properties of compressed electronegative gases', *ibid.*, 104A,, pp. 123–138

HOWELL, A. H. (1939): 'Breakdown studies in compressed gases', *Trans. AIEE*, **58**, pp. 193–204

IEC 26 (1969): 'Specification for new insulating oils for transformers and switchgear'

IKEDA, C., and YODA, B. (1971): 'Effects of electrode surface conditions and dust particles on the electrical breakdown of SF_6 gas', *Elect. Eng.* (Japan), **91**, pp. 67–70

JAMES, A. C., RYAN, H. M., and POWELL, C. W. (1966): 'Electrical strength of compressed air in a uniform field', *Nature*, pp. 403–404

KAO, K. C., and HIGHAM J. B. (1956): 'Electric breakdown of dielectric liquids'. Elect. Research Association Report E/T67

KAWAGUCHI, Y. *et al.* (1970): 'Dielectric breakdown of SF_6 in nearly uniform fields', *Trans IEEE*, **PAS-90**, pp. 1072–1078

KIDD, A. C. (1967): 'Epoxide resin castings and laminates – assessment of temperature limitations be measurements and temperature of deflection at high stress'. Proceedings of the Symposium on High-Voltage Applications of Epoxide Resins, University of Manchester Institute of Science and Technology, pp. 73–78

KNORR, W., MÖLLER, K., and DIEDERICH, K.J. (1980): 'Voltage-time characteristics of slightly non-uniform arrangements in SF_6 using linearly rising and oscillating lightning impulse voltages'. CIGRE, Paris, Paper 15-05

KOK, J. A. (1961): 'Electrical breakdown of insulating liquids' (Phillips Technical Library)

KRASUCKI, Z. (1962): 'Processes leading to discharges in oil-impregnated paper', *Proc. IEEE*, **109B**, pp. 435–439

KRATZENSTEIN, M. G. (1970): 'Impulse breakdown of insulating oil', *Bull Assoc. Suisse Elect.*, **61**, pp. 105–117

KUFFEL, E., GRYZBOWSKI, and McMATH, J. P. C. (1970): 'Breakdown across insulation surface in vacuum under direct, alternating and surge voltages of various wave shape'. Proceedings of the Fourth International Symposium on Discharge and Electrical Insulation in Vacuum, pp. 227–231

LAMBETH, P. J. (1971): 'Effect of pollution on high-voltage outdoor insulators', *Proc. IEE*, **118**, pp. 1107–1130

LEGG, D., and ROTHWELLL, K. (1967): 'Epoxy resin in high-voltage switchgear'. Proceedings of the Symposium on High-Voltage Applications of Epoxide Resins, University of Manchester Institute of Science and Technology, pp. 63–72

MALIK, N.H., and QURESHI, A.H. (1979): 'A review of electrical breakdown in mixtures of SF_6 and other gases', *IEEE Trans.*, **E1-15**, (1)

MATTINGLEY, J. M. and RYAN, H. M. (1971): 'Potential and potential-gradient distributions for standard and practical electrode systems', *Proc. IEE*, **118**, pp. 720–732

MASON, J. H. (1960): 'Discharge detection and measurements', *ibid.*, **112**, pp. 1407–1423

MOLE, G. (1952): 'Design and performance of a portable a.c. discharge detector'. ERA Report V/T 115

MULCAHY, M. J. (1972): 'Effect of particles and fibres on d.c. breakdown voltages in air at atmospheric pressure' in 'Gas discharges'. *IEE Conf. Publ.* **90**, pp. 382–384

MULCAHY, M. J. (1964): Private communication

NELSON, J. K., SALVAGE, B., and SHARPLEY, W. A. (1971): 'Electric strength of transformer oil for large electrode areas', *Proc. IEE*, **118**, pp. 388–393

NITTA, T., SHIBUYA, Y., FUJIWARA, Y., ARAHATA, Y., TAKAHASHI, H., and KUWAHARA, H. (1977): 'Factors controlling surface flashover in SF_6 gas insulated systems'. IEEE Power Engineering Society, Summer Meeting Mexico City

OPPERMANN, G. (1972): 'Die Paschen Kurve und die Gultigkeit des Paschen Gesetzes fur Schwefelhexafluorid'. Internationales Symposium Hochspannungstechnik Technische

Universitat Munchen, pp. 378–385

PALMER, S., and SHARPLEY, W. A. (1969): 'The electric strength of transformer insulation', *Proc. IEE,* 116, pp. 2029–2037

PHILIP, S. F., and TRUMP, J. G. (1966): 'Compressed gas insulation for electric power transmission', 1966 Annual Report, Conference on Electrical Insulation and Dielectric Phenomena, NAS Publication, pp. 100–104

POWELL. C. W., and RYAN, H. M. (1972): 'Breakdown characteristics of air at high temperatures' in 'Gas discharges', *IEE Conf. Publ.* 90, pp. 285–237.

RABINOWITZ, M. (1964): 'New observations on electrical breakdown in vacuum'. Proceedings of the First International Symposium on the Insulation of High Voltages in Vacuum, pp. 119–134

ROBINSON, W. G. (1972): 'Developments in porcelain insulators', *Electrical Rev.,* pp. 435–438

ROHRBACH, F. (1968): 'High voltage insulation problems in electrostatic separators' in ALSTON, L. L. (Ed.): 'High voltage technology' (Oxford University Press)), chap 18

RYAN, H. M. (1967): 'Prediction of alternating sparking voltages', *Proc. IEE,* 114, pp. 1815–1821

RYAN, H. M., and WALLEY, C. A. (1967): 'Field auxiliary factors for simple electrode geometries', *ibid.,* 114. pp. 1529–1536

RYAN, H.M., and WATSON, W.L. (1978): 'Impulse breakdown characteristics in SF$_6$ for non-uniform field gaps'. CIGRE, Paris, Paper 15-01

SALMANG, H. (1961): 'Ceramics – physical and chemical fundamentals' (Butterworths), chap. 12

SCHMID, R. (1967): 'Voltage stress-time behaviour of epoxy resins', Proceedings of the Symposium on High-Voltage Applications of Epoxide-Resins, University of Manchester Institute of Science and Technology, pp. 55–59

SHANNON, J. P., PHILLIP, S. F., and TRUMP, J. G. (1965): 'Insulation of high voltage across solid insulators in vacuum', *J. Vacuum Science & Technology,* 2, p. 234.

SPITZER, F. (1970): 'Prebreakdown and breakdown measurements on transformer oil' in 'Dielectric materials measurements and applications', *IEE Conf. Publ.* 67

STEINBIGLER, H. (1969): 'Anfangsfeldstarken und ausnutzungsfaktoran rotations-symmetrischer electrodenanordnungen in luft', Diss. TH Munchen

STRIVASTAVA, K. D., and TOURREIL C. De (1968): 'Electrical breakdown across ceramic surfaces in high vacuum under d.c. and pulse voltages'. Proceedings of the Third International Symposium on Discharges and Electrical Insulation in Vacuum, Paris, pp. 243–247

TAKUMA, T., WATANABE, T., and KITA, K. (1972): 'Breakdown characteristics of compressed-gas mixtures in nearly uniform fields', *Proc. IEE,* 119, (7), pp. 927–928

TASCHNER, W. (1976): 'Impulse characteristics of a few basic arrangements in SF$_6$ at different gaps and pressures', *ETZ-A,* 97, pp.118-120

TRUMP, J. G., and VAN DER GRAAF, R. J. (1947): 'The insulation of high voltages in vacuum', *J. Appl. Phys.,* 198, p. 327

TRUMP, J. G. (1954): 'Dielectric materials and applications' (Wiley), p. 154

USHIO, T., SHIMURA, I., and TOMINAGA, S. (1971): 'Practical problems on SF$_6$ gas circuit breakers', *Trans. IEE,* PAS–90, pp. 2166–2173

WATSON, A., BELL, W. R., and MULCAHY, M. J. (1968): 'Factorially designed vacuum breakdown experiments'. Proceedings of the Third International Symposium on Discharges and Electrical Insulation in Vacuum, pp. 151–156

WAYE, B. E. (1967): 'Introduction to technical ceramics' (Maclaren and Sons)

ZALESSKII, A.. M. *et al.* (1970): 'Electric strength of SF$_6$ in the field of coaxial electrodes', *Elektrichestvo*, pp. 83–85

ZANON, C. S. (1964): 'An effect of electrode temperature on vacuum breakdown'. Proceedings of the First International Symposium on Insulation of High Voltages in Vacuum, MIT,. pp. 206–214

ZAJIC, V. (1957): 'High-voltage circuit breakers' (Artia Prague), chap. 3

ZEIN EL-DINE, M. E., and TROPPER, H. (1956): 'The electric strength of transformer oil', *Proc. IEE,* **103c**, pp. 35–45

ZOLEDZIOWSKI, S. (1967): 'Breakdown of epoxy resin under impulses of epoxide resins'. University of Manchester Institute of Science and Technology, pp. 43–53

Chapter 13

Cost effective design

C. H. Flurscheim, B.A., Fel.I.E.E.E., C.Eng., F.I.Mech.E., F.I.E.E.

13.1 Cost effective approach to design objective

The development of a new product such as a circuit breaker has implications which should be considered as a whole so that the engineering content of the work can be seen in its proper context, for the objective of producing a new circuit breaker must be to meet the technical needs of power systems with a product that will be competitive and produce a satisfactory financial return over the period the apparatus will be in production. This requires that the design shall be judged not solely by the quality of its performance, nor solely by its cost, but by its 'cost effectiveness', which is a measure of the performance afforded in relation to both the realistic market requirements and cost.

The term 'quality of performance' as used here embraces all those attributes that define the operation of a circuit breaker under service conditions. It must, therefore, be dependent in the first place on the effectiveness of the specification on which the design has been based, including not only the definition of the more obvious ratings, but also the definition of associated characteristics such as the level of reliability, electrical and mechanical safety, the degree of maintenance required, and the flexibility of the design necessary to meet varied requirements. With this background quality of performance is then dependent on the efficiency with which the specification has been implemented in the design and manufacturing process.

In no area of design is the term 'cost effectiveness' more significant than in circuit breakers, for both the level of performance needed on actual networks and the level of performance provided by circuit breakers can vary widely. In the past, excessive variations in both performance and cost of circuit breakers were sometimes due to weaknesses in purchasing specifications, which failed to define adequately the performance level required, or to variations in design interpretations of what was in fact necessary to provide reliable service when the service and network conditions were not adequately appreciated. Even today, with more elaborate specification procedures and more advanced circuit breakers, important differences in design principles and in performance offered can result in significant differences in cost, and comparisons must, therefore, be made on a 'cost effective' basis, in relation to actual installation needs.

Cost here includes a comprehensive interpretation of materials, labour and fixed overheads, the latter embracing charges relevant to manufacturing plant facilities employed, allocated charges for development and special tooling, with circuit breakers a very significant figure, and interest charges on capital required to finance work in progress to achieve the output levels envisaged.

This need for cost effective design draws attention to a basic difference that exists between the development of circuit breakers and that of large transformers or turbogenerators. The latter, although they may be based on standard engineering data, are designed to a considerable extent on a custom built basis to meet specific needs, and it is then mandatory to design with the objective of 'first time' success involving some margin in design parameters to cover residual risks. Circuit breakers, on the other hand, are now nearly always developed as standard apparatus and are proved by prototype testing and cost reduced by value analysis methods before going into production. This affords the opportunity to produce fully tested 'cost effective' products which do not have built-in and unnecessary margins which inflate cost, and it follows that if they are designed on the basis of 'first time' success they may well prove too expensive and be an unsaleable product.

Two major and interdependent areas of decision are required before a development programme can be initiated. First, it is necessary to determine the design philosophy, for it is clearly possible to adopt two approaches to development: 'to lead' or 'to follow'. 'Leading' in design implies investing in an extensive research and development programme which in switchgear may require financial support for many years before any return is obtained. 'Leading' implies research into new systems of arc interruption or into major advances in existing technology, and these are usually, in the early stages, of a character such that the final outcome is difficult to predict, but which if successful would ultimately produce advance designs which will enhance the reputation of their sponsors, provide an economic return on investment and assist in increasing the share of the market captured. 'Following' a design system initiated elsewhere avoids the risks and expense of major research and development programmes, which may be beyond the financial and technical scope of a particular organisation, but will probably involve purchasing of patent licences, and carries with it the implication that the design policy must aim at achieving an adequate share of the market on a profitable basis despite royalties, by very careful attention to all aspects of 'cost effectiveness'. However, some advantages may accrue to a following policy since the designers can learn from the mistakes made elsewhere, and to a minor degree may be able to leapfrog the competition by introducing improvements involving relatively little technical risk, and yet still benefit from the lower overheads associated with the reduced development costs.

Following on a decision on design policy it becomes essential to define the requirements to be met by the circuit breakers to be developed, which requires a dialogue to relate the commercial needs and development concepts that may have pointed the way by establishing new possibilities. If it is decided to adopt a 'leading' policy then some flexibility of approach and financial control is essential in permitting research and advance development to be executed without closely defined objectives. But as soon as such work reaches a stage of proven practicability, the controls necessary become similar to those required with all other developments characterised as 'following' and should become specific. At this stage, the definition should include comprehensive specification together with an assessment of what the development will involve in time and cost, for the best design concepts may prove uneconomic if they are misdirected in their objectives, or proceed without clear directives in front of the designers. Some of the factors requiring consideration in this definition are therefore outlined in the following pages.

13.2 Specification of requirements

13.2.1 Specification of performance and rating
The definition of the development project should include the standard ratings to be met, such as system voltage, current, and short-circuit rating, and should amplify these by specifying other aspects of performance required which affect the design such as total break time, restriking rate to be withstood, overvoltage factors, out-of-phase switching, high-speed reclosing, single-phase switching, point-on-wave closing, capacitor bank switching duty, or the special duty requirements associated with for example, rolling mill or traction service. These requirements should be supplemented by a clarification of the national, international and clients specifications that have to be met. The specification of reliability levels, in terms of failure rates that can be tolerated, can also be an important measure directed towards improving overall cost effectiveness.

13.2.2 Specification of future rating requirements
With circuit breakers, it is quite a usual procedure for the range of current, short-circuit and voltage ratings to be enlarged and for the actual performance limits to be 'stretched' during the life of the equipment. The longer term objectives and possibilities should be declared and considered in the initial stages of development, because a study of these should make it possible to minimise subsequent changes to standardised assemblies.

13.2.3 Specification of special requirements
The range of special requirements which may be necessary with circuit breakers is often numerous and those which are to be taken into account during the design

should be agreed with the commercial interests at the definition stage, as provision for these can often be made at slight cost if they are considered early in the development by provision of suitable interfaces. Nothing is more detrimental to flow production if changes are introduced indiscrimately on specific orders which then involves hold up of details and equipment for modification.

Special components can include auxiliary switches with additional or adjustable contacts, which may require additional drives, special terminal boards requiring additional space, special interlocks, duplicated trip coils, silencers for air-blast circuit breakers, galvanised finish or mounting arrangements such as chassis fitted with wheels.

Special environmental conditions need to be considered because circuit breakers have a very wide range of installation conditions, and it may not be economic to meet them all with a standard design. These can include situations such as arctic climates associated with subzero temperatures and build up of ice; tropical climates associated with exposure to high sun temperature and invasion by insects; high altitudes associated with low air pressures; exposure to extremes in humidity, to fog or salt laden or explosive atmospheres; exposure to blown sand, cement dust, to high flood levels, to vibration or earthquake shock; and suppression of radio interference to especially low levels.

'Environment' also includes situations such as transportation limitations in terms of weight, dimensions and freight cost that apply on the expected delivery routes by sea, rail, road and air; special handling situations that may arise such as rough usage in ports or in service, or exposure during transport at sea, which can affect design requirements in terms of size of components, robustness, corrosion protection, and design for packageing and lifting; and test and erection conditions on site which may be important if the circumstances for carrying these out are unfavourable.

It is then necessary to specify the normal conditions to be met and any special conditions for which standard modifications must be available. The introduction of interfaces or modifications to meet such special requirements will have cost consequences, and it is then a question of achieving a balance between the cost of making provision for those which seem desirable and the consequences arising from omitting those which are considered less important, such as loss of orders or the considerably increased costs which will follow if it is found essential to embody such changes on an *ad hoc* basis.

13.2.4 Specification of production requirements

Estimates are essential for the year-to-year build up of total production, of batch manufacturing levels and for the breakdown into the various forms, if there is to be a choice of systems. This is a difficult task, but it is necessary for guidance in selecting the design basis in relation to alternative manufacturing processes, and the degree of special tooling possible, in addition to providing the basis for the overall financial estimates.

Delivery requirements should also be specified as they are an important technical and commercial aspect of design, for this influences the stock of special materials, details, subassemblies and finished products to be carried and therefore reacts on the design, which must make possible the acceptance of the delivery periods required with an optimum value of stock.

All designs have to be executed in relation to the manufacturing facilities available, and it is therefore important to know if subcontracting manufacture to remote factories is intended. This will require consideration of the manufacturing facilities which will then be available, which may react on the design and the breakdown of manufacturing information instructions in order to facilitate subcontracting of suitable components. Subcontracting factories may not be as sophisticated as those of the parent organisation and manufacturing information may need to be more comprehensively documented, details may need to be capable of being manufactured on simpler machines, or duplicated tooling requirements may be so expensive that the combined total production of some components should be organised for central manufacture, which can again react on the design. The specification should therefore clarify the remote manufacturing facilities that may be involved and the degree of local and central manufacture that is envisaged.

13.2.5 Definition of the development programme

A detailed development programme is then necessary to cover all the principal stages in the development, although these will to some extent overlap. This includes, therefore, the time required for research departments to provide any basic information that may be necessary, and for the engineering departments to provide prototype manufacturing information, the time required for completion of prototypes and development testing and production tooling. For circuit breakers, there is a special need to clarify the prototype testing situation as this includes mechanical endurance testing, short-circuit testing, impulse testing, thermal testing, reliability testing and may require a long period, and it affects, therefore, the date by when the engineering department should be in a position to know the design changes that may then be required.

The time scale for the overall operation will affect costs. If the time allowable is too short it will prove necessary to develop an excessive number of subassemblies in parallel and the interaction of development changes between these is likely to increase the total effort and cost. If the time is long, development work is likely to continue to fill the time available with perhaps unnecessary escalation in cost. There is then an optimum overall programme which should permit efficient development at minimum cost.

Estimates are then necessary for the work load and the cost of the different sections of the development project which together contribute to the assessment of the validity of the project and the charges to be taken against production costs. Separate estimates are advisable for research and engineering, draughting, manufacture of prototypes, type testing, reliability assessments, short-circuit testing, production tooling and special shop facilities required. With circuit breakers, this total is likely to be a very large sum which will form a significant component of the overhead costs which have to be recovered on the product production in the following years. All this information will also aid in subsequent monitoring of progress.

13.2.6 Specification of cost objectives

Preliminary estimates are needed for anticipated manufacturing costs of the project and all its variants, for comparison with commercial requirements. A cost which is sometimes overlooked is the additional cost which will arise during the learning period associated with initial production, since initial manufacturing costs usually exceed estimates during the period when design and manufacturing teething troubles are eliminated and piece rates on the shop floor are established. These additional costs are legitimately a part of the development situation, and are to some extent a reflection of the extent of changes in design philosophy.

Learning costs arise from a number of causes and careful attention to these can reduce their impact. Production circuit breakers are nearly always developed through the prototype procedure, but this does not necessarily ensure satisfactory implementation of production design, because of differences between handmade and tool details, because of the variation in machining tolerances permitted, because of difficulties that will occur in production if the details require new manufacturing processes, because the changes, which have been found essential in the prototype, may not have been adequately recorded and translated into the production design and tooling, and because of problems of assembly, adjustment and testing. Learning troubles will also be increased if the urgency of contract commitments is permitted to require 'production' to be initiated before 'development' has been completed. Nevertheless, since development is likely to be continued so long as time permits, release for manufacture is a matter for very careful judgement to achieve an optimum situation.

The cost and dislocation consequences of these situations can be restricted by careful study of limits and by testing of critical components and assemblies deliberately manufactured to the extreme of the limits permitted; by observing quality assurance procedures during development which will ensure that all changes required by the design office or made by the factory are processed by the correct procedure so that production drawings take such modifications into account; by study of assembly problems during design and prototype construction; by elimination of adjustments which are critical, and which may be made satisfactorily by development staff but which can cause trouble in production; and by systematic assessment during prototype testing of the routine testing procedures that will be necessary in production.

13.3 Cost effective design: Standardisation

With the design policy and definition of objective determined, the design function is then required to convert these ideas into hardware. There are then two important and allied features in the approach to design of circuit breakers that are fundamental to the improvement in 'cost effectiveness'. Power circuit breakers are produced in numbers that may vary from tens to tens of thousand per annum of a type and the most effective way of reducing cost is to ensure the highest degree of standardisation of designs, so that these can be made in the largest possible batch quantities, so that the manufacturing processes to be learnt are kept to a minimum, and so that the maximum possible degree of tooling can be embodied, justified by reduction in variations.

Secondly, to improve the possibilities of standardisation, circuit breakers should be modular in construction, so far as this is possible, so that they can be constructed out of a number of identical assemblies selected as required to meet different ratings. This is not an invitation to produce a wide range of modules, which would defeat the objective, and insistance on restricting variety in modules is as significant a policy as is the modular principle itself. An additional advantage arising from modular construction is that, with some classes of circuit breakers, it will aid in completely separating interruption and insulation to ground functions, which can improve reliability. Modular design will reduce further the number of major items that have to be proved from the point of view of performance and reliability, enabling a more thorough approach to be made to these issues, and will contribute, therefore, not only to economy in production, but to 'cost effectiveness' in its broadest sense.

Where exact standardisation is not possible, then there should be the maximum interchangeability of minor subassemblies such as trip gear units and details, and so far as possible the subassemblies which cannot be identical should at least be similar to reduce manufacturing learning costs and to reduce the variety of problems arising in maintenance by clients.

The design concept of a circuit breaker should therefore be laid out with the objective of dividing it up into self contained subassemblies suitable for modular construction and standardisation across a range of ratings. This will also speed up design by making possible some degree of development of complementary subassemblies in parallel, and it will make possible the assembly and delivery of subassemblies for test, transport and erection, it will simplify ordering on the factory, and improve the accuracy of cost returns from the factory because these will then be related to small assemblies instead of complete circuit breakers.

This approach will also contribute to controlling the total level of work in progress, in manufacture and in stock. For the diversity factor in demand requires increasing total stock with increase in variety of components performing similar functions to meet a given production level.

The optimum stock level is theoretically that which will result in the minimum overall cost of the product in relation to the output and deliveries called for. The larger the batch manufacturing levels the lower will be the manufacturing cost, but at the same time this will result in a higher capital investment in stock and increased risks, if stock control fails to keep in line with changes in demand or obsolescence. The optimum level depends then partly on commercial policy decisions with relation to delivery requirements, and partly on the design, which largely controls the manufacturing cycle through the longer delivery components and materials it demands, and through the complication of individual components which determine the number of consecutive operations needed and the time they will take in passing through the manufacturing cycle.

It may, therefore, prove more economical to substitute smaller or simpler standard items for larger and more complicated items which appear to be less expensive solutions, but which require many sequential operations in the manufacturing cycles. Large castings, forgings and porcelains fall into this category requiring comparative analysis on an overall cost effective basis including the financing of work in progress involved. Thus it may be preferable to manufacture a large porcelain column from several smaller units secured together by epoxy resin or clamped together by cemented flanges, than to make this from a single unit which might otherwise appear to be the most economical solution.

Similarly it may be preferable to use two standardised air receivers, held in stock, than to design and build one larger receiver which is only rarely required, and which would either inflate work in progress if it is stocked or be more expensive if manufactured on an *ad hoc* basis.

Large and complicated components requiring many manufacturing operations in series should also be considered in the light of the fact that they involve higher risks in terms of delays in deliveries of raw materials, longer manufacturing cycles and greater risks of scrap due to error.

There is perhaps no class of equipment in which the number of details can build up to so unmanageable a level as in switchgear unless great care is paid to this subject. It is in attention to standardisation, to modular construction, to similarity, to use of common details that significant improvements in cost effectiveness can be made, quite as important as those obtainable from the follow up study of detail design and manufacturing methods based on the usual 'value analysis' procedure which can be applied to circuit breakers with the same technology as to all other engineering products.

13.4 Cost effective design: Performance and simplication

Another powerful approach to increasing cost effectiveness, which is especially valid with circuit breakers, is through improvement in performance. This embraces up rating the short-circuit current and sometimes the voltage per break and reduction in arc energy and consequential distress which often provides the basis for further cost reduction through simplification, so that these two approaches are often interlocked.

In oil circuit breakers, the long history of development has as yet exposed no evidence that finality of performance has been achieved, and the possibility still exists of making further improvements, perhaps aided by the introduction of new and improved materials. In oil interrupters, reduction in the arc energy is important because of the consequential generation of hydrogen gas, for if this is reduced the problem associated with dissipation of the gas and prevention of reignition are eased, both of which in turn aid in increasing current rating and reducing the volume of material required for the interrupter and for the complete circuit breaker.

An important guide in design envisaging future up rating with any class of circuit breaker is, however, that the initial development should be carried out based on the needs of the lower short-circuit ratings that are required, since experience shows that careful testing and detailed development will usually make it possible to increase the short-circuit rating significantly without increasing the overall dimensions. This does not apply, of course, to the more predictable aspects of design, such as load current or test voltage.

An example of this approach is shown in Fig. 13.1a and b which compares an early 66 kV 1500 MVA oil circuit breaker with a modern design rated at 2500 MVA, the design being based on improved interrupters. From this starting

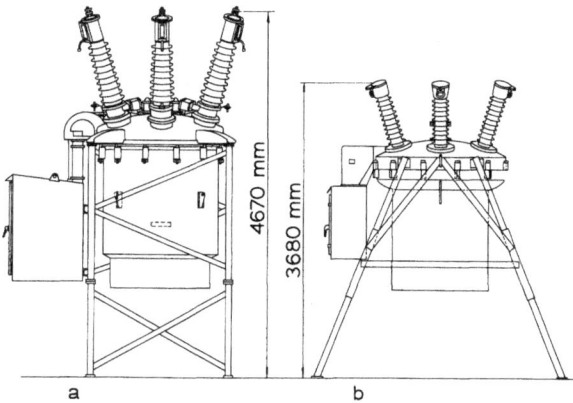

Fig. 13.1 Comparison of 66 kV tank type oil circuit breakers [GEC]

    (a) old design
    (b) new design

point the following were the principal changes:

(a) The smaller interrupter and reduced gas generation permitted the tank to be reduced in diameter, this was accomplished by stepping down from the dome using the outer ring of a steel pressing, the inner cut out of the same ring being used to form the dished bottom of the tank.

(b) The reduction in gas generation allowed the internal insulation phase barriers required in the older tank to be eliminated without the risk of phase-to-phase flashovers.

(c) The oil barrier type bushings with internal current transformers were redesigned with external c.t.s reducing the size of the bushing and the space required, and making c.t. changes easier to carry out, and avoiding the need for oil sealed secondary connections previously necessary.

(d) The dome was simplified by elimination of the welded brackets needed to support the breaker on the frame, the frame itself was reduced in weight by triangulated construction.

(e) A study of the internal mechanism (Fig. 13.2a and b) resulted in the replacement of the earlier lift rod system having three lift rods, a metal cross bar, dual guides and separately mounted energy absorbing devices by a simpler system having a single tubular guide with integral energy absorber, shaped insulation plates to carry the moving contacts and simpler parallel movement linkage.

(f) The earlier design offered both solenoid and spring closing mechanisms (Fig. 13.3a) which had to be dismantled and reassembled and readjusted during erection on site. The redesign (Fig. 13.3b) was standardised with a very simple motor hydraulic mechanism, in which the spring force was held against the hydraulic pressure, release for closing being by means of a dump valve which also disengaged an (unloaded) safety prop. This mechanism was permanently bolted to the underneath of the dome thereby avoiding the need for site assembly and interconnecting framework.

(g) The resultant overall weight reduction was 47% the oil reduction 24%, and since the complications was also greatly reduced the cost reduction was of the same general order.

An example of cost effective design for a 66 kV low oil circuit breaker is shown in Fig. 13.4a and b. In this case the redesign was again based on radical improvement in performance of the interrupter itself, directed towards reducing arc duration and arc length.

The consequential reduction in diameter and gas generation and the availability of higher strength porcelain, then made it possible to eliminate the large SRBP pressure tube required in the older design, replacing this by a small fibreglass tube in the top half of the assembly, which as 'fall out' also improved the dielectric security of the breaker in the open position. The reduction in gas generated made possible a reduction in size of the exhaust gas chamber at the top of the interrupter

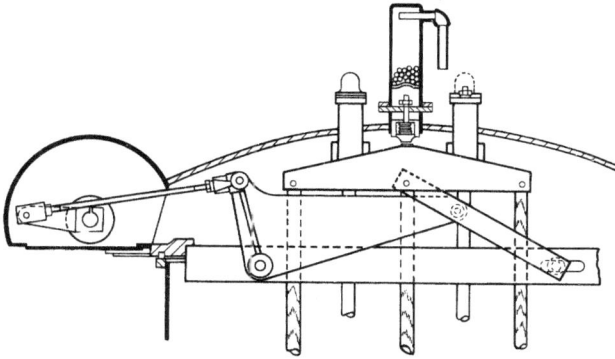

(a)

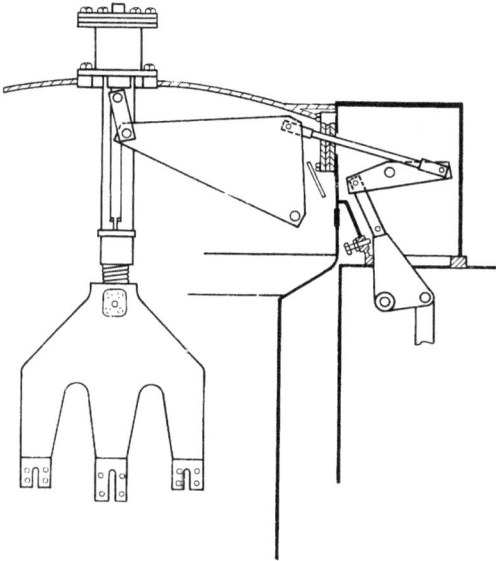

(b)

Fig. 13.2 Comparison of internal mechanisms of 66 kV tank type oil circuit breakers [GEC]

(a) old design
(b) new design

(a)

(b)

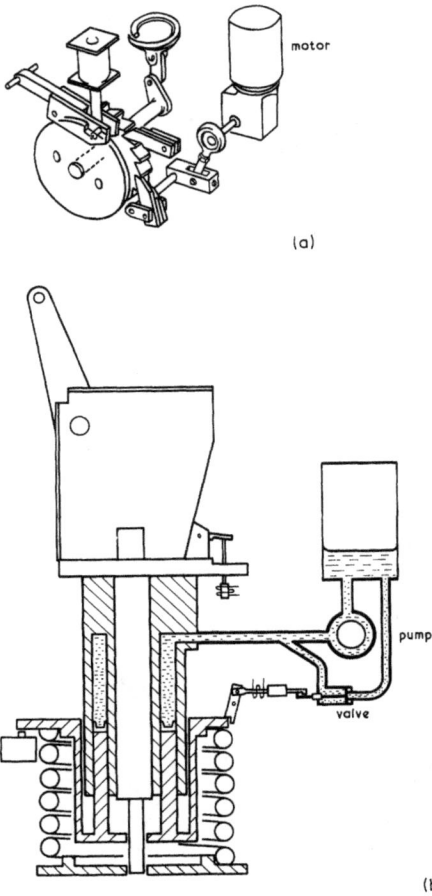

Fig. 13.3 Comparison of operating mechanisms of 66 kV oil circuit breakers [GEC]

   (a) motor wound spring closer (old)
   (b) hydraulic mechanism (new)

assembly, and a detailed study of the current carrying components contributed to reduction in inertia and cost of these details. The overall result was again a very significant reduction in weight and cost.

In gas blast circuit breakers reduction in the rate of flow of compressed air or SF_6 required for a given duty offers cost saving possibilities through reduction in

OLD   NEW

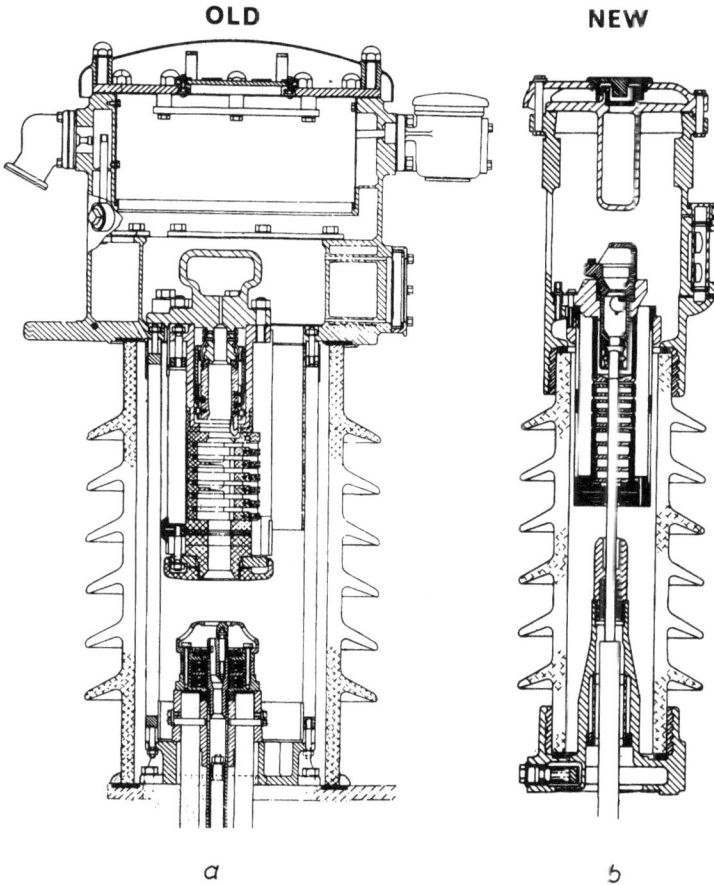

a   b

Fig. 13.4 Comparison of 66 kV low oil circuit breakers [Reyrolle]

    (a) old design
    (b) new design

size of valve gear required or of volume of the storage receivers, or in the energy needed to drive the contact system of puffer type SF_6 breakers, for a given breaking capacity, all of which is then dependent on a better understanding and implementation of the deionisation processes in actual high power nozzle system.

Another approach to reduce cost of air-blast breakers through improved

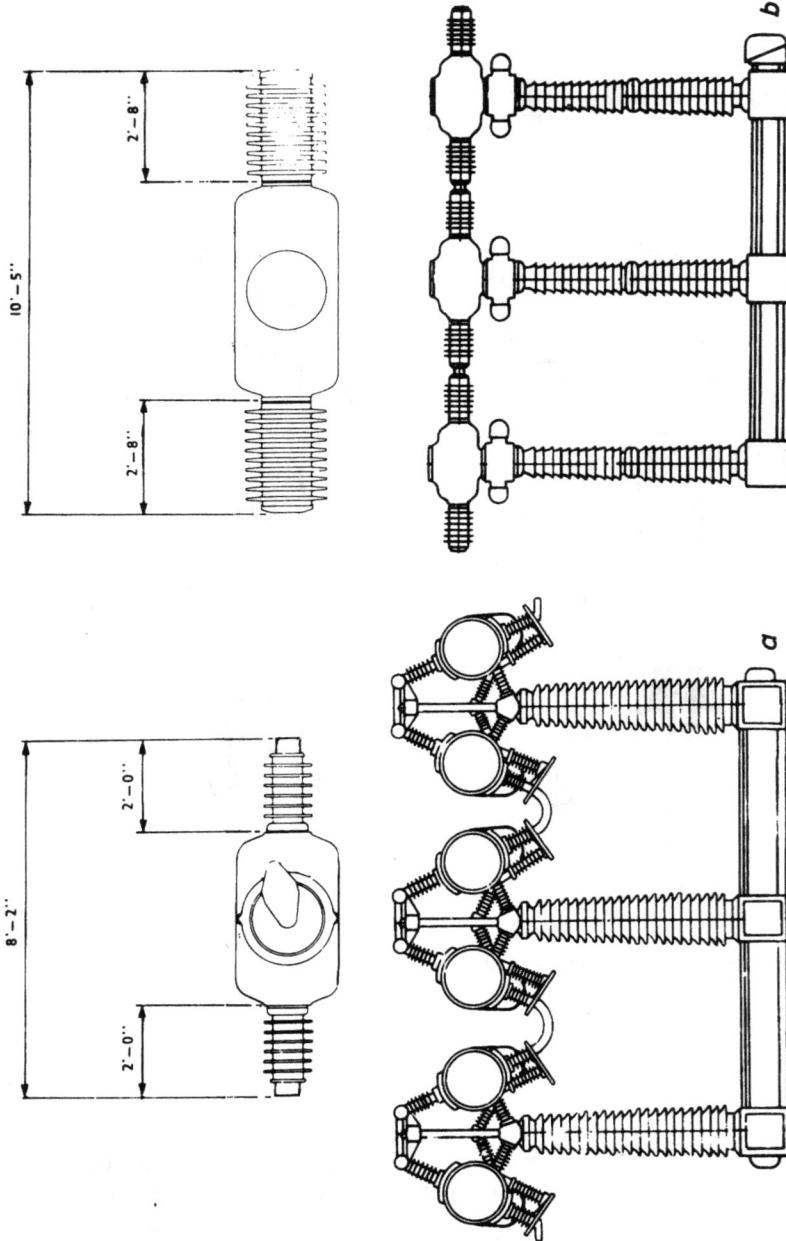

Fig. 13.5 Comparison of 400 kV air-blast circuit breakers [Reyrolle]

(a) old design
(b) new design

performance is to increase the working pressure. Fig. 13.5a and b show the original design of a 400 kV 12 break 35 000 MVA air-blast circuit breaker capable of interrupting short line faults at 10 000 V/μs and operating at 30 atm pressure, and a redesign operating at 60 atm which enabled the voltage per break to be doubled. The individual pairs of interrupters were of course larger, but the objectives were to halve the number of servo valve systems, and to reduce the primary insulation and the complicated structure necessary for the 12 break interrupter system. The total weight of the 3-phase breaker was reduced by 45%, and although working at the higher pressures increased the cost per ton, the overall effect provided a significant cost reduction.

A redesign based solely on reducing complication of mechanism is shown in Fig. 13.6a and b, which compares the original and the redesigned compressed air motor and hydraulic damping system for controlling the sequence switch used in a range of 220–400 kV air-blast circuit breakers, the switch comprising a rotating air break isolator driven by a porcelain column which provided a double air break gap in the open position. The switch turns 60 deg in approximately 0·1 s and requires controlled damping for both the closing and opening movements to restrict shock on the porcelain.

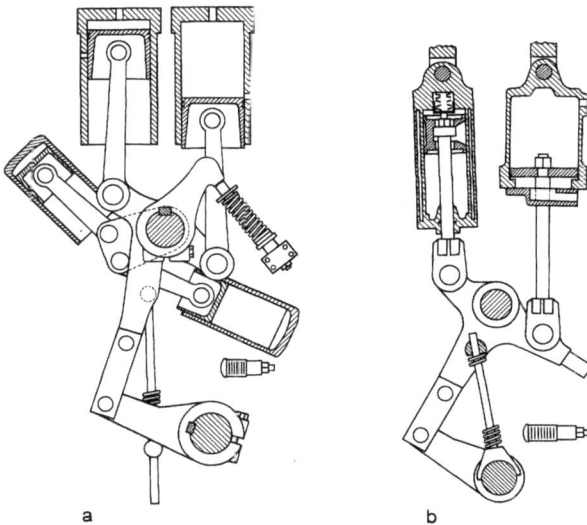

a   b

Fig. 13.6 Comparison of operating mechanisms for a sequential isolator air-blast circuit breaker [GEC]

(a) old design
(b) new design

The earlier design was made with two compressed air driving pistons on one level and two oil damper pistons for closing and opening duties on a lower level, the oil damping system being submerged in an oil tank. The redesign was constructed with both air motor and oil dampers as double acting pivoted cylinders, all mounted in a single plane, the interaction forces being transmitted directly. Although air pipe and oil sealing arrangements were more complicated, the number of air and oil damper components was halved, and the need to transmit high damping torques from the plane of the oil damper system to the plane of the air motor was eliminated. In addition this permitted a simpler bearing arrangement, the two-tier welded steel plate assembly required to carry the bearings in the earlier design being replaced by bearings welded to the air receiver and cover of the mechanism. The damper oil volume was reduced to one quarter, accessibility of the mechanism was improved, and the overall cost factor was reduced by 30%.

Fig. 13.7 gives a comparison of old and redesigned welded mechanism frames for

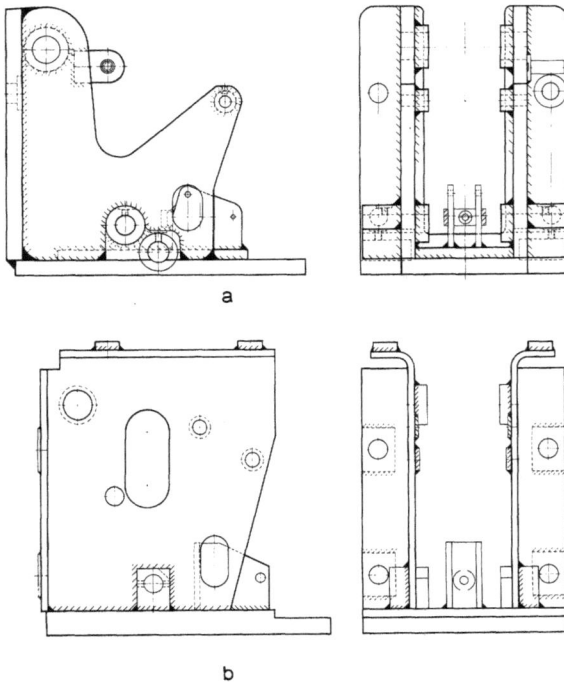

a

b

Fig. 13.7 Comparison of welded steel oil circuit breaker mechanism frame [GEC]

(a) old design
(b) new design

an oil circuit breaker, an example based on using conventional value analysis methods. The base plate was converted from two plates welded together to one thick plate; the back and side plates were changed from three thick plates with considerable flame cutting to two thin plates cropped and bent so as to form also the side plates, incorporating in addition top mounting fixings to permit the mechanism to be bolted directly to the circuit breaker dome; the small subframe was converted from two side plates plus a welded on back strap to a single bent sheet; and a large amount of heavy fillet welding was eliminated and replaced by small fillet welds, the overall cost being reduced by 25%.

Vacuum interrupter tubes, because they are relatively new, have a limited technological background which suggests that there are many possibilities for further improvement in cost effective design based on performance and processing. These might for example be made possible by new approaches to contact cooling, by improved materials, by improvement in dielectric stress control, or by using multibreak construction within a single vacuum enclosure, once the problems of clean assembly for such arrangements can be solved.

13.5 Cost effective design: Use of materials

Circuit breakers employ a wide variety of materials some of which have rather specialised applications in this context. It is fundamental to the use of materials that design should take their special properties into account and the implication is that when new materials are considered a new look is needed as to the form the design should take, both from the point of view of economical use of material and the flexibility the material permits in the design approach. Porcelain, cast epoxy resin, laminated compressed wood, and fibreglass are examples where their properties are particularly important to circuit breaker design.

13.5.1 Porcelain
Porcelain insulation forms a large part of many types of circuit breakers and becomes a major component of cost at the higher voltages (Section 12.5.6). It is then important to understand design factors which control cost effective use of ceramics, involving a process which is amongst the oldest in the world, which has made major advances in recent years, but which is still very largely an art.

The advent of very high-voltage systems has led to the need for correspondingly large insulators. Single piece construction entails high capital expenditure for example in the provision of special kilns, so that the 'on cost' tends to increase significantly with size, and, if the design permits unit construction, economies can be achieved.

Very tall single piece insulators are produced by a jolleying and slip-jointing process or a combination of jolleying, slip jointing and glaze jointing. Often such large insulators are tapered from one end to the other, and this method of construction necessitates the provision of a large number of moulds and also the employment of specialist labour. Construction of the insulators from small units not exceeding about 1·5 m in length, and wherever possible having parallel sides, results in economies of manufacture.

The overall diameter of the insulator and the design of sheds also play a considerable part in the cost of construction. The creepage to be provided is of course determined by pollution conditions, and the creepage length necessary varies in most cases between 15–25 mm/kV of a 3-phase rated voltage. To meet this, a number of plain sheds with small overhang is a better cost effective proposition than the use of a smaller number of deep sheds of 'antifog' design. The two configurations are shown in Fig. 13.8a and b and the antifog b design results not only in a high firing cost, due to the increased overall diameter of the unit, but inhibits the employment of the turning method of insulator shaping due to the complexity of the design and the deep drip rings of the antifog sheds. There is also a higher risk of loss during drying and firing with the more complicated shed

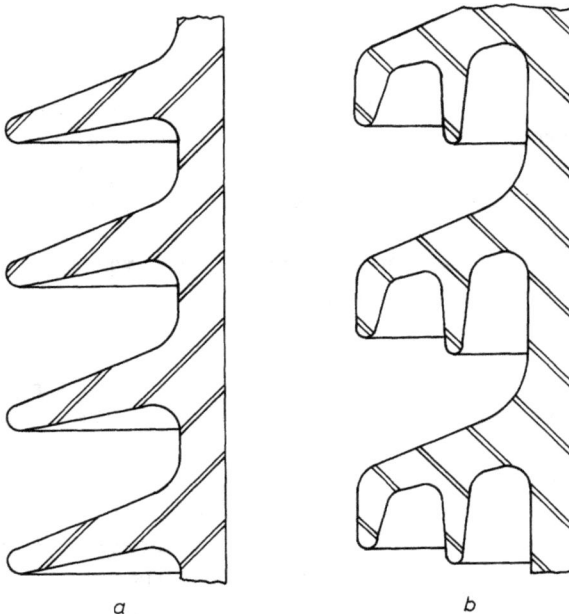

a   *b*

Fig. 13.8 Comparison of porcelain rain sheds [Doulton]

    (a) shed suitable for mechanical production
    (b) antifog shed with drip rings

shapes. It is, therefore, preferable to use a larger number of simple sheds type *a* when the necessary performance can be achieved by this method. Whether antifog or plain sheds are used, the design should always provide for a root thickness of the shed which is graded to the wall thickness to avoid the risk of loss caused by uneven stresses occurring during drying and firing.

The relationship between the height and the diameter of an insulator is also important. Long slender units require special supports during the firing process and even so there is some risk of excessive bending or bowing of the insulator taking place. A ratio of height to body diameter not exceeding 6·1 is normally recommended.

The grinding of porcelain insulators is usually a slow and time consuming operation. Although the 'dressing' of end faces to provide sealing faces can be carried out with reasonable speed, the work entailed in bringing length and diameter dimensions within close tolerances can be very costly. Design should endeavour to accommodate the normal tolerance applicable to ceramic manu-facture, which is between ±2 or 3%, and use closer tolerances only where these are essential in the application of the assembled insulator.

The siliceous porcelain body which over the years has been in almost universal use, has in recent years been supplemented by material having higher alumina content which has an increased cross breaking strength. The higher cost of the high strength material must however be borne in mind when considering the possibility of reducing the size of an insulator for a given specified strength or increasing the stress, particularly if extensive grinding is called for. The increased cost associated with high strength porcelain arises because the materials are more expensive; the preparation of the body materials to the desired degree of possible fineness is also more expensive; the finished article is more dense and therefore all other things being equal the weight of the insulator is greater than for the conventional insulator; and the grinding costs for the high strength material are appreciably higher where fine limits on dimensions are specified.

Where high internal pressures are involved, the stress in ceramic tubes must be calculated by the thick wall theory and take the tensile forces resulting from the end fittings into account. The reduction in strength due to end fittings in such designs can be significant. Where insulators are fitted with metal flanges and the maximum strength has to be developed, in hollow tubes subject to pressure or post insulators which are subject to cantilever loads, the depth of the flange is important and experience has shown that cemented flanges should then have a depth of the order of 40% of the porcelain core diameter. In many applications, however, the final assembly of the insulator is not required to develop this maximum strength, and in these cases the flange does not need to have such a high ratio of depth to core diameter.

Experience has also shown that clamped flange fittings facilitate the develop-ment of maximum strength of porcelain insulators more especially if the external glazing is not destroyed by grinding processes in the region of the flange. This applies particularly to large diameter housings that are sometimes required in circuit breaker assemblies. For most applications, however, adequate strength can be obtained with cemented fittings at more economic cost levels.

An important feature of insulator construction is the glaze finish which serves at least two functions; first, to provide a smooth surface that will reduce the tendency of dirt to adhere and facilitate washing; and, secondly, a compatible design of glaze will increase the unglazed porcelain strength of the insulator by the order of 25%, increasing the cost effective use of the material. Glazing should in general be applied both to the outside and the inside of hollow insulators which are stressed.

Resistance glaze offers opportunities for the future when the extensive trials at present being carried out (1973) to establish the life of these materials have been completed. The adoption of semiconducting glaze will then be of special benefit to the designer of high-voltage insulators because of the voltage control it imposes on the insulator and the resultant economies that should be possible in the size of insulator used. The conducting glaze itself will cost more than the standard for two reasons; first, because material costs are higher, and, secondly, because of the additional process controls which will result in a somewhat higher rejection rate. The control of the actual resistance of the glaze is likely to fall within a tolerance of ±20%, which will present little difficulty for single piece columns, but might suggest the need for selective assembly in multiple unit constructions.

13.5.2 Cast epoxy resin
The introduction of cast epoxy resin has offered very wide possibilities of reassessment in design, and examples are given below.

At voltages above 20 kV special care is necessary to control the axial stress gradients in insulation (Section 12.5.5). The traditional bushing achieves this control by including concentric metal foils. A satisfactory degree of stress control can, however, be obtained by careful design of the external configuration of cast resin insulators so arranged as to ensure maximum spacing between the lines of equipotential where these intercept the profile of the insulator, and Fig. 13.9 shows a 33 kV cast epoxy insulator made in this manner. In this case, the cast epoxy design is clearly simpler than a conventional condenser bushing, but in itself it showed no appreciable cost economy as compared to an s.r.b.p. design. However, the length of the insulation was only one third of that required by a conventional bushing and consequential savings in space in the complete assembly resulted in significant overall cost reduction.

Fig. 13.10 shows a 33 kV outdoor oil circuit breaker designed to take advantage of the possibility of obtaining the necessary clearances in air by using bent bushings mounted vertically instead of the normal approach of using straight bushings mounted at an angle. Here again the bushing itself showed no economy in cost. But the reduction in size of the circuit breaker dome and tank combined with the simplification and standardisation of parts made possible by elimination of angular machining on a number of components, resulted in reduction of overall weight in

EARTH METAL

LINES OF
EQUIPOTENTIAL

CONDUCTOR

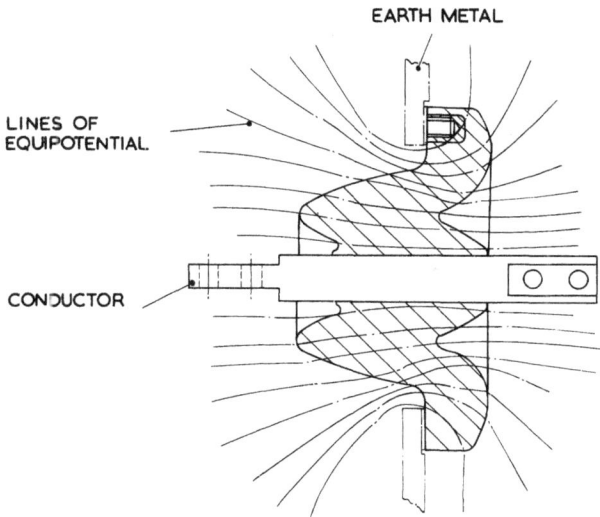

Fig. 13.9 Cast resin 33 kV bushing [GEC]

comparison with the conventional design of 57%, combined with a significant reduction in overall cost.

A third example of cost effective design primarily associated with the use of epoxy resin is given by a comparison of the interrupters of an old and a new 132 kV double-break oil circuit breaker rated at 3500 MVA (Fig. 13.11*a* and *b*). The two designs of interrupters are basically similar in the use of butt type main contacts and sliding contacts at the lower ends, used for making contact with linear low ohmic resistors required for damping and grading purposes.

The overall improvements were in part due to a better arrangement of arc control cross jet plates, which provided increased internal creepage and more effective interaction between the oil and arc plasma, which shortened the arcing time and permitted the plate assembly to be reduced in diameter by 20%. The lower stresses resulting from smaller diameter and the development of cast epoxy technology then permitted a redesign which eliminated a large number of machining and assembly operations.

The s.r.b.p. tube which required rings attached by screwed metal pegs at top and bottom, and a glued on stress shield to protect these pegs, was replaced by a single piece epoxy resin moulding; the bolting ring to secure the interrupter was replaced by integral quick release screw threading in the epoxy body; the resistor contact carrier which was previously screwed in was dropped into the redesigned enclosure from the top and the spring stem guides for the contacts were eliminated,

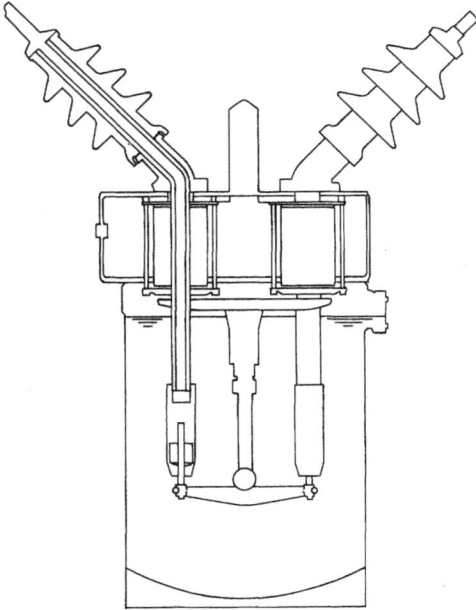

Fig. 13.10 33 kV oil circuit breaker with bent cast resin bushings [Reyrolle]

simplifying also the machining of the current carrying fingers. The cost comparison between the two designs resulted in a reduction of 56%.

13.5.3 Laminated wood and fibreglass operating rods

Another approach to improving cost effectiveness is through control of inertia, as circuit breakers, more especially high performance designs, require very high accelerations of the moving contact system, a problem which can be reduced by the use of new and improved materials. One of the design issues is then to reduce the moving contact mass and the accelerating forces required with consequential reduction in the strength necessary for the insulation and mechanical drive components, which in turn will permit further reduction in the driving forces and mass of material to be moved, saving in the power to be supplied by the operating mechanism and therefore in cost. There are other limits introduced by the forces necessary to close the circuit breaker against short-circuit making current, but reduction in the moving mass will lead to economies throughout the rest of the circuit breaker system, and weight control therefore becomes an important factor in the cost effective equation.

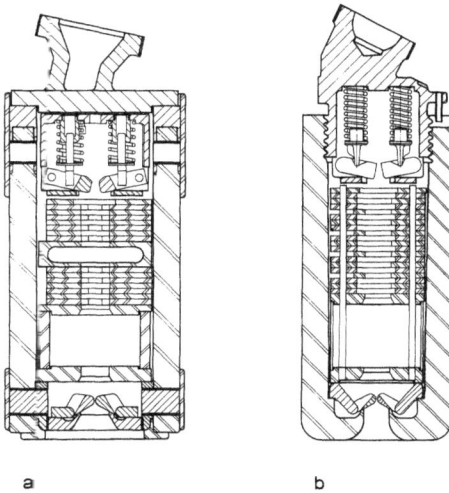

Fig. 13.11 Comparison of s.r.b.p. and cast resin 132 kV crossjet interrupters [GEC]

    (a) old design, s.r.b.p.

    (b) new design, cast resin

A significant proportion of the inertia of the moving system in circuit breakers is attributable to insulation and it is therefore important to design the insulation drive units with the minimum weight compatible with the tension, compression and stiffness requirements. Treated wood, such as ash, has been used extensively in the past, but is variable in both mechanical and electrical properties so that it has been replaced almost entirely by improved materials such as compressed impregnated laminated wood or fibre glass. Because of the variety of detail and applications it is not possible to give an accurate guide, but Table 13.1 gives a comparative picture of the strength and cost of different materials associated with a variety of constructions for the end fittings, per unit strength provided in tension, relating to the systems shown in Fig. 13.12. The single rivet system of Fig. 13.12a is the simplest end fitting, but requires additional location against rock, leading to a more usual construction with multirivets (Fig. 13.12b). Swaged end fittings (Fig. 13.12c) develop a much higher percentage of material strength, but can only be used on round rods and the end fitting process is expensive. Expanded end fittings similar to those used on steel cables (Fig. 13.12d) can only be used with round fibre glass, but is the most effective method for transmitting tension forces on long thin rods of this material. The continuously wound unidirectional fibre glass loop (Fig. 13.12e) develops 100% of the theoretical strength of the material, but it is expensive

Table 13.1 Relative performance of insulating rod systems (approximate)

Type (Fig. 13.12)	Material	Method of fixing	Percentage of material strength developed (of full cross section)	Relative insulation weight per unit tensile strength	Relative insulation cost per unit tensile strength	Relative cost of end fittings per unit tensile strength
a	Laminated compressed wood	single rivet	20	1·0	1·0	1·0
a	Fibreglass sheet	single rivet	20	0·73	0·85	0·88
b	Laminated compressed wood	multi rivet	25	0·80	0·80	1·0
b	Fibreglass sheet	multi rivet	25	0·58	0·68	0·88
c	Laminated compressed wood (machined from square)	swaged	50	0·40	0·5	0·62
c	Fibreglass sheet (machined from square)	swaged	50	0·29	0·43	0·55
d	Unidirectional fibreglass	expanded	80	0·10	0·72	0·53
e	Unidirectional fibreglass endless loop	loop pin	100	0·08	0·90	0·10

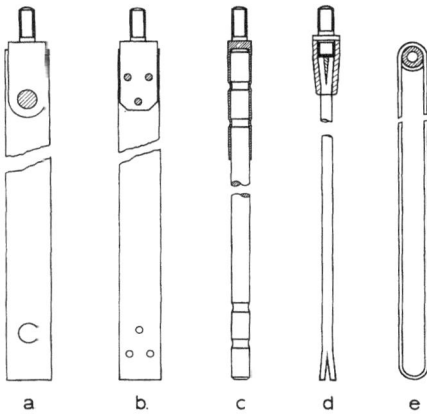

Fig. 13.12 Comparison of circuit-breaker insulation operating rods and end fittings

(*a*) single rivet
(*b*) multirivet
(*c*) swaged
(*d*) expanded
(*e*) endless loop

because of the method of manufacture, requiring separate winding of each individual rod, but this design is attractive for drives that are always in tension because of the simplicity and reliability of the end fittings (British Patent 998391).

Where high tensile forces have to be transmitted, the use of high grade but expensive insulation materials can contribute significantly to cost effectiveness. This may however be modified considerably if compression and stiffness become of importance in the application because of the increased volume of material that must then be used irrespective of tensile stress needs.

13.5.4 Filament wound insulation

Filament winding now offers the possibility of reducing costs of shaped insulation components subjected to high electrical and mechanical stress. An example is the SF_6 circuit-breaker puffer cylinder designed originally using a terylene epoxy cloth cylinder and a separate screwed-in cast epoxy-resin nozzle (as in upper section of Fig. 13.13), which was redesigned with a single-piece wound-terylene filament epoxy component (lower section of Fig. 13.13). This was stronger mechanically and less expensive to produce than the earlier concept, and in addition made it possible to incorporate a cyclo-aliphatic resin layer, necessary for arc protection of the nozzle, as part of the winding process, dispensing with a separate moulding operation necessary with the cast-resin component.

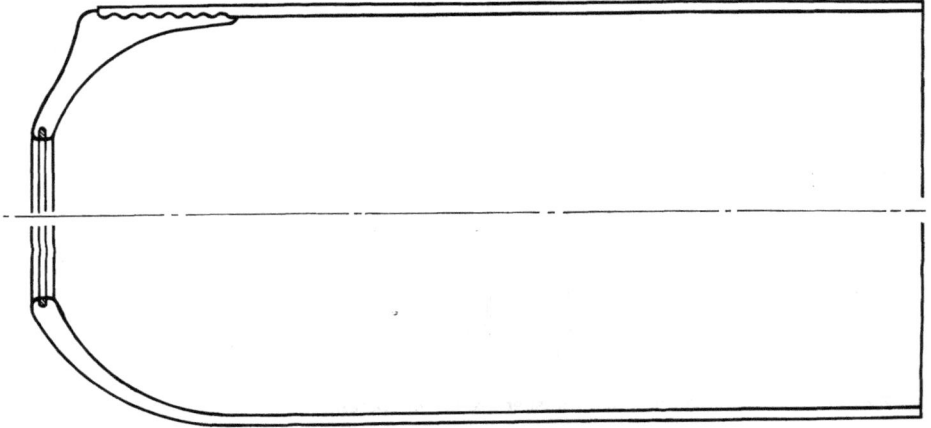

Fig. 13.13   SF$_6$ circuit breaker puffer cylinder. Old and new design (GEC)

13.6 Cost effective design: assembly and test

Circuit breakers have not only to be constructed as a set of parts, but have to be assembled and tested. Study of these processes during the design stage will contribute to the cost effective result by improving accessibility, by avoiding intrinsic difficulties, by subdivision enabling operations to be carried on in parallel, by reduction in space occupied and time required on the factory floor, and by simplifying adjustment and measurement procedures. There is also the issue as to whether production levels will warrant any significant degree of automated assembly or test, both of which should be considered as a normal part of the value analysis procedures.

The more obvious problems are related to inaccessibility, difficulty of placing gaskets in position so that they will stay in place, difficulty in inspection, problems with adjustments more especially if two or more adjustments are interdependent and cannot be finalised on their own, problems of clearances owing to build up in limits requiring selective assembly; risks associated with sequential tightening and torque limitation especially with ceramic components; problems arising with cleanliness, of special importance with hydraulic systems, and dryness, of special importance with SF$_6$ systems. These can in general be resolved by the use of models, logical attention to detail design, by the issuing of appropriate assembly instructions, and by close involvement of the design department with the assembly and test procedures during development.

The cost of assembly and testing complete circuit breakers in a factory and repeating this on site is considerable in the field of high-voltage designs and a careful study of this problem can contribute to cost effectiveness. To avoid the technical need for complete assembly in the factory and still ensure reliability, each subassembly has to be complete in itself, and have clearly definable electrical and mechanical functions, capable of being tested in isolation from the rest of the circuit breaker. This may of course require special test fixtures, with carefully programmed test procedures and limits of acceptable performance, but this can avoid expensive assembly and dismantling, and occupation for a long period of the large area of the factory floor required to accommodate e.h.v. circuit breakers. Performance limits acceptable for each section, such as synchronising errors, operating times, gas flow or leakage rates, contact acceleration, mechanical setting clearances, minimum trip voltage and minimum operating air pressures, should be established so that the need for readjustment or rejection is not a matter for reference to the engineering department.

Design for ease of testing requires that provision is made in the design stage for the measuring instruments, such as travel recorders, pressure gauges, flow meters, oscillographs etc., that will ensure ease of coupling up and accurate results.

The ideal to aim at is to ensure that in the first place assembly and test is not done twice, and then that these functions are not dependent on specialist skills and judgement on the shop floor, not because these are undesirable or not available, but because they are expensive and will vary. If the design can be assembled and tested satisfactorily by relatively unskilled manpower, and if all the relevant features and adjustments lend themselves to straightforward methods of quality assurance, then the design will have gone a long way towards ensuring cost effective achievement.

13.7   Changes in design

The use of value-analysis procedures applied to standardised circuit breakers and their components provides an effective method of extending downwards the trend of production costs, but involves changes in design, and all such changes need to be considered in terms of their possible effect on performance, reliability and certification.

For example, changes in welding methods can lead to fatigue failure, changes in finishes can result in increased wear of sliding seals, quite small changes in mechanism components or inertia can alter the making and latching performance, and minor changes in materials or dimensions in the arcing regions can affect interruption. Any such changes may therefore require re-testing and certification of some aspect of performance.

Changes should also be considered with respect to any dislocation they may cause in production and stocking of parts, and in terms of interchangeability of spare parts required on site, which is important to clients.

In principle, changes which originate from cost reduction, as apart from those necessary to correct any troubles, should be accumulated in the design office and be introduced as a Mark II model at a suitable time in production, together with any recertification this may require. Unless a systematic procedure of this kind is adopted, the disruption in production and testing costs caused by sequential introduction of a series of small changes is likely to exceed the economies to be expected from value analysis activity.

13.7   Acknowledgments

The preparation of this chapter has required information freely given by the following organisations:

Doulton Insulators Limited
GEC Switchgear Limited
Reyrolle/Parsons Limited

Virtual current chopping

M. P. Reece, B.Sc., Ph.D., F.Eng., F.I.E.E.

A.1 Introduction

To distinguish between 'Real' or 'Conventional' current chopping and 'Virtual' current chopping is, on the face of it, easy. As with many phenomena the more deeply the processes are examined, the more blurred the demarcation becomes.

'Real' or 'Conventional' current chopping may be defined as the forced interruption of an alternating current at a time usually just before the natural current zero. The word forced implies that the increasing resistance of the arc in the circuit-breaker (defined as the instantaneous arc voltage divided by the instantaneous arc current) is the cause of the current being 'forced' to zero at a time when it would not be zero if the resistance of the circuit-breaker arc were zero when current is finite.

'Virtual' current chopping may be defined as the extinction of the arc at a current zero which is not the natural power frequency current zero. Virtual current chopping can occur even in an ideal circuit-breaker in which the arc resistance is finite only when the instantaneous current is zero. Such a circuit-breaker cannot cause 'conventional' current chopping.

These definitions imply that any circuit-breaker, however stable its arc, and however free of a propensity to current chop it may be, can still give rise to 'virtual' chopping.

A.2 Conditions for virtual chopping to occur

The definition implies that in the circuit being interrupted or energised in which the circuit-breaker is arcing, some process must occur which causes the current in the circuit-breaker to become zero transiently at a time when the power frequency current in the major current limiting inductive reactance of the circuit is not zero. This can be called the occurrence of parasitic current zeros.

A.3   Causes of parasitic current zeros

There are many causes of parasitic current zeros. Any rapid change of voltage which can induce a high frequency current in the arcing circuit-breaker which is greater in magnitude and opposite in sign to the instantaneous magnitude of the power frequency current in the circuit-breaker must give rise to a parasitic current zero.

Examples are:

A restrike or prestrike in one phase of a circuit-breaker giving rise to a current pulse in that phase or indeed in another phase (by mutual coupling).[1,2]

A lightning stroke which produces a transient current pulse.

The deliberate introduction of a current pulse into a circuit to cause a circuit-breaker to interrupt at other than its natural current zero, for example to interrupt DC with an AC breaker or to make a current limiting AC circuit-breaker.[3,4]

Any switching on one phase of a circuit-breaker which may cause a parasitic current zero in an adjacent phase.[5]

A.4   Likelihood of interruption at a parasitic current zero

The likelihood of a circuit-breaker interrupting a quasi steady current flowing between its contacts when that current is momentarily reduced to zero by a transient current pluse superposed on it depends on two factors; the time constant of the arc, and the transience of the parasitic current zero.

The time constant of the arc depends on the type and design of the circuit breaker, and on the magnitude of the quasi steady current flowing just prior to the parasitic current zero.

The transience of the parasitic current zero depends on the rate of fall of current, and the rate of rise of recovery voltage which appears across the arc if it attempts to interrupt.

Thus interruption is most likely to occur when a circuit breaker with a very short time constant experiences a parasitic current zero of low transience, when it is arcing with a relatively low quasi steady current in the arc.

Conversely interruption is least likely to occur when the circuit breaker has a long arc time constant, when the transience of the parasitic current zero is high, and when the quasi steady arcing current is high.

Two general conclusions can be drawn.

1.   The higher the frequency of the transient oscillatory current, the greater its excess magnitude (the amount by which it exceeds the instantaneous value of the power frequency current), and the higher the instantaneous value of the

power frequency current the less likely is the arc to be extinguished and a virtual chop to occur.

2. The longer the time constant of the arc in the circuit-breaker the less likely is the arc to be extinguished and a virtual chop to occur.

These two general conclusions follow from the nature of the arc interruption process which is described in some detail in Chapter Two.

A.5 Prospective overvoltage generated by virtual chopping

Just as the transient component of the prospective overvoltage produced by a 'conventional' current chop is given generally by $V = IZ$, or commonly in single frequency circuits by

$$V = Ik \sqrt{\frac{L}{C}} \, ,$$

so exactly the same expressions describe the prospective overvoltage generated by a virtual current chop.

The prospective overvoltage due to virtual chopping is often larger than that due to conventional chopping because the quasi steady current interrupted is frequently higher in the case of virtual chopping. In both cases of course, reignition of the arc usually prevents prospective overvoltages being reached.

If repeated virtual chopping occurs in an opening circuit breaker, the increasing electric strength of the widening contact gap may cause the level of over-voltage to increase with time. This has been called 'voltage escalation' by some writers.

A.6 The sensitivity of different circuit-breakers to parasitic current zeros

The sensitivity of common circuit breakers may be listed in increasing order as air magnetic, oil (cross jet pot), oil (oil pumped), SF_6 and air blast, (almost equally sensitive) vacuum (chromium copper contacts) and vacuum (high copper contacts).

It will be appreciated that different designs of the same type of circuit breaker may have different arc time constants, for example an air blast breaker with a long parallel nozzle will have a longer time constant than one with a short convergent/divergent nozzle.

In the case of vacuum interrupters, the minimum current which can maintain a stable dc arc appears to be more important in determining the sensitivity of the interrupter to parasitic current zeros than does the arc time constant.

Thus high chromium alloy contacts, with five times the arc stability of high copper alloy contacts, are much less prone to virtual current chopping than are high copper alloy contacts.

A.7   Forms of virtual chopping

(a)   Chopping in the same phase
Probably the most common example of virtual chopping occurs when a circuit-breaker is interrupting a small current, and chops this current conventionally before the natural current zero. (Thus far this process falls exactly within the definition of conventional current chopping).

As a result of the conventional current chop, a transient recovery voltage occurs (sometimes known as the suppression peak). If the voltage is high enough to cause the circuit-breaker arc to reignite, the local capacitances and inductances will cause a large high frequency damped oscillatory current to flow around the local circuit through the circuit-breaker.

In the early stages of this high frequency damped oscillatory current the rate of change of current may be very high, the circuit-breaker will not interrupt it, and it will keep the arc conducting, thus allowing the initially linear rise of power frequency current in the major circuit inductance.

There are now two possibilities:

> Either the arc will remain conducting for the whole of the next loop of power frequency current, or as the amplitude of the oscillatory current falls, the arc may extinguish at one of the high frequency current zeros.

> If this happens, a 'virtual current chop' occurs. Recovery voltages appear to earth on both sides of the circuit-breaker, which may or may not cause another reignition.

If the arc does not reignite, the transient recovery voltages will subside and the power frequency will remain across the open breaker, with the circuit successfully interrupted. If a further reignition occurs, another damped oscillatory current will flow and the previous phenomena will be repeated, with either further reignitions or final interruption.

(b)   Virtual chopping on closing
Real or virtual current chopping can occur during closing as well as during opening of a circuit-breaker.

If, for example an unloaded transformer is being energised by a circuit-breaker, the following may occur. (It will be appreciated that the details of the processes will depend on the properties of the circuit-breaker and on the circuit parameters).

Considering a single phase earthed system for simplicity, the closing circuit-breaker will spark when the electric strength of the intercontact gap is exceeded by the system voltage. This will give rise to two currents, the power frequency magnetising inrush current, and superposed on this a transient oscillatory current. The circuit-breaker may not interrupt at a parasitic current zero, in which case it will continue to arc, either until the contacts close or until the current zero, whichever comes first. In the former case there will be no chopping phenomena, but if the circuit-breaker is relatively slow in closing, a current zero process will occur almost exactly analogous to that described in para: A.7(*a*). There may be a conventional current chop just before the natural current zero, which then may or may not be followed by a reignition, which may or may not interrupt at a parasitic current zero.

The important difference between the 'opening' and 'closing' cases is that in the first, the dielectric strength of the circuit-breaker is increasing with time, and in the second it is decreasing. This means that the generation of high overvoltages by either the real or subsequent virtual current chopping is on the whole less likely to occur on closing than on opening. Nevertheless, significant overvoltages can occur during closing in certain circumstances.

(c) Chopping in an adjacent phase
In three phase circuits which are being switched (either opened or closed), a reignition in one phase can cause a parasitic current zero in an adjacent phase, by mutual coupling between the phases.[5]

Since the reignitions in one phase usually occur near to the current zero in that phase (often soon after) it follows that the currents in the adjacent two phases will have a magnitude of about ·866 I and should either of these currents be interrupted at a parasitic current zero, the energy stored in the associated inductance, and thus the prospective overvoltage resulting will be much larger than in the originating phase. On occasion virtual chopping can cause the interruption of currents in all three phases simultaneously.

A.8 Practical importance of virtual current chopping

Virtual current chopping, although figuring in the literature and being familiar to specialised power system engineers, does not appear to give rise to many practical problems in electrical power supply systems.

This appears to be largely because those types of circuit-breaker with the shortest arc time constants, namely vacuum and SF_6, both usually have high dielectric strengths and a very rapid recovery of dielectric strength after arc extinction. Thus although they are more capable than oil or air magnetic circuit-breakers of interrupting at parasitic current zeros, they are generally less prone to reignition which generates the parasitic current zeros.

Nevertheless any circuit-breaker can occasionally experience contact separation just before a natural current zero, so it is essential that this will not lead to virtual chopping with generation of destructively high overvoltages.

Only on very rare occasions in the UK does one come across cases of 'adjacent phase' virtual chopping, and here overvoltages generated can be higher, because the power frequency current is higher, the stored inductive energy is thus higher, and also the contact separation is usually larger, with a higher prospective dielectric strength.

Experience in the UK has shown that with common items of plant such as transformers and motors, switched with oil, air magnetic, air blast, SF_6 or vacuum breakers, there has been no significant history of troubles due to virtual chopping.[6] It is worth noting that vacuum breakers in the UK almost without exception use interrupters with high chromium content contacts, with limited sensitivity to parasitic current zeros.

There are, of course, always exceptions, and in special circumstances even oil circuit breakers can interrupt at parasitic zeros. One case at 132 kV has been reported.[7] Other cases are known where frequency triplers have been involved. These cases are however insignificant in practical terms.

Abroad, however, there have been some reports of troubles in the USA when high reactance three phase transformers have been switched with vacuum switches, probably with high copper content contacts.[8]

Other problems relating to virtual current chopping have been reported from various parts of the world, and in one recent case Cornick[9] was able to compare a UK chromium copper contact vacuum interrupter with an overseas made interrupter switching the same circuit. These tests clearly demonstrated the critical importance of the composition of vacuum interrupter contacts to this phenomenon.

References

1. ITOH, T., MURAI, Y., OHKURA, T., and TAKAMI, T. (1972): 'Voltage escalation in the switching of the motor control circuit by the vacuum contactor.' IEEE Winter meeting - New York, N.Y. Paper T72 053-2
2. BOEHNE, E.A., and LOW, S.S. (1969): 'Shunt capacitor energization with vacuum interrupters — a possible source of overvoltage'. *IEEE Trans.*, **PAS-88**, pp 1424-1443
3. GOODWIN, A., REECE, M.P., and SMITHERS, B.W. (1960): 'A preliminary study of direct current interruption by vacuum switches using the artificial current-zero injection method'. British Electrical and Allied Industries Research Association, Report UR: CON HAR EMR/1085
4. GREENWOOD, A.N., and LEE, T.H. (1972): 'Theory and application of the commutation principle for HVDC circuit-breakers'. *IEEE Trans.*, **91**, pp. 1570-1574
5. MURANO, M., FUJII, T., NISHIKAWA, H., NISHIWAKI, S., OKAWA, M. (1974): 'Three phase simultaneous interruption in interrupting inductive current using vacuum switches'. *IEEE Trans.*, **PAS-93** pp 272-280
6. BLOWER, R.W., CORNICK, K.J., and REECE, M.P. (1978): 'The use of vacuum switchgear for the control of motors and transformers in industrial systems'. *IEE Proc.*, International Conference on Developments in Distribution Switchgear, pp. 5-10
7. REECE M.P., and WHITE E.L. (1957): 'Measurement of overvoltages caused by switching-out a 75 MVA 132/33 kV transformer from the high voltage side'. British Electrical and Allied Industries Research Association, Report reference G/T 313
8. PANEK, J., and FEHRLE, K.G.: 'Overvoltage phenomena associated with virtual current chopping in three phase circuits'. *IEEE Trans.*, **PAS-94**, pp. 1317-1325
9. CORNICK, K.J. (1984): 'Switching performance of distribution circuit-breakers'. Private communication

Index

www.ingramcontent.com/pod-product-compliance
Lightning Source LLC
Chambersburg PA
CBHW060417220326
41598CB00021BA/2206